SELECTED
INVERTEBRATE TYPES

CONTRIBUTORS

WILLIAM BALAMUTH
Northwestern University

FRANK A. BROWN, JR.
Northwestern University

JOHN B. BUCK
National Institutes of Health

WILLIAM D. BURBANCK
Drury College

CHAUNCEY G. GOODCHILD
S.W. Missouri State College

LIBBIE H. HYMAN
American Museum of Natural History

MARGARET L. KEISTER
National Institutes of Health

LEWIS H. KLEINHOLZ
Reed College

JOHN H. LOCHHEAD
University of Vermont

MADELENE E. PIERCE
Vassar College

W. MALCOLM REID
Monmouth College

MARY D. ROGICK
College of New Rochelle

TALBOT H. WATERMAN
Yale University

SELECTED INVERTEBRATE TYPES

Edited by

FRANK A. BROWN, Jr.

JOHN WILEY & SONS, Inc.
New York   London

COPYRIGHT, 1950
BY
JOHN WILEY & SONS, INC.

All Rights Reserved

This book or any part thereof must not be reproduced in any form without the written permission of the publisher.

FIFTH PRINTING, JANUARY, 1962.

PRINTED IN THE UNITED STATES OF AMERICA

Preface

During the past few years it has been becoming increasingly evident that someone or some group would soon have to devote time to the preparation of a laboratory practicum of invertebrate zoology to keep our colleges and universities provided with minimum essential published information for proper instruction. At the time of this writing no suitable guide to the laboratory study of our common American invertebrate animals is available. There appeared to be no single larger and more interested group concerned with this problem than the staff of the Department of Invertebrate Zoology, of the Marine Biological Laboratory, Woods Hole, Massachusetts. This task was, therefore, assumed with some assistance from other contributors, including former staff members.

The book differs from most American laboratory manuals in being considerably more informative and being extensively illustrated. In these respects it tends in the direction of the well-known German "Prakticums" in its approach. In these characteristics it will probably be viewed with alarm by many of the older school of invertebrate zoologists. We feel, however, that the day has largely passed when a study of the morphology of invertebrates was an end in itself and hence the student could afford time to do practically a research project on each species studied, being provided only with a specimen and a few pertinent, directed questions. On the contrary, with the tremendous growth of the numerous areas of experimental biology, the subject of invertebrate zoology is being compressed more and more in the average college curriculum. Our students today must obtain, in one-half to one-quarter the time allowed students of an earlier generation, all their basic knowledge of comparative invertebrate morphology. The more efficiently the material is presented, the more ground can be covered in the restricted time available, and hence the more rapidly can they obtain whatever is deemed a minimal foundation.

The present volume assembles in condensed form much of the content of numerous voluminous monographs on the various animals treated, and for a number of species contains the equivalent material though no monograph has yet been published.

The content of the book is probably adequate for a two-year laboratory course, but by a proper selection of types, and in many cases even of organ-systems studied, can be effectively utilized in a course of shorter length. No specific directions are provided as to what student records should be kept; it is felt that each instructor prefers to prescribe these for himself.

It is hoped that the volume will also have a second type of use, namely, introducing many investigators to species they might profitably select for their own experimental investigations. Numerous individuals still untrained in the biology of invertebrates can find among these animals favorable situations or preparations for the solution of basic biological problems. It has seemed to us a sad circumstance that, as this fact is becoming so thoroughly appreciated, the field should be so barren of all published aids. It is hoped that the publication of the volume will result in a rejuvenation in the teaching of this important subject which provides so much promise in the solution of virtually every basic problem in biology and medicine.

Many of the figures are original and are drawn from the actual specimens; all the remainder have been redrawn with more or less modification from the originals. We were very fortunate in having, as one of our members, Miss Marie Wilson, who executed with skill and meticulous care most of the drawings. We wish also to acknowledge our gratitude for the permissions granted by many authors and publishers to have certain of their figures utilized in the preparations of these illustrations. Specific acknowledgments are included in the legends and in the list following this preface.

FRANK A. BROWN, JR.

Woods Hole, Mass.
December, 1949.

Acknowledgments

For permission to reproduce illustrations acknowledgment is due to Mrs. Jane Z. Hegner (*Human Protozoology*, Hegner and Taliaferro) for Fig. 21; C. C. Thomas, Publishers, and Dr. R. R. Kudo (*Protozoology*, Kudo) for Figs. 23, 28, 30A and 30B; McGraw-Hill Book Co. (*The Invertebrates*, Hyman; *Animal Biology*, Wolcott) for Figs. 27, 43, 47, 50, 51, 52, 53, 54, 55, and 175 and 176; The Macmillan Co. (*A Treatise on Zoology*, Part II, Lankester; *A Textbook of Zoology*, Vol. I, Parker and Haswell; *Animal Biology*, Woodruff, 2nd Ed., Copyright 1938) for Figs. 57, 163, and 173; G. P. Putnam's Sons (*Field Book of Ponds and Streams*, by Ann Haven Morgan, Copyright, 1930, by Ann Haven Morgan) for Fig. 58; Cambridge University Press (*The Invertebrata*, Borradaile, Potts, Eastham, and Saunders) for Figs. 59 and 174B; The Blakiston Co. (*A Manual of Common Invertebrate Animals*, Pratt) for Fig. 60; Dr. A. C. Chandler (*Hookworm Disease*, Chandler) for Figs. 135, 136, 137, 138, and 139; Edward Arnold and Co. (*Oyster Biology and Oyster Culture*, Orton) for Figs. 174C and D; General Biological Supply House (*Turtox News*) for Fig. 190; and to Dr. A. Petrunkevitch (*Morphology of Invertebrate Types*, Petrunkevitch) for Fig. 221.

Grateful acknowledgment is also made to the sources of the following figures: Fig. 119, Sommers and Landois, Brumpt, *Précis de parasitologie*, Libr. de l'Acad. Méd., Paris; Fig. 125, A. Meyer, in Bronn's *Klassen und Ordnungen des Tier-reichs*, Abt. 2, Buch 2; Fig. 160, A. M. Marshall and C. H. Hurst, *A Junior Course of Practical Zoology*, 4th edition, G. P. Putnam's Sons; Fig. 162, Carl Vogt and Emil Yung, *Lehrbuch der praktischen vergleichenden Anatomie*, Vol. I, Friedrich Vieweg und Sohn, Braunschweig; Fig. 211, L. Cuenot, *Archives de zoologie expérimentale et générale;* Fig. 232, W. Bateson, *Quarterly Journal of Microscopical Science;* Figs. 154, 155, 156, J. H. Ashworth, Figs. 207, 208, 209, S. T. Burfield, Figs. 212, 213, 223, H. C. Chadwick, all from *Liverpool Marine Biological Committee Memoirs*.

Classification and Table of Contents

Phylum Protozoa   1
    SUBPHYLUM I. PLASMODROMA
       Class I. Flagellata or Mastigophora
          Order 1. Euglenoidina
             Euglena   1
             Peranema trichophorum   3
          Order 2. Chrysomonadina
          Order 3. Cryptomonadina
          Order 4. Dinoflagellata
             Ceratium hirundinella   5
          Order 5. Chloromonadina
          Order 6. Phytomonadina
             Volvox   7
          Order 7. Protomonadina
             Trypanosoma   11
          Order 8. Polymastigina
             Trichomonas   13
          Order 9. Hypermastigina
             Trichonympha   15
       Class II. Rhizopoda or Sarcodina
          Order 1. Amoebozoa or Lobosa
             Amoeba proteus   17
             Entamoeba histolytica   19
             Arcella vulgaris   22
             Difflugia oblonga   23
          Order 2. Foraminifera
          Order 3. Heliozoa
             Actinophrys sol   25
             Actinosphaerium eichhorni   25
          Order 4. Radiolaria
       Class III. Sporozoa
         Subclass I. Telosporidia
           Order 1. Gregarinida
              Monocystis lumbrici   28
              Gregarina blattarum   30
           Order 2. Coccidia
           Order 3. Haemosporidia
              Plasmodium vivax   32

Subclass II. Cnidosporidia
  Order 1. Myxosporidia
  Order 2. Actinomyxidia
  Order 3. Microsporidia
Subclass III. Sarcosporidia
  Order 1. Sarcosporidia
  Order 2. Globidia
Subclass IV. Haplosporidia

SUBPHYLUM II. CILIOPHORA

Class IV. Ciliata
  Subclass I. Protociliata
       Opalina obtrigonoidea   36
  Subclass II. Euciliata
    Order 1. Holotricha
       Didinium nasutum   38
       Coleps hirtus   40
       Holophrya simplex   42
       Prorodon griseus   43
       Paramecium caudatum   44
       Colpoda cucullus   48
       Frontonia leucas   49
       Colpidium colpoda   51
    Order 2. Spirotricha
       Spirostomum ambiguum   52
       Stentor coeruleus   56
       Stylonychia pustulata   59
       Euplotes patella   61
    Order 3. Chonotricha
    Order 4. Peritricha
       Vorticella campanula   63
       Epistylis plicatilis   66
       Carchesium polypinum   66
       Zoothamnium arbuscula   66
Class V. Suctoria
       Podophrya collini   69

Phylum Mesozoa

Class I. Rhombozoa

Class II. Orthonectida

Phylum Porifera   72

Class I. Calcarea or Calcispongiae
  Order 1. Homocoela
       Leucosolenia   72
  Order 2. Heterocoela
       Scypha (= *Sycon*)   74

CLASSIFICATION AND TABLE OF CONTENTS

Class II. Hexactinellida
  Order 1. Hexasterophora
      Euplectella — 76
  Order 2. Amphidiscophora
      Hyalonema — 77

Class III. Demospongiae
  Subclass I. Tetractinellida
    Order 1. Myxospongida
    Order 2. Homosclerophora
    Order 3. Choristida
  Subclass II. Monaxonida
    Order 4. Hadromerina or Astromonaxonellida
      Cliona — 78
    Order 5. Halichondrina
      Halichondria — 78
    Order 6. Poecilosclerina
      Microciona — 80
    Order 7. Haplosclerina
      Spongilla — 81
      Chalina (Haliclona) — 78
  Subclass III. Keratosa
      Hippospongia — 83

Phylum Cnidaria — 85

  Class I. Hydrozoa
    Order 1. Hydroida
      Tubularia crocea — 85
      Hydractinia echinata — 93
      Pennaria tiarella — 96
      Obelia geniculata — 99
      Gonionemus murbachii — 102
      Pelmatohydra oligactis — 104
    Order 2. Milleporina
    Order 3. Stylasterina
    Order 4. Trachylina
    Order 5. Siphonophora

  Class II. Scyphozoa
    Order 1. Stauromedusae
    Order 2. Cubomedusae
    Order 3. Coronatae
    Order 4. Semaeostomeae
      Aurellia aurita — 112
    Order 5. Rhizostomeae

  Class III. Anthozoa
    Subclass I. Zoantharia
      Order 1. Actinaria
        Metridium senile — 119

CLASSIFICATION AND TABLE OF CONTENTS

    Order 2. Madreporaria
          Astrangia danae   127
    Order 3. Zoanthidea
    Order 4. Antipatharia
    Order 5. Ceriantharia
  Subclass II. Alcyonaria
    Order 1. Stolonifera
    Order 2. Testacea
    Order 3. Alcyonaceae
    Order 4. Coenothecalia
    Order 5. Gorgonacea
    Order 6. Pennatulacea
          Renilla köllikeri   131

Phylum Ctenophora   136

  Class I. Tentaculata
    Order 1. Cydippida
          Pleurobrachia pileus   136
    Order 2. Lobata
    Order 3. Cestida
    Order 4. Platyctenea
  Class II. Nuda
    Order 1. Beroida

Phylum Platyhelminthes   141

  Class I. Turbellaria
    Order 1. Acoela
          Polychoerus   141
    Order 2. Rhabdocoela
          Stenostomum   143
    Order 3. Alloecoela
    Order 4. Tricladida
          Bdelloura   145
          Dugesia   148
          Procotyla   154
    Order 5. Polycladida
          Hoploplana   156
  Class II. Trematoda
    Order 1. Monogenea
          Polystomoides oris   158
    Order 2. Aspidobothria
          Aspidogaster conchicola   161
    Order 3. Digenea
          Cryptocotyle lingua   165
          Opisthorchis sinensis (formerly *Clonorchis*)   173
          Gorgodera amplicava   179
          Fasciola hepatica   185
          Schistosoma haematobium   187

CLASSIFICATION AND TABLE OF CONTENTS

Class III. Cestoda
  Subclass I. Cestodaria
    Order 1. Amphilinidea
    Order 2. Gyrocotylidea
  Subclass II. Eucestoda
    Order 1. Tetraphyllidea
        Phyllobothrium laciniatum 190
    Order 2. Lecanicephaloidea
    Order 3. Proteocephaloidea
    Order 4. Diphyllidea
    Order 5. Trypanorhyncha
        Otobothrium crenacolle 194
    Order 6. Pseudophyllidea
        Diphyllobothrium latum 196
    Order 7. Nippotaeniidea
    Order 8. Cyclophyllidea
        Taenia pisiformis 199
        Echinococcus granulosus 204
    Order 9. Aporidea

Phylum Rhynchocoela 209

  Subclass I. Anopla
    Order 1. Paleonemertini
    Order 2. Heteronemertini
  Subclass II. Enopla
    Order 1. Hoplonemertini
        Amphiporus ochraceus 209
    Order 2. Bdellomorpha

Phylum Acanthocephala 214

  Class I. Metacanthocephala
    Order 1. Palaeacanthocephala
    Order 2. Archiacanthocephala
        Neoechinorhynchus emydis 214
    Order 3. Eoacanthocephala

Phylum Aschelminthes 220

  Class I. Rotifera
    Order 1. Seisonacea
    Order 2. Bdelloidea
    Order 3. Monogononta
        Hydatina senta 220
  Class II. Gastrotricha
    Order 1. Macrodasyoidea
    Order 2. Chaetonotoidea
        Chaetonotus brevispinosus 223
  Class III. Kinorhyncha or Echinodera

    Class IV. Priapulida
    Class V. Nematoda
        Order 1. Enoploidea
        Order 2. Dorylaimoidea
        Order 3. Mermithoidea
        Order 4. Chromodoroidea
        Order 5. Araeolaimoidea
        Order 6. Monhysteroidea
        Order 7. Desmoscolecoidea
        Order 8. Rhabditoidea (or Anguilluloidea)
                Turbatrix aceti   225
                Rhabditis maupasi   227
        Order 9. Rhabdiasoidea
        Order 10. Oxyuroidea
                Enterobius vermicularis   234
        Order 11. Strongyloidea
                Necator americanus   237
        Order 12. Ascaroidea
                Ascaris lumbricoides   243
        Order 13. Spiruroidea
        Order 14. Dracunculoidea
        Order 15. Filarioidea
                Wuchereria bancrofti   251
        Order 16. Trichuroidea
                Trichinella spiralis   253
                Trichuris trichiura   257
        Order 17. Dioctophymoidea
    Class VI. Nematomorpha or Gordiacea
        Order 1. Gordioidea
        Order 2. Nectonematoidea

Phylum Entoprocta   261
                Barentsia   261

Phylum Ectoprocta   264
    Class I. Gymnolaemata
        Order 1. Stenostomata
                Crisia   264
        Order 2. Ctenostomata
                Bowerbankia   265
        Order 3. Cheilostomata
                Electra   267
                Bugula   267
    Class II. Phylactolaemata
                Plumatella   269

Phylum Phoronidea

CLASSIFICATION AND TABLE OF CONTENTS

Phylum Annelida   271
- Class I. Polychaeta
    - Order 1. Errantia
        - *Neanthes virens* (formerly *Nereis*)   271
    - Order 2. Sedentaria
        - *Arenicola cristata*   279
        - *Amphitrite ornata*   289
    - Order 3. Myzostomaria
- Class II. Archiannelida
- Class III. Oligochaeta
    - *Lumbricus terrestris*   295
- Class IV. Hirudinea
    - Order 1. Rhynchobdellida
    - Order 2. Gnathobdellida
        - *Hirudo medicinalis*   303
    - Order 3. Pharyngobdellida

Phylum Echiuroidea
- Class I. Echiurida
- Class II. Saccosomatida

Phylum Sipunculoidea   309
- *Phascolosoma gouldii*   309

Phylum Mollusca   318
- Class I. Amphineura
    - Order 1. Polyplacophora
        - *Chaetopleura apiculata*   318
    - Order 2. Aplacophora
- Class II. Scaphopoda
- Class III. Pelecypoda
    - Order 1. Taxodonta
        - *Yoldia limatula*   319
    - Order 2. Anisomyaria
        - *Pecten irradians*   321
    - Order 3. Eulamellibranchiata
        - *Venus mercenaria*   324
        - Fresh-water mussel   334
- Class IV. Gastropoda
    - Order 1. Prosobranchia
        - *Busycon canaliculatum* (formerly *Fulgur* and *Sycotypus*)   336
    - Order 2. Opisthobranchia
        - *Eolis*   344
    - Order 3. Pulmonata

xvi   CLASSIFICATION AND TABLE OF CONTENTS

    Class V. Cephalopoda
        Order 1. Dibranchiata
            Loligo pealeii   347
        Order 2. Tetrabranchiata

Phylum Brachiopoda   357

    Class I. Inarticulata
        Order 1. Atremata
        Order 2. Neotremata
    Class II. Articulata
        Order 1. Protremata
        Order 2. Telotremata
            Terebratella   357
            Terebratulina septentrionales   357

Phylum Onychophora

Phylum Arthropoda   360

SUBPHYLUM I. CHELICERATA

    Class I. Merostomata
        Order 1. Xiphosura
            Xiphosura polyphemus (formerly *Limulus*)   360
    Class II. Pycnogonida
        Order 1. Colossendẹomorpha
        Order 2. Nymphonomorpha
        Order 3. Ascorhynchomorpha
        Order 4. Pycnogonomorpha
    Class III. Arachnida
      Subclass I. Latigastra
        Order 1. Scorpiones
        Order 2. Pseudoscorpiones
        Order 3. Opiliones
        Order 4. Acari
      Subclass II. Soluta
        Order 1. Trigonotarbi
      Subclass III. Caulogastra
        Order 1. Palpigradi
        Order 2. Schizomida
        Order 3. Telyphonida
        Order 4. Kustarachnae
        Order 5. Phrynichida
        Order 6. Araneae
            Argiope aurantia   382
        Order 7. Ricinulei
        Order 8. Solifugae

CLASSIFICATION AND TABLE OF CONTENTS

SUBPHYLUM II. MANDIBULATA
Superclass I. Crustacea
  Class I. Eucrustacea
    Subclass I. Branchiopoda
      Order 1. Anostraca
            Artemia   394
      Order 2. Notostraca
      Order 3. Onychura
            Daphnia magna   399
    Subclass II. Ostracoda
      Order 1. Myodocopa
      Order 2. Podocopa
    Subclass III. Branchiura
    Subclass IV. Copepoda
      Order 1. Mystacocarida
      Order 2. Calanoida
      Order 3. Misophrioida
      Order 4. Monstrilloida
      Order 5. Harpacticoida
      Order 6. Cyclopoida
            Cyclops   406
      Order 7. Notodelphyoida
      Order 8. Caligoida
      Order 9. Lernaeopodoida
    Subclass V. Cirripedia
      Order 1. Thoracica
            Lepas anatifera   413
      Order 2. Acrothoracica
      Order 3. Apoda
      Order 4. Rhizocephala
    Subclass VI. Ascothoracica
    Subclass VII. Malacostraca
     Division I. Leptostraca
      Order 1. Nebaliacea
     Division II. Syncarida
      Order 1. Anaspidacea
     Division III. Peracarida
      Order 1. Mysidacea
            Heteromysis formosa   418
      Order 2. Cumacea
      Order 3. Thermosbaenacea
      Order 4. Tanaidacea
      Order 5. Isopoda
      Order 6. Amphipoda
     Division IV. Eucarida
      Order 1. Euphausiacea
      Order 2. Decapoda
            Crayfishes (and *Homarus*)   422
            Callinectes sapidus   447

Division V. Hoplocarida
  Order 1. Stomatopoda
Superclass II. Progoneata
 Class I. Pauropoda
  Order 1. Pauropoda
 Class II. Diplopoda
  Order 1. Ancyrotricha
  Order 2. Lumacomorpha
  Order 3. Oniscomorpha
  Order 4. Colobognatha
  Order 5. Nematomorpha
  Order 6. Proterospermorpha
  Order 7. Opisthospermorpha
    Spirobolus marginatus 462
 Class III. Symphyla
  Order 1. Symphyla
Superclass III. Opisthogoneata
 Class I. Chilopoda
  Order 1. Geophilomorpha
  Order 2. Scolopendromorpha
  Order 3. Lithobiomorpha
  Order 4. Craterostigmorphora
  Order 5. Scutigeromorphora
 Class II. Insecta
  Order 1. Collembola
  Order 2. Protura
  Order 3. Entotrophi
  Order 4. Thysanura
  Order 5. Odonata
  Order 6. Plectoptera
  Order 7. Orthoptera
    Periplaneta americana 475
  Order 8. Isoptera
  Order 9. Dermaptera
  Order 10. Plecoptera
  Order 11. Embioptera
  Order 12. Corrodentia
  Order 13. Zoraptera
  Order 14. Mallophaga
  Order 15. Anoplura
  Order 16. Hemiptera
  Order 17. Thysanoptera
  Order 18. Neuroptera
  Order 19. Mecoptera
  Order 20. Diptera
    Drosophila melanogaster (larva) 496

CLASSIFICATION AND TABLE OF CONTENTS

    Order 21. Siphonaptera
    Order 22. Trichoptera
    Order 23. Lepidoptera
    Order 24. Megaloptera
    Order 25. Rhaphidiodea
    Order 26. Coleoptera
    Order 27. Strepsiptera
    Order 28. Hymenoptera

Phylum Linguatula

Phylum Tardigrada

Phylum Chaetognatha   505
        Sagitta elegans   505

Phylum Echinodermata   515
  SUBPHYLUM I. ELEUTHEROZOA
    Class I. Asteroidea
      Order 1. Phanerozonia
      Order 2. Spinulosa
      Order 3. Forcipulata
          Asterias forbesi   515
    Class II. Ophiuroidea
      Order 1. Euryalae
      Order 2. Ophiurae
          Ophioderma brevispinum   523
    Class III. Echinoidea
      Order 1. Centrechinoidea
          Arbacia punctulata   528
      Order 2. Clypeastroidea
          Echinarachnius parma   538
      Order 3. Spatangoidea
    Class IV. Holothuroidea
      Order 1. Dendrochirota
          Thyone briareus   541
      Order 2. Molpadonia
      Order 3. Apoda
  SUBPHYLUM II. PELMATOZOA
    Class I. Crinoidea

Phylum Enteropneusta (Hemichorda)   547
    Class I. Balanoglossida
          Saccoglossus kowalevskii   547
    Class II. Cephalodiscida (Pterobranchia)

CLASSIFICATION AND TABLE OF CONTENTS

Phylum Chordata   549

   SUBPHYLUM I. TUNICATA (UROCHORDA)

     Class I. Larvacea

     Class II. Ascidiacea

        Order 1. Aplousobranchia (Krikobranchia)
            Amaroucium constellatum   549

        Order 2. Phlebobranchis (Diktyobranchia)

        Order 3. Stolidobranchia (Ptychobranchia)
            Molgula manhattensis   553
            The Ascidian larva   560

     Class III. Thaliacea

        Order 1. Pyrosomatida
        Order 2. Salpida
        Order 3. Doliolida

   SUBPHYLUM II. CEPHALOCHORDA

   SUBPHYLUM III. VERTEBRATA

Bibliography   565

Index   593

1.

Protozoa

Euglena

By William Balamuth

Euglena (Fig. 1) is included in the Class Mastigophora with numerous other unicellular flagellate organisms. Owing to the loose definition of the class, many diverse types have been grouped together: "plant-like" forms which contain the photosynthetic pigment, chlorophyll, and in which flagella may play a minor role; actively motile green flagellates; actively motile non-pigmented flagellates otherwise closely resembling the previous group; and a wide variety of quite distinct non-pigmented forms which appear "animal-like." In recognition of these contrasting affinities it is conventional to distinguish the Subclass Phytomastigina from the Subclass Zoomastigina. The most reasonable view seems to be that the flagellates are a basic group in which one line (the plant-like Phytomastigina) indicates the transition toward algal plants, whereas the other (the Zoomastigina) marks the level of divergence for animal-like Protozoa—and thus all higher animals.

Euglena belongs to the Order Euglenoidina, which comprises relatively large forms usually of a conspicuously green color. In members of this order one or two flagella are present, arising from the invaginated anterior end of the body, or reservoir. A distinctive carbohydrate reserve called paramylum is segregated by euglenoid flagellates. All chlorophyll-bearing members possess a light-sensitive organelle called the stigma. Several non-pigmented forms (e.g., *Peranema*, described below) are placed in this order because of their basic features.

There are about one hundred species of *Euglena* widely distributed in fresh-water ponds and streams rich in organic matter. Most of them range in length from 50 to 100 μ, although some reach 400–500 μ.

Laboratory cultures are easily maintained in infusions containing milk, or rice, etc. The greatest hazard is to use too rich a medium and thus promote bacterial growth at the expense of the *Euglena*. In observing living material it is always best to prepare several slides at the outset so that normally active individuals can be found later. Proper lighting is a critical factor in studying living protozoa. Avoid overillumination since glare renders the semitransparent bodies invisible. In general, reduce the lighting until it permits maximum definition and yet discloses the full range of colors. At best the flagellum is difficult to see in life owing to its rapid beat, but it can be slowed by viscous substances like tragacanth or methocel. Of temporary stains, iodine solutions bring out the flagellum, while acetocarmine or acetomethyl green delineates the nucleus. Permanent stained preparations disclose most internal detail.

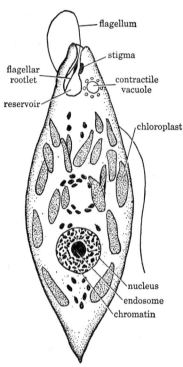

Fig. 1. Generalized organization of *Euglena*, adapted from *E. viridis*.

The body is covered by a plastic, often striated pellicle which permits changes in shape called euglenoid movements. At the anterior end of the body is a tubular inpocketing called the reservoir, which widens out posteriorly. Lateral to the reservoir is a contractile vacuole which is difficult to see. Careful study has shown that it forms by the fusion of several small vacuoles, and it empties into the reservoir.

The single flagellum extends into the body through the reservoir; in its course it bifurcates into two rootlets, one of which bears a swelling. Each rootlet is anchored in a basal granule (blepharoplast) in the wall of the reservoir. During locomotion the flagellum is directed obliquely backward with wave-like motions passing from base to tip, while the body rotates on its long axis and follows a spiral path. Alongside the reservoir is a small reddish body, the stigma, which marks a light-sensitive region. *Euglena* displays clear-cut phototaxis, as is easily demonstrated in mass cultures. It is normally photopositive in weak and photonegative in strong light.

The chlorophyll is contained in special bodies called chloroplasts which vary greatly in number, size, and shape (spherical, rod-shaped, etc.). The paramylum reserves may lie free in the cytoplasm in varied patterns or be attached to chloroplasts. The nutritional requirements of *Euglena* have been carefully analyzed. In the presence of light, chlorophyll acts as in green plants to permit the utilization of carbon dioxide as a source of carbon (phototrophic nutrition). In darkness, or after loss of chlorophyll, an organic source of carbon must be supplied (heterotrophic nutrition). These compounds are absorbed through the pellicle (saprozoic feeding). Specific requirements for nitrogen differ according to the species, but no proof exists that vitamins are required, nor is there any good evidence that *Euglena* ever ingests solid food.

There is a conspicuous nucleus in the posterior or midbody region. In stained preparations of the nucleus a dark, compact endosome can be distinguished from numerous dispersed chromatin granules. During asexual reproduction the nucleus divides mitotically. Chromosomes arise from the dispersed chromatin while the endosome persists and pinches in two. The nuclear membrane remains intact during mitosis, as in most protozoa.

Euglena has a relatively simple life history. No sexual reproduction is known; asexual reproduction occurs by means of longitudinal binary fission. Under unfavorable conditions individuals may round up and form resting cysts, and in some forms fission takes place within cysts. In old cultures masses of individuals may aggregate and form a palmella stage, in which they lose their flagella and become embedded in a gelatinous matrix. With restoration of favorable conditions the individuals may return to the motile phase.

Peranema trichophorum
By William Balamuth

Peranema (Fig. 2) illustrates the difficulty of separating plant and animal flagellates through arbitrary criteria. Its basic organization clearly resembles that of *Euglena* (see below), yet it is colorless, is without a stigma, has two flagella, and ingests solid food through a

PROTOZOA

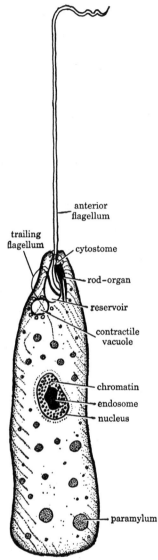

Fig. 2. *Peranema trichophorum.*

mouth opening. There are only two or three species, and the present description refers to the commonest member, *P. trichophorum.*

Peranema is widely distributed in stagnant fresh water, often being found with *Euglena.* The faintly striated pellicle is plastic, permitting extraordinary euglenoid movements; the body may contort into a sphere or even extrude pseudopodium-like extensions in the manner of amoebae. During locomotion specimens vary from 25 to 80 μ long by 10 to 15 μ wide; the body normally has a bluntly rounded posterior end and tapers anteriorly. Two flagella are present. Both flagella are anchored in blepharoplasts in the reservoir wall. One of these is usually not visible; it adheres closely to the striated pellicle and normally does not become free. The other flagellum is very prominent and typically extends straight out ahead of the body. It is usually stiff along most of its length, but the tip can become highly active and flicker in any direction. During forward locomotion the distal tip bends sharply backward. Under ordinary conditions *Peranema* moves upon a substratum in a gliding or creeping manner. More rarely it may move freely in a spiral path in the manner of *Euglena.*

The anterior, flask-shaped reservoir is closely associated with nutrition. A pair of rod-like structures called the rod-organ extends alongside the reservoir and continues a short distance posteriorly in the cytoplasm. Just anterior to the rod-organ a small slit-like mouth (cytostome) opens laterally into the cytoplasm. The details of ingestion are very difficult to observe and have been long disputed; the mouth can become widely distended during ingestion, the rod-organ serving as a support while a food vacuole forms. The food of *Peranema* consists of other smaller protozoans and bacteria, this typically animal-like mode of feeding being termed holozoic.

The nucleus is located in the middle of the body, and in stained preparations the same parts can be distinguished as in *Euglena:* a central endosome surrounded by dispersed chromatin granules. *Peranema* also displays a simple, completely asexual life history, involving longitudinal binary fission as the sole form of reproduction. True resistant cysts have not been described, although individuals may round up and offer some resistance to unfavorable conditions.

Ceratium hirundinella
By William Balamuth

Ceratium (Fig. 3) illustrates well the organization of the Order Dinoflagellata. This is a sharply delimited group of plant flagellates represented in the marine, and to a lesser extent in the fresh-water, plankton. The body of these forms is usually covered with a cellulose wall grooved by a transverse and a longitudinal furrow, each containing a flagellum. Chromatophores of yellow-brown color are typically present, and metabolic reserves include either starch or oil or both.

Of the some eighty species of *Ceratium* one (*C. hirundinella*) is by far the most widely distributed, occurring both in the marine plankton and in many fresh-water ponds and lakes. The anterior and posterior ends of the body are drawn out into long horns. In *C. hirundinella* there are one apical and 2–3 antapical horns, but they have been shown to undergo great geographical and seasonal variation and account for a bodily size range of 100–700 μ. This species is one of the "armored" dinoflagellates and has a fairly heavy cellulose wall composed of closely fitting reticulate platelets. The two furrows intersect on the left ventral surface, and a pair of flagella arises near this point. The transverse furrow (annulus or girdle) divides the body into nearly equal anterior and posterior regions called epicone and hypocone, respectively. The annulus contains a ribbon-shaped flagellum and is interrupted on the midventral surface by a large membranous plate. The broader longitudinal furrow (sulcus) extends obliquely backward and contains a thread-like flagellum which projects beyond the body. The combined flagellar movements produce the characteristic rotating and rolling motion of the body (*dino* = whirling). The transverse flagellum undulates to cause forward and rotational components, while the more whip-like movement of the longitudinal flagellum aids in forward motion and seems to serve also as a rudder.

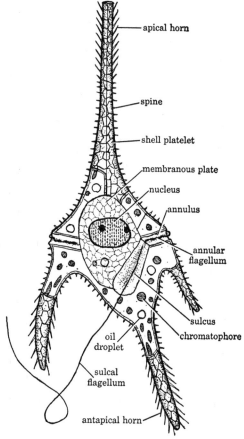

FIG. 3. *Ceratium hirundinella*, semidiagrammatic, with portion of shell removed.

Chromatophores are numerous and scattered, with yellow-brown pigments masking the chlorophyll. In addition to phototrophic nutrition *C. hirundinella* feeds holozoically, ingesting prey by means of pseudopodial extensions from the furrows. Metabolic reserves include both starch and oil droplets distributed through the cytoplasm. A non-contractile vacuolar system called the pusule has been reported but is difficult to demonstrate in *Ceratium*.

The single nucleus is large and central in position. It contains a few conspicuous endosomes surrounded by chromatin granules arranged in persisting rows. Nuclear division is mitotic and involves a large number of chromosome-like bodies, the endosomes being resorbed during the process. The only proved method of reproduction is a binary fission which cuts obliquely through the body, each offspring retaining a portion of the parental shell and restoring the lost elements. Resistant cysts form at times; the protoplasm retracts from the wall and becomes surrounded by a thick cyst wall. These cysts may remain viable for several years.

Volvox

By William Balamuth

Volvox (Figs. 4 and 5) is a unique microorganism which has been known since Leeuwenhoek's time (1700) and has been widely studied because it presents complex colonial organization and sexual differentiation at a relatively primitive level. The Order Phytomonadina, of which it is a member, includes a variety of solitary and colonial forms which usually have 1–4 flagella, are of grass-green color, segregate starch and oils, and undergo both asexual and sexual reproduction. Solitary forms and the trophic members of colonies have essentially a simple structure. *Volvox* stands at an endpoint of colonial differentiation in its large number of individuals or zooids (500–50,000) comprising a colony (coenobium) and in the sharp division of labor between the non-reproductive somatic zooids and the relatively few reproductive zooids.

Of the eighteen recognized species of *Volvox* only two (*V. aureus* and *V. globator*) are widespread in the plankton of small fresh-water ponds and lakes. The mature coenobia are hollow spheres or ellipsoids ranging in diameter from 0.5 to a few millimeters. A single layer of biflagellate zooids occupies the periphery of the gelatinous colonial envelope, while in different species the interior is filled with gelatinous material or water. The reproductive zooids are restricted to the posterior half of the coenobium and are easily distinguished by their much greater size. In any material at hand one should look for asexually produced daughter colonies in various stages of development and for stages of sexual differentiation into gametes and zygotes (see life cycle, below). Each zooid is enclosed by its own gelatinous sheath, which presses against adjacent ones to form polygonal surface markings. In most species protoplasmic strands link adjacent zooids, emphasizing the structural individuality of the whole. These strands

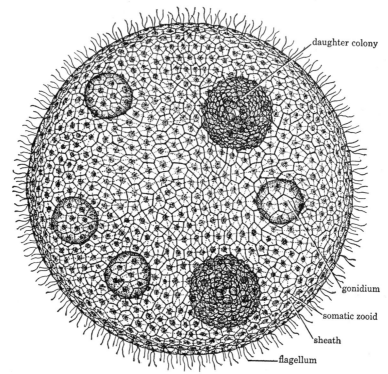

FIG. 4. *Volvox globator,* coenobium containing two daughter colonies and four gonidia.

are difficult to see in permanent preparations but can be demonstrated in living material by the application of dilute aqueous methylene blue. To observe flagella and the internal boundary of the layer of zooids, the focus should be leveled on an optical section of the coenobium. This also will disclose the shape of the zooids in lateral view. If dark-field illumination is available, it will afford a striking demonstration of flagellar activity.

The small size of the somatic zooids makes detailed study difficult. However, their organization closely resembles that of solitary genera (e.g., *Chlamydomonas*), one of which may be available for comparison. Each has two anterior flagella, two or more small contractile vacuoles, and a large cup-shaped chloroplast filling out the posterior half of the body. The nucleus is anterior to the chloroplast and contains a prominent endosome surrounded by dispersed chromatin granules. The

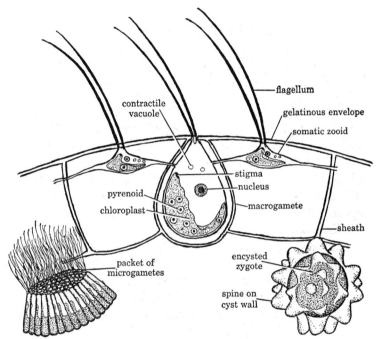

FIG. 5. Small sector of coenobial surface of *Volvox globator*, side view and highly enlarged, to show diagrammatically the stages in sexual differentiation.

nucleus divides mitotically during embryonic development of the coenobium.

Each zooid has a stigma near its external surface; those in zooids of one half of the colony are larger and mark the functional anterior end of the coenobium. The organism's movements are well coordinated in response to light and other stimuli. The combined action of all flagella produces rotation on the anteroposterior axis and a smooth rolling motion (*volvo* = roll) with the anterior pole out ahead.

The nutritional requirements of *Volvox* are incompletely known and presumably are more complex than in most green flagellates. Reports of successful cultivation in completely inorganic media have not been confirmed, although a practical method of maintaining stock cultures has been described recently. This method utilizes animal extracts added to a balanced salt solution (heterotrophic nutrition). Even if inadequate by itself, the chlorophyll mechanism is active, and starch reserves are segregated around the large pyrenoid contained in each chloroplast.

The life cycle approaches those of metazoa in complexity. A small number of special asexual zooids (3–10 or more) called gonidia are set aside in the posterior half of developing coenobia. Each increases to ten or more times the diameter of vegetative zooids and undergoes cleavage by means of longitudinal binary fission to form a complete daughter colony. These offspring become motile in the interior of the parental colony and finally escape when the latter dissociates. The various species of *Volvox* differ in being monoecious (e.g., *V. globator*) or dioecious (e.g., *V. aureus*); gonidia are present in either case. Monoecious forms are protandrous, ensuring the release of male elements before female gametes develop in the same colony. Sexual reproduction involves the union of biflagellate, sperm-like microgametes with spherical, ovum-like macrogametes. The macrogametes closely resemble the gonidia but are about twice as numerous (10–30 or more); they remain in the parental colony until fertilized. The microgametes develop as plate-like bundles in multiples of 16; they are released from the colony as motile units which break up into individual gametes in the vicinity of macrogametes. After fertilization the zygotes acquire a reddish brown cyst wall which is often sculptured, and in this condition enter a dormant period after the parent dissociates and the somatic elements degenerate. If zygotes are available for study, they will indicate the relative size of macrogametes.

It is of interest to note that all phases of the life cycle except the zygote contain the haploid nuclear condition, since meiosis occurs in the diploid zygote before its development into a new colonial generation. The resting zygote occupies the longest period of the life cycle, the colonial phase usually appearing for a brief period only in the spring. The first several generations developing from a zygote produce exclusively asexual coenobia, sexual stages beginning to appear in later generations, thereby evincing gradual attainment of sexual maturity. In several respects *Volvox* suggests a possible transition from unicellular to multicellular organization. Not least of these respects is the sharp division of labor between somatic and germinal elements which results in death of the soma at times of reproduction.

Trypanosoma

By William Balamuth

Trypanosoma (Fig. 6) belongs to the Order Protomonadina, which comprises relatively simple forms possessing 1–2 flagella provided with a few accessory kinetic structures. *Trypanosoma* and its relatives constitute a closely knit family of pleomorphic parasites which are of great medical and economic importance. Members of the genus are found in the blood of all vertebrate groups and are transmitted to them by blood-sucking invertebrates. *Trypanosoma gambiense* causes a fatal sleeping sickness and has decimated large areas of mid- and west Africa. *Trypanosoma brucei* parasitizes African cattle and causes the fatal disease called nagana. Chagas' disease, caused by *T. cruzi*, is restricted to South and Central America and results in a high percentage of fatalities. This last species has been found in insects in the southwestern United States and thus constitutes a potential threat in this country.

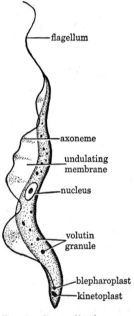

Fig. 6. Generalized organization of *Trypanosoma*.

Since the stages found in the vertebrate host provide the basis of taxonomy, we shall describe the structures of a generalized type as seen in a smear of vertebrate blood. Common frogs and reptiles are good sources of living material, and dried blood smears stained with Wright's or Giemsa's stain are easily obtained. However, the distortions imposed by dry fixation and smearing obscure many internal details, so for careful study it is preferable to combine wet fixation with hematoxylin staining. Trypanosomes range in length from about 15 to 60 μ. The shape resembles that of a curved, flattened blade with one long edge convex, the other concave, and the ends of the body tapering. The convex edge of the body is thrown into irregular rippling folds and is called an undulating membrane; it is essentially a modified flagellum. Along the edge of the membrane runs a stout thread, or axoneme, which is anchored near the posterior end of the body in a minute blepharoplast. Just posterior to the blepharoplast is a small rod-shaped body, the kinetoplast (sometimes erroneously called the parabasal body), which is connected to the blepharoplast by fine fibrils. These two structures usually stain as one. The flexible axoneme continues along the margin of the membrane to the anterior end of the body and then emerges as a free flagellum for a short distance. The flagellum and accessory structures are termed the kinetic apparatus. Locomotor behavior varies from a sluggish wriggling in some species to a lively torpedo-like darting in others (*trypan* = gimlet; *soma* = body).

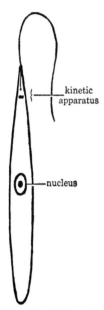

Fig. 7. Schematic diagram of leptomonad stage, to show relative position of kinetic apparatus and nucleus.

The rounded nucleus occupies a central position and in routine blood smears may stain solidly. Special technique (iron hematoxylin) demonstrates that it is actually vesicular, with a large endosome surrounded by a clear space. Chromatin is either embedded in the endosome or adherent to the nuclear membrane.

Food must be absorbed from the blood since no mouth is present. The nutritional requirements have been studied in bacteria-free culture media and are quite complex, probably as a result of adaptation to the special habitat. For example, ascorbic acid and hematin have been shown to be essential growth factors. Special metabolic reserves are not as characteristic as in plant flagellates, but greenish refractile granules often can be seen in living individuals. These are termed volutin and apparently represent reserve nucleic acid. As in most parasitic protozoa, no contractile vacuole is present.

The life cycles of trypanosomes are characterized by form changes in the different sites occupied and by alternation between a vertebrate and a blood-sucking invertebrate (e.g., certain flies, bugs, leeches). In the invertebrate host the entire kinetic apparatus is anterior to the nucleus. The undulating membrane may persist in this prenuclear site (crithidial stage), or it may be replaced by a typical flagellum (leptomonad stage, Fig. 7). The latter is considered to be the ancestral type from which have been derived the other structural patterns in the family.

Trichomonas

By William Balamuth

Trichomonas (Fig. 8) is one of the better-known members of the Order Polymastigina. The ties defining the order are loosely drawn, being based essentially upon the presence of 3–8 flagella (or of multiple

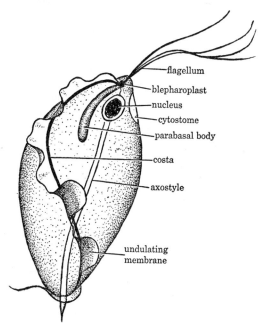

Fig. 8. Generalized organization of *Trichomonas*.

units in one group). Most are parasitic and have accessory kinetic structures, which features are well illustrated by *Trichomonas*. This parasitic genus occurs in a wide range of vertebrate hosts as well as in termites and a few other invertebrates, usually occupying some part of the digestive system. Only one host is involved in each life cycle. About 75 species have been described; 3 of these are present in the human: in the buccal cavity, large intestine, and urogenital system respectively. Common sources for laboratory study include rodents, fowl, frogs. To study living slide material the specimen may be diluted with physiological saline solution (0.6 per cent sodium chloride for cold-blooded, 0.9 per cent for warm-blooded, hosts) and then capped with a coverslip. Most species are quite small (5–20 μ long), and structural details are difficult to make out in life and in routine hematoxylin preparations. The present description has been generalized in order to define the main structural features of the genus.

The body is pear shaped, tapering posteriorly. There are four anteriorly directed flagella, the proximal portions of which frequently become intertwined. A trailing flagellum forms the border of an undulating membrane along the so-called dorsal edge of the body and usually extends freely beyond. *Trichomonas* moves in a characteristic jerky manner, rotating on its longitudinal axis and following an erratic path. During locomotion the membrane is in constant rippling motion, while the anterior flagella lash from side to side, frequently as a whiplike unit. The body is quite plastic, becoming readily deformed in burrowing through debris. All flagella are anchored near the anterior end of the body in a large blepharoplast complex which usually stains as a single mass. Just posterior to the blepharoplast is the prominent nucleus which, according to the species, has either dispersed or concentrated chromatin, lending respectively either a massive or vesicular appearance. The blepharoplast complex serves as the point of insertion for the following structures. A fiber termed the costa, or chromatic basal rod owing to its affinity for hematoxylin, extends in an arc beneath the membrane and marks its line of attachment to the body. The axostyle is a stout, semirigid rod which runs medially through the body and frequently has a spike-like projection from the posterior end. It appears hyaline and non-staining ordinarily, but iodine solutions frequently disclose internal, brownish-staining granules and suggest glycogen. A club-shaped parabasal body, varying in length in different species, extends posteriorly from the blepharoplast and lies in a plane between the nucleus and costa. The parabasal body becomes visible only after special treatment and will not be seen in ordinary material. It has been shown to have a complex internal structure and is not homologous to the structure in trypanosomes often called by the same name. The above system of integrated extranuclear organelles is referred to as a mastigont.

A longitudinal clear area called the cytostome is present anteroventrally, opposite the surface of the body bearing the undulating membrane. Solid food is definitely ingested by many species, apparently through the cytostome as well as other parts of the surface. Saprozoic feeding also occurs.

The life cycle of *Trichomonas* is simple. Cysts do not occur, so that trophic forms must be the transmissive stage. Longitudinal binary fission is the sole means of reproduction. The nucleus divides mitotically. The cytoplasmic organelles are either retained and shared between the two offspring (e.g., blepharoplasts, flagella, costa), with the missing ones being restored by new outgrowth; or certain parental structures are resorbed (e.g., axostyle, all or most of the parabasal body), which requires their complete new formation. During the entire process of reorganization the dividing blepharoplast complex serves as extranuclear centrioles to orient the mitotic spindle and as centers of outgrowth for new components of each mastigont. Many special features of *Trichomonas* are difficult to make out, but the genus can be established conclusively by identifying the undulating membrane, costa, and nucleus.

Trichonympha

By William Balamuth

Trichonympha (Fig. 9) is the most widespread member of the Order Hypermastigina, a group of bizarre, highly complex forms found only in the digestive tract of termites and certain roaches. Distinctive group features include many flagella arranged in rows or tufts, and usually multiple parabasal bodies and axostylar filaments. *Trichonympha* occurs with other flagellates in American termites (*Reticulotermes* in the East and *Zootermopsis* in the West) as well as in the wood-eating roach, *Cryptocercus*, of the Southeast. The genus *Trichonympha* comprises relatively large flagellates, ranging in length from 50 to 360 μ in different species, and they are easily prepared for study. In some cases it suffices to squeeze the posterior abdomen of the host and use the fluid drop extruded from the anus, or the intestine first may be dissected free. In temporary preparations, diluted with 0.6 per cent sodium chloride solution, the flagellates usually become distorted and motionless for a few minutes, but then resume activity. This material is especially favorable for the preparation of permanent iron hematoxylin smears, since the protozoan fauna is concentrated, the specimens are large, and several different types (including trichomonads) occur.

The body is subdivided into three regions: a tapering, flagellated rostrum bearing an apical cap and delimited posteriorly by a transverse fissure; an intermediate bell-like region which bears most of the flagella; and a rounded, non-flagellated posterior region. The numerous flagella (up to 10,000 in some species) are organized into longitudinal rows marked by surface furrows; they differ in length, the longest ones (up to 150 μ) arising most posteriorly and extending over the non-flagellated zone. The individual flagella are anchored in basal

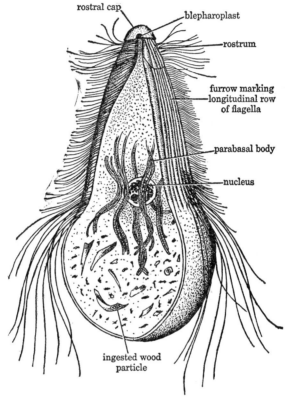

FIG. 9. *Trichonympha collaris*, from the western termite, *Zootermopsis*. (Modified from Kirby, 1932.)

rodlets which in turn are interconnected by subsurface fibrils. At the base of the naked cap is a hemispherical granule. This has been called the blepharoplast because the flagella are ultimately connected to it through fibrils, and because it suggests a blepharoplast in its position and centriolar role during mitosis (see also *Trichomonas*). The parabasal apparatus consists of a large number of slender, sinuous cords which extend posteriorly from the rostral region to the non-flagellated zone. In some forms the cords make contact with the nucleus or even form a basket-like support for it. The single nucleus lies near the middle of the body. The chromatin is widely dispersed in the form of granules, strands, or a permanent spireme, and a peripheral endosome is present. No homologue of an axostyle occurs in *Trichonympha,* although multiple axostylar filaments occur in other hypermastigotes.

Trichonympha is quite active in the normal state. During forward locomotion it rotates unevenly on a longitudinal axis. The rostral flagella are most active, whereas the long body flagella beat slowly if at all. Vigorous vibrations often pass along the body surface in waves and are probably caused by the beating body flagella. The rostrum is very flexible, periodically bending from side to side and proving an effective agent in burrowing through intestinal debris. The food consists mainly of wood particles ingested through the non-flagellated region. It is of interest to note that the flagellate fauna lives symbiotically with its termite host. Experiments have shown that the termites themselves cannot digest cellulose and cannot survive if the protozoa are removed.

It was formerly believed that the life cycle followed the typical animal flagellate plan: longitudinal binary fission as the sole means of reproduction, and occasional encystment as a response to an unfavorable environment. Now, however, complex, undeniable sexual processes are known to occur in several hypermastigotes, including *Trichonympha,* and in one polymastigote. These processes involve gametogenesis and fertilization and are induced by molting of the host, *Cryptocercus.*

Amoeba proteus
By W. D. Burbanck

Amoeba proteus (Fig. 10) is an unusually hardy laboratory animal. Because of their slow movements and relatively large size, amoebae are favorable material for routine laboratory studies. Decaying pond weeds kept in large, shallow crystallizing dishes often yield *A. proteus.* From such a temporary culture, a more stable one may be started by subculturing large active individuals in a boiled-hay infusion medium containing numerous small ciliates and flagellates. Under optimal conditions, animals will frequently be found measuring up to 1000 μ (1 mm.) in length and under intense illumination may be seen readily with the unaided eye. Even when viewed with relatively low magnification the animal is seen to exhibit protoplasmic streaming in one general direction. This streaming is one phase of the phenomenon known as amoeboid motion which results in locomotion of the animal; the rounded cytoplasmic protrusions are called pseudopodia. The relatively blunt pseudopodia in *A. proteus* are referred to as lobopodia. Except in older specimens that have not divided for many days, the upper surfaces of the lobopodia, when viewed with high magnification, typically exhibit an irregular corrugated appearance. This diagnostic characteristic for the species is believed due to secondary cytoplasmic currents flowing above, and parallel or at right angles to, the main pseudopodial stream.

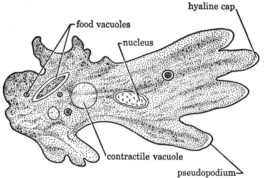

FIG. 10. *Amoeba proteus*. (Redrawn after Schaeffer.)

The outer boundary of the animal is termed the plasma membrane. Beneath it is a visible regional differentiation of the cytoplasm. The ectoplasm is most peripheral and constitutes a thin, non-granular layer. At the physiological anterior end, the end toward which the protoplasm is flowing at any given time, the ectoplasm at the extreme ends of the pseudopodia is thickened into a hyaline cap. Lying beneath the ectoplasm and in contact with it, there is a rather indefinite area of dense granular cytoplasm in which the granules are not streaming or in Brownian movement; this is the plasmagel. The remaining granular cytoplasm, which is actively streaming, is the plasmasol. Plasmagel and plasmasol are merely two physical states of protoplasm; there is a continuous and reversible transformation of plasmasol to plasmagel. Many of the foregoing features are rendered especially clear by study of the animal by dark-field illumination.

Carried along in the stream of flowing endoplasm, the organelles tumble over one another. The large single contractile vacuole lies in the ever-changing posterior end; the rough, discoidal, often indented nucleus, food vacuoles, and crystals are constantly changing their positions. Food vacuoles vary in number and appearance with the kind of food available. The crystals are largely bipyramidal with truncated apices.

Although slow in their movements, amoebae are able to catch the most agile prey living in their environment. Bacteria, *Chilomonas, Monas,* algae, *Colpidium, Paramecium,* nematodes, and even rotifers make up their varied diet. The prey is surrounded by pseudopodia forming a food-cup. As the pseudopodia coalesce in closing the food-cup, a relatively large amount of force is applied, as evidenced by the cutting in two of a *Paramecium* caught in the closing "sphincters." Although an amoeba can ingest from 50 to 100 *Chilomonas* daily, only relatively short-term cultures may be maintained on a pure growth of this flagellate. Ciliates seem to be a necessary part of the diet if stocks of amoebae are to be kept for long periods. After digesting its food, *A. proteus* voids the undigestible residue left in the food vacuole by merely moving away and leaving the waste behind.

Amoeba proteus may survive as long as 20 days without food. During this period of starvation, its mass may be reduced as much as 95 per cent. The animal may survive adverse conditions by producing a protective cyst.

The animals are sensitive to various changes in their external environment; clear-cut reactions may be obtained from such stimuli as light, chemical substances, and touch.

Amoeba proteus usually reproduces by binary fission. Before binary fission of the animal, the body rounds up with numerous pseudopodia radiating in all directions. At a temperature of 30°C., the nucleus of a well-fed individual will undergo mitosis in 21 minutes with the division of the cytoplasm taking place at the telophase of the nuclear division. Both plasmotomy, or breaking up of one individual into many, and sporulation have been described but not completely established.

Enucleated amoebae, or portions lacking a nucleus, will continue to behave normally except for an inability to reproduce and metabolize food properly; consequently, such modified individuals die in a relatively short time.

Entamoeba histolytica

By William Balamuth

Entamoeba histolytica (Figs. 11 and 12) closely resembles free-living Amoebina morphologically, yet biologically it is notable as a widespread, dangerous parasite of the human large intestine. Three other genera (*Endolimax, Dientamoeba, Iodamoeba*) and one other species (*Entamoeba coli*) share the same general site, but they live solely in the lumen and ingest only bacteria and debris whereas *E.*

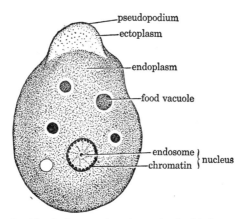

FIG. 11. *Entamoeba histolytica,* trophozoite stained with hematoxylin and showing eruption of clear pseudopodium.

histolytica bores into the intestinal wall and dissolves living tissue (hence the specific name *histolytica*). It is a curious fact that amoebae, apparently indistinguishable from *E. histolytica*, occur in other mammals, lower vertebrates, and some invertebrates, often without evident damage to these hosts. Laboratory turtles, snakes, and frogs are common sources for study. The contents of the lower intestine may be examined by diluting with physiological saline solution on slides. Iodine solutions disclose the nuclei in temporary stains, while iron hematoxylin is preferred for permanent preparations. The present description is based upon study of stained slides of *E. histolytica*.

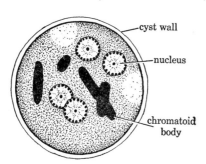

FIG. 12. *Entamoeba histolytica,* mature cyst containing four nuclei and chromatoid bodies (latter usually disappearing by this stage).

The active amoeboid stage, or trophozoite, usually ranges in long diameter from 15 to 25 μ. In fixed material the body is usually oval or rounded, any pseudopodia having been retracted. The cytoplasmic mass is divided into an outer clear zone of ectoplasm and an inner granular to alveolar zone of endoplasm; these zones may not be apparent in stained specimens. They are seen best during the formation of pseudopodia, since these are extruded rapidly and at first consist only of ectoplasm, the endoplasm tending to lag behind. The rapid formation of pseudopodia is correlated to the progressive monopodal motility of this species and is unique among human intestinal amoebae. In other species pseudopodia form more sluggishly and the amoebae move relatively little in random paths.

Entamoeba histolytica is holozoic in part, incorporating erythrocytes and other host-tissue elements into food vacuoles. Bacteria are less frequently ingested than in non-pathogenic amoebae. Organic and inorganic substances are also absorbed saprozoically from the medium. No contractile vacuole is present.

Nuclear structure offers the most stable means of identifying the species of intestinal amoebae. All have a vesicular nucleus, but each displays a different arrangement of chromatin and endosome. In *E. histolytica* there is typically a small central endosome and a relatively fine layer of chromatin granules pressed against the nuclear membrane. Conditions of fixation and staining may be responsible for considerable variation.

The cyst constitutes the other distinctive phase in the life cycle. For reasons not well understood trophozoites in the intestinal lumen regularly round up, eliminate ingested food, and at this stage are termed precysts. These then encyst within a thin resistant membrane. Cysts are spherical bodies usually ranging in diameter from 7 to 15 μ. Nuclear division occurs inside the cyst, so that the mature cyst contains four (rarely eight) minute nuclei with the characteristic species structure. Two kinds of metabolic reserves occur in cysts. Glycogen may be diffusely present in the cytoplasm, appearing as empty vacuoles in hematoxylin smears. In young cysts deeply staining, bar-like masses with rounded ends are present. These are called chromatoid bodies because of their affinity for nuclear stains, but they are not true chromatin.

To summarize the life cycle, there are essentially two stages, trophozoite and cyst, and only asexual reproduction occurs. The trophozoites regularly undergo binary fission marked by a modified nuclear mitosis. The mature cysts represent the transmissive stage. After ingestion by a suitable host or implantation into culture media the multinucleate amoeba actively emerges from the cyst, and a series of fissions restores the normal uninucleate condition.

Arcella vulgaris

By W. D. Burbanck

Arcella vulgaris (Fig. 13) commonly occurs in water taken from bogs or swamps which contains much vegetation. Beside these freshwater habitats, it may also be taken in moist forest soils. Its most conspicuous and characterizing feature is the presence of a light yellow to dark brown shell or test.

There are many subspecies of *A. vulgaris* with a great variety of test size and ornamentation. The test in this species is almost hemispherical and often has a rounded margin projecting like the brim of a hat. The height of the test is typically about one-half the diameter. This latter measurement may be from 30 to 100 μ. Centrally located on the flattened side of the test which is the one normally applied to the substratum is a depression the shape of an inverted funnel which leads to an opening, the pylome. The test is made up of intricately arranged siliceous prisms set in a tectin base which microscopically give the skeleton a reticular appearance.

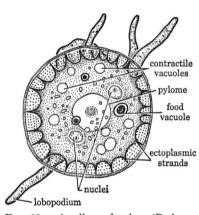

Fig. 13. *Arcella vulgaris.* (Redrawn after Leidy.)

The centrally placed cytoplasm is attached to the inner test wall by ectoplasmic strands. Except for ingested algae usually present, the endoplasm is colorless. Around the periphery there are four or more clear vacuoles; these may fuse to form a single large eccentric one that may be seen to function as a typical contractile vacuole. Two prominent vesicular nuclei lie directly opposite each other. Flowing through the pylome are small finger-like hyaline lobopodia which function as locomotor and feeding organelles.

When a specimen of *Arcella* is reversed, that is, turned so that the top of the test is in contact with the substratum, usually two to five and sometimes up to fourteen dark spots appear simultaneously; they are gas vacuoles. Apparently, they aid the animal in righting itself when reversed in a small volume of water and enable it to float to the surface when reversed in larger volumes. Since the disappearance of the gas vacuoles is correlated with the attachment of the pseudopodia to the substratum, it has been suggested that the gas escapes through them. This hypothesis is further borne out by the fact that small flagellates in the surrounding medium congregate about the pseudopodia. This observation and evidence from indirect chemical tests both suggest that oxygen is the gas present in these vacuoles.

Arcella grows well in infusions which contain *Chilomonas*, green flagellates, and small ciliates. When *Arcella* is kept in such a manner without an excessive bacterial growth, interesting regeneration studies on pieces of the animal may be made.

Asexual reproduction takes place by partial extrusion of the body through the pylome. This extruded mass grows until it equals that of the encapsulated portion. While the cytoplasm is thus enlarging, the two nuclei undergo mitosis and two of the four nuclei migrate into the extruded mass. When the separation of the nuclei is complete, a new light-colored test is formed by the extruded portion, after which the cytoplasm divides to produce two individuals. Apparently the extruded mass may also, on occasion, bud off small individuals, and it has been reported that the parent cytoplasm may also sometimes form many small amoebulae which migrate through the pylome. Whether gametes are formed and unite to produce zygotes is still an unanswered question.

Arcella sometimes leaves its test for unknown reasons and produces a new one. At other times, it may remain within its test and become a resting cyst.

Difflugia oblonga
By W. D. Burbanck

Difflugia oblonga (Fig. 14) possesses a beautifully symmetrical test usually made up of intricately placed sand grains of a definite size range. In certain localities, diatom shells or spicules from freshwater sponges may be used to supplement the sand grains. Often the animals are a vivid green because of zoochlorellae in their cytoplasm.

Difflugia will often be found in material aspirated from small leaf-choked puddles of water. It is also associated with delicate aquatic vegetation although it may be taken from wet soils.

Although there is considerable variation in this species, the test is usually pyriform or flask-shaped, having a neck of variable length with a terminal opening, or pylome. The long axis may measure from 60 to 580 μ, and the width of the bulbous part may be from 40 to 240 μ. At times the width may approach the total length. There may be as many as a half-dozen slender branching, or simple, pseudopodia present at one time. The pseudopodia are usually cylindrical, but become slightly flattened when the animal is actively moving along the substratum.

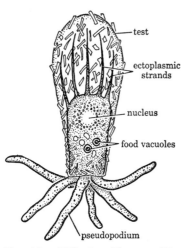

FIG. 14. *Difflugia oblonga.* (Redrawn after Leidy.)

Although the internal organelles are usually obscured by the algal bodies, it is possible to observe them in the few relatively colorless forms which occur in every good culture. The comparatively large nucleus is centrally located in the endoplasm occupying the bulbous part or fundus of the test. Around the periphery of the endoplasm numerous contractile vacuoles and food vacuoles are scattered.

In addition to microorganisms, *D. oblonga* feeds upon filamentous algae such as *Spirogyra*. This is accomplished by making a small hole in the cell wall, allowing the pseudopodia to enter.

Much can be learned about the internal morphology and behavior of the pseudopodia by removal of the test. The test can readily be broken by applying pressure to the cover glass. Naked forms, so obtained, may be kept alive for as long as a week. During this period, small enucleated pieces may separate from the parent animal, and toward the end of the period the uncovered body generally rounds up and dies.

Difflugia typically reproduces by fission. This process results in two individuals, the parent and the offspring, which remain in contact until the test of the new individual is complete. Animals in late stages of binary fission have been mistaken for conjugating forms. At the time of fission, some of the foreign particles which were ingested by the parent are present in that portion of the cytoplasm which is extruded through the pylome to form a new individual.

As with *Arcella*, there have been reports of numerous small difflugians being formed by a single parent, and it may be that some of these function as gametes. The actual significance of such observations will require additional work on the life cycles of both *Difflugia* and *Arcella*.

Difflugia oblonga is very active in the spring and most of the summer, but becomes encysted within its test in the late fall.

Actinophrys sol and *Actinosphaerium eichhorni*

By W. D. BURBANCK

These two species possess radiating ray-like pseudopodia, termed axopodia, and hence are aptly named sun-animalcules or heliozoans. The principal difference between the two is size, with *Actinophrys* being the smaller. In most of the known aspects of their biology, they are quite similar.

These species are found freely suspended or on floating and rooted vegetation in both the marine and fresh-water habitats. *Actinophrys sol* (Fig. 15) may also be found on slides that have been submerged for 2 or 3 weeks. On such slides or in material collected with a fine standard plankton townet, both animals may be seen to exhibit slow, rolling movements effected by the axopodia. In view of the resemblance between flagella and axopodia, it has been thought that there may be some phylogenetic relationship between the heliozoans and flagellates.

The pseudopodia of *A. sol* are relatively longer than those of *A. eichhorni* (Fig. 16). Those of the former species are one to three times the diameter of the body, that is, 40–50 μ; the axopodia of the latter are about equal to the diameter of the body, or between 200 and 1000 μ, although typically nearer the lower length.

The bodies of both animals are highly vacuolated, being almost frothy in appearance. The spherical shape of both species may be distorted by the ingestion of very large food organisms or by a swelling of one or two contractile vacuoles in the periphery. *Actinosphaerium eichhorni* is the more compact with one or two peripheral layers of large, evenly placed vacuoles. In both forms, the vacuoles in the endoplasm have a polyhedral appearance because of the mutual pressure among them. Both heliozoans also have contractile vacuoles that are constant in position and, after filling, contract with such violence that the whole body receives a visible shock. After contraction, the periphery remains indented until the vacuole refills. In *A. sol* the contractile vacuoles empty every minute or two, being more active than those of the larger form. The marine variety of *A. sol*, as with most marine species of protozoans, does not have contractile vacuoles.

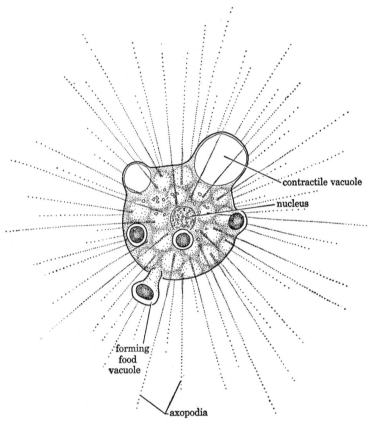

Fig. 15. *Actinophrys sol* feeding on algal zoospores. (Modified after Leidy.)

In spite of their sluggishness, both animals are voracious in their food habits. They ingest algae, algal zoospores, ciliates, and rotifers. Apparently, the axopods are capable of paralyzing small forms, for upon contact, activity of the prey ceases immediately. Rotifers and large *Euglena* are slowed down by contact with several axopodia, whereupon they are trapped by clear protoplasm from the body flowing around them. After digestion, the residual matter is egested.

ACTINOPHRYS SOL AND ACTINOSPHAERIUM EICHHORNI

The ingested food and the vacuolated condition of the body render the nuclei difficult to see. In particularly favorable specimens, or after treatment with dilute acetic acid, the nuclei may be rendered visible. In *A. sol* there is only one, central nucleus. In *A. eichhorni* there may be from one to three hundred. As the cytoplasm of this animal enlarges, the nuclei also enlarge.

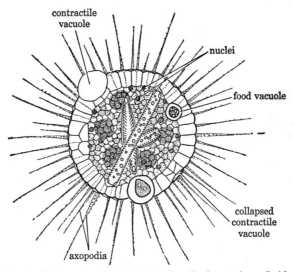

FIG. 16. *Actinosphaerium eichhornii*. (Redrawn from Leidy.)

Both animals usually reproduce by binary fission. *Actinosphaerium eichhorni*, however, also reproduces asexually by budding and by a process of unequal division resulting in pieces with dissimilar numbers of nuclei. This last process is known as plasmotomy.

Sexual reproduction in both animals is very similar despite the difference in nuclear condition. At the beginning of the process, the axopodia are withdrawn and the organism becomes encysted in a case covered with small siliceous spicules; the encysted individual then divides and undergoes reductional divisions to form gametes. After this, there occurs a process of autogamy or self-fertilization. The resulting zygote in *A. sol* divides to form two resting spores from each of which a small individual emerges. In *A. eichhorni* the nuclei of the zygote divide without cytoplasmic division so that the new individual is again multinucleate. *Actinophrys sol* is capable of simple encystment not involving sexual reproduction. This species also at times produces swarm spores which are isogametes.

Monocystis lumbrici *

By W. D. Burbanck

One of the largest and best-known acephaline gregarines is *Monocystis lumbrici* (Fig. 17), which almost always can be found in the seminal vesicles of the earthworms, *Lumbricus terrestris*, *L. rubellus*, and *L. castaneus*.

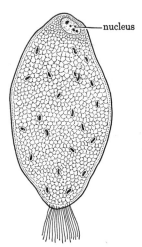

Fig. 17. Trophozoite of *Monocystis lumbrici*. (Redrawn after Hesse.)

Living specimens of this parasite are easily obtained and should be used in this study; moreover, cross sections through the seminal vesicles of infected *Lumbricus* should be used. *Monocystis* is most numerous in the early spring, and examination of earthworms at that time should reveal heavy infections.

To obtain living parasites, an infected worm can be anesthetized by submerging it in 7 per cent alcohol or 5 per cent chloretone. After such treatment, the anterior third of the animal should be slit open along the mid-dorsal line and the body wall reflected. If small pieces of the seminal vesicles, large lobed organs located approximately in segments 9–16, are pinched off with forceps and gently teased in 0.75 per cent NaCl solution, large ciliated-appearing trophozoites (200 by 70 μ) and cysts (160 μ in diameter) will readily be found.

The trophozoite is a parasite of the sperm mother cell and sperm ball of the earthworm. A young trophozoite has very dense and granular cytoplasm. It extracts nourishment osmotically from the sperm ball as it matures, and becomes less dense; cytoplasmic inclusions of paramylum and chromidial granules also become evident with growth. There is a large, well-defined nucleus which internally, in unstained individuals, has irregularly shaped moniliform bodies.

The mature, elongate, mononucleate trophozoite is surrounded by a layer of degenerate sperm cells whose tails superficially give the

* Probably synonymous with *Monocystis agilis*.

trophozoite its ciliated appearance. The cytoplasm of the sperm balls is almost completely used up, but the nuclei remain intact. In addition to the sperm tails, the trophozoite itself has a characteristic tuft of hair-like processes extending from the posterior end of the body. A degenerate, unarmed epimerite is located at the anterior end of the body. Trophozoites are very active, and as they move, their bodies may resemble the shape of an hour-glass, spindle, or top. These shapes and movements may be due to fine longitudinal and transverse fibrils.

As with other gregarines, two mature trophozoites of *M. lumbrici* encyst together in a large, two-layered cyst; this non-sexual union is called syzygy. In the cyst the united trophozoites, now called gamonts or gametocytes, undergo repeated nuclear divisions and ultimately numerous gametes are produced. Since the gamonts do not entirely use up their protoplasm in the formation of gametes, a somewhat vacuolated residual mass is left. Later, the gametes unite in pairs to form zygotes. Male and female gametes come from different gamonts. While still within the cyst, the zygotes secrete spore coats and become smooth, fusiform spores. Within the spore, three nuclear divisions occur and eight nuclei result. In a fully developed spore, each of the eight nuclei, together with a small amount of cytoplasm, is separated into thin, elongate sporozoites.

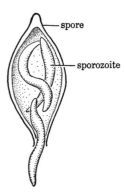

FIG. 18. Sporozoite of *Monocystis lumbrici* escaping from spore. (Redrawn after Hesse.)

Spores of *M. lumbrici* are apparently distributed when the host is eaten by birds, mammals, reptiles, insects, or arachnids; some other species of gregarines in *L. terrestris* escape with the feces or through the dorsal pores of the coelom. Spores do not dehisce in the earthworm, where they were originally formed, nor do they open in the lumen of the gut of the predator. The thick spore coat seems to be able to resist the digestive juices of other animals; consequently, the spores are voided with the feces. However, when these spores are ingested by earthworms, the spore coat is digested and a dehiscence occurs (Fig. 18). The liberated sporozoites migrate through the tissue to the seminal vesicles and transform into exceedingly small trophozoites from 1 to 16 μ in length. The experimental liberation of sporozoites may be demonstrated by examining the gut content of an earthworm previously fed viable spores. Since artificial feeding is a difficult process, better results may be obtained by injecting the spores into the esophagus.

There seems to be no evidence in support of the theory that transfer of *M. lumbrici* from an infected host to an uninfected one occurs during copulation. Also, although most worms are infected with *Monocystis*, their fertility is not greatly impaired, since most of the seminal vesicle is not involved. The worm is apparently able to combat some of the parasites by forming a resistant envelope around trophozoites or by phagocytizing and killing even resistant spores.

Gregarina blattarum
By W. D. Burbanck

The sporozoan, *Gregarina blattarum* (Fig. 19), is an intestinal parasite of cockroaches, such as *Blatta orientalis*, *Periplaneta americana*, and *Blatella germanica*.

Living parasites are best for study; however, cross sections of guts of infected insects may be needed to see all the stages mentioned in this account. Living gregarines may be obtained from roaches by removing the intestinal tract to a microscope slide and splitting it lengthwise. The contents of the tract, including parasites, may be teased free into a drop of either distilled water or Ringer's solution. Plasmolysis is retarded in distilled water, but the parasites remain active for a longer time in the salt solution.

The first stage to be seen will probably be the large, motile, bipartite sporadin or sporont, composed of an anterior protomerite, measuring 100–120 μ, and a posterior deutomerite, measuring 130–450 μ. This lumen-inhabiting stage is dark and dense in appearance, and quite often a spherical nucleus is visible in the deutomerite. Locomotion of a gliding nature on a path of secreted mucus is apparently due to the contraction of longitudinal myonemes. Often two sporadins will be seen in linear association with the protomerite of the second individual or satellite in contact with the deutomerite of the anterior individual or primite. Rarely, one primite may have two satellites of the same or unequal size.

Normally in the host, two associated sporadins will encyst together in a non-sexual union called syzygy. Such an encystment may be induced artificially by placing the associated sporadins in egg albumen. In this medium, the sporadins bend sharply upon each other so that the deutomerite of the primite and satellite are closely applied. The two individuals now begin to rotate, forming a spherical cyst with a hyaline, gelatinous membrane. Later, the cysts become elongate-oval and measure 450–900 μ. The encysted individuals are called gamonts.

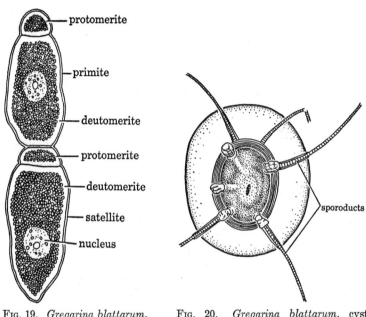

FIG. 19. *Gregarina blattarum*, sporadins in association. (Redrawn after Schneider.)

FIG. 20. *Gregarina blattarum*, cyst with developing sporoducts. (Redrawn after Schneider.)

Cysts are eliminated from the alimentary tract. If they remain moist for several days, gametes develop from the periphery of the gamonts. Next, the gamonts fuse and lose their identity, remaining as a residual mass. The gametes unite in pairs, and spores are formed. After these activities, the cyst produces, on the average, thirteen sporoducts which evert through the gelatinous cyst membrane (Fig. 20). This eversion may be due to an enzymic action. When everted, drops of oil first empty from the ducts. This oil is thought to lubricate the passage for the many spores which follow. Threads, measuring 87 mm. in length and containing as many as 10,000 small spores, 4–8 μ in length, have been observed to issue from the sporoducts.

Within the spore a group of eight sporozoites is formed; either the spore or sporozoite is infective for the cockroach.

A sporozoite, the usual infective stage, upon entering the alimentary tract of the host, penetrates an epithelial cell. Intracellularly, a tripartite trophozoite consisting of an anterior epimerite, a middle protomerite, and a posterior deutomerite is formed. The epimerite remains intracellular as a holdfast, presumably ingesting food. The lumen-forms with epimerites attached are cephalonts. Cephalonts may be free moving in the alimentary tract, but eventually the epimerite is lost and the resulting form is the already mentioned bipartite sporadin.

Plasmodium vivax
By W. D. BURBANCK

Malaria is a disease produced by *Plasmodium*, a type of sporozoan parasite. Four species of *Plasmodium* infect man, while related species occur in other vertebrates. *Plasmodium vivax*, the causative agent of tertian (benign tertian) malaria, is most common in this country. *Plasmodium malariae* causes quartan malaria, *P. falciparum* causes estivo-autumnal (subtertian or malignant) malaria, and the rarer *P. ovale* causes another tertian form of malaria. Since an extensive literature on the epidemiology, life history, and clinical effects of the disease is readily available, only a brief summary of the life history together with a description of the stages of *P. vivax*, which may be demonstrated in the laboratory, will be included here. The other three species of *Plasmodium* affecting man are so similar to *P. vivax* that, with minor modifications, the following account will apply also to them.

Life History. Exceedingly small spindle-shaped organisms, the sporozoites, are injected into the human host by a bite of an infected female mosquito belonging to the Genus *Anopheles*. The parasite does not immediately establish itself in the red blood corpuscles but goes through an exoerythrocytic stage of development before penetrating the erythrocyte. After this stage, which has not been fully described for human parasites, the animal enters a red blood corpuscle, where it feeds upon corpuscular cytoplasm. Young intracellular forms, called trophozoites, feed, grow rapidly, and exhibit considerable amoeboid activity. Since a red-stained nucleus of the trophozoite lies eccentrically in blue-stained cytoplasm containing a colorless vacuole, the organism resembles a signet ring and is called the ring stage. As the organism enlarges, it becomes less active and its nucleus divides into twelve to twenty-four chromatin masses in a characteristic process known as schizogony or segmentation. The parasite is now known as a schizont or segmenter. Next, division of the schizont results in the formation of twelve to twenty-four small intracorpuscular parasites, the merozoites. When the corpuscle ruptures, the merozoites, together with darkly pigmented hemozoin granules, are liberated into the blood stream of the human host. When merozoites are released in large enough numbers (150,000,000), a typical paroxysm, consisting of chills followed by fever, occurs.

Released merozoites usually attach to and penetrate another erythrocyte in which they continue the trophozoite-schizont-merozoite asexual cycle. However, some of the merozoites, because of factors which are not fully understood, develop into two types of gametocytes, the microgametocytes and the macrogametocytes. If blood containing the gametocyte stages is ingested by an infective mosquito, the sexual cycle of *Plasmodium* may be completed in the insect host. Within the stomach of the mosquito, the microgametocyte, in a process known as exflagellation, produces about eight flagella-like microgametes which burst out of the parent protoplasm and lash about. After a less conspicuous maturation, the macrogametocyte is ready to receive the microgamete. Fertilization results in the formation of a motile, elongated cell known as an oökinete (vermicule). This stage burrows into the stomach wall of the mosquito, becoming an oöcyst which, after feeding on host tissue, enlarges and produces by nuclear division, within 10–20 days, hundreds of spindle-shaped sporozoites. After this process of sporogony, the sporozoites break from the oöcyst into the hemocoel through which they eventually find their way into the salivary glands to remain until injected into a new host.

Stages Found in Human Blood Smears. Typical infected red blood corpuscles may be studied from thin blood smears stained with some modification of Romanowski's polychrome method. Giemsa's and Wright's methods are commonly employed. Only slides with a high percentage of infected erythrocytes should be used by inexperienced observers. It will be necessary to use the oil immersion objective, and differentiation between infected cells and leucocytes must be made.

Trophozoite or Ring Stage (Fig. 21A, B). Infected erythrocytes, often somewhat atypical in shape, are usually larger than normal cells.

34　PROTOZOA

Within an erythrocyte, the signet-ring form is usually central or submarginal, having a red nucleus, a vacuole surrounded by blue cytoplasm, and may have fine pseudopodia. Two parasites are occasionally found in a single cell. Bright-red-staining granules, known as Schüffner's dots, may appear within the erythrocyte. As the tropho-

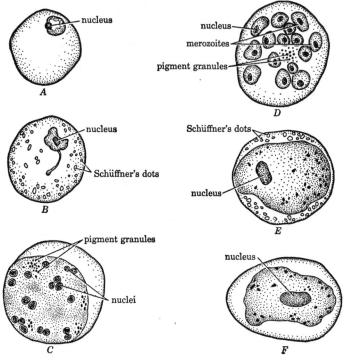

Fig. 21. Stages of *Plasmodium vivax* as seen in smears of human blood. (Redrawn from Hegner and Taliaferro, after Norris.) *A*. Trophozoite in ring stage. *B*. Older trophozoite. *C*. Schizont. *D*. Formation of merozoites. *E*. Macrogametocyte. *F*. Microgametocyte.

zoite grows, small yellowish brown pigment granules appear in the cytoplasm. They become more abundant and darker as the parasite grows.

Specific diagnosis depends upon minor differences which may be detected in this and later stages. In *P. malariae* and *P. falciparum*, there are no Schüffner's granules, and infected erythrocytes are not enlarged. *P. falciparum* frequently produces a multiple infection of the erythrocyte, and many parasites lie near the margin of the red corpuscle. Although *P. ovale* causes enlargement of the erythrocyte and Schüffner's granules are present, it is often found in distorted, oval-shaped corpuscles which also possess a ragged edge.

Schizont (Fig. 21C). The nucleus, which was a single mass in the growing trophozoite, has divided into a number of masses, the typical number being sixteen in *P. vivax*. The cytoplasm of the parasite shows various degrees of separation into strands and particles, and pigment tends to aggregate in isolated clumps.

Merozoite (Fig. 21D). In the mature schizont twelve to twenty-four merozoites (1.5 μ in diameter), each having a nucleus and a small amount of cytoplasm, will be found. The pigment is concentrated in one or two clumps. At this stage, practically the entire erythrocyte is filled with the parasite.

Macrogametocyte (Fig. 21E). Although occasionally macrogametocytes are present in early infections, more typically they appear later than the trophozoites. Young macrogametocytes resemble full-grown trophozoites, but the pigment is more regularly scattered through the cytoplasm and the cytoplasm is more homogeneous and lacks vacuoles. The regular outline of the parasite is circular or oval. Full-grown macrogametocytes (9–10 μ in diameter) are usually larger than mature trophozoites.

The gametocyte stages of *P. falciparum* are sausage-shaped and lie within elongated erythrocytes. Since this stage is more easily recognized in this species than in *P. vivax*, it is usually figured as the typical gametocyte stage for the genus.

Microgametocyte (Fig. 21F). The microgametocyte (7–8 μ in diameter) contains less cytoplasm than the macrogametocyte and stains more lightly. The nucleus is more diffuse and larger, often assuming a stellate or spindle shape.

Stages Found in the Mosquito. Microgametes and macrogametes: If gametocytes are numerous in the blood, the process of exflagellation of the microgametocyte may be demonstrated on a slide by drawing out some blood into citrated physiological saline solution. Stained smears may also be obtained from supply houses. The microgamete or flagellum is seen to have a nucleus located somewhere along its length, and sometimes a structure resembling an undulating membrane may be seen.

Little change is usually noted in the macrogametocyte before fertilization, but extrusion of polar bodies has been described.

Oökinete. Oökinetes formed from the union of a microgamete with a macrogamete may be demonstrated in permanent slides, or smears may be made of the infected stomach of a mosquito. This elongated body glides along in a manner which has been described as similar to gregarine movement.

Oöcyst. The oöcysts appear as tubercle-like enlargements in the muscular layer, or projections from the outer epithelial wall, of the stomach of an infected *Anopheles* mosquito. Stained sections will demonstrate the nuclear divisions which accompany sporogony or sporulation.

Sporozoites. Sporozoites are best demonstrated from smears made from salivary glands of infected mosquitoes. They are elongated filaments somewhat thickened in the center. A central, darkly stained nucleus is surrounded by lighter cytoplasm.

Opalina obtrigonoidea

By W. D. BURBANCK

The opalinids, intestinal parasites in Anura, have been useful in understanding the world-wide distribution and relationships of their hosts. These astomous Protozoa have also been of interest because they may be living representatives of animals, which, in the distant past, gave rise to the Euciliata or true ciliate Protozoa.

Opalina obtrigonoidea (Fig. 22) is found in the rudimentary rectal caecum of two common frogs, *Rana pipiens* and *Rana palustris*. It is also found in as many as seventeen other species of frogs and toads.

Living opalinids, obtained from the recta of pithed adult or larval infected amphibian hosts, should be studied, if at all possible. However, stained whole mounts also may be used as a substitute or to supplement the study of living organisms.

The body of *O. obtrigonoidea,* as seen in amphibian Ringer's solution, resembles a scalene triangle and is from 400 to 500 μ in length by about 180 μ in width. The anterior end of the body is very much flattened dorsoventrally, while the posterior part of the body is rounded. The pellicle, although thin, is relatively tough and raised into sharp, longitudinally spiralling folds between which parallel rows of very long cilia protrude. Seen dorsally, the anterior right-hand corner is called the apex. Near this, a pair of stout fibrils, the falx, runs along the anteroventral surface. A furrow in the pellicle, lying between the very large falcular cilia, is sometimes intrepreted to be a vestigial cytostome. Since the falcular fibrils are connected by other fibrils to the basal granules of the cilia, the falcular fibrils are thought to have a neuromotor function.

The basal granules of the cilia lie directly beneath the pellicle. Internal to these is an alveolar ectoplasm. The endoplasm is gelatinous and does not flow from cuts made in the body. Within the endoplasm, numerous evenly spaced monomorphic nuclei are found. Scattered among the nuclei are even more numerous spherules of unknown function. In the posterior end of the body, there are a so-called "excretory pore" and "bladder."

Opalina may be cultured under anaerobic conditions by using sterile technique. Pütter's fluid, consisting of 100 parts of 8 per cent solution of sodium chloride, 5 parts of 30 per cent solution of Rochelle salts, and 400 parts of distilled water, has been recommended as a stock solution in this technique. The medium is boiled and is then subjected to suction to remove all the dissolved air. To this fluid, powdered egg white, in the proportions of 0.25 gm. per 200 cc., is slowly added to prevent flocculation. Opalina washed in sterile Ringer's or Pütter's fluid are then placed in 9 cc. of the nutritive mixture contained in a 15-cc. sterile tube. A sterile one-hole rubber stopper is placed in the culture tube, and the final seal is made by placing a small glass plug in the hole. Under such conditions, animals multiply over a period of about 25 days before transfer to fresh medium is necessary. During this period, the opalinids may be studied in the culture tube with the aid of a dissecting microscope.

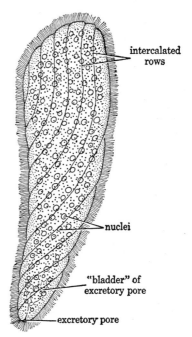

FIG. 22. Opalina obtrigonoidea. (Compiled from Metcalf and from Cosgrove.)

Opalinids reproduce both asexually and sexually. Binary fission may be transverse, longitudinal, or oblique. In the spring, numerous small individuals with one or several nuclei are produced. These encyst and are voided in the feces of the host. The cyst, covered with a resistant hyaline membrane, settles to the bottom of the particular body of water where young tadpoles of various species feed. When ingested by tadpoles, the parasites excyst and by a series of longitudinal divisions become differentiated into macro- and microgametes. The microgamete swims tail first and appears to be very sticky. It eventually fuses with a macrogamete to form a zygote. The latter grows into an adult opalinid as its host metamorphoses or forms an infectious cyst which is voided with the feces of the host.

Didinium nasutum

By W. D. BURBANCK

Didinium nasutum (Fig. 23) is an extremely voracious ciliate. Its enormous appetite is satisfied by a diet of such large ciliates as *Paramecium, Frontonia, Colpoda, Colpidium,* and *Tetrahymena*. In nature, it lives in the same freshwater habitat as its prey, and, cultured artificially, it flourishes in various infusions capable of supporting its food organisms.

Normally, *D. Nasutum* is 80–200 µ long and barrel-shaped, but, when distended with food, is almost oval. A well-fed animal appears very dark and dense, whereas a starved one becomes vacuolated and almost transparent.

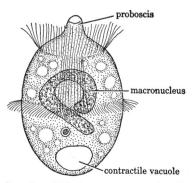

FIG. 23. *Didinium nasutum*. (Modified after Kudo.)

The anterior end of *Didinium* is prolonged into a tube-shaped proboscis or seizing organ made up of parallel trichocyst-like bodies which give the organ a hyaline appearance. A mouth opening or cytostome is at the tip of the proboscis. Two girdles of cilia, also called pectinelles, encircle the body. One ring of cilia is close to the base of the proboscis, while the other is about one-third of the total length from the posterior end. Often, just before fission, animals are seen with four rings of cilia.

Within the endoplasm, a horseshoe-shaped nucleus may often be seen in unstained specimens or by staining with acetocarmine. A contractile vacuole and cytopyge are both in the extreme posterior end of the animal. An active cyclosis carries solid wastes to the cytopyge opening, where they are voided.

In its erratic hopping, bobbing, spinning movements, *Didinium* reminds one of a child spinning and bouncing on a pogo stick or a figure skater whirling or pirouetting on one skate. In *D. nasutum*, the "stick" or "skate" is the protrusible proboscis or seizing organ which is used by the animal to test every object with which it comes in contact. If the proboscis contacts a soft animate object, it becomes fastened. The cytostome may then open to engulf a whole large protozoan, or the contents of a smaller one, such as *Tetrahymena*, may be sucked through the unexpanded proboscis. Contact with the seizing organ may also break *Tetrahymena* or *Colpidium* into pieces so that they cannot be ingested. With either *Paramecium caudatum* or *P. aurelia*, the most frequently eaten food of *Didinium*, the cytostome is opened widely and the proboscis withdrawn which pulls the prey through an opening large enough to engulf it.

When *Didinium* is introduced into cultures of large *Paramecium multimicronucleatum*, many of the latter are ingested. Moreover, peculiar truncated pieces of paramecia may be observed; they are the living fragments of *P. multimicronucleatum* that proved too large for didinia to engulf completely after individuals had been attacked. In its attempt to ingest these large ciliates, *Didinium* may distend so much that it ruptures and dies, thus liberating its crippled victims. Large paramecia have also been observed to throw off masses of trichocysts when attacked. Small didinia are sometimes successfully warded off with this defense mechanism, and they swim away with *Paramecium* cytoplasm clinging to their proboscides. In attacking *Stentor*, *Spirostomum*, hypotrichs, rotifers, and each other, didinia are unsuccessful. In the last case, however, two animals may cling together for a short time.

Not only must didinia have plenty of food ciliates, but these ciliates must be in good condition before they will support normal growth. A healthy *Didinium* and its progeny will ingest eight large paramecia in one day.

In a rapidly growing culture of *Didinium*, transverse fission occurs at the rate of three to four divisions per day. In finger-bowl cultures, numerous pairs of dividing animals may be seen to pull apart suddenly. If the parent animal has been well fed, the daughter cells may again divide even though their food supply is inadequate.

Although asexual binary fission is the chief means of reproduction in *D. nasutum*, sexual reproduction, as evidenced by the conjugation of two mature individuals, also occurs.

Under adverse conditions nearly all the didinia in a culture will encyst after settling to the bottom of a culture or some other substratum covered with a bacterial zoogloea. In this somewhat gelatinous material encysting animals spin round and round. Finally, they become quiescent, having formed a cyst wall several layers thick. For some unknown reason, cultures of *Didinium* do not always encyst under suboptimal conditions and animals may continue to swim until they round up, become vacuolated and feeble, and die. Since nuclear reorganization takes place while animals are encysted, this process may be followed by collecting and staining encysted forms at spaced intervals. Encysted *D. nasutum* kept in small vials of spring water have remained viable up to ten years.

Excystment of *Didinium* is also interesting to observe. Initially, in this process the endoplasm begins an intense cyclosis within the cyst. Later, the cyst wall becomes thinner, and an internal vacuole becomes larger. Suddenly, the animal breaks through the greatly stretched and barely visible cyst wall and swims free. Excystment may be induced by adding a heavy concentration of well-fed food ciliates to cultures containing viable cysts; filtrates made of the medium in which the food ciliates have lived will also have the same effect.

Coleps hirtus

By W. D. BURBANCK

Coleps hirtus (Fig. 24) is a widely distributed fresh-water ciliate with an ability to tolerate a wide range of temperature, oxygen, salinity, and acidity. Furthermore, *C. hirtus* possesses "armor" which protects it from Suctoria and *Didinium* known to feed on other small ciliates.

Because of its ability to survive a wide variety of environmental conditions, *C. hirtus* may be collected in many dissimilar habitats during all seasons of the year. It is most abundant, however, in the warmer months. *Coleps* does not grow well in the laboratory on hay infusions, but small jars containing growing *Oscillatoria* and various filamentous green algae will support this ciliate for several weeks.

The barrel-shaped body of *Coleps hirtus* is brownish, is slightly more convex dorsally, and is 40–65 μ long by about 20–30 μ wide. In the alveolar layer of the ectoplasm beneath the pellicle, the body is covered with armor plates which are modified into oral teeth at the anterior end, and caudal spines at the posterior end. The plates are organic in composition and are arranged in four anteroposterior girdles of twenty plates each which encircle the animal like the staves of a barrel. Each plate bears elevated rectangular areas that give the surface of the ciliate the appearance of being divided into more or less symmetrical squares. These details may be confirmed by allowing a 20 per cent solution of sodium carbonate to run under the coverglass to dissociate the armor plates. Shorter cilia are located between the longitudinal plates, and longer ones, important in locomotion, occur around an anterior cytostome. The ciliate swims at a moderate speed and rotates on its long axis in a counterclockwise direction. The slight dorsoventral asymmetry may be the cause of a wobbling of the anterior end.

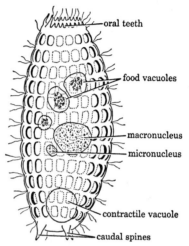

Fig. 24. *Coleps hirtus*. (Modified after Noland.)

The fact that *Coleps* congregates around the decomposing bodies of small aquatic animals may indicate that this ciliate is a scavenger or it may be eating bacteria present under such conditions. An anterior cytostome is capable of great expansion, enabling the animal to ingest large protozoans and green flagellates. At times, the endoplasm may be so filled with large numbers of food vacuoles that the body shape is distorted.

A centrally located spherical macronucleus and micronucleus are present. A contractile vacuole is posterior, opening by a lateral pore. A subterminal cytopyge is also present.

Coleps hirtus has never been seen to conjugate, although individuals of other species of the genus do conjugate by a fusion of their oral ends.

Transverse binary fission of this form is interesting. The central elongated dividing area is free of plates, a situation which obtains even after the anterior and posterior parts have separated. As it takes a while for the plates to form on the transparent halves, the two daughter cells look superficially like acorns projecting from their cups.

Holophrya simplex

By W. D. Burbanck

In sewage-free fresh, brackish, or salt water covered by a rich bloom of algae, the small ciliate, Holophrya simplex (Fig. 25), can usually be found in the surface scum. Although inconspicuous and relatively simple in structure, its voracious food habits make it a significant link in the ecological food chain.

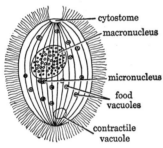

Fig. 25. Holophrya simplex. (Redrawn after Kahl.)

Holophrya simplex varies greatly in size; a given individual is capable of great elongation and contraction. However, an approximate size of 34 μ long and 18 μ wide may be given. These variations, coupled with the ability of this ciliate to ingest large amounts of algae and microscopic crustacea, result in great distortions of the oval body.

Superficially, H. simplex might be confused with Prorodon, for the two forms have the same general shape, movements, and environment. However, the former has no trichites forming a conspicuous oral basket. In fact, the terminal cytostome of H. simplex is evident only when food is being ingested. This species of Holophrya typically has eighteen to twenty rows of cilia. It is possible to demonstrate this fact by the Klein silver-line technique. By this method, the basal bodies or granules, representing the origins of cilia, may be stained.

Like the cytostome, the cytopyge of H. simplex is evident only when material is passing through it. Both the contractile vacuole and its pore are posterior. There is a large spherical central macronucleus and a small ellipsoidal micronucleus adjacent to it. At times the endoplasm may be so filled with food vacuoles that the internal organelles are obscured in not only the living but also the stained specimen.

Asexual reproduction is accomplished only by binary fission. Unlike Prorodon, no temporary division cysts are formed. However, somewhat gelatinous resting cysts are typical for this holophryan.

Like Prorodon, conjugation occurs by fusion of the oral regions and nuclear exchange takes place through the endoplasmic bridge.

Prorodon griseus

By W. D. Burbanck

In temporary rain pools such as hog wallows, *Prorodon griseus* (Fig. 26) is often found in large clumps at the surface of the water. Even when the pools dry up, sediment taken from the bottom and placed in water may yield great numbers of excysted *P. griseus* within 24 hours.

During the trophic, free-swimming state *Prorodon* is somewhat variable in shape, but, when fully extended, it is between 165–200 μ in length and 40–60 μ wide. The body is ellipsoidal, with the anterior end slightly wider than the posterior. At times animals may double or triple their volumes due to the very large amount of food which they ingest. In the expanded state, the ciliate rotates on its long axis, propelled smoothly and rapidly by fine, dense cilia which uniformly cover the body. At the anterior end the cilia are somewhat longer than elsewhere.

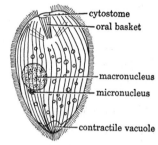

Fig. 26. *Prorodon griseus*. (Redrawn after Kahl.)

The ectoplasm of *P. griseus* contains no trichocysts or other inclusions except for twenty to thirty deeply embedded, large, parallel, rod-like trichites. These form a highly distensible oral basket around the subterminal cytostome.

Although the endoplasm is usually grayish, *Prorodon* may occasionally become green because of the presence of numerous intracellular, symbiotic zoochlorellae. The food vacuoles are large due to water taken in with the food. Food is digested quickly, aided by cyclosis and interrupted periods of churning which are said to occur in the cytoplasm. As the food vacuoles become reduced in size, indigestible particles are voided through a cytopyge at the posterior end.

The voluminous macronucleus is usually located centrally and is either spherical or elongated. It has a prominent nucleolus, and a compact pear-shaped micronucleus which is attached externally to its nuclear membrane. Behind the nuclei, at the extreme posterior end, is the large contractile vacuole. It is supplied by canals converging from all parts of the body.

Although some trophic forms are to be found in cultures at all times, the animals spend most of their time in one of two kinds of cysts. When taken from dried mud, animals excyst from relatively small thick-walled protective cysts. Almost immediately after excysting, the shortened animals conjugate, remaining fused at their anterior ends for about 4 hours. If, during this period, conditions become adverse, conjugation may continue in a large temporary thin-walled cyst formed by the two united individuals. During conjugation, there is an exchange of nuclear material. Interestingly enough, animals in conjugation are seldom collected from natural habitats.

Binary fission takes place after the ex-conjugants expand, lengthen, swim freely, and feed. Again, if conditions are altered too radically, fission may occur within a temporary cyst. Both the new oral baskets and the contractile vacuoles are formed before the nuclei divide. The macronucleus divides before the micronuclei. Nuclear division takes about an hour and a half and is completed before the new individuals separate. It is, therefore, possible to prepare stained mounts showing nuclear division by fixing animals at the time when the division furrow first becomes visible. In laboratory cultures, free-swimming, dividing animals will generally be found in large numbers until about 5 hours after conjugation, when they again form protective cysts.

Macronuclear reorganization takes place only in the permanent protective cyst and requires at least 4 days. Both the cilia and the oral basket are lost during this process, but the contractile vacuole and micronucleus remain essentially unchanged. After the macronucleus is reorganized, the lost parts are reformed, and, if conditions are favorable, the animal is ready for excystment.

Paramecium caudatum

By W. D. Burbanck

Of all the ciliates, probably the best known are the various species of the genus *Paramecium*. One of the largest and longest known of these species is the cosmopolitan *Paramecium caudatum* (Fig. 27). Its widespread distribution is surprising, for there is no definite proof that this or other species of the genus form resistant cysts. Since *P. caudatum* lives only in isolated bodies of fresh water, methods of dispersal that have been suggested are by means of the feet of aquatic birds and by way of droplets of water carried by the wind from one body of water to another.

For general laboratory use, P. *caudatum* will grow satisfactorily on a boiled-hay infusion medium to which wheat or rice grains have been added. It has also been maintained on several complex sterile media, the simplest of which is yeast juice. Other less uniform but very satisfactory methods of growing paramecia include a monofloral suspension of the bacterium, *Pseudomonas ovalis*, and a suspension of pure, living yeast.

Paramecium caudatum is moderately large, being 180–300 μ long by 45–75 μ wide. The body is covered by a semi-rigid pellicle which maintains the characteristic slipper shape of the animal. This thin, tough covering has a surface pattern of hexagonal depressions which may be demonstrated by allowing animals to dry on a microscope slide in a suspension of India ink. Cilia project through the pellicle in the center of each hexagonal area. All the cilia are of uniform size except at a somewhat pointed posterior end of the body, where they are longer, and in the cytopharynx, where they form an undulating membrane.

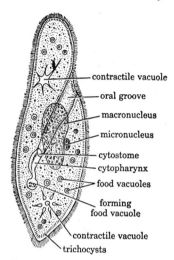

FIG. 27. *Paramecium caudatum*. (Modified after Hyman.)

In the peripheral ectoplasm, underlying the pellicle, are basal granules of the cilia and trichocysts, thousands of tiny rod-shaped, glassy bodies oriented perpendicularly to the surface. Electron microscopy has shown that each trichocyst consists of a minute oval body with a hood-shaped cap covering a pointed tip. After chemical or mechanical stimulation, trichocysts are extruded by a rapid uptake of water. The rapidity of elongation during extrusion suggests that a cylindrical membrane covering the undischarged trichocyst is preformed in the resting trichocyst. The extruded trichocyst consists of a sharply pointed tip and a spindle-shaped shaft which projects through the pellicle into the surrounding water. Some evidence has been adduced to show that the tangle of trichocysts discharged around a predator is effective in driving it away. However, it is likely that the usual function of the trichocysts is to anchor the animal temporarily during feeding, only a few being used at a time for this purpose.

When observed from the oral side, a prominent channel, the oral groove, is seen to extend from the right side of the anterior end diagonally across the animal to the left posterior half, where it becomes closed over at a point called the cytostome to form a cylindrical cytopharynx or gullet, which extends back to about the posterior third of the body, where it ends in the endoplasm. Bacteria, small flagellates, and organic detritus are funneled into the oral groove as a consequence of the action of the body cilia and the undulating membrane. The food vacuoles are formed by pinching off the elongated watery sacs which grow successively in endoplasm of low viscosity at the bottom of the cytopharynx. The growth and pinching off of the vacuoles are probably brought about by the contraction of endoral fibrils in the endoplasm posterior to the cytopharynx. The food vacuoles in *Paramecium* after leaving the cytopharynx pass rapidly on a fixed course toward the posterior end of the body and slowly move on a varied course to the cytopyge. During the rapid movement, due to the action of the endoral fibrils, and the slow movement, due to cyclosis, the food vaculoes undergo a regular series of changes in size and in the appearance of the food within. The fact that digestion occurs within these vacuoles shows that digestive enzymes must be secreted into them by the endoplasm. The origin of the changes in acidity in the vacuoles is less certain, both because the colorimetric methods used to measure hydrogen-ion concentration are unreliable, and because evidence has been put forward which relates the changes in acidity to changes in the water content of the vacuoles. Finally, undigested or indigestible material is egested through a fixed opening, the cytopyge, which lies a short distance posterior to the cytostome.

Paramecium caudatum normally has two contractile vacuoles, one about a fifth of the body length from the anterior end, and one a similar distance from the posterior end. Each is fed by six or more radiating canals in the endoplasm, which enlarge during the systole of the vacuole, then discharge into it as it grows. Each vacuole empties to the outside through a pore which is constant in position, but invisible except at the moment of discharge. The contraction of the vacuole is almost instantaneous. The posterior vacuole contracts at a slightly greater frequency than the anterior. The contractile vacuole is generally agreed to function in eliminating the osmotic water which is constantly entering through the pellicle and food vacuoles. Thus, paramecia put into hypertonic solutions show a reduced rate of contraction of the contractile vacuole. The process is one requiring energy, as indicated by the fact that paramecia deprived of oxygen swell up and burst because of their inability to eliminate the osmotic water.

Paramecium caudatum has a massive macronucleus which usually has an indentation along the side or end. In this depression lies a single, fairly large, compact micronucleus. Division of both the macro- and micronucleus of *P. caudatum* may be easily observed by staining with aceto carmine at the time the animal undergoes binary fission. A more complicated nuclear picture is obtained, however, if two animals are in conjugation, or if autogamy is taking place. In all of these cases of nuclear reorganization, the macronucleus will be seen in a fragmented condition.

As in five other species of *Paramecium*, *P. caudatum* contains several "sexually" differentiated classes. In *P. caudatum*, five such classes have thus far been described. Individuals from each of the classes or varieties will not conjugate with members of any of the other four varieties. Each variety in this species is made up of two mating types. Animals belonging to one type will not conjugate with members of their own type, but they will conjugate with those of the second type in their variety.

Paramecium is excellent material for the study of simple behavior because of its predictable reactions to various stimuli. When an individual collides with an object, it backs up; this is an avoidance reaction brought about by a reversal of the ciliary beat. After backing off, it swims forward again in a slightly different direction and by such successive trials may continue until it has passed around the obstacle. In a dense culture of actively swimming paramecia, small drops of very weak acetic acid introduced into the water "trap" many of these ciliates. Because of their natural roving movements, paramecia enter a drop of weak acid, and at a point where they pass into the surrounding water they give an avoidance reaction. Apparently, the change of environment from the optimal condition provided by the weak acid acts as a stimulus when the animal comes in contact with the non-acid water at the edge of the drop; thus, paramecia are prevented from escaping from the drop and large numbers of animals may accumulate in a single drop. When placed in a cylinder of "hay-tea" or spring water, paramecia exhibit a negative geotaxis and may be found in a ring around the top of the vessel after a short time—cultures of paramecia may be concentrated in this way. Paramecia introduced into small troughs of water, with a temperature gradient from 10 to 38°C., will accumulate at their optimal temperature of 25–26°C. Here again, the final position of the animals is due to a series of avoidance reactions.

The reaction of *Paramecium* to an electric current is different from that produced by any other agent. When the current is weak, animals swim toward the cathode, whereas under the influence of stronger currents the ciliates swim toward the anode. These results are not due to normal "trial and error" activities typical of *Paramecium*; but rather, under the influence of both constant electrical currents and induction shocks, *Paramecium* responds as though it were forced to behave as it does. The reaction to electricity, therefore, is apparently a purely laboratory product.

Colpoda cucullus
By W. D. BURBANCK

Because of its widely scattered cysts, *Colpoda cucullus* (Fig. 28) is one of the most easily obtained protozoa. Unheated infusions of dried hay, moss, and soil almost always yield active individuals in about 24 hours.

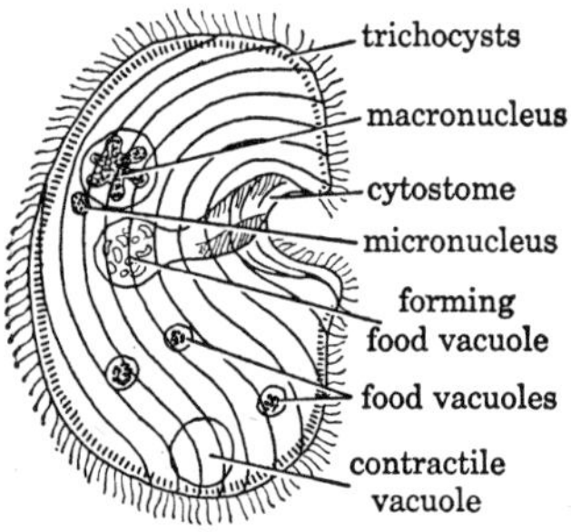

Colpoda cucullus is a moderately large holotrichous ciliate. Average individuals are 80 μ in length, while well-fed specimens measure up to 120 μ. The shape of the animal superficially resembles a flattened kidney bean. A clearly defined, diagonal groove is located on the aboral surface extending from the right side of the posterior third of the body to the middle of the left side where it forms a prominent notch. The groove continues for a short distance on the aboral surface to terminate in an oval cytostome lying near the left side. Cilia are borne on the right side of the short cytopharynx, and rows of cilia on the left side of the same organelle give that area a cross-hatched appearance. These cilia sweep bacteria and other small organisms into a yellowish endoplasm, thus forming numerous dark food vacuoles. There are delicate cilia all over the body of the animal; they are somewhat more prominent on a keel-like area with a series of eight to ten indentations which lies anterior to the cytostome.

Embedded in the ectoplasm are small clear trichocysts which have a doubtful function in this species. Within the endoplasm, near the center of the extreme posterior end of the body, the contractile vacuole may be seen. Its contents are voided through a dorsal pore. The

FIG. 28. *Colpoda cucullus*, oral view. (Redrawn after Kahl.)

fairly large macronucleus and small micronucleus lie on the right side of the animal just anterior to the level of the cytostome.

Conjugation is seldom observed in artificial cultures, but when it does occur there is a fusion of the preoral regions of both animals. Asexual reproduction takes place in thin-walled division cysts. Usually, there are two divisions, resulting in four individuals. While in this type of cyst, part of the endosome of each of the four nuclei is extruded. It is thought that this reorganizational process may in some way take the place of the nuclear changes which occur during conjugation.

The earlier-mentioned resting cysts are covered with a thick-walled protective covering. In such protective cysts, the quiescent animal may remain for long periods of time. Animals in such cysts have remained viable even when subjected to temperatures of liquid air ($-180°C$.) for $12\frac{1}{2}$ hours, $70°C$. for 26 hours, and $106°C$. for 1 hour.

Frontonia leucas

By W. D. Burbanck

Frontonia leucas (Fig. 29) is a large, common ciliate capable of rapid movements. Although usually colorless, populations of green animals are found which contain symbiotic zoochlorellae.

Seen from the aboral side, *F. leucas* is shaped like a left-moccasined foot. The anterior end is somewhat flattened dorsoventrally, whereas the posterior end is more cylindrical. In clone cultures, the average length is 150–350 μ and the width is 70–150 μ. Individuals from wild cultures have been reported to be twice the maximum sizes given here.

Frontonia is apparently omnivorous, for it can subsist on bacteria, algae, amoebae, arcellae, and even rotifers. The usual food, however, seems to be *Oscillatoria*, diatoms, and other algae. On occasion, filaments of *Oscillatoria* two or three times the length of the animal may be seen rolled up in the endoplasm. It is difficult to grow *F. leucas* in hay infusion cultures, but it can be maintained successfully for a week or two in small jars containing growing cultures of the above algae.

Except in the oral region, which deserves special notice, the animal is uniformly covered with cilia. Embedded in the ectoplasm are numerous fusiform trichocysts, usually arranged at right angles to and in contact with the pellicle. As seen with a high power of the microscope, the trichocysts give the edge of the body a serrated appearance and, in surface view, appear as small circles. Medially, they extend into the fluid endoplasm. When pressure is applied to the coverglass or irritants are run under it, the trichocysts suddenly become transformed into long hair-like filaments that project from the whole surface of the body.

Of all the structures thus far mentioned, the complex cytostome is probably the most characteristic of the genus. Located ventrally in the anterior fourth or third of the body, the cytostome proper is shaped more or less like a heart with its point toward the anterior end. Seen from the oral surface, the cytostome is closer to the right side. From this side a very narrow postoral groove extends almost to the posterior end of the body, then swings toward the opposite side at its posterior end. Because this groove can open widely, the animal is able to eat long filamentous algae and such sizable food as rotifers.

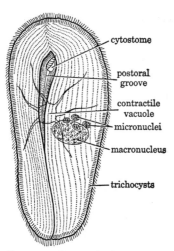

FIG. 29. *Frontonia leucas*, oral view. (Modified after Bullington.)

Very fine cilia form a boundary along the right side of the cytostome proper and extend the length of the postoral groove. On this same side in the cytostome proper there is a complex, three-layered undulating membrane. On the opposite side, there are rows of semimembranous cilia. The innermost cilia of this group extend across the cytostome into the groove.

Another unusual organelle of this infusorian is a very large, centrally located contractile vacuole. This vacuole is fed by eight to ten canals, some of which may be branched at their distal ends. The pore of the vacuole opens through the aboral side of the body.

At approximately the same level of the contractile vacuole is a large oval macronucleus. Either embedded in this nucleus, or near it, are the micronuclei. These are usually not less than three or more than five in number. During conjugation, the macronucleus fragments into small rounded bodies. Also, during conjugation, the complex oral apparatus seems to disappear while the animals are joined by their oral surfaces. The ex-conjugants, however, reform the apparatus as soon as they separate. Also, during binary fission, the oral apparatus is reabsorbed and is formed anew by each of the daughter cells.

Colpidium colpoda

By W. D. Burbanck

Colpidium colpoda (Fig. 30A) has been reported from fresh water, sea water, and soil. The size in sea water is not as great (45 μ) as in fresh water, and the anterior end of the marine form is straighter than that of the fresh-water one. These ciliates are especially numerous in fresh water contaminated with sewage. Apparently, under such conditions, *C. colpoda* feeds mainly on the enormous numbers of bacteria present.

Colpidium is roughly pear-shaped with the anterior end somewhat drawn out and twisted to the right. The animal habitually lies with the cytostome to the right. A slightly concave side, conventionally called the ventral side, is in contact with the substratum, whereas the opposite, or dorsal, side is somewhat convex. The twisted appearance is further enhanced by a diagonal line, a preoral suture, sculptured in the pellicle, which, when seen from the oral surface, curves from right to left and terminates in a small triangular cytostome at the posterior end of the depressed area. Under

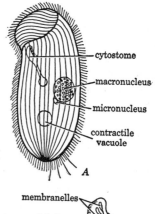

Fig. 30. A. *Colpidium colpoda*, oral view. (Redrawn from Kudo, after Kahl.) B. *Colpidium colpoda*, cytostome. (Modified from Kudo, after Furgason.)

normal conditions, *C. colpoda* is 90–150 μ in length and one-third to one-half as broad as long. In old cultures, starved animals become very small.

For a protozoan of such apparent simplicity, the cytostome is quite complex. The cytostome is bordered along its right margin by an undulating membrane supported by an ectoplasmic extension, and along its left by three membranelles. The shape and structure of the cytostome are of considerable taxonomic importance because they distinguish this ciliate from other closely related genera (Fig. 30B).

If specimens are prepared according to the Klein silver-line technique, basal bodies of the cilia will be rendered visible and fifty-five to sixty meridional rows of these basal granules can be counted.

Typically, there is also a postoral row or meridian. As in other holotrichous ciliates, the body of *C. colpoda* is completely covered with functional cilia of uniform length. At the posterior end, however, there are also some very long, stout cilia, the presence of which may account for the ability of the animal to make the quick turns and gyrations that it exhibits from time to time.

Colpidium colpoda has a centrally located, round or ellipsoidal macronucleus and a small micronucleus. As yet, there is no evidence of any nuclear reorganization other than the occasional extrusion of small granules from the nucleus. There is a simple, dorsal, contractile vacuole in the posterior third of the body. The cytopyge, also posterior, is ventral.

During 10 years of observation of animals grown in hay-tea, no conjugation or encystment was observed. Reproduction under these laboratory conditions, therefore, appears to be only by binary fission.

This ciliate has been grown successfully on suspensions of twenty-five different species of bacteria. With pure cultures of bacteria, washed animals, and an aseptic technique, the highest division rates have been obtained when bacteria of the Family Enterobacteriaceae were used as food for *C. colpoda*.

Strangely enough, this ciliate does not seem to grow on a non-living non-particulate medium even though representatives of several closely related genera, and even species in the same genus, will grow well under such conditions. Although growth is reduced to one division every 48 hours, bacterially free animals will grow in a sterile non-living medium made up of supernatant 5 per cent autoclaved brewers' yeast added to a suspension of the bacterium, *Pseudomonas fluorescens*, which has been heat killed at 60°C.

Because of its constant predictable division rate on bacterial suspensions and non-living media, *C. colpoda* is excellent material for experimental work. It may be used in such population studies as an analysis of the growth curve, intra- and interspecific competition studies, and problems of nutrition.

Spirostomum ambiguum
By W. D. BURBANCK

The unusually large ciliate, *Spirostomum ambiguum* (Fig. 31), is often found in shallow, quiet ponds among fallen tree leaves which are undergoing decomposition and giving off hydrogen sulfide. During the day, the animals take cover beneath the leaves, but at night they may be found swimming freely. Numerous tests of the hydrogen-ion concentration of both natural habitats and artificial cultures indicate that *S. ambiguum* can tolerate a range of pH from 6 to 8 with an optimum at 7.4.

Unlike many other ciliates which flourish in simple infusion cultures, *S. ambiguum* does not grow well in shallow, wide-mouthed culture dishes. Apparently, in such a medium, it requires a relatively large volume with a small open surface. Such an environment may be provided by the use of standard bacteriological culture or test tubes. Two grains of wheat boiled in aquarium water and allowed to stand for 2 or 3 days will provide a heavy bacterial growth to which the spirostoma may be successfully transferred if the volume of the added culture, the inoculum, is not too small. A less easily controlled but more natural medium may be made by boiling some of the leaves found in the pond where the animals were collected. Both these methods are useful in acclimatizing the animal when it is being transferred from wild to artificial culture, but very vigorous cultures may be established, even in shallow dishes, if a tablespoon of cow manure is added to a liter of timothy-hay wheat-grain infusion. When such cultures show signs of dying out, an addition of more manure rejuvenates them. In any event, the optimal pH of 7.4 should be maintained for successful results.

There are several gross morphological structures that are particularly noticeable. The peristome, seen from the dorsal side, i.e., with cytostome on right, occupies two-thirds of the right side of the ciliate. It turns obliquely to the left at its posterior end. Beneath the delicate pellicle, there is a furrowed ectoplasm which is arranged in conspicuous parallel stripes running obliquely to the

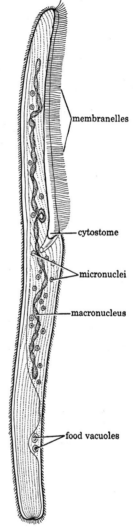

Fig. 31. Oral view of *Spirostomum ambiguum*. (Redrawn after Stein.)

left from the anterior end. Beneath each furrow, next to the basal bodies of the cilia, are contractile ectoplasmic fibrils, the myonemes. When *S. ambiguum* gives an avoidance reaction, the myonemes contract, causing the stripes to appear like spirals around the shortened body of the animal. In addition to these structures, a very large, posterior contractile vacuole and a long, beaded, refractile, centrally located macronucleus are very evident.

At the extreme anterior end, at a point from which the myonemes radiate, the peristomial membranelles begin their spiral. As seen from the dorsal aspect, these membranelles lie on the lower right margin of the long oral groove. At the cytostome they turn abruptly to the left, passing through it into the cytopharynx. Although the cytopharynx has no undulating membrane, there are rather specialized cilia occupying the side opposite the membranelles. Such an arrangement of cilia and membranelles enables this large ciliate to ingest only bacteria and small flagellates. The rest of the body is covered with uniform cilia except for the slightly longer caudal ones which are reported to be sensitive to touch.

Interesting comparisons can be made between food and suspended carmine particles fed to the animal. In both cases, these substances are wafted to the cytostome by the membranelles, and at the end of the cytopharynx balls of food or carmine are formed. If the balls are food, endoplasmic flow, cyclosis, takes them first forward to the extreme anterior end, and down the side opposite the cytostome. From here they are carried around the feeding canal and contractile vacuole and finally voided from the cytopyge, which lies in the middle of the truncated posterior end. On the other hand, if the balls are of carmine, shortly after leaving the end of the cytopharynx, they are shunted to the opposite side, where they are summarily carried to the cytopyge and voided.

The very large posterior contractile vacuole is somewhat pear-shaped, and when distended, it is almost as broad as the animal, but normally not more than one-eighth of the total length. The single tributary canal runs from the anterior end of the body. Just before the vacuole empties, the tributary or feeding canal closes. Under normal conditions, the vacuole contracts at intervals of less than 10 minutes and voids its contents from a pore at the posterior end. Apparently, a complete emptying of the contractile vacuole involves the contraction of the adjacent body wall, for the posterior end of the animal appears compressed after the vacuole is emptied. Both the canal and vacuole reform in their original positions.

In contrast to the very conspicuous beaded or moniliform macronucleus, the micronuclei are very small and difficult to observe even when successfully stained. There is no correspondence between the number of lobes in the macronucleus and the number of micronuclei near them. Sometimes as many as five micronuclei may be found near a single lobe, and the total number of micronuclei exceeds the ten to fifty lobes of the macronucleus. Because of the large size of the macronucleus, *S. ambiguum* forms good material for studies in protozoan regeneration. When pieces of the macronucleus are included, even fragments one-fiftieth to one-seventieth of the original volume regenerate successfully.

Binary fission in this animal is a long and complex process taking 7–9 hours for completion. Individuals about to divide always seem to have the cytostome located at about the middle of the body. The first evidence of division is seen in the posterior half of the animal where new peristomial membranelles are formed and start to move in an uncoordinated manner. Later, the macronucleus loses its lobated appearance and undergoes several movements and modifications before division.

Before division, the micronuclei enlarge and the tributary canal of the contractile vacuole enlarges near the center of the body. When cytoplasmic constriction has progressed sufficiently, a new terminal pore is formed and begins to function immediately. By the time the two daughter cells are ready to separate, the enlarged macronuclear mass, which has again become vermiform, divides into two almost equal parts. Individuals which have just undergone division may always be recognized by the fact that the cytostome is near the posterior end and the macronucleus is unlobed. Both these juvenile conditions are soon rectified as the animals mature.

In conjugation, one member usually fuses to the peristomium of the other slightly posterior to the anterior end. Apparently, neither animal feeds just before conjugation nor during the union; consequently, the endoplasm becomes quite clear. While the animals are still united, commissures between the lobes of both macronuclei break, resulting in their fragmentation. After a union of 60–72 hours, the animals separate with the macronuclei being reformed in each exconjugant.

Stentor coeruleus

By W. D. Burbanck

Stentor coeruleus (Fig. 32) is a large ciliate that possesses a sky-blue pigment, stentorin, which is one of the most indestructible of all animal pigments. When fully extended, *S. coeruleus* is between 1 and 2 mm. in length, although some individuals in a state of extreme extension may reach the unusual size of 4 mm. The attached and extended forms best demonstrate the trumpet-shaped form of the body. As a swimming form, *S. coeruleus* assumes an oval or pyriform appearance.

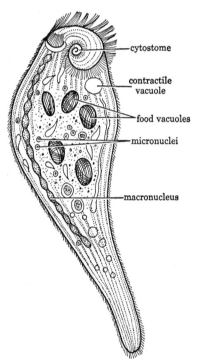

Fig. 32. *Stentor coeruleus* somewhat contracted. (Modified after Stein.)

The blue stentor may be collected in both streams and ponds. In streams contaminated with sewage, it may be very abundant. When brought into the laboratory, much debris should be brought with the animals. Large hay-infusion cultures may be established in liter graduates. Under such conditions, the animals have a pyriform appearance and remain free-swimming rather than attaching to the sides of the vessel.

The body of the animal is covered with cilia except on several small areas. These cilia are arranged in definite rows along colorless stripes which alternate with blue granular ones. At the proximal or posterior end of the animal, the body is free of cilia and produces ramose pseudopodia which act as a holdfast when the body is extended. The holdfast is capable of anchoring the animal to solid objects and to the surface film.

Cilia on the proximal region of the body are longer than those on the anterior or distal end. Certain cilia on the body appear to be

STENTOR COERULEUS

somewhat more rigid and may perhaps have a tactile function. On the transverse frontal field, the cilia are the shortest, and in the region of the cytopharynx, where the granular stripes are the narrowest, the ciliary rows are correspondingly closer together. The frontal field is bordered by a single row of membranelles which carry food in a clockwise path (viewed from the peristomial end) to the buccal pouch, which is a depressed area of the frontal field spiralling down to the cytostome. The actual sunken portion of the pouch is so placed that the membranelles and cilia create a great vortex which brings food into this "hopper." Here, a remarkable process of selection and rejection of food takes place. In some cases, the ingestion of certain materials may be avoided by closing the pouch, while in others, if undesirable food does enter, it can be spewed out. Not only can selection be made between living food and non-living carmine particles, but even between two species of flagellates belonging to the same genus. The stentor may ingest living food ranging all the way from small flagellates to rotifers, for even the latter cannot escape when once caught in the powerful vortex. Other species of *Stentor* may be eaten, and cannibalism within the same species is a common occurrence.

When placed on a depression slide and studied microscopically, *S. coeruleus* reveals great similarity to *Spirostomum*. The stentor may be orientated by considering the part of the circumference of the frontal field nearest the cytostome as the ventral side. The beaded or moniliform nucleus with its fifteen to twenty lobes will then be seen to lie on the left side. Micronuclei lie very close to the lobes. During the process of both fission and conjugation, the macronucleus may assume a vermiform or rounded appearance. Unlike *Spirostomum*, the contractile vacuole is anterior in position; however, it is similar in having one radiating canal. Both this hydrostatic organelle and its opening are to the right of the cytostome. In a normal animal the contractile vacuole may be seen to empty at intervals of about 3 minutes at a temperature of 21°C. Residual material in the food vacuoles after digestion may be defecated through a definite anal pore or cytopyge near the opening of the contractile vacuole, although waste may be voided also through other areas on the surface of the body. The process of defecation seems to involve endoplasmic contraction.

Stentor coeruleus is capable of both rapid and slow contractions. The first are apparently due to the action of special myonemes, while the second seem to be a result of general protoplasmic activity. The myonemes are located beneath the colorless stripes; these myonemes are closer together where the pigmented stripes are narrower. In the midregion of the body on the ventral side, the myonemes ramify and one myoneme may have several branches. As in *Spirostomum*, the myonemes are in the inner or alveolar layer of the ectoplasm.

Fission in the stentors offers excellent material for a study in morphogenesis. The first inkling of an approaching division is the development of a second adoral zone in the midventral region in the area of the ramifying myonemes. The oblique division which ultimately takes place requires about 7 hours at 17–20°C. to go to completion, with 2½ hours of this time actually devoted to the cytoplasmic constriction.

After the new adoral zone has appeared, the sequence of events is as follows: pushing out of the new adoral zone and frontal field from their initial longitudinal orientation into a transverse one; cytoplasmic constriction; elongation of the distal half (the parent zooid); and actual separation. Attending these external changes are alterations in the macronucleus and the formation of a new contractile vacuole just below the level of the new adoral zone. This description is greatly oversimplified, and a careful scrutiny of the living material will yield a satisfying experience to the observer.

Conjugation of the stentors is infrequent in laboratory cultures. This process takes place in old cultures in which the animals are in a semistarved, food-vacuole-free condition. Usually, there is a marked difference in the size of the partners. They are attached to the substrate at some distance from each other; the stentors bend toward one another and a small area of the periphery of their adoral fields becomes fused. If disturbed, they are able to swim freely. The macronucleus will be seen to be rounded up and, for the first time, the micronuclei are visible in the living condition. This relatively rare phenomenon occurs usually in late spring or early fall.

Because of its large body and nucleus, *S. coerulus* is excellent material for regeneration studies. When mutilated, even though several organelles may be undamaged, it undergoes an extensive dedifferentiation followed by redifferentiation of a new set of organelles.

The animal has been shown to exhibit an interesting behavior in response to carmine particles. If an attached animal is subjected to a continuous stream of carmine particles in suspension, a series of reactions may be observed. At first, the attached animal merely bends away; then, it reverses its ciliary beat; next, it contracts; and, if the treatment is continued long enough, it will break its attachment and swim away.

Stylonychia pustulata
By W. D. Burbanck

One of the most prevalent ciliates to be found in infusion cultures made from natural waters or soils is the hypotrich, *Stylonychia pustulata* (Fig. 33). As compared with the holotrichous ciliates, its

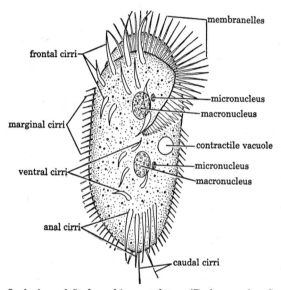

Fig. 33. Oral view of *Stylonychia pustulata*. (Redrawn after Summers.)

swimming movements are neither as graceful nor as rapid. The flat, rigid body moves erratically over the substratum, clambering about on ventral cirri that provide the animal with a rather awkward-appearing, creeping type of locomotion.

Belonging to the same order, Spirotricha, as *Stentor* and *Spirostomum*, *S. pustulata* has also a well-developed adoral zone which spirals in a clockwise direction (viewed from the peristomial end). The body is rounded at both ends with the anterior end narrower than the posterior one, thus giving the animal an elongated-oval shape. Except for monsters that will be mentioned later, the usual length is about 150 μ.

The arrangement of membranelles and cirri is both diagnostic of the Genus *Stylonychia* and characteristic for the Suborder Hypotricha.

Several genera of hypotrichs have the ventral cirri arranged with eight cirri in front of the anterior nucleus in rows of three-three-two. Posterior to these and more or less scattered in the central region are five ventral cirri. Behind these, arranged parallel to each other and projecting beyond the posterior end, are five anal cirri. Unique to *Stylonychia* are three caudal cirri with the middle cirrus longer than the other two. Projecting beyond the sides of the animal are a variable number of marginal cirri. All these cirri are made up of fused cilia. During the artificial manipulation of placing the animals on a slide and more or less compressing them under a coverglass, the cirri may fray out into a shaving-brush-like dissociation of their component cilia.

Most striking and complex is the adoral wreath of membranelles which starts a little to the right side of the animal's anterior end and spirals to the left, continuing slightly beyond the level of the anterior nucleus. This can best be seen in ventral aspect. Membranelles are made up of short rows of fused cilia. Both the sculpturing of the peristomial region and the action of the powerful membranelles direct food into the cytostome. Apparently, the membranelles are also important locomotor organelles, and the marginal cirri opposite them seem to be enlarged to compensate for the powerful action of the membranelles.

Across from the membranelles on the other side of the peristomium is located an extensive undulating membrane. The gathering of food is accomplished by the two last-mentioned organelles, while its concentration and movement into the endoplasm are accomplished by cytoplasmic fibrils which extend far up the peristomium. At the cytostome, some turn in as postesophageal fibrils which are most important in concentration of bacteria. The vibrations of these fibrils apparently rotate the forming food vacuole. The coordinated movements of, and the fibrillar connections among, the membranelles, oral fibrils, and perhaps anal cirri point to the presence of a coordinatory or neuromotor system.

Because of its arrangement of membranelles, cirri, undulating membrane, and fibrils, *Stylonychia* is capable of utilizing a wide variety of foods. A heavy concentration of bacteria such as *Pseudomonas ovalis*, green flagellates, small ciliates such as *Tetrahymena* and *Colpidium*, or even cysts and young of its own kind is adequate. The cannibalistic habit is prevalent only when there is a scarcity of other foods. The cannibalistic forms are monstrous, being several times the average size. The animals grown on *Colpidium colpoda* are intermediate in size between those grown on bacteria and the cannibalistic forms.

Stylonychia pustulata has two large prominent macronuclei and a single micronucleus lying on the oral side of each macronucleus. Lateral to the posterior end of the adoral wreath, at the same level as the anterior nucleus, is the contractile vacuole. The appearance of the endoplasm is dependent upon the kind and size of food ingested. When large ciliates or its own kind are ingested, the endoplasm may become almost opaque and the body greatly distorted.

Stylonychia normally reproduces by binary fission, the animal constricting transversely.

The cytoplasm of *Stylonychia* is more rigid and dense than that of the holotrichous ciliates and thus is excellent material for regeneration studies. The animals may be cut with delicate needles or modified razor blades, or mutilated by centrifugation. Even if a small external organelle is disturbed, a complete reorganization will take place. In such cases, it is interesting to watch the rapid formation of a whole new field of ventral cirri and the casting off of the old organelles. Apparently, the presence of a macronucleus but not a micronucleus is necessary for fragments to regenerate. Dedifferentiation and regeneration of the anterior external organelles also take place after binary fission.

Stylonychia pustulata has proved itself to be excellent for experimental studies upon the active individual because nuclear reorganization or endomixis which affects the division rate occurs only during encystment. With a pH of 7.6 and a temperature of 26°C., using a mixture of foods containing bacteria, ciliates, and yeast, a maximal division rate of between four and five divisions per day may be obtained. Selection of large and small individuals over a period of several generations will lead to remarkably stable races of large and small animals.

Conjugation in *S. pustulata* takes place readily with a union of the animals in the peristomial region. Maintenance of ex-conjugants and a comparison of their division rate with that of non-conjugants has led to the conclusion that, at least in these ciliates, the process of conjugation does not have a stimulating effect.

Euplotes patella

By W. D. Burbanck

Euplotes is one of the best-known genera of ciliates. *Euplotes patella* (Fig. 34) is the best-defined species of a group of patella-like hypotrichs which includes the species *E. eurystomus*, *E. woodruffi*, and *E. aediculatus*. All these species have at one time or another been confused with *E. patella* although they are all about one-third larger than that species.

Euplotes patella is found in both fresh and brackish water. It has a subcircular to elliptical outline with the dorsal side convex and the ventral one concave. On the dorsal side, there are six distinctive ridges

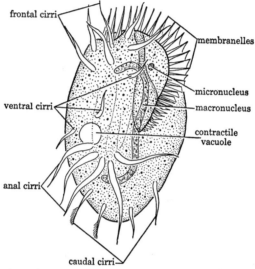

FIG. 34. Oral view of *Euplotes patella*. (Modified after Stein, from Pierson.)

with corresponding rows of sensory bristles embedded in rosettes of granules. The average size is 91 μ in length by 52 μ in width.

On the ventral side, the cirri are arranged in a characteristic fashion. There are six frontals arranged in two rows, three scattered ventrals, five anals arranged in parallel rows, and four caudals which at times have a frayed appearance. Still observing the animal from the ventral side, thirty-four to forty-five membranelles will be seen to be arranged in a spiral from the right to the left side of the animal around a comparatively small peristomial field. On the right, there is a somewhat raised peristomial lip which overhangs part of the field. This elevated portion apparently prevents food from being swept across the animal instead of into the cytostome.

The fact that *E. patella* does not have an undulating membrane or very small endoral fibrils is compensated for by the adoral wreath's curving through the cytostome into the funnel-shaped cytopharynx, so that posterior membranelles perform the task of moving food through the cytopharynx. At its internal end, the cytopharynx opens into a large postpharyngeal sac.

The dense endoplasm contains a large C-shaped macronucleus which is distinctive when compared with the other closely related species. The small micronucleus lies near the macronucleus close to the membranelles in the cytopharynx. Seen dorsally, the contractile vacuole lies near the right side at a level with, or slightly anterior to, the anal cirri. Depending on the temperature and the salinity of the medium, the vacuole empties at intervals of 40–70 seconds. The obscure cytopyge is located close to the pore of the contractile vacuole.

By careful use of the Klein silver-line technique, a definite system of pellicular fibrils may be demonstrated. Since the fibrils connect with bristles, this system is thought to be sensory. The fibrils connecting the anal cirri and membranelles make up the neuromotor system. This system is believed to coordinate the movements of these locomotor organelles. These deeper fibrils of the neuromotor system may be seen by using Heidenhain's iron alum-hematoxylin staining technique.

Because of its relatively simple food-gathering apparatus, *E. patella* flourishes only on food of comparatively large size. Rye, wheat, timothy-hay, or split-pea infusions in which there is a heavy growth of *Chilomonas paramecium* are excellent media for the growth of this hypotrich. Neither single nor mixed bacterial cultures give good results.

Along with various species of *Paramecium, E. patella* has definite mating types. There are six of these, and animals from one type will conjugate with those of the other five types, but not with members of its own type. Before conjugation, nuclear reorganization and replacement of locomotor organelles take place, and after conjugation the eighteen ventral cirri are again replaced.

Unlike *Stylonychia, E. patella* does not form cysts. Binary fission is accompanied by the formation of a new set of ventral cirri for each daughter and a new peristomium for the posterior daughter. The anterior daughter cell retains the old peristomium with no apparent modification.

Vorticella campanula
By W. D. Burbanck

When microscope slides are submerged in lakes having a good deal of rooted vegetation or if submerged plants from rivers or streams are collected and examined, the large social peritrichous ciliate, *Vorticella campanula* (Fig. 35), will often be found in large groups with its attached stalks coiling and uncoiling. This species is easily recognized because it has an endoplasm filled with numerous conspicuous refractile granules and it lives only in uncontaminated waters.

Although this peritrich is usually campanulate or bell-shaped, it is capable of great individual variation in both shape and size. *Vorticella*

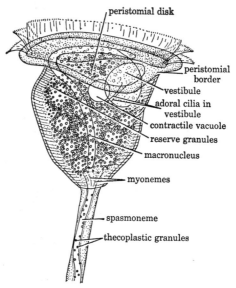

FIG. 35. *Vorticella campanula*. (Modified after Noland and Finley.)

campanula varies in size up to 157 μ in bell length by 99 μ in width, with the length of the stalk varying from 50 to 4150 μ.

As viewed from the peristomial end, the adoral zone spirals in a counterclockwise direction toward the cytostome. This is the reverse of the condition in *Stentor*. Upon close examination, the pellicle will be seen to have indistinct circular striations as though, phylogenetically, the body had once been ciliated. Apparently this may have been the case, for basal bodies can be demonstrated by the Klein silver-line technique.

The adoral zone lies in a groove between the peristomial disk and peristomial border or lip. Marginal to an area where the disk is raised, there is a sunken area, the vestibule, which on its outer border distorts the otherwise circular form of the lip. On the elevated portion, the cilia of the adoral zone circle one and one-half times in a counterclockwise direction when viewed from the peristomial end. The bases of these cilia are fused. There are three rows. The inner circlet is double and the cilia stand erect and are in constant undulation, while the outer circlet is single, inclining over the lip, and acts as a guide for food carried toward the vestibule. As the rows of cilia approach the vestibule, the two inner rows separate from the outer one. These two inner rows enter the vestibule together along its inner wall with the outer row of cilia passing along the outer wall of the vestibule and becoming fused into a strong undulating membrane.

The cytostome is found at the bottom of the vestibule leading into a delicate cytopharynx which is collapsed except when distended with food. Water and food passing into the cytostome flow down the vestibule between the two inner ciliary rows and the outer one. A return current of water, into which the cytopyge and pore of the contractile vacuole empty, passes between the inner cilia of the double circlet and the inner wall of the vestibule.

In spite of the fact that the endoplasm of *V. campanula* is almost opaque because of the continuously moving reserve granules, it is still possible to see its organelles. Even in unstained specimens the long vermiform macronucleus is visible. To see the small micronucleus, it is necessary to stain with acetocarmine or any of the more permanent stains. Near the attachment of the stalk, myonemes are prominent. The contractile vacuole is about 7 μ in diameter when fully distended and lies between the macronucleus and the vestibule, emptying into the latter about once every 6 seconds. At times, the endoplasm may be almost full of food vacuoles. Observations indicate that they follow a fairly definite course before egestion.

Attached to the bell-shaped body is a long stalk which contains a very large highly contractile myoneme or spasmoneme spiralling down its center. Upon contraction, it gives the appearance of an unstretched spring, and the myonemes in the peristomial border pucker over the peristomial disk. On the sheath of the spasmoneme there are some characteristic bodies called thecoplastic granules which are useful taxonomic structures. On the stalk proper, bacteria may often be found living commensally.

If kept in wide shallow dishes after collection, *V. campanula* will survive for some days provided bacterial growth does not become too great. *Vorticella convallaria*, which is similar in appearance and size, multiplies well in cultures with heavy bacterial growth.

Since rotifers and two free-swimming ciliates, *Amphileptus* and *Lionotus*, eat *V. campanula*, the fortunate observer may be able to watch their predacious activities.

By reducing both oxygen and food, it is possible to induce encystment. At excystment, individuals called telotrochs, with a posterior wreath of cilia in addition to their oral ciliation, swim rapidly and erratically until they attach and form stalks. In binary fission, which occurs by a longitudinal division, one of the daughters always becomes a telotroch. Under adverse conditions, individuals may develop the posterior wreath of cilia and swim to a more favorable location.

Several generations after excystment, conjugation may occur. An individual may undergo a series of divisions resulting in numerous small microconjugants. These are considerably smaller than telotrochs and swim in an even more erratic manner, stopping frequently as they hit the bases of the bells of attached individuals. Eventually, the microconjugant fuses with a macroconjugant which is outwardly like the ordinary vegetative zooid, but which has undergone nuclear modification.

Epistylis plicatilis, *Carchesium polypinum*, and *Zoothamnium arbuscula*

By W. D. Burbanck

The three colonial peritrichous ciliates, *Epistylis plicatilis* (Fig. 36), *Carchesium polypinum* (Figs. 37, 38), and *Zoothamnium arbuscula* (Figs. 39, 40) may be collected on submerged microscope slides or on vegetation under conditions similar to those for *Vorticella*. The three may be easily differentiated on the basis of the structure of their stalks. *Epistylis* has no spasmonemes and is non-contractile; it possesses a regular dichotomously branching stalk. *Carchesium* has complex spiral spasmonemes interrupted at each bifurcation, permitting the contraction of single zooids. *Zoothamnium* has a single branching spasmoneme connecting all the zooids. In the last case, the whole colony contracts as a unit.

All three animals have zooids that bear a striking resemblance to *Vorticella*. The zooids of *Epistylis* are elongate and conical with an adoral zone bearing four bands of cilia; those of *Carchesium* are conical and campanulate; and *Zoothamnium* is also conical-campanulate to spherical. All three of these forms are extremely large and are readily visible to the unaided eye. *Zoothamnium* is the largest of the three with a height up to 6 mm., while *Carchesium* and *Epistylis* may be 4 and 3 mm. respectively.

Of the three species, only *E. plicatilis* is not found in sea water; however, there are species of this genus found in that habitat. If animals are attached to slides, they may be kept for some days when placed in large aquaria or tanks in which a fresh supply of either sea or chlorine-free fresh water is introduced, depending upon normal habitat.

All three of these peritrichs form resting cysts, and cysts of *Amphileptus* may also be present in a colony of *Epistylis* since this free-moving ciliate eats *Epistylis* and encysts on stalks of its prey. Suctoria and amoebae also parasitize *Epistylis*, and both *Amphileptus* and *Lionotus* feed on *Carchesium* and *Zoothamnium*.

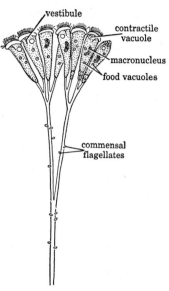

FIG. 36. *Epistylis plicatilis*. (Modified after Kent.)

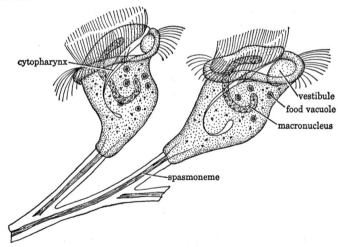

FIG. 37. Zooids of *Carchesium polypinum*. (Modified after Kent.)

68   PROTOZOA

Because of its unusual size and ability to regenerate, *Zoothamnium* may be used for interesting studies in reorganization. There is in this form a definite physiological gradient which resembles that found

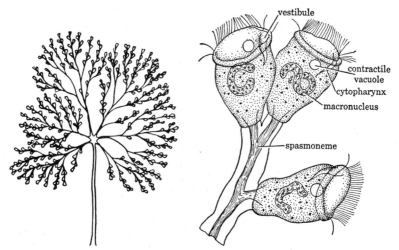

FIG. 38. Colony of *C. polypinum*. (Modified after Kent.)

FIG. 39. *Zoothamnion arbuscula*, zooids. (Modified after Furssenko.)

in the Metazoa. Apical zooids exert control over those located at a lower level. If the apical zooid is cut off, a lower one will assume the dominant position and influence.

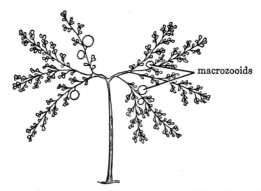

FIG. 40. Colony of *Z. arbuscula* with macrozooids. (Modified after Furssenko.)

All three peritrichs have macroconjugants and microconjugants similar to those found in *Vorticella*. They also have telotrochs, which in *Zoothamnium* are very large macrozooids called ciliospores. In

E. plicatilis, ciliated "embryos" form within parent zooids and escape through spout-like apertures on the bell of the parent.

Podophrya collini

By W. D. Burbanck

Among the many forms often appearing in hay-infusion cultures seeded with pond water is the primitive suctorian, *Podophrya collini* (Fig. 41). Its large size and unique type of reproduction both set it apart from other species of the same genus.

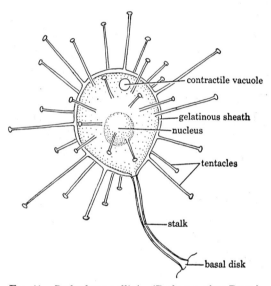

Fig. 41. *Podophrya collini*. (Redrawn after Root.)

The globular body of *P. collini* is, on the average, from 40 to 50 μ in diameter. Well-fed individuals may be 118 by 102 μ, and starved ones, 22 by 18 μ. Generally, the stalk is about 50 μ in length and is attached to the substratum by a basal disk. The presence of 30–60 slightly knobbed tentacles gives the animal a pin-cushion-like appearance. The tentacles are arranged evenly over the body and point in all directions except toward the region of attachment. They are in constant movement and within a minute may extend or contract from 35 to 70 μ in length.

A large macronucleus 25–35 μ in diameter is located centrally in the endoplasm with 3–12 small micronuclei close by or scattered near the periphery of the body. A large contractile vacuole also lies near the periphery. There is no cytostome; material is taken directly from the prey through the fairly rigid tentacles. Both the tentacles and body are covered with a gelatinous sheath.

The chief foods of *P. collini* are *Paramecium aurelia* and *caudatum*. Species of *Colpidium*, *Colpoda*, *Glaucoma*, and *Halteria* are also always seized when they come in contact with the tentacles. Such vigorous forms as *Frontonia*, *Bursaria*, *Vorticella*, and *Didinium* may be occasionally eaten but are usually successful in tearing themselves free, even though some of their cytoplasm may have been sucked from their bodies. *Podophrya collini* does not seem able to seize such forms as *Stentor*, *Spirostomum*, and *Coleps* which have a tough outer covering. Although flagellates such as *Chilomonas* and *Bodo* strike the tentacles, they apparently do not stimulate the seizing reaction. This also seems to be true for ciliated podophryean embryos, which frequently collide with the tentacles but are not eaten.

The feeding of *P. collini* is the result of fortuitous collisions of ciliates with the extended tentacles. As soon as one tentacle comes in contact with the prey, additional tentacles join in holding and eating it. One to several hours are required to ingest a large *Paramecium*.

After *P. collini* has eaten and increased enormously in size, an interesting process of invagination associated with reproduction takes place, resulting in an internal tube communicating to the outside by a pore. Cilia are formed, projecting into the lumen of the cavity, while the macronucleus divides and additional contractile vacuoles appear in the parental cytoplasm. At this time, the external opening enlarges and the invaginated portion everts, with the new macronucleus, contractile vacuoles, and food vacuoles flowing into the newly forming ciliated embryo. After this evagination, a constriction takes place which results in the separation of the ciliated embryo from the parent. The flattened, kidney-shaped embryo subsists on the material stored in the food vacuoles received from the parent and exhibits ceaseless activity. During its free-swimming period, the large ciliated embryo (about 70 by 30 μ) may undergo binary fission.

As a result of such internal budding and subsequent fission of swimming embryos, one parent may produce from five to fifteen embryos. After such activity, the parent's volume often is reduced to a size smaller than that before prereproductive feeding.

When tentacles are broken or tangled with *Paramecium* trichocysts, the body of *P. collini* may erupt with cytoplasm streaming out of the ruptured area. The resulting spherical mass becomes ovoid; cilia then form and the transformed individual may swim away, leaving the shrunken body and stalk behind, or if the basal disk gives way, they may be dragged along behind.

After swimming for several hours, the embryo comes to rest on either its anterior or posterior end. About 10 minutes later, tentacles which are immediately functional grow out at the unattached end of the body. One of the contractile vacuoles enlarges, while the other two become smaller and finally disappear. The cilia are resorbed into the body. About a half hour after coming to rest, a refractile spherule pushes out at the extreme end of the animal, immediately adhering to the substratum as the basal disk of the future stalk. Within the next 2 hours, tentacles, body, and stalk take on the normal adult appearance.

In a smaller but somewhat similar suctorian, *P. fixa*, conjugation takes place between two adjacent stalked adults by fusion of their bodies. Although this phenomenon may occur in *P. collini* it has, as yet, not been described.

When food is not available over several weeks, the podophryean becomes smaller and smaller, finally undergoing encystment. During this process, the gelatinous sheath which covers both body and tentacles greatly thickens. The tentacles degenerate at their tips, forming elongated knobs. These degenerated tentacles are rather easily broken off, but may persist for as long as 4 months.

2.

Porifera

Leucosolenia

By W. D. Burbanck

The small, delicate sponge, *Leucosolenia* (Fig. 42), grows in shallow water below low tide mark, on large boulders and jetties. In spite of its apparent frailty, this sponge is found where wave action is intense and does not live in calm water.

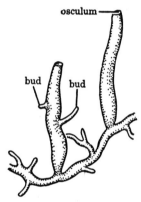

Fig. 42. A colony of *Leucosolenia*.

The colony of *Leucosolenia* consists of clusters of slender, whitish, tubular individuals joined together irregularly. It is difficult to define an individual because of the budding type of growth typical of representatives of this genus. Individual branches may be up to 25 mm. in height. Examination with the unaided eye reveals that there is a simple circular opening, an osculum, at the distal end of each mature tube. Carmine suspended in sea water containing a living individual is carried away from the osculum instead of into it. The opening serves, therefore, as a vent or exhalant siphon rather than a mouth.

Favorable sections made with a razor, or better, stained permanent slides of a cross section of *Leucosolenia*, help explain the structure of and water circulation through the animal. It is now seen that the water issuing from the osculum must have entered the internal cavity, or spongocoel, of the animal through the many fine pores in the body wall. The pores in *Leucosolenia* differ from those in more complex types of sponges by being intracellular. These modified epidermal cells are aptly named porocytes (Fig. 43). Although the chief function of these specialized cells is to admit water, they can ingest such organisms as diatoms. Usually, the cells are described as resembling a doughnut in shape; actually, they more closely resemble a short piece of hollow macaroni.

Pieces of the macerated living sponge may exhibit independent and erratic movements. Under high magnification, one of the pieces which has come to rest may be seen to be made up in part of uniflagellated

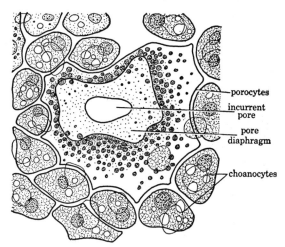

Fig. 43. Porocyte of *Leucosolenia* with surrounding choanocytes. (Redrawn from Hyman.)

cells, the choanocytes. Around the bases of the flagella of some of these cells, hyaline collars project from the cell bodies, but in other choanocytes their collars may be retracted. The beat of the flagella of the choanocytes pumps water through the sponge. Another equally important function of the choanocytes depends on the ability of the cell body to engulf small particles of food around the outer edge of the base of its collar. The food is then digested, and its derivatives are passed on to wandering amoebocytes. Finally, the choanocytes take an active part in the nourishment of the ova and sperm and facilitate their union.

The general physiological condition of the animal determines the activity of the flagella, and the flow of water is controlled also by the porocytes and by a perforated membrane which is stretched across the osculum. There is no coordinating nervous system, the individual cells responding to stimuli as independent effectors. Sponges with this porocyte-spongocoel-osculum arrangement are typical of the simplest Porifera and are called ascon sponges.

Many sponges are maintained in a turgid semirigid condition by the pressure of the current of water propelled through the system of pores and canals. To some extent this may be true of *Leucosolenia*, although the chief means of support appears to be small interlaced calcareous rods or spicules. Those which are elongate and pointed at each end are the monaxon type, while those with three rays are the triaxon type. Found less frequently are tetraxon spicules with rays radiating in tetrahedral arrangement.

Spicules dissected from the living specimen may show amoebocytes which secrete spicules and are called calcoblasts. This is an appropriate name, for when an acid is added to the spicules, an effervescence follows which indicates that the spicules are composed of calcium carbonate. The monaxon spicules may be seen to be formed by two calcoblasts working in cooperation, while the triaxon type is formed by three.

Leucosolenia, as a typical ascon sponge, represents a stage through which the more complex sycon and leucon sponges pass during their development. Because of this and other resemblances, reference will be made to this simple sponge during the discussion of the more complex ones which follows.

Scypha (= *Sycon*)
By W. D. BURBANCK

Scypha [*] (Fig. 44) is a small marine sponge belonging to the Class Calcarea. It resembles *Leucosolenia* and occupies the same shallow water habitat where wave action provides the animals with an adequate supply of food and with well-oxygenated water.

In both size and anatomical complexity, *Scypha* surpasses *Leucosolenia*. Individuals of *Scypha* are 20–25 mm. in height and 5–6 mm. in diameter. Asexually reproducing animals may have one or two sizable buds attached to the base of the parent. Branching, however, is by no means as extensive as in the ascon sponges.

In *Leucosolenia*, the single tubular spongocoel is lined with collar cells. In *Scypha* and other sycon-type sponges, there is still a tubular

* Formerly misnamed *Grantia*.

SCYPHA (= SYCON)

spongocoel, but the choanocytes are located in many regularly spaced radial canals which radiate from the spongocoel. Perhaps, therefore, the increased mass of the syconoid sponges has resulted from this adaptation, which, in proportion to the mass of the body, presents a far larger number of choanocytes to the food-laden water than occurs in ascon sponges. Longitudinal sections through both the living and dried specimens reveal the openings of the radial canals into the spongocoel. These are the apopyles or internal ostia. In cross sections, it can be seen that the outer ends of the radial canals end

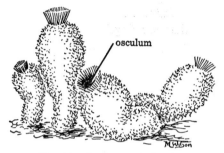

FIG. 44. A cluster of *Scypha*.

blindly. Careful scrutiny may also reveal tiny openings between individual collar cells in the layer of choanocytes, the prosopyles. Externally, the animal possesses over its entire surface innumerable pores or ostia. In *Leucosolenia*, the pores are intracellular; in *Scypha*, they are intercellular. Water entering the ostia goes directly into blind pouches, incurrent canals, which have no choanocytes but are lined with epidermis. The incurrent canals are of approximately the same size and shape as the radial canals, and interdigitate among the latter in the body wall. The prosopyles mentioned earlier connect the incurrent and radial canals. The water propelled by the choanocytes flows from the former into the latter. Finally, it flows through the apopyles into the spongocoel and out through a relatively large osculum at the distal end of each individual.

Around the osculum of *Scypha* there are giant monaxon spicules arranged to form a funnel-shaped collar. With this exception, the spicules are morphologically and developmentally very similar to those in *Leucosolenia*. The spicules of *Scypha* are arranged in a definite manner in respect to its canal system. At their distal, blind ends, the radial canals fuse with the epidermis to form a cortex. This cortex has monaxon spicules partly projecting beyond its surface, giving the animal a bristly appearance. Along the radial canals, triaxon spicules are arranged in a pattern with one point directed toward the distal end of the canal. Finally, around the apopyles, next to the spongocoel, tetraxon spicules occur along with the triaxon spicules.

Scypha is a hermaphroditic organism, in which spermatozoa and ova arise from specialized amoebocytes called archaeocytes. In macerated living *Scypha* and in stained sections, sperm spheres and segmenting eggs are frequently found. Swimming amphiblastula larvae are common during the summer months on the northeastern Atlantic coast of the United States. Stained whole mounts of the larvae show plainly that the flagellated cells are smaller and more numerous than the non-flagellated ones. When the larva finally comes to rest, an invagination takes place and an osculum breaks through at an end opposite a point of attachment. The flagellated larval cells develop collars and become functional choanocytes. At this time, *Scypha* closely resembles the simple ascon sponge.

Euplectella

By W. D. Burbanck

In the United States the elegant glassy sponge, *Euplectella*, is popularly called "Venus's flower basket" (Fig. 45). It is usually seen as dried specimens or as intricately latticed skeletons. In either case, little may be ascertained about the organization of this sponge from such material.

The curved, rigid structure of the individual is thought to be an adaptation to the slow, constant water current found at depths from 500 to 5000 meters in waters off the western Pacific islands.

Most of the four- and six-rayed siliceous spicules are interlaced and fused at their tips to form a three-dimensional or dictyonine type of network. An oscular sieve strengthens the terminal opening of the sponge, while projecting ledges of spicules wreathing the vase add still greater rigidity. The openings, or parietal gaps, in the meshwork of spicules, although

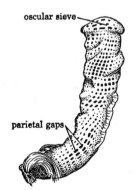

Fig. 45. *Euplectella*. (Modified from Pratt.)

connecting with the spongocoel, are apparently not part of the canal system.

Since *Euplectella* occurs on slimy oceanic oozes, the root spicules are adapted to anchor the sponge to its unstable substratum. Projecting from the surface of dried uncleaned specimens are the feather-duster-like spicules or floricomes. Localized regionally are other types of spicules.

In spite of the great complexity of the siliceous skeleton, the position of the osculum, spongocoel, flagellated chambers, and incurrent and excurrent canals strongly resemble the simple sycon-type and leucon-type arrangement found in the Calcarea.

Hyalonema

By W. D. Burbanck

Hyalonema (Fig. 46) is a hexactinellid sponge found along the New England coast. It lives in water only 10–15 meters deep.

Superficially, *Hyalonema* looks like a ball of glass wool on a stick. The "stick" or columella is made up of a root tuft of very long twisted spicules. These spicules constitute a holdfast for the sponge. The middle part of the columella commonly has symbiotic polyps attached to it, while the proximal end projects into the body of the sponge.

In addition to possessing siliceous, six-rayed spicules similar to those of *Euplectella*, *Hyalonema* has large and small amphidisks surprisingly similar in appearance to those of some of the fresh-water sponges. Extending from all over the surface are small, branching, five-rayed spicules. These resemble small Christmas trees on cross-shaped bases.

When the upper surface of the sponge body is depressed, the resulting cavity may be termed a spongocoel since the excurrent canals open into it; but where the surface is extended into a gastral cone by upward projection of the columella, no spongocoel exists.

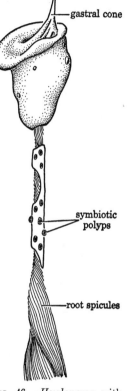

Fig. 46. *Hyalonema* with symbiotic coelenterates attached. (Redrawn from Hyman.)

Cliona, Halichondria, Chalina

By W. D. Burbanck

Since the three leucon-type sponges, *Cliona* (Fig. 47), *Halichondria* (Fig. 48), and *Chalina* (Fig. 49) are all in the Subclass Monaxonida,

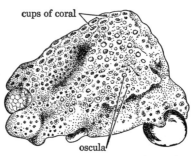

Fig. 47. An encrusting colony of *Cliona* on a piece of coral. (Redrawn from Hyman.)

it is not surprising that they resemble each other in various ways. However, their spicules, gross anatomy, and ecology differ in so many respects that a comparison of the three genera is valuable.

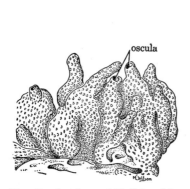
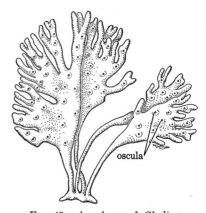

Fig. 48. A colony of *Halichondria*.   Fig. 49. A colony of *Chalina*.

When pieces of these sponges are boiled in 5 per cent potassium hydroxide, the spongin is dissolved, leaving the siliceous spicules in the residue. Study the spicule types and identify them with the aid of figures (Figs. 50, 51, 52).

In shallow water both *Cliona* and *Halichondria* form low encrustations on rocks, while in deep water their growth is massive. Often clam shells, particularly of *Venus*, are completely riddled by *Cliona*.

FIG. 50. Oxea spicules of *Halichondria* and *Chalina*. (Redrawn from Hyman.)

FIG. 53. Isochelas spicules of *Microciona*. (Redrawn from Hyman.)

FIG. 51. Spiraster spicules of *Cliona*. (Redrawn from Hyman.)

FIG. 54. Toxa spicules of *Microciona*. (Redrawn from Hyman.)

FIG. 52. Tylostyle spicules of *Cliona*. (Redrawn from Hyman.)

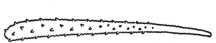

FIG. 55. Acanthostyle spicules of *Microciona*. (Redrawn from Hyman.)

The broken surface of fragments of the shells shows extensive tunnels eroded by the sponge. Each tunnel has an osculum slightly raised above the mass of sponge encrusting the surface of the shell. The color of *Cliona* is light yellow, and the substance of the sponge is very compact. It has a sulfurous odor. *Halichondria*, on the other hand, even when taken from deep water, is soft and is brownish yellow in appearance with high delicate oscula through which water pours with more vigor than for specimens of either of the other two genera. Lastly, the branching yellowish brown spongin-laden flesh of *Chalina* perforated with many oscula has resulted in its picturesque common name of "dead-man's fingers." Except when broken from its stalk and washed ashore, this sponge is not found in shallow water.

Microciona

By W. D. Burbanck

This large spectacular sponge is found in two forms. In shallow water it forms a soft, thin encrustation on rocks and shells, while in deeper water the colonies growing on similar objects become massive and occur up to 15 cm. in height. Although the siliceous spicules prepared by boiling pieces of the animal in 5 per cent potassium hydroxide are interesting because of the several types present (isochelas, toxas, acanthostyles) (Figs. 53, 54, 55), the most unusual characteristic of this animal is its tremendous capacity to reassociate, regulate, and regenerate.

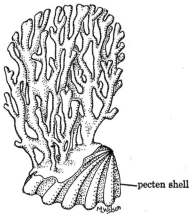

Fig. 56. *Microciona* attached to a *Pecten* shell.

A day or two suffices for interesting studies of cell reassociation in *Microciona* (Fig. 56). Examine microscopically cells which have been thoroughly dissociated by forcing a piece of the living animal through bolting cloth (heavy silk used in processing wheat flour) onto slides supported by Syracuse dishes in fingerbowls of clean sea water. Apparently, only choanocytes, nucleated archaeocytes, and granular amoebocytes are able to survive such strenuous treatment. During reorganization, small filopodia may be seen to extend from some cells to others. Upon contact these filopodia withdraw, pulling cells together into small clumps. After several days, flickering choanocyte flagella may be seen under oil immersion within even tiny cell aggregations.

In early August, swimming larvae may be observed in preparations of macerated adult specimens in sea water.

Spongilla

By W. D. BURBANCK

Spongilla (Fig. 57) is probably the best known of fresh-water sponges. Species in this genus are found in both ponds and streams. Often these sponges are various shades of green because of algae living symbiotically within them. *Spongilla* usually grows on submerged sticks and plants with the colony exhibiting more branching than is found in the other genera of fresh-water sponges.

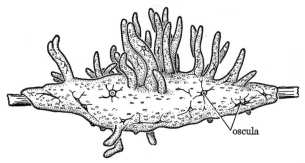

FIG. 57. A colony of *Spongilla* growing on a submerged twig. (Redrawn from Lankester, after Waltner.)

As with most sponges, spiculation is the best means of identification. The siliceous spicules of *Spongilla* are embedded in spongin and are elongated and pointed at both ends; smaller types are generally spiny. As with many fresh-water sponges there are resting bodies or gemmules protected by amphidisks. These peculiar, dumbbell-shaped spicules are absent in *Spongilla*, although typical body spicules are present in the protective, chitinous covering of the resting bodies.

In late summer and early fall, *Spongilla* produces gemmules which remain quiescent throughout the winter. If the gemmules are transferred to aquaria, small sponges producing a vigorous flow of water will develop and live for a short time.

Although spicules protect both the body and the apertures of *Spongilla*, a well-known odoriferous condition of these sponges may also have a protective significance.

Spongilla is an example of a rhagon sponge. Beside the relatively large oscula through which vigorous streams of water may be seen to issue, other smaller apertures or dermal pores are scattered over the very thin surface. Through these, water passes from the outside into large subdermal cavities; thence, it flows through incurrent canals

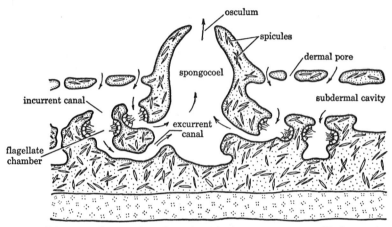

Fig. 58. Diagram of a section through a fresh-water sponge. (Redrawn from Morgan.)

to numerous flagellated chambers (Fig. 58) which finally, by way of excurrent canals, carry water to the spongocoel and through the osculum.

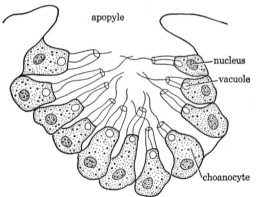

Fig. 59. A flagellated chamber of *Spongilla*. (Redrawn from Borredaile and Potts, after Vosmaer.)

An unusual free-swimming larva is characteristic of *Spongilla*. In most sponges, the larva after settling upon a suitable substratum undergoes invagination and metamorphosis during which processes the external flagella become the internal choanocytes. This does not happen in *Spongilla;* rather a cavity develops within the larval body while it is still surrounded by maternal tissue, and there is a differentiation of spicule-forming cells or scleroblasts, of the minute choanocytes characteristic of this genus (Fig. 59), and of amoebocytes. The flagellated larva, after hours or days of swimming, attaches itself by its anterior end and undergoes development of its canal system while exhibiting no invagination; it utilizes those choanocytes already produced internally by the larva. The external flagellated cells used in larval locomotion are not retained.

Hippospongia

By W. D. Burbanck

The commercial sponge, *Hippospongia* (Fig. 60), occurs in Florida on rocky bottoms at depths up to 10–15 meters. It is a typical horny sponge. Individuals may live to be 50 years old, gradually becoming

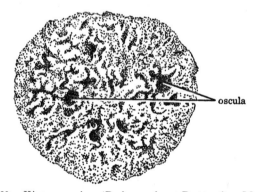

Fig. 60. *Hippospongia.* (Redrawn from Pratt, after Moore.)

extremely large and massive. As seen by the large number of oscula, a single sponge is made up of few to many individuals. Because of its numerous orifices and cavities, *Hippospongia* harbors a host of small commensal crustacea and worms.

Hippospongia resembles the monaxonid sponges except for its lack of siliceous spicules. The spongin skeleton of *Hippospongia* is very extensive, as can be ascertained by an inspection of the cleaned skeleton. Even in the commercially prepared sponge the raised oscula are evident. The water-imbibing qualities of these sponges may be understood by examining a small piece of the cleaned sponge. The spongin strands are separated by such small spaces that very large quantities of water are held because of capillarity.

The exterior of the living sponge is covered by a dark non-living membrane almost the consistency of thin, tanned kidskin. Unlike the situation in many other Keratosa, foreign objects do not adhere to the epidermis. Internally, numerous flagellated chambers may be seen with the dissecting microscope. Water enters a chamber by a prosopyle and leaves it by a specialized, elongated canal called an aphodus.

3.

Cnidaria and Ctenophora

CNIDARIA

Tubularia crocea

By T. H. WATERMAN

Tubularia crocea (Figs. 61–65) is a marine colonial hydroid. It occurs in dense tufts of irregularly and sparsely branched long stems bearing hydranths and arising from a tangled mass of prostrate stolons. This species is common in shallow water, particularly on wharf pilings, from Arctic regions to Cape Hatteras and from British Columbia to San Diego. Other, fairly similar species occur on both coasts of North America and in Europe.

Either preserved or living material may be used in studying this animal, although the latter is to be preferred since experimental as well as morphological observations may be made on the live animals. Even for purely structural examination only specimens which have been fixed in an expanded condition will be satisfactory.

In preparing *Tubularia*, and in fact most other cnidarians, for preservation it is necessary to anesthetize the animal before introducing the fixing fluid. This may be done most effectively by gradually introducing magnesium sulfate into the dish in which the animal, maximally expanded, has been placed. One method of doing this is to suspend a small cloth bag of crystals of this salt in the sea water over the specimen so that, as the magnesium goes into solution, it will gradually diffuse out and into the water near the animal. Care must be taken that the effect of the salt is sufficient to prevent contraction before fixation is attempted. Animals so fixed in the expanded state may then be dissected or stained and prepared in any of the standard ways for histological study.

CNIDARIA AND CTENOPHORA

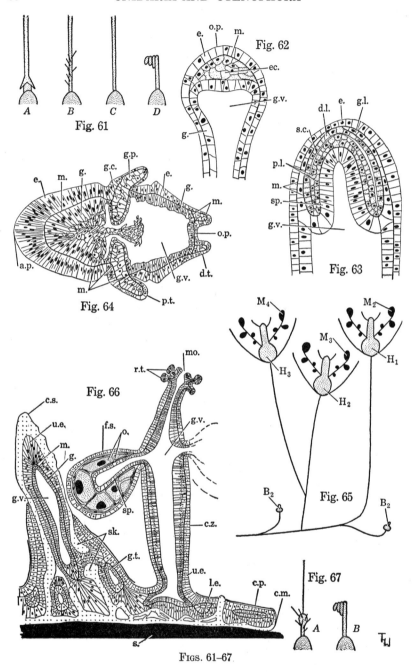

Figs. 61–67.

FIG. 61. Nematocyst types of *Tubularia*. (Data from Weill.) *A*. Stenotele, *B*. Basitrich. *C*. Atrich. *D*. Desmoneme.

FIG. 62. Young gonophore of *Tubularia*, showing early stage of ectocodon development. This structure, which is endodermal in origin in this species, gives rise to the sex cells as well as the subumbrellar and manubrial epidermis. $\times 500$. (After Liu and Berrill.)

FIG. 63. Older gonophore of *Tubularia*, showing the formation of a manubrial structure (spadix), subumbrellar cavity, and further development of the ectocodon. $\times 500$. (After Liu and Berrill.)

FIG. 64. Longitudinal section of an actinula of *Tubularia*, the larval stage which emerges from the female gonophore in this species. (After Lowe.)

FIG. 65. Growth pattern of a colony of *Tubularia*, showing the relations between polypoid and medusoid structures.

FIG. 66. Section through part of a colony of *Hydractinia*, showing a mature female gonozoid and the relation between the coenosarc and the skeleton. $\times 60$. (Compounded from papers of Colcutt and of Ballard.)

FIG. 67. Nematocyst types of *Hydractinia*. (Data from Weill.) *A*. Microbasic heterotrichous eurytele. *B*. Desmoneme.

ABBREVIATIONS FOR LABELS ON FIGS. 61–67

a.p. : aboral pole
B_2 : bud (second generation)
c.m. : colony margin
c.p. : crust of periderm
c.s. : chitinous spine
c.z. : column of female gonozooid
d.l. : distal layer of ectocodon (subumbrellar epidermis forms from this layer)
d.t. : distal tentacle
e. : epidermis
ec. : ectocodon
f.s. : female sporosac
g. : gastrodermis
g.c. : endodermal cushion
g.l. : gastrodermal lamella (radial and ring canals form in this layer)
g.p. : endodermal plug
g.t. : gastrodermal tube
g.v. : gastrovascular cavity
H_1 : hydranth (first generation) = primary zooid of colony

H_2 : hydranth (second generation)
H_3 : hydranth (third generation)
l.e. : lower epidermis
m. : mesolamella
M_2 : medusoid (second generation)
M_3 : medusoid (third generation)
M_4 : medusoid (fourth generation)
mo. : mouth
o. : ova
o.p. : oral pole
p.l. : proximal layer of ectocodon (gametes and manubrial epidermis form from this layer)
p.t. : proximal tentacle
r.t. : reduced tentacles
s. : substrate
s.c. : subumbrellar cavity
sk. : skeleton
sp. : spadix
u.e. : upper epidermis

As with many other sessile organisms, interesting early stages of development and growth for hydroids like *Tubularia* may be obtained (in proper season) by exposing a rack of clean glass slides in shallow water near rocks or piles where the adults are growing. Examination of these surfaces from time to time will reveal their gradual colonization by large numbers of forms, among which may be found the specific ones in question.

A colony of *Tubularia* consists of feeding individuals (hydranths), reproductive individuals (gonophores), and stems and stolons that are shared by the whole colony, whose parts they interconnect. These latter communal structures are referred to collectively as coenosarc.

The hydranths, which altogether comprise the trophosome, or feeding part of the colony, are large for hydroid polyps and are borne terminally on long slender stems (pedicels) which may branch occasionally to give rise to other hydranths. Each individual polyp of *Tubularia* has the sac-like radially symmetrical body plan characteristic of the Phylum Cnidaria. The hydranth, or trophozooid, consists of a narrow neck attaching it to the stem and a larger, conical region bearing two whorls of solid, filiform tentacles. The distal, shorter tentacles arise from the hypostome (or manubrium) and surround the terminal mouth. The proximal tentacles, which are longer, arise at the base of the widest portion of the hydranth, the gastric region.

The axis of symmetry is the oral-aboral axis and runs from mouth to pedicel of the zooid. The colonies of *Tubularia*, as contrasted to the radially symmetrical individuals of which they consist, are not symmetrical but form irregular tufts whose shape and size are determined by age and growing conditions.

The only body cavity of *Tubularia*, as in other Cnidaria, is the digestive (gastrovascular) cavity which is thus a coelenteron. This is the basis for the name Coelenterata, frequently applied to this and other (Porifera, Ctenophora) animal groups. The "mouth" of the hydranth is the only external opening of the gastrovascular cavity. The latter is continuous, however, with the cavities of the stems and stolons which are tubular. Thus the coelenteron of each individual of the colony is interconnected with that of all others.

The body wall of *Tubularia*, like that of other cnidarians, is three-layered, consisting of two epithelia: the gastrodermis, which lines the coelenteron and develops from the embryonic endoderm, and the epidermis, which covers the outer surface of the body and develops from the embryonic ectoderm. Between these epithelia lies the mesoglea, which is of ectodermal origin. As in other hydroid polyps, the mesoglea of *Tubularia* is a thin acellular mesolamella.

These three layers of the hydranth body wall are continuous with similar elements of the coenosarc of the colony's stems and stolons. Outside the living layer of the coenosarc is the periderm (= perisarc), a chitinous transparent layer secreted by the epidermis. In *Tubularia* and the other gymnoblastean hydroids this periderm stops just proximal to the neck of the zooids instead of forming a cup-like hydrotheca around the hydranth as in calyptoblasts, such as *Obelia*.

Histologically the tissues of *Tubularia* are well differentiated, although the general structure of the body attains only the tissue, rather than the organ, level of development. Of foremost interest are the cnidoblasts, cells which contain the nematocysts. These are diagnostic of the whole phylum, although *Tubularia* has types of these organelles characteristic of its own group. Of the wide variety present in the phylum *Tubularia crocea* has four: desmonemes, atrichs, basitrichs, and stenoteles (consult Hyman, 1940, pp. 382 ff., for details of nematocyst classification). These may be studied by the methods suggested in the section on *Metridium*.

Nematocysts and their associated intracellular structures comprise a remarkably complex effector mechanism. When mature, they are so oriented in the distal part of the cnidoblast that any prey contacting the surface in which they lie causes them to discharge sticky or poisonous threads of minute size. These entangle or pierce the food animal. By their means *Tubularia* can hold and subdue relatively large, active prey, as experiments with living hydroids and small plankton animals will readily demonstrate.

Another interesting cnidarian cell type to be found in *Tubularia* is the epithelial cell which is partly differentiated into a second kind of cell. Thus there are epitheliomuscular cells in the epidermis and nutritive-muscular cells in the gastrodermis. These cell types have their basal portions broadened and flattened so that they can contain myofibrils paralleling the epithelial surface; the distal parts may remain epithelial or be digestive, depending on their location. These partly contractile elements have frequently been considered as primitive forerunners of completely differentiated muscular cells, but other interpretations of the facts are quite possible.

Prepared histological sections are desirable for examining these details most effectively. However, a number of interesting features may be observed by teasing out fragments of fresh tissue of *Tubularia* and examining them microscopically under moderately high power.

Stain various pieces by drawing a drop or two of different vital dyes under the coverslip. One per cent sea-water stock solutions of methylene blue, bismarck brown, methyl green, and neutral red are suitable for this purpose. Add just enough dye to color the water lightly. Observe the tissue and remove the dye when the tissue is lightly stained. Such facts as the distribution pattern of the nematocysts, the structure of the mesoglea, and the cell types of the epidermis and gastrodermis can be readily determined in this way.

No special respiratory structures or pigments are present in *Tubularia*. This is in keeping with the unusually large surface to (living) volume ratio found in most cnidarians because of their essentially epithelial organization. However, in this species, as in other large hydroids, like *Corymorpha*, the internal surface area of the stem is increased by certain structural devices. A slice of *Tubularia* stem cut crosswise will indicate the arrangement in this hydroid.

Furthermore, flagellar currents are present internally. They circulate the fluid in the gastrovascular cavity to various parts of the body. Muscular movements of the hydranths also aid this circulation. The effectiveness of these currents may be readily demonstrated by introducing a small amount of a suspension of fine carmine or India ink particles. By this means the rate of transport of gastrovascular fluid as well as any patterns of flagellar currents may be determined.

Amoeboid wandering cells which actively migrate through the body or are passively carried by movements of the gastrovascular fluid also provide means of transport in *Tubularia's* body.

No specially organized excretory system is present in this species.

Feeding and Digestion. *Tubularia* eats small animals which happen to come in contact with the tentacles. Caught and paralyzed by the nematocysts, the prey is brought to the mouth by bending of the tips of the tentacles inward and contraction of the hypostome toward the food. The prey is then engulfed by the mouth and passes into the coelenteron. Large food particles are prevented from passing into the stem by a cushion of gastrodermal tissue which constricts the body cavity at the base of the hydranth.

Typical feeding reactions can be elicited in hydranths with small bits of *Mytilus* meat or juice. Experiments should be carried out to determine whether actual contact with the food is essential to evoke these responses and whether the zooid distinguishes among food, filter paper soaked in clam juice, and plain filter paper. This can be done with fine forceps by offering the food to large hydranths under the binocular microscope. The results obtained will depend among other things on the nutritional state of the hydranth, as repeated trials of the same procedure will demonstrate.

Digestion is partly extracellular. Ingested food masses, which may be surprisingly large, are rapidly disintegrated in the body cavity, and the resulting fragments are engulfed by cells of the gastrodermis. Digestion is completed intracellularly.

Nervous System. *Tubularia* has a nervous system which despite its diffuse and primitive organization is made up of discrete neurons like those of higher forms. The neurons are bipolar or multipolar, and their cell bodies are scattered through the system instead of being aggregated into ganglia or nuclei. Hence there are no true nerves or nerve tracts, which are cable-like strands of axons without accompanying neuron cell bodies. Instead the nerve cells are loosely organized into epidermal and gastrodermal plexuses interconnected by fibers coursing through the mesoglea. Neurosensory and neuroeffector junctions are formed by various neurons in these plexuses. Note that the neurons are discrete cells; the concept that they form a syncytial nerve net is erroneous. Synaptic junctions are thus present between them, although such junctions may not be morphologically differentiated.

Some of these features of the nervous system may occasionally be observed in *Tubularia* which have been vitally stained with methylene blue as described above.

Their arrangement may also be determined by testing the reactions of the hydranths to a fine needle or brush touched to various parts of the surface of zooids and stems. Such study can indicate presence or absence of discrete pathways of conduction, directionality of impulses over these, facilitation, spread of excitation, areas of greatest sensitivity, and so on.

In addition to the neurons, neurosensory cells, epithelial sensory cells, and perhaps free nerve endings are present, but no specific sense organs occur in *Tubularia* or any other of the polypoid coelenterates.

Skeletomuscular Systems. Except for the periderm which forms a transparent chitinous exoskeleton there are no skeletal structures in *Tubularia*. The muscular system is epitheliomuscular in nature. The general orientation of its contractile fibrils is longitudinal in the epidermal elements and circular in those of the gastrodermis. The muscular system thus consists roughly of two cylinders of muscle one inside the other.

Some of the contractile elements, particularly in the tentacles, are believed to be independent effectors. In that case they are like the cilia, nematocysts, and some of the gland cells of this form in being independent of nervous control.

The details of the arrangement of the muscles of *Tubularia* have to be studied in histological sections, although the observations on cell types and feeding reactions discussed above will have yielded some useful information on this point.

The movements carried out by *Tubularia* with its muscular system are limited to expanding and contracting reactions and feeding behavior since the adults have no locomotor ability. The reaction patterns during feeding have already been studied.

Reproduction. By asexual processes of plant-like budding *Tubularia* colonies give rise to more hydranths, and when mature these zooids produce the sexually reproducing entities, the gonophores. The latter are borne in clusters springing from the hypostome just above the proximal whorl of tentacles. In these gonophore clusters the distal buds are the oldest. Thus, from the base to the tip of the cluster, a summary of gonophore development may be observed. It starts with an outpushing of epidermis, mesolamella, and gastrodermis, then proceeds to the formation of a gamete-producing individual of the colony. Its structure is quite different from that of the hydranth, however.

A study of the comparative anatomy and comparative life cycles of other hydroids indicates that the gonophore is an incompletely developed medusa which is never set free from the colony. Typically the cnidarian medusa is a bell- or saucer-shaped free-swimming creature with tentacles and a mouth on the lower or subumbrellar surface of the disk (see the sections on *Obelia, Gonionemus, Pennaria,* and *Aurellia*). Despite the distinction between such a medusoid type and the polyp, or zooid, their basic structural organization is closely similar, and in many coelenterate species both forms occur in the life cycle.

In *Tubularia*, however, the medusoid individuals remain attached to the colony by an aboral stalk. Their anatomy reflects the typical tetramerous symmetry of hydromedusae. Tentacles are reduced to four knob-like buds. As in all coelenterate life cycles where medusoid stages or structures are produced, these reduced medusae of *Tubularia* produce the gametes. The species is dioecious. Sex cells develop from endodermal cells and are shed into the subumbrellar cavity. Eggs remain in the female gonophore until fertilization and part of the larval development have occurred. Thus the planula, the ciliated, mouthless stereogastrula typical of many cnidarians, never becomes free-living. Instead it metamorphoses into a polypoid, tentaculate, actinula larva, which is then set free. The latter larva is of unusual interest since it appears to be closely similar in structure to the stem form of coelenterate from which all present-day members of the phylum evolved.

With a pair of fine needles tease out some of the gonophores of *Tubularia* and observe as much as possible of the stages in its life history. Note that even in this species belonging to the simplest of the true metazoan phyla the gametes have already become differentiated into the large non-motile egg and small, flagellated, swimming sperm typical of most of the animal kingdom. It is also worth while to allow some actinulae to settle down and attach in a finger bowl of sea water (renewed from time to time). Each becomes a primary zooid and, if fed, will begin to form the typical stolonial colony, thus completing the life cycle.

The factors involved in growth and differentiation of *Tubularia* have been widely studied by means of regeneration experiments. The interstitial cells which form cnidoblasts also are active in regenerating hydranths. Some simple experiments of this sort may easily be carried out. If clean, vigorous-looking polyps are used, regeneration takes place rapidly. Determine the extent and type of regeneration in lengths of stem which have neither hydranth nor proximal parts present. See if the proximal and distal ends behave similarly, and test whether a fine thread tightly tied around the middle of the piece affects this. Also test the influence of a terminal hydranth on proximal regeneration with and without such a thread. Further interesting results can be obtained by varying the length of stem left attached to an isolated hydranth or by removing the latter's tentacles.

Hydractinia echinata

By T. H. WATERMAN

Hydractinia (Figs. 66, 67) is a small, colonial, marine hydroid of stolonial habit. It forms a spiny encrusting mat of stolons from which the polymorphic zooids arise on unbranched pedicels. This hydroid is common in shallow water on rocks, wharf pilings, and especially gastropod shells inhabited by hermit crabs. Its geographic range is from Arctic regions to Dry Tortugas on the Atlantic coast.

For collecting this coelenterate the shell-houses of hermit crabs are a convenient and common substrate. It is necessary to identify the species, however, since *Podocoryne* and *Stylactis*, which are quite closely related, are also common on shells from similar habitats.

In living colonies the polymorphic zooids of *Hydractinia* are relatively easy to identify, but unless fixation has preserved the polyps in fully expanded condition, killed material is almost useless for this purpose. The discussion on the preparation of *Tubularia* for study is equally pertinent here.

External Structure. The stolons of this species fuse to form a network of gastrodermal tubes covered above and below by sheets of epidermal epithelium resulting from coalescence of the epidermis of neighboring individual stolons. A secreted layer of periderm attaches the colony to the substratum and also projects from the colony's surface in the form of jagged spines which are a diagnostic structural feature. There is no periderm around the zooids, which emerge from this stolonial base. Five types of persons occur in the colonies, although not all in one, since the species is dioecious. The feeding hydranths, or trophozooids, bear one whorl of filiform tentacles at the base of a conical hypostome. They are larger than gonozooids. The latter have a short rounded hypostome and no tentacles, although knobs, each containing large nematocysts and resembling vestigial tentacles, surround the base of the hypostome.

The gonozooids bear the sexually reproducing individuals of the colony. In *Hydractinia* the medusoid stage is even more suppressed than it is in *Tubularia*. The gonophores of the present species are reduced to simple evaginations of a columnar stalk, the blastostyle. Such rudimentary gonophores are called sporosacs and bear eggs or sperm, depending on the sex of the colony. These structures are borne on the stem of the gonozooid proximal to the reduced tentacles.

Similar in general structure to the reproductive persons of the colony are the spiral zooids. They lack gonophores, however, are more elongate, and frequently recurve on themselves, whence the name. They are presumed to be specialized defensive zooids.

In colonies of *Hydractinia* growing on snail shells inhabited by hermit crabs, there may be found near the margins of the shell still another type of polyp, the tentaculozooid. These are even more elongate than spiral zooids, also lack tentacles, and in addition have no nematocyst knobs. These polyps may be sensory in function.

To study the polymorphic colonies fragments of snail shell encrusted with *Hydractinia* provide a convenient source of material. Note that both sexual and asexual reproduction within the colony is restricted to certain zooid types. Except for the structure of the stolonial mat and its spines, the basic plan of *Hydractinia's* organization is closely similar to that of *Tubularia*. Techniques described above for that hydroid can profitably be employed in this case to study the differences in tissue organization and distribution of cell types in the various forms of the *Hydractinia* zooid. Note that only two types of nematocysts occur in this species: (1) desmonemes and (2) microbasic heterotrichous euryteles.

Feeding and Digestion. Repeat on *Hydractinia* the experiments on feeding behavior suggested for *Tubularia*. In the present form it is of interest to note any differences in responses to food and mechanical stimulation when these stimuli are applied separately or together to the different types of zooids present. By this means information should be obtained on whether or not the spiral zooids and tentaculozooids have special functions.

The frequent occurrence of *Hydractinia echinata* on gastropod shells inhabited by hermit crabs has led many zoologists to believe that this relationship is of some commensal significance. Feeding by both animals, however, is apparently quite independent, and either can flourish without the other. The relationship is thus merely a facultative one.

Nervous System and Sense Organs. The possible function of tentaculozooids as sensory polyps is a point that should be studied further. Evidence for greater concentration of sensory and nerve elements in these zooids could be obtained by appropriate staining techniques (e.g., *intra-vitam* methylene blue) and by observations on reactions like those suggested for *Tubularia*.

Skeletomuscular System. The specialized peridermal developments of the encrusting part of *Hydractinia* colonies have already been mentioned. The movement and behavior patterns of the zooids have also been observed. Note that movement and contractile ability of the zooids are extensive in this species, partly because of the absence of periderm on the zooids and their stalks.

Reproduction. A free-living medusoid, as in *Tubularia*, is absent from the life cycle of *Hydractinia*. However, in the two other genera of the Hydractinidae, *Podocoryne* and *Stylactis*, free medusae are produced.

About the only internal detail that can be observed in the sac-like gonophores of *Hydractinia* is a simple structure, homologous with the manubrium of a complete hydromedusa. This simple columnar structure is the blastostyle referred to above. On this are borne the sex cells which are actively shed when ripe, through muscular contraction of the walls of the gonophore.

Several interesting experiments may be done on the spawning of this hydroid. Since the species is dioecious, it would obviously be an advantage if both male and female colonies spawned at the same time in any given neighborhood. This does indeed occur in nature; the spawning is synchronized by some photosensitive mechanisms. Only when it is exposed to light for a certain period, after a preceding conditioning period in the dark, does a colony shed its gametes. Using ripe colonies of both sexes, determine by exposure to experimentally controlled periods of light and dark what the necessary sequence is for spawning. Check these results with spawning time in control colonies just exposed to normal night-day light sequence. Note whether time of day influences ability of a given light sequence to induce spawning. Muscular elements of the gonophores are involved in these reactions. This may be demonstrated by repeating some of the experiments in calcium-free sea water, since muscle cannot contract in the absence of this cation.

After spawning, early cleavage stages, planula larvae, and initial colony formation may be seen if fertilization has been effected. Evidence for the indeterminate nature of the early cleavages in *Hydractinia*, as is characteristic of the phylum, can be adduced from the subsequent development of isolated blastomeres, teased apart with fine needles.

Regeneration in this species, while basically like that of *Tubularia*, has an additional feature of importance; polypoids are readily regenerated, but fragments of each type will grow back only their own kind of zooid. Experiments on this hydroid are more difficult to carry out, however, because of their considerably smaller size.

Pennaria tiarella

By T. H. WATERMAN

This (Figs. 70, 71) is a colonial marine hydroid which forms a loose bushy growth with regular pinnate branches. The colonies may reach 15 cm. in height. Unlike the two preceding hydroids, this form has a free medusa stage in its life cycle. It is common in shallow water on piles, rocks, and seaweed from the Bay of Fundy to the Gulf of Mexico.

Some of the techniques and data given for the preparation of other hydroids will also be found pertinent to the present species and useful for its study.

External Structure. *Pennaria* hydranths have large conical hypostomes bearing scattered capitate tentacles and a basal whorl of filiform tentacles. Just distal to the latter the gonophores are borne. They develop into relatively large medusae with four knob-like tentacles. Spawning may occur while the medusae are still attached to the polyps, but they may break free first. The species is dioecious.

Growth in *Pennaria* colonies is monopodial, i.e., the growth zone is just proximal to the terminal hydranths of the main stem and its branches. After an internode of stem growth has occurred in this region, a hydranth buds off alternately on one side, then the other, of the stem. If it is to give rise to a branch, this subterminal hydranth develops a growth zone just proximal to itself and then proceeds to form the branch just as the main stem has been formed by the primary zooid. The latter is thus the oldest polyp, and the terminal zooid on the branch nearest the base of the colony is the next oldest. The youngest obviously are the penultimate hydranths of the stem and branches.

General Body Organization. Many of the basic structural features of *Pennaria* are not strikingly different from those already observed in other gymnoblastean hydroids. The capitate tentacles of this species are of interest, however, for the marked aggregation of nematocysts in their tips. These may be more readily observed if the tentacles are stained with methyl green or methylene blue. The largest will be found to be stenoteles. In addition there occur here the other three types of stinging cells present in *Pennaria:* (1) desmonemes, (2) basitrichs, and (3) microbasic mastigophores. Suggestions for their study will be found in the section on *Metridium*.

Details of the other various cell types making up the tissues of this hydroid may be demonstrated by the following maceration technique (which may also be applied to other coelenterates):

1. Place the fresh tissue to be studied in a sea-water solution of equal parts of 0.05 per cent osmic acid and 0.2 per cent acetic acid for 3 minutes.

2. Wash in several changes of 0.1 per cent acetic and leave in the latter 12 hours.

3. Wash for a few minutes in water.

4. Stain with acid carmine and mount in glycerine.

5. If cells are not sufficiently isolated, separate them further by tapping the coverglass gently.

The above procedure provides only a temporary mount; for permanent preparations a different stain and dehydration must be used. This technique should clearly demonstrate the various cell types present, and such details as the muscle fibers of the hydromedusa may be observed for the first time in the series of hydroid types presented in this book.

The medusoid form is bell- or thimble-shaped and, as mentioned above, has four reduced tentacles present on its margin. The outer surface, from whose apex an aboral stalk attaches it to the hypostome of the parent hydranth, is called the exumbrellar surface; the inner surface, the subumbrellar, bears the mouth centrally at the end of a tubular pendant structure like a bell clapper. This is called the manubrium. Extending from the margin of the bell towards the oral-aboral axis is a shelf-like double layer of epidermis, the velum, which partly closes off the subumbrellar cavity from the exterior. It functions in these small medusae to produce a sort of jet propulsion when the bell contracts. Both ex- and subumbrellar surfaces consist of epidermis.

The gastrovascular system comprises a central gastric cavity and four radial canals which join a ring canal running around the margin. The exumbrellar layer of mesoglea is much thicker than the mesolamella characteristic of the subumbrellar layer and the polypoid colony, although in the hydromedusae it is still an acellular layer.

The analyses and suggestions made for *Tubularia* and *Hydractinia* apply generally to the *Pennaria* polyp as well. Further comment and observations on hydromedusae will be given for a form, *Gonionemus*, more suitable for detailed study.

Reproduction. In *Pennaria* the medusoid stage, which reaches a size of only 2 mm. or thereabout, usually becomes free living but has only a brief period of pelagic life, during which it feeds little if at all. Growth of the medusa, maturation of the gametes, and sometimes even their release occur while the gonophore is still attached to and dependent on the polyp colony. Thus, although less suppressed, the medusoid stage is only somewhat less so than in *Tubularia*, for example. An examination of early cleavage stages is of interest because of the remarkably irregular arrangement of the blastomeres. These gradually become organized into a morula, and further development is more typical.

Obelia geniculata
By T. H. WATERMAN

This species (Figs. 68, 69) is a colonial, marine hydroid which grows on short, unbranched, zig-zag stems arising from a prostrate stolon. The hydranths, on short pedicels, arise alternately from the "shoulders" which occur at the nodes of the stem. Free-living medusae occur in the life cycle; they develop in a gonangium characterized by an axial collar. This hydroid is common in shallow water, on piles, rock, and seaweed. Its range is the Atlantic coast from Arctic regions to the Gulf of Mexico and the Pacific coast from Southern California to Oregon. Other species differing from *geniculata* only in growth habit and structural details occur on both North American coasts and in Europe.

External Structure. Each hydranth of *Obelia* bears a single whorl of filiform tentacles at the base of the hypostome, which swells distally into a subspherical structure. A hydrotheca, part of the periderm, forms a bell-shaped cup around the base of the hydranth. At the base of this hydrotheca the polyp rests on a shelf-like projection which constricts the cavity within the periderm at this point. Gonophores which develop into free discoidal medusae are borne on modified polyps called blastostyles. Each of the latter is enclosed in a peridermal protective structure that in this case is a vase-shaped gonotheca constricted terminally by a collar. Gonotheca, blastostyle, and gonophores together form a gonangium. Gonangia arise in an axillary position between the simple stem and the pedicels of hydranths.

The growth of the *Obelia* colony is sympodial; i.e., each new hydranth arises as a bud from the stem just proximal to the next youngest polyp. The gonangia arise as secondary buds in the axils so formed. Thus the youngest hydranths are the most distal ones, and the stem of the colony, which in monopodial forms is all produced by the primary zooid, here results from the composite effort of all the hydranths it bears. Each internode of the stem is contributed by one hydranth.

General Body Organization. Most of the important characteristics of the *Obelia* trophosome (the feeding zooids collectively) have been mentioned above. One further prominent feature of the periderm should be observed, namely, that stems, pedicels, and gonangial stalks

100 CNIDARIA AND CTENOPHORA

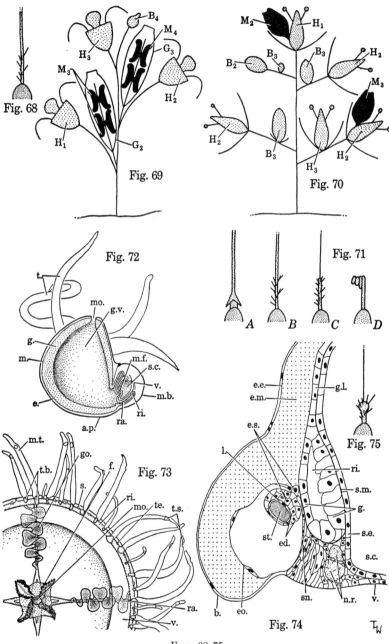

FIGS. 68–75

FIG. 68. Nematocyst type of *Obelia:* basitrich. (Data from Weill.)
FIG. 69. Growth pattern of a colony of *Obelia*, showing its sympodial structure and the relation between gonangia and hydranths.
FIG. 70. Growth pattern of a colony of *Pennaria*, showing its monopodial structure and the relation between polypoid and medusoid stages.
FIG. 71. Nematocyst types of *Pennaria*. (Data from Weill.) *A*. Stenotele. *B*. Basitrich. *C*. Microbasic mastigophore. *D*. Desmoneme.
FIG. 72. Longitudinal section of polyp of *Gonionemus*, showing medusa bud on the column. ×50. (After Joseph.)
FIG. 73. Oral view of developing medusa of *Gonionemus*. ×6. Preserved specimen.
FIG. 74. Radial section of the bell margin of a young medusa of *Gonionemus*, showing details of a statocyst. ×550. (After Joseph.)
FIG. 75. Nematocyst type of *Gonionemus:* microbasic heterotrichous eurytele. (Data from Weill.)

ABBREVIATIONS FOR LABELS ON FIGS. 68–75

a.p. : aboral pole
b. : margin of medusa bell
B_2 : bud (second generation)
B_3 : bud (third generation)
B_4 : bud (fourth generation)
e. : epidermis
ed. : ectodermal cells around lithocyte
e.e. : exumbrellar epidermis
e m. : exumbrellar mesoglea
eo. : ectodermal wall of statocyst
e.s. : endodermal stalk cells of lithocyte
f. : fimbriated lip
g. : gastrodermis
G_2 : gonangium (second generation)
G_3 : gonangium (third generation)
g.l. : gastrodermal lamella
go. : gonad
g.v. : gastrovascular cavity
H_1 : hydranth (first generation)
H_2 : hydranth (second generation)
H_3 : hydranth (third generation)
l. : lithocyte

m. : mesolamella
M_2 : medusoid (second generation)
M_3 : medusoid (third generation)
M_4 : medusoid (fourth generation)
m.b. : medusa bud
m.f. : manubrium forming
mo. : mouth
m.t. : marginal tentacle
n.r. : nerve rings
ra. : radial canal
ri. : ring canal
s. : statocyst
s.c. : subumbrellar cavity
s.e. : subumbrellar epidermis
s.m. : subumbrellar mesoglea
sn. : sensory epithelium
st. : statolith
t. : tentacles
t.b. : tentacular bulbs
te. : tentacle bud
t.s. : tentacular suckers
v. : velum

are all annulated by ring-like constrictions. These presumably add flexibility to the colony, but their significance is not clear. Only one type of nematocyst has been found to be present: basitrichs.

Like the anthomedusa of *Pennaria*, the leptomedusa of *Obelia* has a marked tetramerous symmetry. It has twenty-four marginal tentacles when liberated from the gonangium, four radial canals on which the sex cells are borne, and eight closed statocysts around the bell margin. At least in young specimens, when the medusa is not actively swimming, the bell is often everted so that the manubrium projects apically from the umbrella. The more usual relationships are assumed, however, when the animal swims by the pulsations of the disk.

The gonophores are released from the gonangium as complete, active, small medusae. They do not, however, have mature gametes. Hence, an appreciable period of pelagic existence is necessary before spawning can occur. Of the hydroids studied so far in the series in this book, *Obelia* has the most complete and relatively most independent sexual stage of the group. In the next form to be considered, however, the medusoid is definitely the predominant generation while the polypoid is a small solitary form.

A special type of asexual reproduction may occur in *Obelia* when the water temperature exceeds 20°C. In this case buds, which would normally become gonangia, break free of the colony, settle down, and by stolon-like growth start a new colony.

Gonionemus murbachii
By T. H. Waterman

The hydroid of this coelenterate species (Figs. 72–75) is a minute (1 mm.), solitary polyp with four tentacles. The medusoid stage is much larger (2 cm. in diameter). It is a free-swimming hydrozoan jellyfish with sixteen or more marginal tentacles and four radial canals along which the gonads are located. The latter have a characteristic ruffled appearance. The medusa occurs in shallow water attached to seaweed (particularly the eelgrass, *Zostera*) by adhesive disks on the tentacles. Before the wasting disease of *Zostera* nearly wiped out the species, *Gonionemus* was common in certain localities in Vineyard and Long Island sounds. Another species, *G. vertens*, is locally abundant on eelgrass from Puget Sound to Alaska.

External Structure. As far as known, the polypoid never develops beyond a minute hydra-like creature with four tentacles, each with a solid core of gastrodermis. Gonophores are borne in gymnoblastean fashion as buds on the column of the hydroid. The medusa of *Gonionemus* is a typical tetramerous hydromedusa with a short manubrium, bearing four brief frilled oral lobes. Special pads of adhesive tissue occur distally on a sort of knee on each tentacle. They may be studied by placing an appropriate piece of tentacle under a coverslip on a slide and examining them in a drop of glycerine under high magnification. At the same time note the nearby batteries of nematocysts and the fact that the medusan tentacles are hollow in this species.

General Body Organization. As in other hydromedusae, the gastrovascular system of *Gonionemus* consists of a central gastric cavity into which the mouth opens and from which radiate four canals. The peripheral ends of these radial canals are joined by a ring canal which runs around the margin of the bell. These gastrovascular canals are tubes of gastrodermis embedded in the mesoglea. The latter forms only a thin layer like the hydroid mesolamella between the gastrodermis and subumbrellar surface but is a thick gelatinous stratum between the exumbrellar epidermis and the gastrodermis. Between the radial canals the gastrodermis forms a thin sheet of flat epithelial cells (one layer thick) like a part of the gastrovascular system whose cavity had been obliterated and walls collapsed to a single layer of tissue. This sheet is called the gastrodermal lamella.

In *Gonionemus*, as in other medusae, note that the gelatinous bulk of the animal consists of mesoglea. The epidermis and gastrodermis are simple epithelia as they are in a polyp like *Tubularia,* but the mesoglea is greatly thickened as described above. Vital staining or permanent fixed and stained preparations will demonstrate that there are few, if any, cells present in the mesoglea of *Gonionemus;* this is characteristic of Hydrozoa but this intermediary layer is cellular, at least in part, in the Scyphozoa and Anthozoa.

At the bases of the marginal tentacles are swellings termed tentacular bulbs; they contain part of the gastrovascular cavity and apparently function in digestion and also serve as sites of nematocyst formation. These bulbs in addition have some sensory functions, but they are not photosensory ocelli as frequently stated. However, sense organs in the form of statocysts do occur in the bell margin between the tentacle bases.

Since *Gonionemus* is sometimes difficult to obtain in the living state, observations on the behavior and physiology of medusae had best be made on more readily available animals. The behavior of the *Gonionemus* medusa is, however, interesting on several counts. First, the chemical stimulation of food apparently excites the animals to greater activity. Random swimming movements, poorly, if at all, oriented to the stimulus, are increased. Second, this medusa has been described as actively fishing. To do so it swims upwards to the surface, turns over, and sinks downward with its tentacles trailing. Both these feeding activities are worthy of notice since many medusae display feeding reactions only after they actually come in chance contact with food.

Reproduction. Note that in *Gonionemus*, as in Scyphozoa and trachyline medusae, the hydroid stage of the life cycle is much reduced. The polyp can reproduce asexually by budding off frustules, which in turn develop into more polyps, but each individual remains solitary and bears gonophores on its pedicel. The medusae produce eggs and sperm in gonads dependent from the radial canals, although the germ cells are believed to originate in the ectoderm. The zygote gives rise to a ciliated planula which swims about, settles down on its aboral end, and metamorphoses into the polypoid stage. When ripe, living material is available, all these early stages may readily be observed in the finger bowls in which the medusae are kept.

Pelmatohydra oligactis

By T. H. Waterman

The hydras are solitary, large (up to 20 mm.), fresh-water hydroids, occurring in ponds, lakes, and sluggish streams. They reproduce sexually and asexually. There is no trace of a medusoid stage in the life cycle. *Pelmatohydra oligactis* (Figs. 76–81), the brown hydra, is common in northern United States, east of the Rockies; it is also of widespread occurrence in Europe.

For identification and study of hydras it is almost essential to culture the animals for a considerable period. Thus the observation of an embryonic theca, critical for taxonomic purposes, will require that specimens be brought into the sexual condition. In order to do this, secondary cultures of *Cyclops*, *Daphnia*, or similar forms must be maintained to feed the polyps.

External Structure. This common brown hydra, like the other members of the family, is organized into a columnar body, basally differentiated into an adhesive disk and distally bearing the mouth in the center of the inconspicuous hypostome. The latter is surrounded by a single whorl of elongate tentacles, which when extended are three to four times as long as the column. Usually there are six or less of these tentacles. *Pelmatohydra* is differentiated from *Chlorohydra* by its lack of the gastrodermal zoochlorellae responsible for the green color of the latter genus. *Pelmatohydra* is differentiated from *Hydra* by the division of its column into a proximal narrow stalk and a distal thicker body. The regions of the column wall are distinct not only in their diameters but also in the absence from the stalk of glandular cells freely distributed in the gastrodermis of the body region. Note that there is no periderm. Epizooic ciliate protozoans belonging to the genera *Kerona* and *Trichodina* are frequently found living on the outer surface of hydras as ectocommensals.

General Body Organization. Basically the structure of the hydras is close to that of the typical hydroid polyp, consisting as it does of a three-layered sac whose cavity, the coelenteron, opens to the exterior by a mouth surrounded by tentacles. It is peculiar in organization, however, in possessing hollow tentacles whose cavities are in direct connection with the coelenteron. Some other hydroid polyps have hollow tentacles, but their lumina are isolated from the main body cavity.

It is noteworthy that, although the hydras have, like the other cnidarians, attained only a tissue level of differentiation (as opposed to the organ differentiation of higher animals), the tissues are specifically differentiated. This may be demonstrated by turning the column of hydra inside out so that the normal relations of epidermis and gastrodermis are reversed. A blunt bristle or fine glass rod may be used to accomplish this by pushing the base of the column up through the mouth. In those polyps which survive without directly turning back to their normal arrangement, it will be found that active migration of epidermal and gastrodermal cells restores them to the original relation by a remarkably rapid cellular reorganization. Such hydras may be feeding and behaving normally again after 2 days. Despite this striking ability to reorganize their normal structure, hydras are unable to reform themselves if dissociated into isolated cells as certain sponges like *Microciona* are well known to do. Small fragments (one-fiftieth of a polyp) will regenerate, but they must consist of at least a few cells of both epidermis and gastrodermis.

106 CNIDARIA AND CTENOPHORA

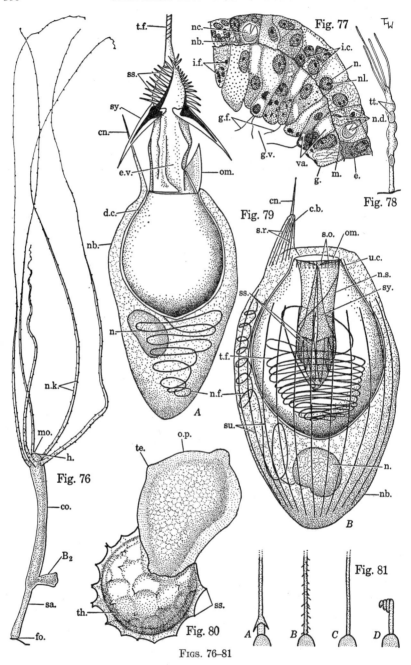

Figs. 76–81

FIG. 76. Aquarium specimen of *Pelmatohydra* with tentacles well extended. ×6.
FIG. 77. Detail of cross-section through digestive region of the column of hydra. ×60. From fixed and stained material.
FIG. 78. Male specimen of *Pelmatohydra* with ripe gametes. (After Hyman.)
FIG. 79. Undischarged (*B*) and discharged (*A*) stenotele of hydra. ×1700. (After Schultze.)
FIG. 80. Young hydra hatching. ×430. (After McConnell.)
FIG. 81. Nematocyst types of hydra. (Data from Weill.) *A.* Stenotele. *B.* Holotrich. *C.* Atrich. *B.* Desmoneme.

ABBREVIATIONS FOR LABELS ON FIGS. 76–81

B_2 : bud (second generation)
c.b. : cnidocil base
cn. : cnidocil
co. : stomach region of column
d.c. : discharged nematocyst capsule
e. : epidermis
e.v. : everted shaft of nematocyst
fo. : pedal disk
g. : gastrodermis
g.f. : gastrodermal flagella
g.v. : gastrovascular cavity
h. : hypostome
i.c. : interstitial cells
i.f. : ingested food particles
m. : mesolamella
mo. : mouth
n. : nucleus
nb. : cnidoblast
nc. : nematocyst

n.d. : nematocysts developing
n.f. : cnidoblast filament
n.k. : nematocyst knobs
nl. : nucleolus
n.s. : nematocyst shaft
om. : operculum
o.p. : oral pole
sa. : stalk region of column
s.o. : supporting rods of operculum
s.r. : supporting rods of cnidocil
ss. : spines
su. : supporting rods of cnidoblast
sy. : stylet
te. : tentacle bud
t.f. : tubular filament
th. : theca
tt. : testes
u.c. : undischarged capsule
va. : vacuoles

Histological study of the various regions of the body either in macerated tissue or in prepared sections will provide further insight into hydra's organization. Techniques suggested for *Tubularia* and *Pennaria* may be used to advantage here also. A maceration procedure recommended for this particular form is the following:

1. Place a hydra on a glass slide and pipette away as much of the surrounding water as possible.
2. Invert this slide over a wide-mouthed vial so that the polyp is exposed to the vapor from the fluid contained therein. This fluid is made up of equal parts of 40 per cent formaldehyde, glacial acetic acid, and 95 per cent alcohol.
3. Remove after 10 minutes; add 2 drops of water and then pipette off; then add another 2 drops of water and pipette off.
4. Add a drop of 40 per cent glycerine and tease out the tissue with fine needles. Add a coverglass.

By the use of these different techniques a variety of cell types will be observed. Basically the polyp consists of a columnar epidermal epithelium and a flagellated gastrovascular epithelium separated by an acellular intermediate mesoglea, called a mesolamella as in other hydroid polyps. Of particular interest are certain cells which are specialized for more than one function. Thus there are epitheliomuscular cells in the epidermis and nutritive-muscular cells in the gastrodermis. Both have their basal portions broadened and flattened so that they can contain myofibrils; the distal parts of these cells remain epithelial or may be digestive in function, depending on their location. Such cells as these with dual special structure and functions are known in but few cases in the whole animal kingdom; the partly contractile, partly impulse-conducting cells found in certain regions of the body in Nematoda are one example found outside the Cnidaria.

Another interesting cell type occurring in hydras is found in the basal disk. Here there are epidermal gland cells able to secrete an adhesive material which cements the animal to the substratum. In addition, this region can also produce a bubble of gas which causes the hydra to float inverted just below the surface film of the water.

There are four different types of nematocysts present in hydra. Techniques for discharging and studying these organelles are given in the section on *Metridium*. The kinds of nematocysts present in hydra are, in order of their size, stenoteles (which are the largest), holotrichs, atrichs, and desmonemes (which are the smallest). Their distribution is characteristic and different for each of the four kinds. No nematocysts occur in the foot, or basal disk, and all types occur in abundance on the tentacles. Only holotrichs are found on the hypostome, while the column has stenoteles and occasionally holotrichs.

Despite their more general final distribution all these nematocysts originate mostly in the distal three-fourths of the column. From there the epidermal interstitial cells which differentiate into cnidoblasts migrate or are carried by gastrodermal wandering cells to their final location in various parts of the body.

In the discharge of a nematocyst, mechanical pressure releases the covering operculum, thus initiating the series of changes which leads to the release of the nematocyst filament. This is an eversion of a pre-existing, internal, tubular structure. The latter lies, before discharge, coiled within the capsule of the nematocyst. The wall of the hollow filament and that of the capsule are continuous. In the process of discharge the previously involuted tube is ejected from the capsule by being turned inside out. Thus the innermost part of the filament in the capsule becomes the outermost, distal end on discharge.

The occurrence of four different types of stinging cells in hydras and their parallelism to the kinds of nematocysts found in the marine hydroids, such as *Tubularia,* suggest that these fresh-water coelenterates are specialized descendants of the latter. From this point of view then the complete absence of a medusoid stage may be considered part of the very general trend in sessile and benthonic fresh-water animals to suppress free-living planktonic stages in their life cycle. Certainly there are widespread parallel cases of more or less complete suppression of the medusa in various marine Hydroida. Thus the notion that hydra represents a primitively simple, ancestral, diploblastic, coelenterate type seems no longer tenable to the majority of workers in the field, although it is widely entertained in the elementary textbooks.

There are no noteworthy features of *Pelmatohydra's* respiration. It should be pointed out, though, that in the closely related *Chlorohydra* the presence of intracellular green algae in the gastrodermis is of respiratory importance. When photosynthesis is occurring, the hydra is able to survive in low-oxygen environments unavailable to hydras without zoochlorellae. Differences in behavior relative to light and oxygen can be readily observed between hydras with and without these intracellular plants.

As in other cnidarians the gastrodermis of hydra is not only a secretory and absorptive tissue but is also concerned with transportation. Most, or all, of the gastrodermal cells are flagellated, with one to five flagella on each. The transport of developing cnidoblasts from their site of origin to their final location by wandering cells of the gastrodermis has already been mentioned. Ingested food may also be carried from the digestive region of the column to all parts of the body by the breaking free of cell masses or fragments containing food, followed by their distribution through peristalsis and flagellar currents. In other body regions these masses or fragments disintegrate and with their contained food are engulfed by gastrodermal cells.

No specific structural or functional systems concerned with disposal of metabolites are known in this form. In species containing zoochlorellae, these plants, like the symbiotic intracellular algae of corals, will function, to some extent at least, as excretory organs for the animal. This is possible because certain waste products of the hydra's protein metabolism will be essential nutrient materials for the algae and hence rapidly used up.

Feeding and Digestion. These functions in hydra conform to the basic coelenterate pattern. Chance contact of the prey with the outspread tentacles of the polyp provides a sort of coarse filter-feeding mechanism. *Pelmatohydra oligactis* is remarkable for the length of its fishing tentacles which are in this respect more like those of some medusae than the usual relatively short ones of other polyps. Spontaneous contractions and expansions of the column and tentacles occur and have been presumed to represent feeding behavior. Recent observations, however, have not confirmed the widely expressed opinion that this spontaneous activity becomes intensified in hungry hydras.

Prey which contacts the tentacles is entangled, held, and paralyzed into submission by the activity of the nematocysts. Muscular movement of the tentacles then conveys the food to the mouth for ingestion. Hydranth feeding is carnivorous and voracious, the body and mouth being notably expansible for large prey. Swallowing, as in many other animals, is aided by mucus produced by peristomal gland cells. This secretion apparently acts, like enterokinase, to stimulate the activity of the exocrine digestive cells, since living animals are not digested if introduced directly into the enteron. Normally, however, rapid extracellular partial digestion is followed by absorption and completion by intracellular digestion, as is typical for the phylum. The activity and enzymes involved may be studied by the same techniques described for *Tubularia* and *Metridium*.

Nervous System and Sense Organs. Morphological and physiological studies of the nervous system and behavior of hydras indicate their similarity to the same phenomena in other coelenterates. Thus spontaneous activities are limited, and coordination is weak and diffuse. Isolated parts of the body are able to carry out a number of their activities as if they were still part of the whole organism. The physiology of the neuromuscular system may be studied by means of experiments like those described for *Metridium* and *Aurellia*. In addition *intra-vitam* staining with methylene blue may occasionally demonstrate parts of the nerve net of this polyp.

No special sense organs are present.

Muscular System. Epitheliomuscular cells with the myonemes of their expanded basal parts provide the contractile elements of hydras. These, as well as the other cell types, may be most readily studied in macerated tissues, prepared as described above (p. 108).

No skeleton or periderm is produced by the hydras.

Reproduction. Since a medusoid stage is completely lacking in hydra's life cycle, gametes are produced directly by the polyp. This absence of a gonophore, or even vestige of one, is believed not to be a primitive hydroid characteristic but rather to be an endpoint in the trend towards reduction of the medusa which, as mentioned above, apparently occurred more than once in the evolution of the Hydrozoa. In *Pelmatohydra oligactis* the sexual phase is induced by lowered water temperatures and hence would in nature occur seasonally in the fall. In the laboratory, however, it may often be produced at any season merely by lowering the temperature of the culture vessels for 2–3 weeks. Eggs and sperm are produced in different individuals, although a number of other hydra species are monoecious.

The sex cells are of epidermal origin, apparently arising from interstitial cells. The gonads, which are mere aggregates of gametes rather than true organs, develop just beneath the epidermis of the column distal to the stalk. Ova become large when mature, equalling the column in diameter. Their maturation may be divided into two phases. First, there is a period during which one interstitial cell grows at the expense of its neighbors. The latter disintegrate and are absorbed by the developing egg. Then there is a period in which yolk is elaborated by the single, surviving oöcyte which by this time has developed radiating, elongate, finger-like processes. Material for the yolk appears at this stage to be absorbed from the gastrodermis by this growing egg cell. When maturation is complete, the egg becomes spherical and breaks through the overlying epidermis. It remains attached to this point for several days and must be fertilized during this period if development is to proceed. A chitinous shell, which may have low spines on its surface, is secreted around the zygote. The latter drops off the female parent shortly and adheres to the substratum. Cleavage is equal and holoblastic; gastrulation occurs by delamination and ingression to form a stereogastrula; at hatching the young hydra has a mouth, column, and tentacles and grows directly into the adult form. The ciliated, larval, free-living planula typical of most cnidarians does not occur in hydra's life cycle.

The male *Pelmatohydra oligactis* is smaller than the female and is difficult to identify as a member of the species. The spermatozoa develop in aggregates under the epidermis through which they erupt when mature. Unlike some other hydras, this one has no nipple on the testis to serve as an exit for the sperm.

Asexual reproduction normally occurs by the familiar budding of new polyps and their separation from the base of the body region of the column. Usually fission does not occur, although both longitudinal and transverse division of hydras may take place in the process of regaining normal organization after abnormal development such as that induced by cuts and grafts.

Aurellia aurita

By T. H. WATERMAN

The sexually reproducing stage of this scyphozoan (Figs. 82–89) is a large (up to 30 cm. in diameter) marine medusa with four frilled oral arms, short marginal tentacles, and alternate branched and unbranched radial canals connected peripherally by a ring canal. The polypoid form is reduced to a solitary trumpet-shaped scyphistoma, about 5 mm. high. This species is common on the Atlantic coast of North America from Greenland to the West Indies in continental waters. The same, or closely similar, species is cosmopolitan in all warm and temperate oceanic regions.

External Structure. The symmetry of the *Aurellia* medusa is strongly tetramerous. The mouth is square in shape, and from its four corners the oral arms arise. The four radii so delineated are called the perradii; at 45° to these are four interradii, and halfway between per- and interradii are eight adradii. The gastric pouch is four-lobed with each of its outpocketings lying along an interradius. From it arises a definite pattern of branched and unbranched radial canals which distally join a common circular ring canal. The perradial canals, arising between the gastric pouches, and the interradial canals, arising from the distal end of each pouch, show a characteristic branched pattern. At the ends of their center branch are niches in the margin of the medusa which scallop the bell into eight lobes. In the eight indentations thus formed complex sense organs called rhopalia occur. They will be discussed below. In the adradii run eight unbranched canals, two of which arise from each of the four gastric pouches. The gonads appear as four horseshoe-shaped aggregates in the floor of these pouches. Beneath them are shallow cavities, the subgenital pits, in the subumbrellar surface. These do not, however, appear to be functionally connected with the gonads or gamete discharge.

The oral arms are large and gelatinous, being about as long as a radius of the bell. Along their subumbrellar surface runs a groove, the two edges of which are convoluted and bear many short tentacles. The remaining tentacles of the medusa are marginal in their position on the disk and are simple, short, and numerous. They arise at the peripheral edge of the exumbrellar surface and are hollow structures. Nematocysts occur on their surface in aggregates like broken rings. Only two of the numerous known types of these stinging cells occur in *Aurellia:* (1) atrichs, and (2) microbasic heterotrichous euryteles. For further discussion of nematocyst structure and function see the sections on *Metridium* and *Pelmatohydra*.

General Body Organization. The basic structural plan and main types of cells and tissues of the *Aurellia* medusa are similar to those of the other cnidarian studied. The mesoglea, however, is unlike that of hydrozoans in its cellular character and in its much larger volume differs from the homologous layer in polypoid coelenterates.

The staining and maceration techniques described for *Tubularia* and *Pennaria* are also useful in the study of *Aurellia* if cell and tissue details are to be observed under moderate magnification.

The body organization of the polypoid stage of *Aurellia*, the sessile scyphistoma, is rather different from that of the hydroids. In this scyphozoan there are four septa, called tenioles, which divide the gastric cavity into four perradial pouches. Ectodermal invaginations of the subumbrellar surface, the subumbrellar funnels, push into these septa. It will be seen that the interradial position of these invaginations of the polypoid stage is the same as that of the subgenital pits of the medusa. Note, however, that the gastric pouches of the latter are not perradial, as are those formed by the septa, but are interradial in position.

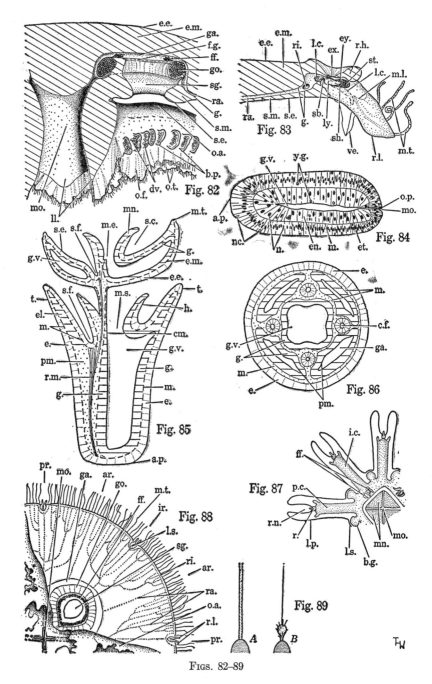

Figs. 82–89

Fig. 82. Interradial section through oral-aboral axis of an *Aurellia* medusa to show relation between gastrovascular system, subgenital pits, gonads, and oral arms. ×2. From preserved material.

Fig. 83. Radial section of the margin of an *Aurellia* medusa to show the relations of a rhopalium to the gastrovascular system and tentacles. ×4. From preserved material.

Fig. 84. Longitudinal section of a planula of *Aurellia*. ×290. (After Hein).

Fig. 85. Schematic longitudinal section of a strobila of *Aurellia* which has released all but its last ephyra. The part of the figure to the left of the oral-aboral axis represents an interradial section, that to the right a perradial one. ×35. (After Percival.)

Fig. 86. Schematic cross section of *Aurellia* through the distal column of a scyphistoma to show the relation between septa, subumbrellar funnels, and gastrovascular system. ×35.

Fig. 87. Quadrant of a young ephyra of *Aurellia*. ×15. From preserved material.

Fig. 88. Quadrant of a young medusa of *Aurellia*. Note that the oral arms have been somewhat displaced to show underlying structures more clearly. ×1. From preserved material.

Fig. 89. Nematocyst types of *Aurellia*. (Data from Weill.) *A*. Atrich. *B*. Microbasic heterotrichous eurytele.

ABBREVIATIONS FOR LABELS ON FIGS. 82–89

a.p. : aboral pole
ar. : adradius
b.g. : bud of adradial gastrovascular canal
b.p. : brood pouches
c.f. : cavity of subumbrellar funnel
cm. : columella
dv. : developing embryos
e. : epidermis
e.e. : exumbrellar epidermis
el. : endodermal core of tentacle
e.m. : exumbrellar mesoglea
en. : endoderm
et. : ectoderm
ex. : exumbrellar sensory pit
ey. : eyespot
ff. : gastric filaments
f.g. : floor of gastric pouch (gastrodermis plus mesolamella plus subumbrellar epidermis)
g. : gastrodermis
ga. : gastric pouch
go. : gonad
g.v. : gastrovascular cavity
h. : hypostome
i.c. : interradial canal of gastrovascular system
ir. : interradius
l.c. : gastrovascular canal running into rhopalial lappet
ll. : lips
l.p. : primary lappet
l.s. : developing margin of bell in Fig. 87; lithostyle in Fig. 88
ly. : lithostyle
m. : mesolamella

m.e. : mouth of ephyra
m.l. : marginal lappet
mn. : manubrium
mo. : mouth
m.s. : mouth of strobila
m.t. : marginal tentacle(s)
n. : cell nuclei
nc. : nematocysts
o.a. : oral arm
o.f. : marginal frill of oral arm
o.p. : oral pole
o.t. : tentacles of oral arm
p.c. : perradial canal of gastrovascular system
pm. : septum
pr. : perradius
r. : rhopalium
ra. : radial canal(s)
r.h. : rhopalial hood
ri. : ring canal
r.l. : rhopalial lappet(s)
r.m. : retractor muscle of septum
r.n. : sensory niche
sb. : subumbrellar sensory pit
s.c. : subumbrellar cavity
s.e. : subumbrellar epidermis
s.f. : subumbrellar funnel
sg. : subgenital pit
sh. : eyespot with pigment cup
s.m. : subumbrellar mesoglea
st. : statolith
t. : tentacle
ve. : velarium
y.g. : yolk granules

Aurellia does not have these septa or funnels present in the medusa; they disappear during the metamorphosis of the strobila into the ephyra (mentioned below). The subgenital pits of *Aurellia* and the possibly homologous subumbrellar funnels of other Scyphozoa increase the surface area of the body especially in the region of the gonads. It has been suggested that this is of respiratory significance, but experimental support for this notion is lacking. The highly branched gastrovascular system of this medusa is obviously of importance for respiration, since it permits the circulation of coelenteric fluid in direct contact with the exterior through the mouth to all parts of the gastrodermis.

Even in a fairly massive jellyfish like *Aurellia*, the active tissues are still epithelial in their organization, and the mesoglea, which is thick, contains relatively minute amounts of active protoplasm. Entire jellyfish probably contain only about 1 per cent organic matter, largely protein. In some species that have been analyzed about 96 per cent has been found to be water. The statements in the literature that marine jellyfish may contain 99.8 per cent water are not correct, since even sea water itself is somewhat less than 97 per cent water usually! However, it has been shown for *Aurellia* that in a brackish environment of 0.6 per cent salinity, in which this medusa sometimes lives, it may consist of 98 per cent water. Thus with such large proportions of water overall oxygen requirements would be expected to be small per unit of fresh weight.

The coelenteric fluid of *Aurellia* is moved about in the gastrovascular cavity by the swimming contractions of the disk and movements of the oral arms. In addition flagellar currents flow distally through the straight radial canals and centrad in the branched ones. A suspension of fine carmine or India ink particles injected into the coelenteron and ring canal will demonstrate this. It is of interest to determine the pattern and speed of particle movement in the various channels by such means.

As mentioned for *Tubularia*, wandering cells also play a transport role, perhaps nourishing and acting as scavengers in the regions of the mesoglea which are too remote from epidermis or gastrodermis for diffusion to be effective. As in hydroids, there are no structural specializations for excretion present in *Aurellia*, and the physiology and biochemistry of this function are imperfectly known.

Feeding and Digestion. The oral lobes of *Aurellia* with their tentacles play a major part in its feeding. As is characteristic of the scyphozoans, gastrodermal tentacles (gastric filaments) occur on the

AURELLIA AURITA 117

wall of the coelenteron. In *Aurellia* they lie in the floor of the gastric pouches just central to the gonads. These tentacles bear nematocysts and hence function in quieting prey which may be brought into the gastric cavity still alive. They also contain gland cells which secrete most of the digestive enzymes for extracellular digestion. Furthermore, intracellular digestion probably also occurs in the gastric filaments. These various phenomena may be observed to advantage if some copepods or other zooplankton stained with neutral red or bromothymol blue are introduced into the coelenteron or even just onto the oral lobes of a living medusa.

Also the specific activity of the enzymes can be tested either by making extracts of various tissues like the gastric filaments and then measuring their activity or by introducing drops of olive oil, pieces of denatured egg white, grains of starch, glycogen, etc., onto the surface of the gastrodermis and observing whether or not they are affected by enzymes.

Nervous System and Sense Organs. As in the hydrozoans the nervous system consists of subepithelial diffuse nerve plexuses. In *Aurellia* there is a concentration of neurons near each rhopalium, forming a sort of ganglion, and radial strands of the network running towards the oral-aboral axis. On the subumbrellar surface the nerve net extends onto the oral arms and tentacles. Conduction of impulses via this nervous system has been measured at 23 cm. per sec. for *Aurellia* (cf. 60 cm. per sec. in some slow mammalian non-myelinated nerve fibers). If a doughnut preparation is made by removing the margin (including rhopalia) and the center of the bell, an impulse once started around the ring in a given direction will circulate around for several hours without fatiguing or changing its amplitude.

The specific sense organs of *Aurellia* are the eight marginal tentaculocysts, the rhopalia. These structures which are presumably specialized tentacles consist of a club-shaped lithostyle which has an equilibratory function, upper and lower sensory pits, and an ocellus. The rhopalium itself is enclosed in a hood and two flanking pairs of tablike lappets. To observe the gross structure of this organ place an appropriate piece of the umbrellar margin in a watch glass under a dissecting microscope.

The frequency of contraction of the medusa's disk is greater under certain conditions when the illumination is increased. A demonstration of the photosensory function of the ocelli may be obtained by a controlled experiment of this sort on medusae with and without rhopalia.

Experiments on the importance of the marginal sense organs have shown that they, or regions of the nerve net near them, normally act as pacemakers for the rhythmic contractions of the medusa. This can be observed when all but two rhopalia are removed from an *Aurellia*. Swimming movements can be seen to originate near the rhopalia. In such a medusa, too, compensatory responses to tipping of the bell are mediated by the rhopalia; these disappear if the latter organs are removed. Note that, phylogenetically speaking, the marginal sense organs of medusae mark the first appearance of an organ in the animal series. With their exception Cnidaria in general have reached only as far as a tissue grade of construction.

Muscular System. Subumbrellar muscle of ectodermal origin provides the main contractile tissue of *Aurellia*. Radial as well as circular fibers are present. Histological examination of fresh or macerated tissue will demonstrate the type of cell structures involved.

The behavior of the neuromuscular system of *Aurellia* may be readily studied with a tissue strip prepared from the ring described above by a radial cut. Stimulation of such a strip produces a contraction which lasts about 1 sec. The refractory period is nearly as long as this, so that summation or fusion of twitches does not take place. Facilitation does occur, however, and after the first response, subsequent responses grow stronger for the first few. Demonstrate these facts by conducting some simple experiments of this type.

Reproduction. The gametes of *Aurellia* are shed into the gastrovascular cavity and passed out through the mouth. In female medusae the developing larvae may be retained for a considerable period in the grooves of the oral arms. Examine *Aurellia* specimens for evidence of such brooding. The gastrula develops by invagination; the planula attaches to a substratum by its anterior, aboral end and metamorphoses into a scyphistoma. The polypoid structure of this stage has been outlined above. It winters over the first year and may give rise asexually to other scyphistomae. Then it begins distally to differentiate a serial array of medusae (strobilization), forming as many as a dozen or more. As these develop, the most distal one becomes free first and swims away as an ephyra, followed seriatim by the others.

The ephyra is a young medusoid form with the characteristic tetramerous symmetry well developed. The sixteen marginal lappets of the disk and the eight sensory recesses for the rhopalia are much exaggerated at this stage in comparison with the adult medusa. Note that the rhopalial recesses are only about half as deep radially as the clefts between rhopalia. The ephyra gradually matures into the medusa. Study a series of the stages in the life cycle of *Aurellia* so that both their individual structure and their interrelations are understood.

Metridium senile

By T. H. WATERMAN

This sea anemone (Figs. 90–92) is a large, sessile, marine polyp. It has a thick columnar body and very numerous short tentacles. There is no trace of a medusoid stage in the life cycle. It occurs commonly on rocks from tide pools to a depth of 90 fathoms. In geographic range *Metridium* is found from Labrador to New Jersey as the variety *M. dianthus* (= *M. marginatum*). This form also occurs on the coast of Europe; another variety, *M. fimbriatum*, if it actually belongs to the same species, extends the range to the Pacific.

External Structure. *Metridium* possess typical anemone organization with attached base, upright column, and oral disk with central mouth and marginal tentacles. The column is differentiated into two portions, a distal, thin-walled, short capitulum and a proximal thick-walled scapus. These two regions are separated by a groove and collar. The wall of the scapus is perforated by small openings called cinclides, and the whole animal is firmly, but not immovably, fastened to the substratum by the pedal disk. Normally the cinclides are closed, but when the anemone contracts on being disturbed, water and acontia may be discharged from these openings. The mouth is an elongate slit, which causes the symmetry to deviate from radial. It opens into a coelenteron through a throat-like actinopharynx. The latter has one to three flagellated grooves, siphonoglyphs, which run longitudinally from the mouth to the lower end of the pharynx. Note that if only one siphonoglyph is present, the symmetry of the oral disk is reduced to bilateral.

General Body Organization. The features of *Metridium's* basic structure can be studied by cutting cross sections of an expanded specimen through the pharynx and below the pharynx. This may be done effectively with a large pair of scissors. If the resulting section is not a smooth one at right angles to the longitudinal axis, trim it with the scissors to make it so. Study the various structures revealed by this means. Similarly a longitudinal cut through a specimen will

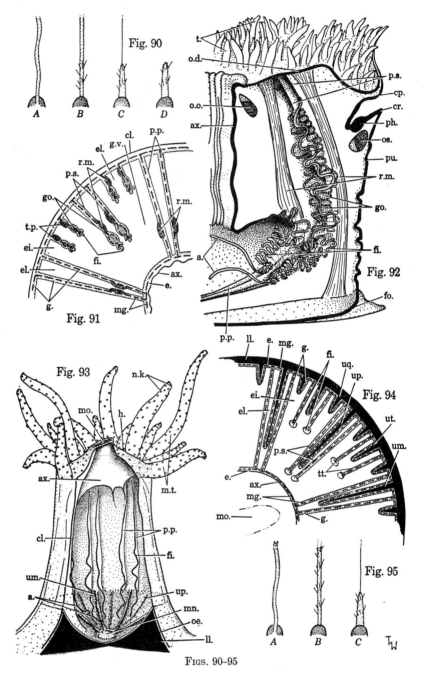

FIGS. 90–95

FIG. 90. Nematocyst types of *Medridium*. (Data from Weill.) *A*. Spirocyst. *B*. Basitrich. *C*. Microbasic mastigophore. *D*. Microbasic amastigophore.

FIG. 91. Schematic cross section through a sector of the column in *Metridium* in the region of the actinopharynx to show relations between septal pairs, retractors, filaments, and gonads. ×2. From preserved material.

FIG. 92. Stereodiagram of *Metridium* to show the basic structure relations within the adult anemone. ×2. From preserved material.

FIG. 93. Cutaway view of a single polyp of *Astrangia* to indicate its general structural features. ×5. From life.

FIG. 94. Schematic cross section through a quadrant of a polyp of *Astrangia*. Note that skeletal parts and the pharynx as diagrammed would be seen in the same cross section only when the polyp is contracted into the corallite. The cycles of septa in this diagram are considerably more developed than in the specimen of Fig. 93, in which only primary septal pairs and primary and secondary sclerosepta are present. ×5. From preserved material.

FIG. 95. Nematocyst types of *Astrangia*. (Data from Weill.) *A*. Spirocyst. *B*. Holotrich. *C*. Microbasic mastigophore.

ABBREVIATIONS FOR LABELS ON FIGS. 90–95

a. : acontia
ax. : actinopharynx
cl. : column wall
cp. : capitulum
cr. : collar
e. : epidermis
ei. : exocoele
el. : endocoele
fi. : septal filament(s)
fo. : pedal disk
g. : gastrodermis
go. : gonads
g.v. : gastrovascular cavity
h. : hypostome
ll. : corallite
mg. : mesoglea
mn. : columella
mo. : mouth

m.t. : marginal tentacles
n.k. : nematocyst knobs
o.d. : oral disk
oe. : coenosarc
o.o. : oral ostium of primary septum
os. : marginal ostium of secondary septum
ph. : sphincter muscle
p.p. : primary septal pair
p.s. : secondary septal pair
pu. : scapus
r.m. : retractor muscles of septa
t. : tentacles
t.p. : tertiary septal pair of *Metridium*
tt. : tertiary septal pair of *Astrangia*
um. : primary scleroseptum
up. : secondary scleroseptum
uq. : quaternary scleroseptum
ut. : tertiary scleroseptum

demonstrate a number of additional details which are not obvious in cross sections.

It will be observed in such preparations that, as in the basic body plan of Scyphozoa and other Anthozoa, the coelenteron of *Metridium* is divided by a series of longitudinal partitions. These septa originate as folds of mesoglea covered with gastrodermis from the wall of the column and project inward toward the oral-aboral axis. Septa which reach and join the actinopharynx are called primary. There are usually six pairs of these septa in *Metridium*. Those which insert near the siphonoglyph are called the directive septa; if there are two siphonoglyphs or more, each pair of primary septa associated with a siphonoglyph is called a directive pair. Between the pairs of primary septa are successively shorter cycles of secondary, tertiary, and quaternary pairs.

On one surface of the septa are longitudinal retractor muscles. These occur on the sides of the pair which face out in the case of the directives; they are on the surfaces facing in on the other septa. Since the part of the coelenteron enclosed between the members of a pair of these partitions is called an endocoele and that part not so enclosed an exocoele, the retractors of the directives are exocoelic in position while those of the remaining septa are endocoelic.

At their oral end the primary and secondary septa of *Metridium* are each pierced by ostia; each primary has two, one marginal and one axial in position. The other septa have only a single marginal one. These perforations keep the chambers of the coelenteron between the septa in communication. Below the actinopharynx they are in communication also. Where the axial edges of the septa are not attached to the actinopharynx, they are thickened to form the septal filaments which basally become free as the thread-like acontia. The latter bear nematocysts and can be partly extruded through the cinclides and mouth. This usually happens when the anemone, on being disturbed, contracts strongly and rapidly. Gonads are borne longitudinally on the septa parallel to the filaments but further away from the edge of the septum. These sex cells do not occur on the directives and usually not on the primary septa in *Metridium*.

Unfortunately irregular arrangement of both siphonoglyphs and septa is common in *Metridium* so that many specimens deviate from a simple, typical plan of organization. Thus specimens with one siphonoglyph predominate, although those with two are not uncommon. *Metridium* specimens with no siphonoglyph or with three also occur, but rarely. Deviations of siphonoglyph pattern will obviously lead to some abnormalities in septal arrangement also.

In *Metridium*, as in other anthozoans, the mesoglea of coelenterates reaches its highest degree of differentiation. In addition to its non-cellular elements it contains stellate amoebocytes and connective tissue cells. Maceration, teasing, and staining procedures described for *Tubularia* and *Pennaria* may be used to advantage in studying this tissue.

Definite respiratory currents are maintained by inward beating of the flagella borne in the siphonoglyphs. The direction of this beat is the reverse of that found over most of the surface of the disk and actinopharynx under normal conditions. It seems likely that the elongation of the mouth, and hence a major deviation from radial symmetry, arose as an adaptation to isolate a special inward flowing respiratory current. The presence of such a circulation is apparently correlated with the restricted access to the coelenteron and the heavy tissue structure of anemones compared with most cnidarians.

The gastrovascular fluid of *Metridium* is circulated by the activity of the siphonoglyphs and gastrodermal flagella. Note that the ostia of the mesenteries and the well-developed musculature of the column and septa also aid in circulation, the former passively, the latter actively. The effectiveness of these various structures aiding in the circulation may be readily observed with the aid of fine carmine or India ink suspensions.

There is good reason to suspect, on the basis of work done on corals, that certain areas of the septal filaments are the main site for excreting waste materials from the body. However, as in other coelenterates, the biochemistry of metabolites and their final fate are not well known for *Metridium*.

Feeding and Digestion. Because of its large size and abundance, these phenomena have been extensively studied in *Metridium*. Food is caught by the nematocysts on the tentacles and oral disk, if it chances to come in contact with them. It is conveyed to the mouth either by ciliary-mucous currents or by muscular movement of the tentacles.

Using *Mytilus* juice and meat, sand grains, filter paper, and filter paper soaked in *Mytilus* juice, determine the reactions of the tentacles, oral disk, and lips to stimulation by these objects. The reversal of ciliary beat to be observed on the edge of the mouth is an example of the dual application of ciliary currents both to cleansing the disk (centrifugal currents) and transporting food particles to the mouth (centripetal currents).

The progress of extracellular and intracellular digestion may be studied by various techniques. In a young active *Metridium* the following may be tried: (1) measure the time necessary for extracellular digestion of small animals or meat fed to the anemone; (2) if the food has been stained, observe its intracellular fate from the coloration of gastrodermal cells; (3) test the digestive ability of the anemone by noting digestion, or its absence, for specific proteins, fats, or carbohydrates; (4) measure the digestive activity of aqueous extracts made from various specific tissues such as tentacles, septal filaments, and acontia. In experiment 4, controls must be run with boiled extracts in which the enzymes would be destroyed.

Metridium is a good coelenterate in which to study the stimuli which cause the discharge of the prey-catching nematocysts. Isolated tentacles observed under the microscope in depression slides are suitable test material. If care is taken, the following facts may be demonstrated: (1) a human hair touched to the tentacle surface will cause many nematocysts to discharge; (2) a clean (flamed before use) glass rod of comparable size does not cause discharge unless the stimulus is violent; (3) food juices introduced with a minimum disturbance do not evoke discharge; (4) in the presence of food juices or other substances such as saliva (1:10 in sea water) mechanical stimulation by a clean glass rod will evoke discharge. Therefore mechanical stimulation aided by a lowering of the threshold brought about by chemical means seems to be the effective cause for the discharge of nematocysts. The specific nature of the sensitizing chemical substance is unknown but may be lipoidal. Without its synergistic effect the stinging cells will not discharge in their normal manner.

To observe more details of the nematocysts draw a fresh *Metridium* tentacle across a slide on which some saliva has been dried. Stain the slide with methylene blue or acid fuchsin. Another useful procedure is to take a small piece of tentacle, oral disk, or whatever region is to be examined and place it on a microscope slide. Then with a sharp scalpel cut the tissue into fine fragments. Considerable detail may be observed without staining, in which case a coverslip should be added with a minimum of fluid on the triturated tissue; if staining is desired, the above stains may be used first and then a coverslip added. Note that best results will be obtained when the tissue is pressed out into very small, thin fragments between the coverglass and the slide. The main features of nematocyst structure, the capsule, filament, and surrounding cnidoblast cell, may be readily seen in such preparations. Note, however, that structural details, such as the arrangement of bristles on the nematocyst thread, are often near the limit of resolution of the light microscope even under oil immersion; hence considerable care must be taken in studying such minutiae. Nevertheless, the fact that from a morphological point of view the cnidoblast is probably the most complicated metazoan cell known and in addition is, in some cases, of systematic and phylogenetic importance makes the study of these organelles of considerable general interest.

In *Metridium* the following four types of nematocysts occur: (1) spirocysts, (2) basitrichs, (3) microbasic mastigophores, (4) microbasic amastigophores (for details of terminology consult Weill, 1934, or Hyman, 1940, p. 383). Compare these nematocysts of *Metridium* with those which can be found in any other available Cnidaria. It should be observed that in *Metridium* nematocysts, as in those of other anthozoans, there is no obvious cnidocil, such as occurs prominently as the so-called trigger of the stenoteles of hydras (Fig. 79), for example.

Nervous System and Sense Organs. With the exception of an aboral tuft of stiff cilia present in the larval stage there are no specific sense organs in *Metridium* or, in fact, in any Anthozoa. Their nerve net is basically of the same type as that of other coelenterates. Conduction has been found to be very slow; in some regions of the body, for example, conduction at a velocity of 10 cm. per sec. or less occurs. However, there are certain aggregates of neurons which form transmitting pathways conducting at rates 10 or more times faster than the slowest regions. Even at this more rapid rate, however, the velocity is about 1 per cent that occurring in the fastest vertebrate myelinated neurons.

The organization of this neuronal network can be determined histologically. On the other hand, it is perhaps of greater interest to demonstrate the relationships physiologically. This may readily be done as follows: (1) test various parts of the body for sensitivity to mechanical and electrical stimulation; (2) note the effects of strength of stimulus on its spread to remote parts of the anemone; (3) make a series of cuts in the tentacles, on the margin, on the column, and other parts to see how they restrict the transmission of excitation. Tongue-like strips cut out of these regions will also help identify any directional properties of the system.

Muscular System. The muscular system is well developed. Epidermal musculature, which is not prominent in any part of the body, is restricted in most forms to the tentacles and the oral disk; gastrodermal musculature on the other hand is well developed. It consists of circular muscles in the tentacles, oral and pedal disks, and column as well as strong longitudinal retractors and some radial fibers in the septa. The septal retractors are the main muscles which contract the anemone.

There is no skeleton in anemones.

Note that, although *Metridium* is in general sessile, it can move slowly about by means of contractile waves which pass over the pedal disk.

Neuromuscular responses of *Metridium* differ from those of medusae, such as *Aurellia*, in two important respects. (1) No response is obtained with a single stimulus applied to the column, but if it is followed soon enough by a second, a contraction occurs. Such facilitation will continue to build up at a rate dependent on the number and frequency of stimuli until a maximum tension is reached. Then when stimulation stops, facilitation, as tested by single test stimuli, slowly disappears. (2) The refractory period is appreciably shorter than the slow twitch of the muscle; hence mechanical summation of tension is a prominent feature of the responses of the system. These facts may be readily demonstrated by stimulating the column of a *Metridium* with shocks of various frequencies and intensities, meanwhile recording the resulting tension developed. Note that the anemone's neuromuscular system differs from that of the medusa discussed above by having a twitch that is tonic rather than clonic in character and by having synapses or neuromuscular junctions in the system which require more than one impulse or series of impulses to effect transmission.

Reproduction. The life cycle of *Metridium* is simple. The gonads are borne on the septa as in other anthozoans, but in this species they are restricted generally to the incomplete septa as mentioned above. Gametes, which originate from gastrodermal interstitial cells, mature in the mesoglea. Zygotes develop into bilaterally symmetrical, free-swimming, ciliated planulae. These begin to metamorphose into anemones, then settle down and attach at their aboral end.

Asexual reproduction occurs commonly in *Metridium*. Pedal laceration is the predominant method. In this process the anemone may move away from its point of attachment, leaving behind it parts of the pedal disk and septa. These regenerate a new anemone in the old site, and the parent redevelops a pedal disk in a new location. Or, alternatively, pieces of pedal disk may be constricted off from its margin. These may move off a slight distance, then regenerate and reorganize a complete anemone. Such a procedure will account for the not uncommon occurrence of a large *Metridium* surrounded by a whole "brood" of small anemones produced by this asexual means. Laceration of these sorts may be quite irregular, hence the frequency of modifications in the basic septal and siphonoglyph arrangement in this species.

Astrangia danae

By T. H. WATERMAN

Astrangia is a stony coral forming small colonies (up to 10 cm. in diameter) encrusting stones and shells in shallow water. It is thus not a reef-building coral. The zooids protrude from closely spaced calcareous skeletal cups into which they may be withdrawn when contracted. Unlike the majority of shallow-water corals, the polyps of *Astrangia* remain more or less expanded in daylight. No medusoid stage occurs in the life cycle. *Astrangia danae* (Figs. 93–95) is the only littoral stony coral (Order Madreporaria) that is found on the Atlantic coast of the United States north of the Carolinas. It is common from Florida to Cape Cod. A fairly similar species occurs on the coast of southern California.

Astrangia is moderately hardy. If kept in aquaria of clean sea water and fed on small animals or bits of meat, it may survive for considerable periods. As with most animals, live or fresh material is preferable for the study of this species, but fixed specimens are quite adequate for analyzing growth habit and even for examining polyp morphology, provided that the colony has been killed in the expanded state.

External Structure. The polyps of *Astrangia* when expanded are 8–10 mm. high and 4–5 mm. in diameter. Their color is usually translucent whitish punctuated on the tentacles and oral disk by denser flecks caused by knob-like concentrations of nematocysts. Frequently the animals appear faintly pinkish or greenish, especially when contracted, because of algae growing in the skeleton. Sometimes, too, green or brown colonies of the coral occur; in these the gastrodermis will be found to contain zooxanthellae, algae living within the polyp's cells.

A ring of hollow filiform tentacles with rounded tips arises around the margin of the oral disk. There are usually twelve larger tentacles (in two cycles of six) alternating with twelve smaller ones. In older individuals there may also be some additional tentacles belonging to an incomplete fourth cycle. In the center of the oral disk is the mouth, which is normally elongate and slit shaped. When feeding reactions are provoked, however, the mouth may gape widely and become circular rather than narrowly elliptical in outline. Similarly, in different phases of activity, the oral disk, bearing the mouth centrally and the tentacles marginally, may be either flat or raised into a projecting conical hypostome.

The calcium carbonate skeleton, or corallum, of *Astrangia* can best be studied in cleaned and dried specimens. Like that of other madreporarians, it is secreted by a special portion of the epidermis of the polyp. This skeleton consists of a small amount of matrix which encrusts the rock or other substratum and from which project circular cups or corallites. Each of these contains the base of a coral animal and tends to become polygonal in outline where the polyps are crowded together. A corallite consists of a cup-shaped theca whose bottom is called the basal plate. From the walls of the theca a system of longitudinal partitions, termed sclerosepta, projects radially. These sclerosepta are arranged in three or four cycles. Three of them are complete when the animal is mature, but the fourth usually is developed only in one sector of the cup. Thus in a regularly organized corallite of *Astrangia* there are six primary, six secondary, twelve tertiary, and a few quaternary sclerosepta. This septal pattern reflects the basic hexamerous symmetry typical of the whole Subclass Zoantharia to which the corals belong. However, the regularity of this pattern may be partly obscured in *Astrangia* when a fourth cycle of sclerosepta begins to appear, since fusion of septa in the other three cycles may then occur.

The sclerosepta push in the body wall of the column and pedal disk of the polyp so that folds of it project inward between the gastrodermal-mesogleal septa, as shown in Figs. 93 and 94. From this relationship it is clear that the skeleton of a coral like *Astrangia* has at least three structural functions, namely: (1) support of the basal regions of the polyp, (2) protection for the contracted animals, (3) increase of gastrodermal surface by the folds of body wall covering the sclerosepta.

General Body Organization. The internal anatomy of an *Astrangia* polyp is like that of other anthozoans in possessing an actinopharynx, opening like a throat from the mouth into the body cavity, and septa, partitioning the coelenteron into partly isolated compartments. As shown in Fig. 93, the actinopharynx, instead of being long and tubular, as in *Metridium* (Fig. 92), is in this coral short and umbrella-shaped with the mouth opening into its apex. The longitudinal septa are shallow and occur in pairs, as they do also in *Metridium*, in contrast with *Renilla*.

In *Astrangia* there are six pairs of primary septa which are perfect, i.e., their central margins reach and fuse with the actinopharynx in the distal body region where this latter structure is present. In addition there are six pairs of secondary septa, and in mature individuals some tertiary pairs. The first and second cycles of tentacles arise from the oral disk in the spaces between the members of a pair of these internal partitions. Thus they are endocoelic in position. Since there are usually more cycles of tentacles than septa, some of the tertiary and quaternary tentacles are exocoelic.

As mentioned above, the sclerosepta of the skeleton are also topographically related to the internal septa and hence the tentacles. Each septal pair has an endocoelic scleroseptum, and there is one cycle of exocoelic sclerosepta which apparently have maintained this position by continual splitting of the original secondary cycle as more septal pairs arise.

As in other Anthozoa and Scyphozoa the coelenteric septa consist of three layers: an outpushing of mesoglea from the body wall, covered on both sides and marginally by a layer of gastrodermis. Like those of *Metridium*, the free edges of the septa of *Astrangia* are thickened to form septal filaments. Embryologically these develop, at least in part, by downgrowths of stomadeal ectoderm. These filaments are considerably longer than the basal edge of the septa on whose margins they lie, so that their course is a conspicuously convoluted one.

During most of their length these filaments are circular in cross section (Fig. 94) as they are in *Metridium* below the level of the actinopharynx. Near the base of the *Astrangia* polyp they project into the body cavity as free acontium-like filaments. These filiform structures may be extruded through the proximal part of the body wall of the polyp and in live colonies may frequently be seen writhing slowly in the shallow grooves between the individuals. This phenomenon is apparently comparable to the extrusion of acontia by *Metridium* when it contracts violently.

Note that in this colonial organism the coenosarc, instead of consisting of a series of tubular stems and rhizomes, as it does in *Tubularia, Obelia, Pennaria,* and the other colonial hydrozoans, consists of a double sheet of tissue separated by space continuous with the gastrovascular cavities of the polyps. Both of the sheets of tissue on either side of this space consist, in turn, of epidermis, mesoglea, and gastrodermis with the latter lining the space. Thus the thin translucent layer of living material interconnecting the polyps of a colony of *Astrangia* is made up of this complex laminated structure.

The nematocysts in this species (Fig. 95), which are borne in the gastrodermis of the acontia and in knob-like aggregates in the epidermis of the tentacles and oral disk, are of three kinds: (1) spirocysts, found only in Zoantharia, and rather different in structure from all other nematocysts; (2) holotrichs; (3) microbasic mastigophores. (Consult Hyman, 1941, p. 383, or Weill, 1934, for details of nematocyst classification.) Techniques useful in the study of the structure and function of the nematocysts of *Astrangia* will be found in the account of *Metridium*.

Feeding and Digestion. According to the best evidence, all corals are specialized carnivores feeding on plankton animals. As in the typical coelenterate pattern nematocysts, tentacles, mucus, and ciliary currents aid *Astrangia* in capturing and engulfing its food. Relatively large animals are paralyzed and then swallowed with surprising speed by the very distensible mouth. Again, as is typical of the phylum, digestion occurs in two phases, a rapid extracellular proteolytic phase, followed by phagocytic ingestion, and intracellular completion of the process. The gastrodermal region most active in both secreting enzymes and intracellular digestion is that of the septal filaments.

These various processes in *Astrangia* may be conveniently studied by feeding the polyps on crabmeat mixed with litmus powder. This will be readily swallowed. The distribution of the dye and its changes of color will demonstrate the fate of this food as well as the pH existing during the various phases of its digestion. The transparent tissues of this species make it a particularly favorable organism for this purpose.

Although *Astrangia* is one of the very few shallow-water corals which are not always infected with zooxanthellae, algae which live intracellularly in the gastrodermis, it is sometimes so infected. Interestingly enough an experimental infection can be produced by feeding this coral on crabmeat with which the appropriate single-celled algae have been mixed (Boschma, 1925). As in *Chlorohydra* the presence of photosynthesizing algae in the tissues may affect the coral's respiration. However, the fact that *Astrangia* normally has no such algae in its cells and that even those corals which do always have zooxanthellae get along quite adequately in the dark shows that this relationship is not critical in the physiology of the individual coral animal or colony. Some careful experiments indicate, however, that the absorption of the waste products of the coral's protein metabolism by the activities of the algae may be of sufficient excretory importance to be of advantage to the coral reef or population as a whole.

The muscular system of *Astrangia*, like that of the other corals, is considerably reduced. Very little epidermal musculature is present, and even the gastrodermal longitudinal retractors of the septa, although they can withdraw the polyp into its theca, are quite inconspicuous structures.

Reproduction. As in other anthozoans the gonads of *Astrangia* are borne on the septa back of the septal filaments. The free-swimming ciliated planula, already beginning to form its longitudinal septa, settles down on its aboral end and develops into the primary polyp of the colony. Subsequent individuals are apparently formed not by fission of pre-existing polyps but rather by the formation of a new mouth in the marginal coenosarc, followed by the gradual organization of the septal, scleroseptal, and tentacular cycles typical of the species. Margins of colonies should be studied for details of this method of asexual development. If ripe, living colonies are available, the gametes and the planula may also be studied to advantage.

Renilla köllikeri

By T. H. WATERMAN

Renilla (Figs. 99–101) is a colonial, benthonic, marine alcyonarian with direct development through a planula to a primary zooid. A medusoid stage is completely lacking, as it is in all the other Anthozoa. A division of labor among the zooids for circulation and feeding results in a characteristic dimorphic colony. The latter consists of a discoidal rachis (6–7 cm. in diameter) and a finger-like peduncle (5–6 cm. long); it lies embedded in the substrate above which only the zooids, all limited to the superior surface of the rachis, are visible. The polyps are colorless or whitish, while the rest of the colony is violet or amethyst in color. *Renilla köllikeri* occurs in the eulittoral zone in southern California; another quite similar species, *R. amethystina*, has a more southerly range on the Pacific coast of North America, while *R. reniformis* occurs in shallow water in the West Indies and on the Carolina coast.

132 CNIDARIA AND CTENOPHORA

External Structure. The colonies are bilaterally symmetrical and bear zooids only on the superior surface of the rachis. The latter is broad and fleshy and has a reniform outline. From the notch (hilum) of the disk the root-like, unbranched peduncle arises. The so-called dorsal track, which is in the position of the midrib of a leaf on the superior surface of the rachis, is free from polyps. The zooids are of two kinds: autozooids and siphonozooids. Over a hundred of each occur on the superior surface of the rachis.

The autozooids are typical anthozoan polyps in their basic construction. In the center of the flattened oral disk of each is the mouth which opens into an actinopharynx. Marginally the oral disk bears eight tapering, pinnate, hollow tentacles, as is characteristic of the Alcyonaria. Only part of the column of the polyp rises above the level of the rachis; this external part is called the anthocodium. The rest of it is buried in the common tissue of the colony, the coenenchyme. These autozooids are feeding polyps, while the siphonozooids are specialized individuals which act as pumps to drive a current of water into the colony.

The siphonozooids are without tentacles and occur in clusters on the surface of the rachis between the bases of the anthocodia. The siphonoglyph, a flagellated epidermal groove of the actinopharynx, is particularly well developed in these polyps and is the means whereby they pump water into the colony. Lining these siphonoglyphs are unusually tall, columnar, epithelial cells bearing long flagella. Part way along the dorsal track a particularly large siphonozooid is to be found; this is a special exhalant polyp.

The remainder of the colony is covered with epidermis continuous with that on the surface of the external parts of the polyps.

General Body Organization. Each polyp has eight perfect septa dividing the distal coelenteron into eight chambers continuous with the cavities within the tentacles. Below the level of the actinopharynx the septa bear filaments, and their margins approach the lateral wall of the coelenteron so that basally there is a large common gastrovascular cavity into which the eight interseptal cavities open. In autozooids, retractor and radial septal muscles are present. Gonads, as well, are borne on the septa. Siphonozooids not only lack tentacles but also retractors and do not bear sex cells.

The walls of the anthocodia have a typical cnidarian three-layered structure: epidermal and gastrodermal epithelia separated by mesoglea. Only one type of nematocyst, the atrich, is present in *Renilla*, as in the whole subclass to which it belongs. The parts of the zooids within the colony consist only of gastrodermis embedded in a mesogleal matrix of the coenenchyme. Between individual polyps a meshwork of gastrodermal tubes, called solenia, branches through this mesoglea, connecting the coelenterons of all the members of the colony. Some such arrangement would obviously be necessary with separate feeding and pumping polypoids, mutually dependent on circulation between one and the other.

As in the anthozoans in general, the mesoglea is a cellular mesenchyme. It gives rise in *Renilla* to skeletal spicules which lie in the coenenchyme. It does not, however, form an axial skeleton in the peduncle, as it does in the stems of many of the other alcyonarians such as gorgonians.

In the peduncle two large tubes are present, a superior and an inferior canal. The superior canal bears the exhalant siphonozooid on the dorsal track, and both canals have their lumina connected with the solenia and body cavities in the rachis. These arrangements with the aid of peristaltic contractions of the walls of the peduncle are used to adjust the amounts of water present in the canals. Thence they control the degree of expansion and the turgidity of the colony. In many other Pennatulacea, like the sea pens, the peduncle's muscular activity is used for movement and burrowing, but *Renilla* does not burrow or move about much.

The foregoing general features of *Renilla* may be studied by a variety of the techniques already suggested for *Tubularia, Metridium,* and others: dissection, vital staining, maceration, injection of carmine, and so on.

No special organs for respiration are present in *Renilla*. However, the development of siphonozooids suggests that in rather large fleshy colonies of this sort an active circulation and renewal of gastrovascular fluid are advantageous; this circulation would obviously be of respiratory importance.

The movement of water through the various cavities and channels of the colony is produced primarily by the siphonoglyphs of the siphonozooids, as has been mentioned above. This primary circulation is aided by flagellar currents established in various other parts of the gastrovascular system, such as the interseptal cavities of the anthocodia. In living specimens these should be carefully analyzed and timed by means of the movement of carmine or India ink particles.

No particular information is available on excretion in *Renilla*, and no structural specializations for this purpose have been found.

134 CNIDARIA AND CTENOPHORA

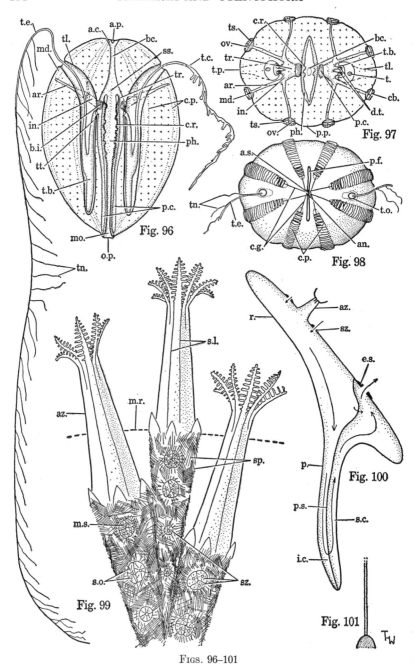

Figs. 96–101

PLEUROBRACHIA AND RENILLA 135

FIG. 96. Longitudinal section through the tentacular plane of *Pleurobrachia*. ×1.5. From preserved material.

FIG. 97. Cross section through an equatorial plane of *Pleurobrachia* transecting the stomach. ×1.5. From preserved material.

FIG. 98. Aboral view of *Pleurobrachia* to demonstrate relations between apical sense organs, comb-plate rows, and polar fields. ×1.5. From preserved material.

FIG. 99. Margin of the superior surface of the rachis of a colony of *Renilla*, showing three autozooids, many siphonozooids, and numerous skeletal spicules embedded in the coenosarc. ×15. (After Eisen.)

FIG. 100. Schematic axial section through the rachis and peduncle of a colony of *Renilla* to demonstrate its general structure and the circulation pattern in the gastrovascular system. (After Parker.)

FIG. 101. Nematocyst type of Renilla: atrich. (Data from Weill.)

ABBREVIATIONS FOR LABELS ON FIGS. 96–101

a.c. : anal canal
an. : anal pore
a.p. : aboral pole
ar. : adradial canal
a.s. : apical sense organ
az. : autozooid
bc. : aboral canal
b.i. : beginning of interradial canal
cb. : comb plate
c.g. : ciliated grooves
c.p. : comb plate rows
c.r. : convoluted pharyngeal ridge
d.t. : tentacular canal (double)
e.s. : exhalent siphonozooid
i.c. : inferior canal of peduncle
in. : interradial canal
md. : meridional canal
mo. : mouth
m.r. : margin of rachis
m.s. : mouth of siphonozooid
o.p. : oral pole
ov. : ovary
p. : peduncle
p.c. : pharyngeal canal(s)

p.f. : polar field
ph. : pharynx
p.p. : pharyngeal plane
p.s. : porous septum of peduncle
r. : rachis
s.c. : superior canal of peduncle
s.l. : lines marking position of septa within column
s.o. : septa of siphonozooid
sp. : spicules in coenosarc
ss. : stomach
sz. : siphonozooid
t. : tentacle
t.b. : tentacle base
t.c. : tentacle partly contracted
t.e. : tentacle extended
tl. : tentacle sheath
tn. : tentacle branches
t.o. : opening of tentacle sheath
t.p. : tentacular plane
tr. : transverse canal
ts. : testis
tt. : tentacular canal

Feeding and digestion are essentially the same in *Renilla* as in other coelenterates. Feeding experiments are particularly interesting if they are carried out to elucidate two phenomena: (1) transportation of food to the coenenchyme and siphonozooids; (2) any coordination of feeding activities between neighboring autozooids, since there appears to be little nervous correlation between persons of the colony.

A diffuse nerve net is present, as in most other polypoids. Conduction of impulses over this system is very slow: 7 cm. per sec. (cf. 60 cm. per sec. in some slow mammalian non-myelinated fibers). Instructive experiments may be carried out by testing the relative sensitivity of various parts of the colony and the spread of response as a function of the kind of stimulus used (mechanical, electrical, etc.) and its intensity. Contraction of anthocodia, and bioluminescence, which can be evoked in epidermal granules, are probably the best responses to observe in such work.

No special sense organs are present.

The only skeletal elements present in *Renilla* are the mesenchymal calcareous spicules previously mentioned. Their deep coloration is responsible for the amethyst or violet color of the radius and peduncle. The contractile elements of the colony are mainly gastrodermal in the anthocodia and also in the peduncle. Epitheliomuscular elements and free muscle fibers lying beneath the originating epithelia are both present.

Reproduction. *Renilla* is dioecious. Sex cells originate from endodermal interstitial cells and mature just under the gastrodermis of the septa of autozooids (with the exception of the asulcal septa which are the two well-flagellated septa attached to the actinopharynx at a point opposite the siphonoglyph). The zygote develops into a stereogastrula and thence into a planula. The latter settles down and metamorphoses into the primary zooid, which produces the peduncle, and then by lateral budding the auto- and siphonozooids. These gradually develop between them the characteristic disk-shaped coenenchyme of the rachis.

CTENOPHORA

Pleurobrachia pileus

By T. H. Waterman

Pleurobrachia (Figs. 96–98) is a pelagic ctenophore without any attached or polymorphic forms in its life cycle. Like all other ctenophores, it is exclusively marine in habitat. In general appearance and behavior these comb-jellies, or sea-walnuts, are reminiscent of coelenterate medusae but differ sharply from the latter by completely lacking the nematocysts diagnostic for Cnidaria. For this reason the phylum has sometimes been called Acnidaria, but, since all other metazoans also lack intrinsic nematocysts, this is hardly a unique feature. The name Ctenophora for the phylum is more appropriate, since eight rows of comb-plates occur only in this group of animals.

In size *Pleurobrachia* averages 15–20 mm. in diameter with tentacles 15–20 times this length. It may appear in coastal waters during the summer in enormous swarms. This species occurs from Greenland to Long Island; the same, or closely similar, ctenophores are found on the Pacific coast of North America, in the Antarctic, and in the South Pacific.

Pleurobrachia is preferably studied in the living state, although preserved material is moderately satisfactory.

External Structure. *Pleurobrachia* is ovate or spheroidal in shape. An oral-aboral axis is established by the mouth and an aboral sense organ. A symmetrical pair of long tentacles which are retractile into tubular sheaths marks one plane of symmetry, the tentacular plane, through the longitudinal axis. At right angles to this plane another oral-aboral plane of symmetry is delineated by the pharynx, the pharyngeal plane. *Pleurobrachia* is thus biradially symmetrical, as are the rest of the ctenophores.

As for nearly all adult members of the phylum, the most characteristic external structures are the ciliated comb-plates, which are arranged longitudinally in eight meridional rows. These are about 45° apart on an equatorial plane and lie on adradii in the sense that this term is used for *Aurellia;* the tentacular and pharyngeal planes are perradial in position.

Each comb-plate consists of a more or less fused row of cilia arranged like the teeth of a comb. Successive plates are spaced so that the free ends of the cilia of the next more aboral comb overlap, by about one-third, the base of the next more oral one. The comb-plates are the locomotor organs of *Pleurobrachia*. It is of particular interest that ciliary activity is able to provide effective movement in so large an animal. The close coincidence of the specific gravity of the animal with that of its environment makes this possible.

The tentacles which are used for capturing food are very extensible. Although they may be completely withdrawn into the body, they are capable of extension to a length many times body diameter. Simple, numerous, short branches in a single series occur throughout the length of the tentacles. On the surface of these branches are knob-like swellings containing a special cell found only in ctenophores. These are lasso cells or colloblasts, elaborate cells which on contact cause small animals to adhere to the tentacles. If living *Pleurobrachia* are available, a study of the effective stimuli for colloblast activity should be made along the lines of the experiments suggested for the nematocysts of *Metridium*.

General Body Organization. The basic structural plan of *Pleurobrachia* is like that of other ctenophores and coelenterates in so far as the body consists of two epithelial layers separated by a mesoglea which is an ectomesoderm. The mesoglea is cellular in *Pleurobrachia* and forms a thick gelatinous collenchyme similar to the mesoglea of scyphozoan medusae like *Aurellia*. However, the formation within ctenophoran mesoglea of mesenchymal muscle cells independent of either epidermis or gastrodermis is an advance in tissue differentiation not achieved by any cnidarian. Longitudinal muscles which contract the tentacles lie within the mesoglea of the solid core of *Pleurobrachia's* tentacles. This musculature is particularly well developed at the base of the tentacle within the tentacle sheath. Some of these details can be observed by vital staining and maceration techniques described for *Tubularia*, *Pennaria*, and *Hydra* among the coelenterates.

The gastrovascular system of *Pleurobrachia*, like that of many scyphomedusae, forms a complex pattern of branching canals. Starting at the oral end, the mouth leads into a pharynx of ectodermal origin. This is laterally compressed so that its flat surfaces determine the pharyngeal plane as mentioned above. About two-thirds of the way from oral to aboral poles the pharynx opens into the stomach. From this part of the system arise two pairs of canals lying in the tentacular plane and an axial canal running to the aboral pole. This last canal is the aboral one. Just beneath the aboral sense organ it divides into four short channels—the anal canals, two of which terminate in anal pores opening to the outside. One of the pairs of canals originating from the stomach is the pharyngeal, which runs orad close beside the pharynx. The other pair of canals runs a short distance laterally before dividing in the tentacular plane to give rise to the tentacular canals, which course along the base of the tentacular sheath. It then redivides twice in an equatorial plane to give rise on one side to four T-shaped meridional canals, each of which underlies a row of comb-plates. The pattern of canals will become more obvious when Figs. 96 and 97 and an actual specimen are correlated with this description. Injection of carmine or India ink suspensions into the mouth or other parts of the system will aid the visualization of these relationships.

In *Pleurobrachia* the pharyngeal and meridional canals end blindly, but in some of the more highly developed ctenophores, like *Mnemiopsis*, the ends of these canals are looped together, forming a more elaborate pattern derived from the basic one of the cydippid type, here illustrated by *Pleurobrachia*.

Respiration in *Pleurobrachia* is aerobic, and oxygen consumption is low. Note again, however, that the bulk of the animal is a watery jelly of mesoglea, as in cnidarian medusae. No special respiratory organs or pigments are present, but the ciliary currents maintained in the gastrovascular canals will obviously be of importance in the transport of respiratory gases, as well as nutrients. Food particles or injected dyes should be used to determine the speed of transport and the directions of current flow in the various parts of the gastrovascular system.

Little is known about excretion in *Pleurobrachia*. The only observed function of the anal canals and pores is to void indigestible material from the gastrovascular canals.

Feeding and Digestion. The fishing tentacles and their colloblasts have already been mentioned. With their aid *Pleurobrachia* obtains its purely carnivorous diet of small zooplankton. Particularly when swarming, the voracious feeding of these ctenophores may deplete the plankton in their neighborhood. Obviously this may be of considerable commercial importance if the plankton involved consists of oyster, certain teleost, starfish, or *Teredo* larval stages. Digestion occurs in two stages: rapid extracellular digestion in the pharynx, followed by intracellular digestion within the widespread gastrodermal cells which first phagocytize the brei resulting from the extracellular process. In living specimens the following observation can be made profitably:

1. Fishing activities of the swimming animals.
2. Time course of phases of digestion, determined by feeding on vitally stained plankton.
3. Digestive ability by force-feeding specific substances like starch and olive oil.
4. Digestive enzymes and their distribution by making extracts; compare various parts and test their digestive activity.

Nervous System and Sense Organs. A subepithelial synaptic nerve net like that of coelenterates is apparently present in *Pleurobrachia*. It is particularly concentrated just below the rows of comb-plates, whose activity is under nervous control. This control not only governs the frequency of beat but may also reverse its effective direction. Instructive experiments may be performed by making superficial and deep cuts in the comb-plate rows and observing the resulting changes in activity patterns.

At the aboral pole of *Pleurobrachia* is a special sense organ consisting of statoliths borne on four fused bundles of cilia. The whole statocyst is enclosed beneath a transparent dome. From it radiate out eight ciliated bands which go to the aboral ends of the comb-plate rows. Experimental section of these bands and removal of the statocysts should be tried; note the resultant effects on ciliary coordination, locomotion, and equilibrium.

Muscular System. The mesenchymal muscles of *Pleurobrachia* have already been referred to. Histologically they consist of unstriated fibers. In a living specimen the distribution and orientation of these contractile elements may readily be determined by stimulating various parts of the body and tentacles electrically. As with other ctenophores, no skeleton is produced by *Pleurobrachia*.

Reproduction. *Pleurobrachia* is a typical ctenophore in being hermaphroditic. Sex cells of endodermal origin lie in the gastrodermis of the meridional canals. When ripe, they are discharged through the mouth. Thus both sex organs in a structural sense and gonoducts are lacking in this ctenophore, as they are in all cnidarians.

Fertilization occurs in the sea water, into which the gametes are shed. Cleavage of the egg is holoblastic and is peculiar in that by the third division it has already established the biradial symmetry of the adult. Early development is definitely mosaic in character. The micromeres which begin to appear with the third cleavage give rise to ectoderm; gastrulation occurs by epiboly of these micromeres and by invagination; mesoderm origin is not known with certainty but is probably ectodermal. Direct development into a young, typical cydippid occurs. In the higher orders of ctenophores the young stages pass through a cydippid stage closely similar to this. For this reason, among others, the Cydippida (the order to which *Pleurobrachia* belongs) are considered the most primitive group in the phylum.

4.

Platyhelminthes

Polychoerus

By L. H. HYMAN

General Remarks. Although members of the Order Acoela are not uncommon in the littoral zone of ocean shores in colder waters, usually not enough specimens of any one species are available for class study. One species, *Polychoerus caudatus*, was formerly common along the Massachusetts coast in certain localities but has disappeared in recent years; however, it may recur. A similar form, *Polychoerus carmelensis*, occurs on the California coast. A remarkable ectocommensal acoel, *Ectocotyla paguri*, is not uncommon on hermit crabs on the Maine coast; this acoel has a plicate pharynx and a caudal adhesive disk. The Acoela live in shallow waters on the surface of the bottom muck or among stones and algae. They have carnivorous habits and feed on small organisms that are ingested through the mouth (and pharynx, when present) into the interior mass of mesenchyme, where digestion occurs. *Polychoerus* will be studied when available.

External Features. Examine a specimen in a watch glass of sea water; note the bright orange or orange-red color, the flat, oval, changeable shape, the rounded anterior end, and the posterior notch bearing a little tail filament. (Additional tails are common in the Californian species.) Observe the method of locomotion. After observing the natural appearance of the animal, mount one on a slide, and cover it with a coverglass supported by a ring of Vaseline; this permits the coverglass to be gradually pressed down on the animal. The animal is completely covered with cilia, which may be seen by focusing on the body margin. Near the center of the animal on the ventral surface is the circular mouth leading into the interior mesenchyme by way of a short tube that constitutes a type of pharynx known as simplex or simple.

Internal Features. About one-fifth of the distance from the anterior end, the statocyst is noticeable in the median line; this is a hollow sphere containing a granule. The statocyst is embedded in the brain, which may be distinguishable. Two or more eyes occur near the statocyst in the Californian species but are wanting in *P. caudatus*. *Polychoerus*, like other Acoela, lacks a digestive tube; hence the interior is filled with mesenchyme (also called parenchyma). The visible structures in the mesenchyme belong to the reproductive system, which is hermaphroditic. Definite gonads are wanting in the Acoela; the sex cells develop *in situ* from mesenchyme cells and lie in longitudinal tracts in the lateral parts of the body with the sperm masses lateral to the eggs. Yolk glands are lacking in the Acoela, but in *Polychoerus* and a very few other genera a nutritive tissue that differs in appearance from the rest of the mesenchyme surrounds the tracts of eggs. Definite oviducts are wanting, but poorly defined sperm ducts are usually present, consisting at first of channels in the mesenchyme and becoming more distinct as they approach the penis. In the median line behind the largest eggs is a rounded body, the seminal bursa, bearing a number of projecting nozzles of a hard yellow material. Each nozzle is the outlet of a sacciform space in the bursa where sperm received at copulation are stored and from which later sperm are emitted through the nozzle into the mesenchyme, eventually reaching the eggs. The seminal bursa opens ventrally in the median line by the female gonopore, through which sperm are received into the bursa at copulation. Behind the seminal bursa is another rounded body, the penis, to which the ripe sperm converge by the pair of sperm ducts mentioned above. The penis is a muscular body with a central lumen and opens ventrally by the male gonopore, a little posterior to the female gonopore. The Acoela lack a nephridial system.

Breeding Habits. The breeding habits of the Acoela are not well known. In the best-investigated cases, including *Polychoerus*, the animals copulate after a simple courtship and each discharges sperm into the seminal bursa of the partner. Direct injection of sperm into the mesenchyme ("hypodermic impregnation") possibly occurs in some Acoela. The eggs seem to be laid either through the mouth or by rupture of the ventral body wall; they are enclosed in cakes of gelatinous secretion and hatch in a short time to young worms.

Stenostomum

By L. H. Hyman

General Remarks. The rhabdocoels, of which *Stenostomum* is an example, are common inhabitants of fresh-water pools, ponds, and lakes, and also occur in the littoral zone of oceans; a few are terrestrial in humid locations. Fresh-water rhabdocoels are obtained by collecting vegetation, pond debris, and bottom muck, but it is seldom possible to collect sufficient numbers of one species for class study; hence it is more practical to increase the numbers of collected specimens by growing them in cultures, as can easily be done for species of *Stenostomum*, *Macrostomum*, *Microstomum*, *Mesostoma*, and other common genera. *Stenostomum* can be found in almost any pond and is readily grown in protozoan-type cultures but has the disadvantages of small size and lack of sex organs. The rhabdocoels are in general carnivorous, capturing small animals or feeding on fresh corpses. Whereas the majority are free-living, there are some entocommensals, probably tending to parasitism, that inhabit the coelom or digestive tract of other invertebrates, especially echinoderms and mollusks; and one or two true parasites such as *Fecampia*, living in crustaceans and undergoing a large amount of parasitic degeneration.

Study of *Stenostomum*. Note characteristics of the *Stenostomum* culture; the animals appear as white rods gliding about. Mount a few on a slide in a drop of culture water without cover, and with low power note their shape and movements. *Stenostomum* reproduces rapidly by fission, and the larger animals consist of a chain of two to several zooids. These are indicated by fission planes and undergo complete differentiation before breaking from the chain. Try to determine the order of development of zooids in the chain by their degree of differentiation (it is not a simple linear order), and if interested consult references. Cover some *Stenostomum* with a Vaseline-ringed coverglass and study details under higher powers. *Stenostomum* is completely ciliated, and the action of the cilia is usually easily seen along the body margins. Note in the surface of the animal bundles of little rod-shaped bodies, the rhabdites; their function is uncertain but it is usually supposed that they discharge to form a mucous material. Observe that the body consists of a thin surface epidermis, an interior digestive tube, and a mesenchyme filling the space between these two. Just behind the anterior tip observe on each side a ciliated pit; this functions as a chemoreceptor. The epithelium of the bottom of the pit is in contact with the brain, a large quadrangular or trapezoidal body. Shortly behind the brain is seen the extensible mouth leading into the oval pharynx. The pharynx is of the simplex type, being simply a muscular tube; externally it bears gland cells, the number and arrangement of which differ in different species. The pharynx leads by a constriction into the tubular intestine that extends nearly to the rear end of the animal and usually contains food remnants. The rear end forms a pointed tail in some species, especially the common *S. tenuicauda*. *Stenostomum* is peculiar in having only one protonephridial system, whereas most rhabdocoels have two. This opens by a nephridiopore near the posterior end, from which the tubule may be followed forward into the head, where it curves back on itself and proceeds posteriorly again, getting gradually more tenuous. The flame bulbs are not usually detectable.

The feeding of *Stenostomum* may usually be observed by placing food with some animals in a watch glass and watching under low powers. Suitable food consists of *Paramecium* that has been damaged by strong centrifuging, crushed small invertebrates such as *Daphnia*, or minute bits of liver or flesh.

There are in the United States a number of different species of *Stenostomum* that differ in body shape, pharyngeal glands, and other details. Identification to species is difficult and usually involves long cultivation.

Reproduction. *Stenostomum* and a few other rhabdocoels (*Catenula, Microstomum*) reproduce rapidly by fission and are seldom seen in the sexual state. *Stenostomum* usually becomes sexual in the autumn, concomitant with a decline in fission. The sex organs are located near the anterior end and consist of one or more ovaries without any ducts or other accessory parts and a testis opening near the head by a duct provided with a cirrus. Hypodermic impregnation appears to be the rule. Each ovary ripens one egg per season, and this is fertilized inside the parent body, then escaping by rupture. All other rhabdocoels reproduce exclusively by the sexual method, and hence all grown specimens are provided with an hermaphroditic reproductive system, usually of complicated structure. Copulation occurs and the fertilized eggs are laid singly, each enclosed in a shell or capsule. *Mesostoma* and related genera produce two kinds of eggs, thin-shelled subitaneous or summer eggs that are self-fertilized and develop at once within the maternal body to young worms, and thick-shelled cross-fertilized dormant eggs with delayed development. Many rhabdocoels of small pools have definite reproductive cycles correlated with the seasons.

Bdelloura

By L. H. Hyman

General Remarks. The Tricladida, of which *Bdelloura* is a member, are commonly termed planarians and are readily recognized by the tubular plicate pharynx directed backwards and the three-forked intestine of which one branch runs anteriorly from the pharynx to the head and the other two turn posteriorly and extend to the posterior end. The triclads or planarians exist as three ecological and morphological groups: marine, fresh-water, and terrestrial. The marine triclads occur sparingly along ocean shores in colder waters or as ectocommensals on other animals. The fresh-water or paludicolous planarians are common everywhere in temperate regions in ponds, lakes, streams, and springs. The terrestrial triclads or land planarians are represented by a large number of species in the humid forests of tropical and subtropical regions and by a few species in humid woods in the temperate zones. In the United States there are a few endemic land planarians and several exotic species that occur in greenhouses and conservatories and may become established in gardens in California, Louisiana, and Florida.

External Features. *Bdelloura candida,* representing the marine triclads, has pronounced epizoic habits, being found only on the king-crab, where it inhabits the leg bases and gills; it is not, however, parasitic but feeds on small organisms brought in by the host's movements. Obtain a *Bdelloura* by scraping one from limulus and place it in a watch glass of sea water. Compare the shape with that of other planarians with which you are familiar. Note the form of the head, the absence of auricles, and the number of eyes. Observe the adhesive caudal disk at the rear end of the animal used in clinging to the host; this is an adaptation for an epizoic life and is wanting in most other triclads. This disk, which is supplied by numerous adhesive gland cells, is continuous with an adhesive band or zone that encircles the animal on the ventral surface near the margin (see Fig. 102). Note the use of the caudal disk and adhesive margins in locomotion.

Pharynx Protrusion. The relations of the pharynx to the pharyngeal cavity are the same as in fresh-water planarians. Those who have not seen pharynx protrusion in the latter may attempt to cause its protrusion in *Bdelloura*. This can be done by placing the worm momentarily in fresh water.

Internal Anatomy. Place a *Bdelloura* on a slide in sea water and cover with a large piece of slide whose border has been Vaselined. Add enough sea water to fill completely the space between the two slides and gradually compress, absorbing excess water, until the worm is flat and quiet. Observe as many systems as possible; internal structures become clearer as the worm gets more compressed, but one cannot expect to see everything on every specimen. Some specimens may show features that others do not.

FIG. 102. Scheme of the adhesive zone of *Bdelloura candida*. (Redrawn after Wilhelmi.)

The worm is completely covered with cilia except on the adhesive areas; their action can usually be seen by focusing on the body margins. The pharynx, pharyngeal cavity, and three-branched intestine are like those of fresh-water planarians; if not familiar with them read directions for the latter. The mouth opening in the rear end of the pharyngeal cavity is sometimes determinable. The nervous system, appearing as light strands, is often distinguishable. Beneath each eye is a white area that constitutes a cerebral ganglion. From each ganglion nerves go forward and laterally and a ventral cord proceeds to the posterior end, giving off lateral branches at regular intervals. Note the marginal nerve into which the lateral branches go and the confluence of the lateral cords at the rear end in front of the adhesive disk, into which numerous nerves are given off. The protonephridial system can sometimes be seen in the anterior part of the worm after long compression. It consists on each side of a zigzag tubule into which flame bulbs open. Some students will succeed in seeing the flicker of one or two flame bulbs in lateral anterior regions; do not confuse them with ciliary flicker on the body surface.

The reproductive system cannot really be understood without the study of stained whole mounts and serial sections, but much can be seen after the animal has been compressed for some time. Figure 103 will help on some points. The testes are numerous rounded or oval bodies between the outer ends of the intestinal branches; they begin at about the level of the ovaries and extend posteriorly to a level behind the penis. The yolk glands are somewhat similar bodies to the medial side of the testes. There are two sperm ducts that run backwards along either side of the pharynx and enter the penis, which is the conical body seen in the midline behind the pharynx. The penis lies in a cavity, the genital antrum (or atrium), to which its relations are the same as those of pharynx and pharyngeal cavity. The delicate connections of the testes with the sperm ducts are not demonstrable. The female system of *Bdelloura* and a few other marine triclads is

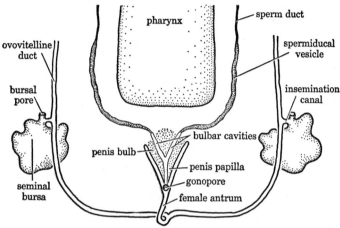

FIG. 103. Copulatory apparatus of *Bdelloura candida*. (Modified after Wilhelmi.)

peculiar in the separation of the sperm-receiving part from the rest of the system. There are two sperm-receiving sacs, called seminal bursae, one to either side at about the level of the penis. They are conspicuous objects in compressed worms. Each opens ventrally by a short canal and a pore (the pore is demonstrable in worms mounted ventral side up) and receives sperm at copulation. The ovaries are a pair of small rounded bodies situated between the second and third or third and fourth side branches of the anterior intestinal branch. They are generally difficult to see. From each an ovovitelline duct (not demonstrable) passes backwards to a point behind the genital antrum. As each passes the seminal bursa of that side it receives from the bursa a delicate insemination canal (not demonstrable) by way of which the sperm reach the eggs. The two ovovitelline ducts unite to a short common duct that enters the rear end of the genital antrum.

The eggs of *Bdelloura* (and all other triclads) are inclosed with yolk cells in a tough-walled cocoon. This cocoon is formed in the genital antrum from droplets emitted by the yolk cells. In *Bdelloura*, these cocoons are oval brown objects fastened by a stalk to the gills of limulus which should be examined to see if cocoons are present.

Syncoelidium pellucidum. *Syncoelidium pellucidum* is another marine triclad epizoic on limulus but is much less common than *Bdelloura*. Some students may find that they have specimens of *Syncoelidium* rather than *Bdelloura*. *Syncoelidium* is only about one-third the length of *Bdelloura* but is much more favorable for study because of its transparency. Particularly the nervous system stands out with diagrammatic clarity and shows the ventral and lateral cords with their regularly arranged transverse connections. Differences from *Bdelloura* consist in the lack of a definite caudal adhesive disk, the confluence of the two posterior intestinal branches behind the pharynx, and the larger, fewer testes. Otherwise, directions for *Bdelloura* may be followed.

Dugesia

By L. H. HYMAN

General Remarks. The fresh-water triclads, here exemplified by *Dugesia* and *Procotyla*, are common inhabitants of fresh waters in temperate zones and can be collected, frequently in large numbers, by looking under stones along the shores of ponds and streams, by placing masses of pond vegetation in pans with water, whereupon the planarians will come to the surface, or by baiting the edges of streams and springs with strips of raw beef. Planarians are easily kept in the laboratory in covered crocks or pans which should be kept cool. They should be fed three times weekly with thin strips of liver or small pieces of earthworms, mealworms, clams, or similar food; after each feeding all remnants of food must be removed and the water changed. Planarians, especially members of the Genus *Dugesia*, are easily fixed in the extended condition by dropping on them 2 per cent nitric acid as they are crawling extended in a small amount of water. They should then be immediately flooded with the fixative (saturated mercuric chloride in 0.7 per cent sodium chloride solution) and after half an hour washed thoroughly in water and treated as usual for whole mounts. Application of iodine is usually necessary to remove excess mercuric chloride. By this method beautiful whole mounts of planarians that have been fed on colored food (see below) can be obtained.

DUGESIA

External Features. The members of the Genus *Dugesia* (formerly called *Planaria* or *Euplanaria*) have triangular heads with conspicuous auricles and two eyes, and tapering bodies colored brown to black. The most common species, *Dugesia tigrina* (formerly *Planaria maculata*), occurs throughout the United States in ponds and streams; it is brown above, usually more or less spotted, and often with a light mid-dorsal stripe. Another common species is *Dugesia dorotocephala* (includes *agilis*), larger and of more or less uniform dark brown to black coloration, found in springs and spring-fed waters.

Examine a planarian in a watch glass of water. Note the pronounced bilateral symmetry, dorsoventrality, and cephalization. Observe the coloration (compare dorsal and ventral surfaces), eyes, projecting auricles, and shape of the head. The auricles are highly sensitive to external factors. Note the gliding form of locomotion, due to a combination of ciliary action and slight muscular waves; observe the behavior of the head and auricles during locomotion. In some specimens a white dash may be noticeable across the base of the auricles; this is the auricular groove, a ciliated groove of sensory nature used in testing the water. Note the habit of the animals of clamping themselves to the glass when attempts are made to pick them up. This clamping is accomplished by an adhesive zone that encircles the ventral surface near the margin and serves as the outlets of glands secreting an adhesive substance. Note the mucous trail left by the animal as it crawls; this mucus is secreted by a tract of gland cells opening on the ventral surface and is important in locomotion to furnish purchase for the cilia. In fresh-water planarians, the cilia are usually limited to the ventral surface and are lacking from the adhesive zone.

Digestive and Other Systems. The digestive system consists of the mouth, pharynx, pharyngeal cavity, and intestine. The mouth can usually be seen as a small white spot on the ventral surface a little behind the middle of the worm. The planarian may be mounted in a little water on a slide which is then turned so that the ventral side is up, and this side then examined for the mouth with a hand lens or under low powers. To see the digestive tract small light-colored specimens may be mounted in water between two coverglasses ringed with Vaseline at the margin; this can then be placed on a slide either side up. Pressure may be gradually applied if necessary. The pharynx is readily seen as a white muscular tube in the pharyngeal cavity, to the anterior end of which it is attached. The attached end of the pharynx leads into the intestine, which at once forks into three branches, one extending forward in the median plane to the head, and the two others curving backwards, one to either side of the pharyngeal cavity, and proceeding to the posterior end of the worm. Each main branch has numerous side branches or diverticula, and these are often united by longitudinal connections paralleling the main branches. Just how clearly the intestine will show up on the preparation depends on whether it contains food and so has a different color from the other tissues. It is therefore preferable to study the intestine in specimens that have been fed on colored food. Planarians starved several days may be put into a dish of water with a blood clot or with bits of liver or animal flesh that have been thoroughly puddled in a thick paste of carmine or carbon in water. The planarians must not be disturbed while feeding or they will leave the food without having completely filled the intestine. Several hours should elapse before any pressure is put on such planarians or before they are fixed as directed above for the making of whole mounts. After such feeding every small branch of the intestine stands out clearly. Whole mounts of planarians that have been fed in this way may also be used in place of live specimens.*

To see the action of the pharynx in living planarians, the latter, after being starved for a few days, should be placed in a dish with small bits of animal flesh. Note the behavior of the planarians, indicating a perception of the presence of food at a distance, the path of the planarian to the food, and the position assumed by the planarian to hold down the food. After the animal is thoroughly attached to the food, quickly but gently lift up one side and note the white pharynx inserted into the food. The pharynx will also usually be protruded in the presence of meat juices and can be made to protrude by applying to the animal such injurious chemicals as dilute acid.

The main parts of the nervous system can usually be seen in whole mounts or in small light-colored specimens pressed between slide and coverglass or, best of all, in white planarians. In the Genus *Dugesia*, the brain has the form of an inverted V, with the limbs surrounding the eyes and the rest extending forward parallel to the head margin. From the brain numerous branches extend forward and laterally to the head margin and auricles, and a main pair of ventral cords proceeds posteriorly about one-third of the distance in from the margin. In some planarians, but not in *Dugesia*, there are also dorsal and

* Such whole mounts are not specimens that have been "injected," as often claimed by supply houses, but are specimens that have been fed on colored food.

lateral longitudinal cords. Numerous transverse connections occur between whatever cords are present.

The protonephridial system is impractical to see, although flame bulbs are sometimes detectable in the lateral regions of the head in greatly pressed specimens. The reproductive system cannot really be understood except by the study of serial sections, although some parts of it may be visible on stained whole mounts.

Histological Structure. Transverse sections through the pharynx and through some region outside the pharynx are desirable for study. Note that the ventral surface is flat, the dorsal surface arched. The section is clothed externally with a cuboidal epithelium, the epidermis, taller dorsally than ventrally. Deeply staining rods, the rhabdites, are conspicuous in the epidermis; determine on which surface they are most abundant. In good preparations, stained with eosin, a small area devoid of rhabdites and cilia is noticeable on each side on the ventral surface near the margin. This is the adhesive zone mentioned above and is the outlet of eosinophilous glands. To the inner side of the epidermis are thin strata of muscle fibers. The interior of the section is filled with mesenchyme, gland cells, and dorsoventral muscle fibers and contains sections of the intestine. The latter appear as hollow, rounded masses, composed of tall, bulging cells containing large vacuoles and various granules and spheres. In specimens that have fed recently, these intestinal cells are filled with food balls of various sizes. So-called gland cells, really protein reserve cells, triangular cells filled with uniform spherules, may be seen among the regular intestinal cells in good preparations. In sections through the pharynx, the latter appears in the center of the section as a circular body enclosed in a circular space, the pharyngeal cavity. The pharynx is of complicated histological construction, consisting of the following layers, from the surface to the lumen: outer epithelium, outer longitudinal muscle layer, outer circular muscle layer, outer zone of gland cells, main nerve plexus, inner zone of gland cells, inner longitudinal muscle layer, inner circular muscle layer, lining epithelium; all these layers are crossed by radial muscle fibers. The inner and outer epithelia are not typical cellular layers, and their nature is not too well understood. The gland-cell layers are evident only with proper staining, as with Mallory's triple stain. The ventral nerve cords are seen (best in sections anterior to the pharynx) as a pair of pale masses just to the inner side of the ventral epidermis about one-third the distance in from the body margin.

Asexual Reproduction. Members of the Genus *Dugesia* regularly reproduce by fission. Fission also occurs in some other genera but is absent in the majority of triclads. One is seldom fortunate enough to witness the fission process, but in a *Dugesia* culture the products of fission are usually easily recognized. Fission takes place behind the pharynx. The anterior product is recognizable by the truncate posterior end, lacking the usual pointed tail; it moves about and behaves like other planarians. The posterior product consists of the small tail end and remains motionless attached to the container until regeneration occurs. Fission can be induced by cutting off the heads of planarians and is also favored by transferring the worms to clean containers as the mucous accumulations on the walls retard fission.

Regeneration. *Dugesia* and some other genera regenerate very well, but many triclads have poor powers of regeneration. Specimens of *Dugesia* may be cut up in various ways, the pieces kept in finger bowls of clean water (avoid fresh tap water), and the regeneration process watched. Regeneration requires several days. Attempts to produce double heads or tails by longitudinal splits will generally fail unless such splits are kept constantly open during the first 24 hours.

Sexual Reproduction. Sexual reproduction is often of rare or sporadic occurrence in planarians that reproduce regularly by fission or fragmentation. In planarians that lack asexual means of reproduction, all individuals develop a reproductive system as they attain mature size, and fresh-water species usually breed at some definite time of the year. Planarians are hermaphroditic but do not fertilize their own eggs, and copulation with mutual insemination is necessary to produce fertile eggs. During the breeding season any one individual copulates repeatedly and lays a cocoon every few days. The cocoons hatch in about 2 weeks into tiny worms.

Dugesia tigrina exists in sexual and asexual strains; the latter never or rarely develop a reproductive system; the former breed annually in late spring and early summer. If sexual specimens are brought into the laboratory at this time, copulation and laying of cocoons can be frequently witnessed. The cocoons are brown to black spheres (reddish when fresh) stuck to the container by a stalk; they hatch to tiny worms in about 2 weeks.

The anatomy of the reproductive system is complicated and not suitable for class study, as a complete set of serial sections of a sexual worm would be required for each student. Further, the details of the copulatory complex differ in every genus and species, as the anatomy of this complex constitutes a main taxonomic character. Figure 104 must therefore serve in place of first-hand study and must be understood to apply only to *Dugesia tigrina*. In triclads in general there are usually numerous testes; in *Dugesia* the testes are small round bodies located in lateral longitudinal tracts along the body length. Each connects by means of a minute sperm ductule with the main sperm duct of its side. As these sperm ducts proceed posteriorly, they widen into sinuous tubes, the spermiducal vesicles, that act as storehouses for ripe sperm; these vesicles can be seen in living sexual specimens by turning these ventral side up as white

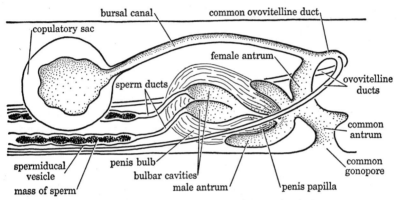

FIG. 104. Copulatory apparatus of *Dugesia tigrina*.

tubes along either side of the pharyngeal cavity. There is one pair of ovaries, usually found in an anterior position shortly behind the level of the auricles. From each an ovovitelline duct proceeds posteriorly close to the sperm duct and collects yolk cells from the yolk glands. The yolk or vitelline glands are clusters of large, deeply staining cells between the intestinal diverticula throughout the worm; they connect by minute ductules with the ovovitelline ducts into which they pour some of their cells during breeding. Sperm ducts and ovovitelline ducts converge to the copulatory apparatus situated behind the pharynx. This consists of male and female parts, both of which open into a common chamber, the genital antrum, that opens below by the common gonopore. The gonopore can be seen in live sexual specimens as a conspicuous white spot in the midventral line shortly behind the mouth. The male part of the copulatory apparatus is the penis, which in general consists of two parts, the penis bulb, a rounded or oval musculoglandular body embedded in the mesenchyme, and the penis papilla, a muscular conical projection into the anterior or male part of the antrum. The sperm ducts, either with or without union to a common duct, pass through the penis bulb and enter the base of the penis papilla, through which a single duct proceeds to the tip. In *Dugesia tigrina*, the sperm ducts enter the penis bulb separately and within it each enlarges somewhat. The female part of the copulatory apparatus consists of a sac, the bursa copulatrix or copulatory sac, that lies between the penis bulb and the rear end of the pharyngeal cavity. It is quite large in *D. tigrina*. From it a canal, the bursal canal, passes backwards above the penis, curves ventrally, and enters the rear or female part of the antrum. In *D. tigrina*, the two ovovitelline ducts, after joining to form a very brief common duct, enter the bursal canal on its curve. Eosinophilous glands, the cement glands, discharge into the terminal parts of the ovovitelline ducts and into the antrum.

In copulation, the penis papilla of each partner emerges by elongation through the gonopore and inserts into the bursal canal of the partner, into which it discharges a mass of sperm. The sperm are held for a short time in the copulatory sac but very soon leave this sac and migrate up the ovovitelline ducts until they come to the beginnings of these ducts adjacent to the ovary where the sperm are held until used. The ovary does not open directly into the beginning of the ovovitelline duct, and the ripe eggs must pass through an intervening membrane. As they do so they are fertilized and then proceed down the ovovitelline ducts in company with yolk cells.

The eggs of triclads are laid in capsules or cocoons that are formed in the male antrum. Here the eggs and yolk cells assemble and become surrounded by a capsule composed of a hard protein material originating from droplets in the yolk cells. The cocoon, containing several fertilized eggs and thousands of yolk cells, is laid through the gonopore with the aid of the secretions of the cement glands which also produce the stalk. The cocoon, pale and soft at first, soon darkens and hardens.

Procotyla

By L. H. Hyman

General Remarks. The broad white planarian, *Procotyla fluviatilis* (formerly erroneously identified as a European species, *Dendrocoelum lacteum*), is common throughout the eastern United States, except in the far south, in ponds, streams, and springs, but seldom

PROCOTYLA   155

occurs in large numbers. It is impractical to keep in the laboratory for any length of time, as it will feed only on live prey such as small crustaceans. It has poor powers of regeneration and is therefore unsuitable for this purpose. Its lack of pigment and sluggish habits make it a satisfactory object for study.

External Features. Obtain a specimen in a watch glass of water and examine with a hand lens or low powers, noting body shape and comparing with *Dugesia*. Observe particularly the shape of the anterior end, best described by the adjective truncate. Observe in the center of the anterior margin the adhesive organ, characteristic of the entire Family Dendrocoelidae, to which *Procotyla* belongs. This organ consists of a sucker-like cushion of glandulomuscular nature used in locomotion and in the capture of prey. Note slightly developed auricles to either side of the adhesive organ and the eyes, usually somewhat irregular in number and arrangement. Most specimens will show the digestive tract, clearly outlined on the white background of the body. In about the center of the body, a white oblong area represents the location of the pharynx, from which the three-branched intestine leads; one branch goes forward in the median line and the other two curve posteriorly and proceed backwards alongside the pharynx and copulatory apparatus. Note numerous slender side branches or diverticula from the intestine, and observe their forking in the periphery. The long white area behind the pharynx is the copulatory apparatus, which consists chiefly of a long and powerful penis bulb. Observe the flexible body margins and the way they are employed in locomotion. The body margin has an adhesive zone as in *Dugesia*. Poke the animal at the rear end and note use of the adhesive organ in hurried type of locomotion.

Feeding Behavior. To see the feeding behavior, the animal should be placed in a finger bowl with plenty of water, and some *Daphnia* added. If the animal is hungry and kept undisturbed, the feeding behavior will probably be seen by some students. The planarian seems to detect the proximity of prey by the water disturbance created by the latter. If a *Daphnia* passes near the planarian, the latter will lunge quickly at it and close the adhesive organ over it. The prey is then passed along the body until the pharynx can reach it.

Internal Features. Mount a worm in a drop of water, cover with a Vaseline-ringed coverglass, and examine with moderate powers. The parts of the digestive tract and other features mentioned above will now be more clearly seen. Much can be detected of the reproductive system, best under low magnification. The ovaries are a pair of small rounded bodies easily seen between the intestinal diverticula less than half the distance between the eyes and the pharynx. Posterior to them, the smaller testes can be seen crowded between the intestinal diverticula. Behind the pharynx the long oval penis bulb is the most conspicuous part of the copulatory apparatus. Between its anterior end and the rear end of the pharynx is seen a vesicle, the copulatory sac. The duct from this, or bursal canal, may be detectable in some specimens running backwards above or to one side of the penis bulb. From the rear end of the penis bulb the short pointed penis papilla may be seen projecting posteriorly. For other details of the reproductive system, a reference should be consulted.

Breeding Habits. *Procotyla fluviatilis* is incapable of asexual reproduction, and hence all individuals develop a reproductive system as they approach mature size. Sexual maturity is attained in late fall and early winter; copulation and laying of cocoons occur as described under *Dugesia*. After the breeding season, the reproductive system degenerates, although the penis bulb may remain noticeable for some time. It is not definitely known whether the same individual again forms a reproductive system as cold weather approaches, but it is presumable that this is the case.

Hoploplana and Other Polyclads

By L. H. Hyman

General Remarks. Polyclads, of which *Hoploplana inquilina* (originally called *Planocera inquilina*) is an example, are common inhabitants of the ocean littoral, under stones, among seaweeds, and on pilings. In general they are larger and broader than other Turbellaria, being generally of a flat, oval shape, although some are long and slender. While many are of dull white to brown coloration, some exhibit bright colors and patterns; those of tropical and subtropical waters are usually larger and more brightly colored than those of temperate regions. In contrast to other Turbellaria, polyclads regularly have numerous eyes (see Fig. 105) and numerous intestinal branches radiating from the pharynx to the periphery. There are numerous small ovaries and testes in lateral regions, but yolk glands are wanting. The polyclads are generally of sluggish habits, but some can swim by undulating the body margins. Although the majority are free-living, a number occur in constant association with other invertebrates, especially mollusks, crustaceans, and tunicates. Often this association means that the other animal is the regular food of the polyclad. The polyclads are in general carnivorous. Because of their thick, opaque bodies, polyclads are not very favorable for study in the live state but a few are relatively transparent; there is also always the difficulty of obtaining an adequate number of specimens of one species.

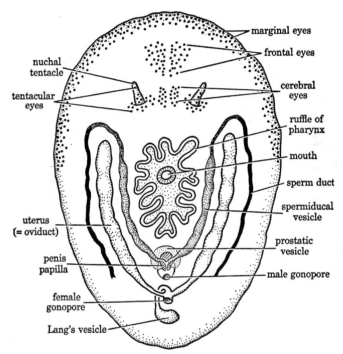

FIG. 105. Composite diagram, showing some morphological characters used in polyclad taxonomic keys.

Study of *Hoploplana inquilina*. This is a relatively small, fairly transparent polyclad that lives as a commensal in the mantle chamber of the large snail *Busycon*, where it seems to feed on the ejecta of the snail. It can be caused to crawl out of the snail if the latter is kept out of water for some hours. Note *Hoploplana* clamped motionless to the walls of the container; they have pale, thin, flat, rounded to elliptical bodies. Mount one as usual under a Vaselined cover. Note in the mid-dorsal region somewhat back from the anterior end a pair of pointed tentacles, each with a cluster of tentacular eyes at its base. In front of these are two more irregular clusters of eyes, called cerebral eyes. Note the absence of marginal eyes. The central part of the body is occupied by the conspicuous ruffled pharynx opening below by the mouth, a circular opening at about the center of the pharynx. The digestive branches radiating from the pharyngeal region to the periphery are more or less visible (actually the pharynx opens into a main intestinal tube above it and the branches radiate from this). The nervous system is readily seen especially in the smaller specimens; the white oval brain is located between the tentacles and from it a network of white nerve branches may be followed.

The reproductive system is hermaphroditic, as usual in flatworms. The sperm ducts filled with sperm are readily seen alongside the posterior part of the pharynx; they lead to a round mass, the prostatic vesicle (reservoir of gland secretions), in the midline just behind the rear end of the pharynx. This vesicle bears a curved thorn or penis stylet, which constitutes the whole of the penis. Behind the prostatic vesicle is seen the female gonopore to which leads a pair of oviducts, called uteri in polyclads as they act as reservoirs of ripe eggs. These uteri course along the sperm ducts, just lateral to them. The ovaries and testes are small round bodies scattered in the peripheral regions. *Hopoplana* will usually lay eggs in the containers; the eggs are laid in a spirally wound gelatinous ribbon.

Other Polyclads. If other polyclads are available, compare them as to size, shape, presence or absence of tentacles, number and arrangement of eyes, color pattern, shape and location of the pharynx, and arrangement of the intestinal branches. On the New England coast, the most common species are the following: *Notoplana atomata*, of a spotty brown coloration with an arched penis stylet seen by putting the worm under pressure; *Stylochus ellipticus*, cream, brown, or olive with a light mid-dorsal band posteriorly; *Stylochus zebra*, a large, handsome worm, with a pattern of chocolate and flesh cross bars, that is found free or in snail shells occupied by hermit crabs; and *Euplana gracilis*, a small, brown polyclad resembling a planarian, easily recognized by the few eyes in two bands.

Polystomoides oris

By C. G. Goodchild

In general, monogenetic trematodes are ectoparasitic on the gills or skin of aquatic vertebrates and have a direct life cycle without the interpolation of intermediate hosts. A well-known exception to this generalization is *Polystoma integerrimum*, which inhabits the urinary bladder of European frogs and has a complicated life cycle involving an alternation of generations. The posterior suckers of this frog parasite are so conspicuous and mouth-like that an early worker

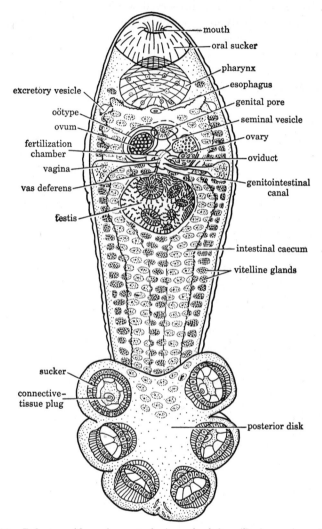

FIG. 106. *Polystomoides oris,* ventral view of adult. (Redrawn from Paul.)

mistook them for mouths and named the worm *Polystoma*. Other polystomes occur on the gills or in the urinary bladders of amphibians and in the mouths or urinary bladders of turtles. *Polystomoides oris* frequents the mouth cavity of the widely distributed, fresh-water painted-turtle, *Chrysemys picta*, from which it extracts blood, epithelial tissues, or tissue exudates as food. It maintains a tenacious hold on the oral mucosa by sharp hooks and by six large suckers borne on a posterior muscular disk termed the cotylophore.

In a living worm or in a preserved whole mount (Fig. 106) notice the elongated oval body and the cordiform posterior disk. The suckers and hooks are rather constant in size, shape, and placement and are, therefore, very useful taxonomically. The suckers are complicated structures set in the parenchyma of the cotylophore and are arranged in three anteroposterior pairs. Each sucker has a fluted, cuticular framework and a heavy, muscular wall. At the base of the sucking cup is a connective-tissue plug or button to which retractor muscles are attached. The cuticular wall of the cup supports the sucker, and a powerful suction is created by muscles including the basal button retractors. The action is similar to that of a pump where a suction is created by the piston moving in the rigid cylinder. A pair of large hooks lies between the posterior suckers, while smaller hooks may be found in association with all the suckers.

Integument, Musculature, and Parenchyma. A somewhat plicated cuticula invests the body and is underlain by a thin hypodermis and a dermomuscular sac consisting of an oblique layer sandwiched between external circular and internal longitudinal muscular layers. Dorsoventral muscle bundles traverse a mesenchymatous parenchyma which is like that of other trematodes (cf. *Cryptocotyle*).

Digestive System. An apical mouth is surrounded by a large oral sucker thought not to be fully homologous to the oral sucker of digenetic trematodes because it lacks a complete external limiting membrane. A large, muscular pharynx lies immediately behind the oral sucker. These parts are lined with inturned cuticula, as is a short esophagus which leads into bifurcated gut caeca. The caeca are initially provided with lateral diverticula, but a short distance posteriorly these disappear and the caeca proceed to the posterior part of the body proper as smooth tubes.

Excretory and Nervous Systems. The protonephridial excretory system is difficult to see even in living whole mounts. Flame bulbs are present; from these structures increasingly larger, coalescing tributaries finally lead into paired excretory canals which in turn open from small excretory vesicles to the dorsal surface. The excretory pores open lateral to the foremost gut diverticula. A brain is located dorsal to the esophagus, and three pairs of nerves proceed anteriad and posteriad from the brain.

Reproductive Systems. Although the worm is a hermaphrodite, cross fertilization is apparently obligatory because eggs are not produced by a single unmated worm. The male and female systems will be described separately.

Male Reproductive System. A large, ellipsoidal testis is located near the middle of the body. From the anterior edge of the testis a vas deferens runs to a short, distended seminal vesicle which opens into a genital atrium located immediately behind the bifurcation of the gut caeca. The genital pore is made conspicuous by an encircling ring of about twenty-seven cirrus spines which are used to hold the worms together during cross fertilization.

Female Reproductive System. The ovary is comma-shaped and is smaller than the testis. It, as well as most of the remaining reproductive organs, lies anterior to the testis. From the ovary a convoluted oviduct passes mediad and presently enlarges to form a fertilization chamber equivalent to a seminal receptacle. Joining this dilation are the right and left vaginae; these tubes are horizontally placed and open on either side by funnel-shaped vaginal pores. A short genitointestinal canal passes from the dilation to the left caecum. Functionally this canal appears to be equivalent to Laurer's canal in the Digenea and affords a means of eliminating excess eggshell material or sperm. Many ovoid vitelline glands are uniformily scattered in the postpharyngeal body region. Yolk material produced by these glands is transported by a system of ducts to a large oötype. The oötype connects posteriorly to the fertilization chamber and functions to elaborate the egg. There is never more than one large (0.25 × 0.18 mm.) ovoid egg in the oötype at any one time. There is no uterus, and the egg is discharged in an unsegmented condition directly from the oötype into the genital atrium. From this chamber it passes to the outside through the genital pore.

Life Cycle. The egg settles to the bottom of the pond or stream, undergoes cleavage, and hatches, in about a month (at laboratory temperatures), as a small, ciliated, free-swimming juvenile measuring only 0.275 × 0.065 mm. This juvenile enters the mouth of another turtle and grows to sexual maturity in about 1 year.

Aspidogaster conchicola
By C. G. Goodchild

Aspidogastrid trematodes occur in mollusks, fishes, and turtles, and a juvenile one, *Stichocotyle*, is often recovered from the rectum of the American lobster.

162 PLATYHELMINTHES

Aspidogaster conchicola (Fig. 107) is a widespread parasite of fresh-water mussels belonging to the Family Unionidae. It is found chiefly in the pericardial cavity and kidneys of its host and often occurs there in masses of twenty or more individuals. The taxonomic

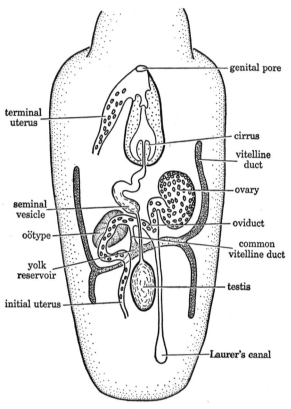

FIG. 107. *Aspidogaster conchicola*, dorsal view of the male and female reproductive systems. (Redrawn from Stafford.)

position of the group to which this parasite belongs is in question because the worm has only a single host in its life history, as in the Monogenea, but its body form is different from theirs. It is not, however, an unquestioned digenetic form because it lacks an alternation of generations and has, in the egg, a fully formed embryo which is an adult in miniature. This problem of classification has been solved by erecting an intermediate order, the Aspidobothria, for the group

ASPIDOGASTER CONCHICOLA

In living or preserved specimens of *Aspidogaster conchicola* note an oval, alveolated sucking disk on the ventral side and a spindle-shaped dorsal region containing most of the body organs. These regions are separated by a groove which marks the position of an internal septum (missing posteriorly) which almost completely divides the body. An anterior neck of the dorsal region, bearing a transversely oval mouth and oral sucker apically, is very mobile and can be extended far forward, waved laterally or reflected over the back. The slightly overhanging posterior body part is less changeable and usually assumes a blunt, conical shape.

The large, ovate sucking disk is a powerful organ of attachment. It is composed, in large worms, of nearly 120 hexagonal sucking compartments, termed alveoli, which are bounded by approximately 30 transverse and 3 longitudinal ridges. The peripheral ridge of the disk is crenated, and transverse ridges connect opposite marginal indentations. Small retractile sensory papillae, the marginal organs, occur at the junction of marginal and transverse ridges.

Integument. A thin, non-spinous cuticula covers the body and inturns to line the terminal parts of the excretory and genital ducts and the digestive canal. Beneath this cuticula several muscle layers, including an outer circular, a separated longitudinal, two layers of mixed diagonal, and another inner layer of circular fibers have been reported. Other non-integumentary muscles run in bundles from the dorsal to ventral surfaces and occur in the walls of internal organs.

Parenchyma. A mesenchyme with intercellular fibers and fluid-filled cavities fills all the spaces among the body organs. This tissue is cut across by the trough-shaped muscular septum previously mentioned. Two layers of muscles, an upper stratum of transverse and a lower longitudinal one, could effect a diaphragm-like motion of this septum to force circulation of the parenchymatous fluid.

Digestive System. The mouth cavity is surrounded by a muscular region which functions as an oral sucker and is morphologically better differentiated in younger specimens than in older ones. The mouth cavity connects directly to a strong, compact, elliptical pharynx which is an efficient food-sucking organ. Behind the pharynx, there is a simple sac-like intestine lined with columnar cells and possessing outer circular and inner longitudinal layers of muscle fibers.

Excretory System. The protonephridial excretory system has been thoroughly studied by many workers. Two funnel-like excretory pores open at the posterior end of the body. These pores connect to a subjacent, transverse chamber from which two lateral excretory canals are given off to follow a rambling course anteriorly; at the level of the pharynx they recurve and extend to the posterior end. The recurrent vessels receive tributaries decreasing in size and branching in a regular fashion. A flame bulb is found at the end of each terminal branch; moreover, ciliated tufts occur at various points on the lining of the recurrent vessels.

Reproductive Systems. This worm is a true hermaphrodite with the ability of self- or cross fertilization. The genital pore is located ventromedianly in the groove separating the neck and ventral sucking disk. Beneath this pore is a common genital atrium into which the terminus of the uterus and a cirrus open.

Male Reproductive System. A median testis lies in the infraseptal compartment slightly behind the middle of the body. A vas deferens leaves the gonad and continues forward until it joins an expanded, curved seminal vesicle which in turn connects to the pointed, eversible cirrus located in a cirrus pouch.

Female Reproductive System. A folded, unpaired, small ovary lies in the ventral compartment anterior and lateral to the testis. A compact, zigzag ovarian stem leads away posteriorly from the main, distended ovarian mass and unites immediately with a ciliated oviduct locally expanded to form the oötype. Three additional female ducts occur in this general area. A blindly ending Laurer's canal runs posteriorly; a common vitelline duct transports yolk material from six or more vitelline glands clustered around the lateral vitelline ducts; and finally, there is the uterus, which in this region may contain spermatozoa and function as a seminal receptacle. In sexually mature individuals, the uterus is tightly recurved and filled with eggs containing embryos in all stages of development. Initially, it runs posteriad and then, turning forward, follows a tortuous course to the genital atrium which it joins by the uterine pore.

Life History. The juvenile of this worm becomes the adult by a gradual, ametamorphic growth. The newly hatched juvenile has anterior, middle, and posterior body regions each of which contains the rudiment of an adult structure. Anteriorly a large oval sucker and a pharynx are present; in the middle body region a cylindrical intestine and two excretory bladders are prominent; while posteriorly an internal sac-like rudiment of the sucking disk is present. The juvenile does not swim and has very limited powers of locomotion. Older juveniles gradually acquire more and more adult characters with the ventral sucking disk undergoing the most striking change.

It starts as the small, invaginated, posterior pocket and becomes the extensive ventral creeping and attaching organ already described. The method of transfer to new hosts is unknown, but may involve water currents to transport the passive juveniles.

Cryptocotyle lingua
By C. G. GOODCHILD

The digenetic trematode, *Cryptocotyle lingua,* is a widely distributed intestinal parasite of fish-eating birds and mammals. It occurs commonly in European and North American loons, gulls, and terns, and often may be found in individuals captured far inland. Young nestling or fledgling birds are usually heavily infected, and a single bird may yield hundreds of worms. Rats, cats, and dogs are suitable mammalian hosts.

To study the morphology of the several distinct stages in the life cycle of this worm the following should be available: an adult worm from the vertebrate host; a mature miracidium contained in the egg; a redia and cercaria from the liver of an infected snail; a free-swimming cercaria; and a metacercaria encysted in the outer surface of a fish. A median sagittal section of the adult worm will help to clarify the arrangement of organs in a unique ventrogenital pit.

Material for study of the adult should include preserved whole mounts or living worms obtained from a recently killed host. The elliptic worms are small (1 mm. long by 0.4 mm. wide) and light brown in color. When a living fluke is located, remove it from the small intestine by gently scraping the mucosa with a scalpel. Mount the worm on a slide in normal avian saline solution, and after carefully compressing it by withdrawing excess fluid from the edge of the coverglass with absorbent paper, seal the edges of the coverglass with Vaseline to prevent evaporation. Other temporary, vitally stained preparations should be made in a similar way by adding a small drop of dilute neutral red (1:5000 in distilled water) to the avian saline.

THE ADULT (FIG. 108)

Integument. The cuticula, which is secreted by the underlying macrocellular hypodermis, is thin and transparent. The anterior half of the worm is covered with small, scale-like spines readily seen in profile at the margins; the spines gradually decrease in size posteriorly.

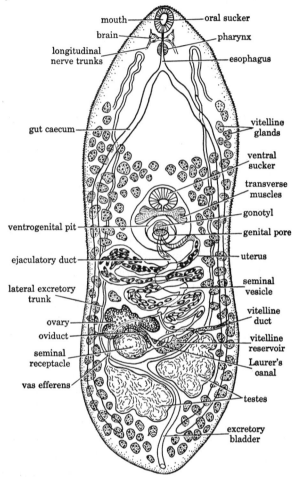

FIG. 108. *Crytocotyle lingua*, ventral view of adult. (Modified after Jägerskiöld.)

Muscular System. This system, in most trematodes, consists of several definite peripheral layers under the hypodermis and scattered bundles of muscles running from the dorsal to the ventral body surfaces. Usually, the outermost layer is circular, beneath which is a thin group of diagonal muscles, and finally there is an innermost layer of longitudinal muscles. These layers, as well as the dorsoventral muscles, are not conspicuous in whole mounts of *Cryptocotyle*.

An oral sucker and pharynx are muscular feeding organs and have their muscle cells arranged chiefly in a radial fashion.

Near the middle of the body is an invaginated, muscular, ventrogenital pit used during the sperm interchange of copulation. The ventral sucker in this worm is small and, in surface view, forms an anterior, crescent-shaped, internal portion of the pit. The muscular fibers of the ventrogenital pit are complexly arranged. Radial muscular fibers are present in the wall of the acetabulum while transverse muscles occur near the outer lips of the pit. These muscles enable the structure to function during copulation as a genital sucker.

Parenchyma. An internal body space between the organ systems is filled with a mesenchymatous tissue called parenchyma. This tissue consists of stellate cells between which are fluid-filled, small, intercellular cavities and a network of fibers. Move the slide until a clear marginal region of the worm is in focus, and then turn carefully to high power. Study the shape and arrangement of the cells and fibers which constitute the chief supporting tissue of the worm. Muscular movements cause the intercellular, parenchymal fluid to ebb and flow over all the body cells. Food and waste-product exchanges, similar to those effected by the tissue fluid of higher forms, are made by the intercellular fluid.

Digestive System. A circular mouth opens ventrally, slightly behind the anterior tip of the body. An oral cavity, the internal space of the oral sucker, connects behind to a very short prepharynx followed in turn by the ovoid muscular pharynx. A slender esophagus leads posteriorly from the pharynx to the point of bifurcation of the gut caeca. The gut caeca are narrow and simple and extend to near the posterior end of the body, where they end blindly.

Reproductive Systems. Most trematodes are hermaphroditic, each worm possessing complete male and female reproductive systems. Moreover, while self-fertilization often occurs and a single worm can produce normal larvae, cross fertilization is probably more common. In *C. lingua,* copulating adult worms are often observed.

As previously noted, the ventral sucker and the genital sinus are located within the ventrogenital pit, a median invagination of the body wall (Fig. 109). The genital sinus receives the ends of both the male and female reproductive ducts which pass through the dorsal part of a papilla-like accessory structure, the gonotyl. This gonotyl is acorn-shaped with a broad, cup-shaped base and a conical tapering tip, and on being protruded during copulation serves to hold the worms together as well as to align the genital pores to facilitate sperm exchanges.

Male Reproductive System. The two testes are lobed, somewhat diagonally placed, and located in the posterior part of the worm. In uncompressed specimens, the lobation may be indistinct and the testes somewhat separated. Fine efferent sperm ducts, the vasa efferentia, which are usually difficult to see, run forward from the testes and

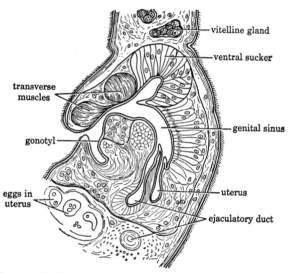

FIG. 109. *Crytocotyle lingua*, median sagittal section of the ventrogenital pit of the adult.

unite, behind the genital pit, to form a vas deferens. The sperm duct widens posterior to the genital pore, forming a sperm-packed seminal vesicle, and finally terminates in an ejaculatory duct which then opens into the genital sinus.

Female Reproductive System. The female reproductive system is more complex and accordingly more difficult to interpret. Ova are produced in a single, lobed ovary, lying in front of the testes usually on the right side. An oviduct runs medioposteriad from the ovary and, after proceeding a very short distance, receives a duct from a large ovoid or spherical seminal receptacle usually packed with sperm and located closely behind the ovary. From this junction, where fertilization occurs and which marks the end of the oviduct and the beginning of the uterus, a short Laurer's canal runs dorsally to open at the surface near the midline at the level of the ovary. It has been postulated that excess sperm and other reproductive materials are eliminated through this canal. Between the ovary and seminal receptacle will be found a vitelline or yolk duct which meets a similar duct from the opposite side in an expanded vitelline reservoir, filled with granular vitelline material, located just ahead of the anterior testis. The vitelline ducts drain many ovoid vitelline glands occupying the postesophageal region of the body. The fertilized egg now passes in the uterus alongside the vitelline reservoir and receives yolk material through a short, common yolk duct from the reservoir. An oötype, an egg-fashioning organ found in many trematodes, is not present in *C. lingua*. The uterus occupies most of the space between the posterior sex organs and the ventrogenital pit.

As the eggs proceed through the convoluted uterus toward the genital pore, they increase in size, form brownish eggshells probably from some of the vitelline materials, and undergo the first few cleavages. The eggs finally pass into the genital sinus and are discharged into the small intestine of the host to be carried with the feces to the outside.

Excretory System. Near the posterior end of the body on the dorsal surface is an excretory pore. Running forward from the pore and bending between the two testes is a clear, vesicular excretory bladder. Slightly ahead of the anterior testis the excretory bladder bifurcates to form short anterior branches. Two main excretory vessels are laterally placed and follow a wavy course anteriorly from the lateral arms of the bladder; at the level of the esophagus they bend back upon themselves. The main lateral trunks have smaller tributary vessels which ultimately connect to the terminal units of the excretory system, the flame bulbs or flame cells. Examine the clear margin of the worm under high power and find small flickering cells. Now study a single cell with the oil-immersion objective and note the cell body with its nucleus and the vibrating tuft of cilia which constitutes the "flame" of the flame bulb. Some workers call this organ complex an osmoregulatory system because they consider it to function only in removing excess water.

Nervous System. The central nervous system consists of a ganglionic concentration near the anterior end and several main longitudinal nerve trunks proceeding posteriorly. Living specimens stained with vital dyes such as methylene blue or neutral red often show the whole extent of the nervous system. The cerebral ganglia, constituting the bilobed brain, are connected by a narrow transverse nerve commissure in the dorsal prepharyngeal body region. From the brain,

170 PLATYHELMINTHES

smaller lateral nerves pass anteriorly to the oral sucker, and two conspicuous main trunks, each with many fine peripheral nerves and interconnective commissures, run outside the gut caeca to the posterior portions of the body. In stained whole mounts the compact reproductive organs usually obscure the nerve cords posteriorly.

LARVAL STAGES

Miracidium and Sporocyst. Cleavage and development of the larva continue for about 10 days after the egg reaches the outside. A free-swimming miracidium does not occur in this life cycle; rather, the fully embryonated egg must be ingested by a snail, usually the edible periwinkle, *Littorina littorea*, which is exceedingly common along parts of the North Atlantic coasts of Europe and North America.

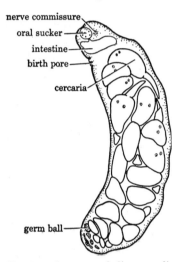

FIG. 110. *Cryptocotyle lingua,* redia from the liver of *Littorina littorea.* (Redrawn from Stunkard.)

The miracidium, as seen in fully embryonated eggs, is ovate and about as long as the egg. It is covered, except on the anterior tip, with long spirally arranged cilia. Several large glands with apically opening ducts occur in the central portion of the larva.

After being eaten, the infective larva hatches and invades the tissues of the snail host. Small oval to vermiform sporocysts have been recovered in the prehepatic lymph spaces adjacent to the digestive tract. Sporocysts typically are simple, sac-like larvae which lack digestive tracts and which, in turn, produce other larvae, the rediae.

Redia. Rediae occur in great numbers in the snail host's liver, which is altered in color and texture as a result of the mass infection. Crush an infected snail and tease the soft, yellowish liver in molluscan Ringer's solution. After noticing the tremendous numbers of larval trematodes in the liver, mount several of the small, colorless, sausage-shaped rediae on a slide and examine under high power. Locate a small, anterior oral sucker and a short, sac-like intestine, restricted, in a mature redia, to the anterior end (Fig. 110). On the ventral side immediately behind the oral sucker is a birth pore provided with small enclosing lips through which cercariae can emerge. A nerve commissure is located on the dorsal side of the intestine immediately behind the oral sucker. A double excretory system with ramifying internal ducts and flame bulbs and with lateral excretory pores, which open about one-third of the body length from the posterior end, may be apparent in favorable specimens.

The interior of the redia is filled with developing cercariae. In the posterior end, germinative cells give rise to small, undifferentiated germ balls which, as they move anteriorly, slowly develop into tailed cercariae. Study, under high power, the posterior end of a carefully compressed specimen and note the character of the germinative cell-wall layer. Also examine the germ balls and note their resemblance to early cleavage stages and morulae in other forms. Now trace the development of the cercaria by examining selected stages. Notice the formation of a tadpole-shaped body, a thin caudal region, and heavily pigmented eyespots.

Cercaria. The cercariae are not fully mature when they emerge through the birth pore. As they migrate through the host tissues, they complete their maturation by developing cuticular spines, a tail fin, and granules in gland cells.

Locate a favorable cercaria (Fig. 111) which still retains its tail and examine it under high power. Note the resemblance between the cercarial body and the adult.

The cercaria has several larval specializations which enable it to find and penetrate its next host. The two conspicuous eyespots composed of black pigment granules are located on the dorsal side of the developing esophageal commissure and function as light-sensitive organs. The larvae are positively phototactic and will collect invariably on the light side of the container. They also react immediately to shadows by swimming actively in the direction of the shadow-casting agent; this reaction would serve in nature to locate and attach to fish, their next host. The tail, which is the organ of locomotion, is long, muscular, and provided with dorsal and ventral fins each apparently stiffened by "bristles." Examine a bristle critically and note that it is merely an optical illusion formed by light refraction at a crest in the homogeneous fin. The center of the larva contains about eighteen cephalic-gland cells with excurrent ducts opening near an anterior oral sucker. The gland cells are filled with a granular secretion, readily visible in neutral-red-stained larvae. This secretion is undoubtedly histolytic and dissolves the flesh of the fish beneath the spot of cercarial penetration. Located peripherally are many small cystogenous gland cells whose secretions probably aid in the formation of the metacercarial cyst.

The larval morphological specializations are transitory and cease to function at the time of metacercarial formation; however, the following cercarial structures are permanent and persist through the adult

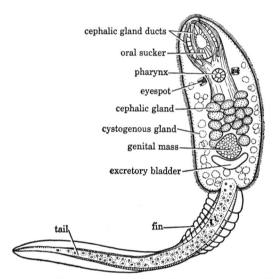

FIG. 111. *Cryptocotyle lingua*, mature, free-swimming cercaria. (Redrawn from Stunkard.)

stage. The cercarial sucker and an excretory system, consisting of posterior bladder, ducts, and flame bulbs, are visible in most larvae. A pharynx, a group of cells near the eyespots, and gut caeca are not well formed. A group of cells immediately anterior to the excretory vesicle is believed to give rise to all the genital organs.

Metacercaria. Free-swimming cercariae readily penetrate the skin and especially the fins of fishes. Most fishes of the intertidal zone, particularly sluggish swimmers such as the cunner, *Tautogolabrus adspersus*, are naturally infected with these metacercariae. Examine a cunner and note the peppering of black dots in the fins and flesh; each spot indicates the presence of a metacercarial cyst which is made more evident by encircling black pigment cells formed by the host.

Cercarial attack, penetration, and encystment may be studied by placing a 3-mm. square of healthy fin from a stunned cunner in a watch glass together with several cercariae concentrated in a small quantity of sea water. Observe, with the binocular microscope, the random swimming of the cercariae and locate cercariae crawling on the surface of the fin. Determine whether there is any positive chemotaxis or whether the finding of the fin by the cercariae is purely chance. When the cercarial tail is shed and active penetration of the fin is begun, transfer the vessel to the stage of the compound microscope and observe the method of penetration and encystment.

The postcercarial changes may be studied by releasing a metacercaria from its cyst. Carefully remove a well-formed cyst from the fin and transfer it to a drop of normal saline on a glass slide. With sharp needles open the cyst and remove the worm. Note the absence of cercarial organs and the continued development of adult organs. As you study the metacercaria, compare its anatomy with that of the sexually mature adult and that of the cercaria, both previously studied.

Opisthorchis sinensis
By C. G. Goodchild

The Chinese liver fluke, *Opisthorchis sinensis*, formerly known as *Clonorchis sinensis*, was first described in 1875 from a Chinese hospital patient whose liver, on post-mortem, was found to be enlarged and turgid and from whose distended bile ducts many small digenetic trematode worms were recovered. Since that time the full medical importance of this worm has been slowly realized, and its epidemiology, therapy, morphology, prevention, and life history have been studied.

The adult worm (Fig. 112), available as a stained whole mount, should be examined under low power to determine its size, shape, and general morphology. The normal size of this worm ranges from 10–25 mm. in length by 3–5 mm. in breadth. Note the rounded, tapering posterior end and the elongated, gracefully tapering anterior end, terminating bluntly with the oral sucker. The anterior one-fourth of the body, which contains an oral sucker, pharynx, brain, and the beginning of two gut caeca, is set off from the posterior body region by a slight marginal notch. At the notch level will be found a ventrally opening genital pore slightly anterior to a rather small ventral sucker. The larger posterior body region contains the majority

174 PLATYHELMINTHES

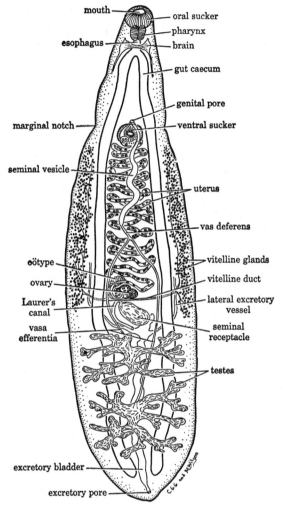

FIG. 112. *Opisthorchis sinensis*, dorsal view of adult.

of the body organs including the complete reproductive systems, the major portion of the excretory system, and most of the gut caeca.

THE ADULT

Integument and Muscular System. A thin, minutely wrinkled cuticula which is devoid of spines overlies the hypodermis and the peripheral muscular layers of the body.

OPISTHORCHIS SINENSIS

Parenchyma. The space between the organ systems in the interior of the body is filled with a mesenchyme called parenchyma. This tissue consists of irregularly shaped cells between which is a matrix of fluid and intertwining fibers.

Digestive System. A large transversely oval mouth opens, slightly subterminally, on the ventral side from a buccal cavity which is surrounded by the well-defined oral sucker occupying most of the space of the extremity. Posterior to the oral sucker is a conspicuous ovoid pharynx which often slightly indents the posterior border of the sucker. The sucker and pharynx, which are provided with radiating muscle fibers, serve for attachment, abrasion of the bile ducts, and aspiration of loosened food particles. A short esophagus, approximately equal to the pharynx in length, joins the pharynx and the two mildly distended gut caeca. The oral cavity, pharynx, and esophagus are lined with a thin, inturned cuticula, while the gut caeca, which reach nearly to the posterior end of the body, are lined with a vacuolated columnar epithelium. This epithelium serves to secrete enzymes which digest the aspirated food. The food consists of blood corpuscles and a mucin-like exudate produced by laceration of the inflamed mucosa of the bile duct.

Reproductive Systems. This hermaphroditic trematode possesses well-defined reproductive organs and a common genital pore.

Male Reproductive System. In large, well-stained specimens, two deeply branched, tandem testes are visible posteriorly. From the main body of each testis arises a thin vas efferens. This duct proceeds forward to a point slightly anterior to the ovary where it joins its mate from the other testis to form a vas deferens. This sperm duct soon enlarges into a sperm-filled seminal vesicle. A narrow ejaculatory duct connects the vesicle to the genital atrium which opens at the ventral genital pore. A cirrus pouch, prostate, and cirrus are lacking in this worm.

Female Reproductive System. A small, slightly lobed, unpaired ovary lies in the median line slightly anterior to an oblong or ellipsoidal sperm-filled seminal receptacle. The seminal receptacle can be easily located because it lies just ahead of the anterior testis. Anteriorly the seminal receptacle gives off a short duct to which initially is connected Laurer's canal and which terminally joins a short oviduct. Laurer's canal curves posteriad around the receptacle and also runs abruptly dorsally to open by a small surface pore located between the anterior testis and seminal receptacle near the midline. The oviduct, after receiving the sperm-carrying duct of the seminal receptacle, proceeds forward a short distance and receives a common vitelline duct which adds particulate vitelline material to the ova. The yellowish vitelline material is elaborated in many small, extracaecal vitelline glands located in the middle third of the body and is transported by vitelline ducts. Locate small vitelline ducts among the glands and see their junctions with transverse vitelline ducts running mediad. The union of the transverse ducts forms the common vitelline duct already mentioned. The oviduct, after receiving this last duct, then penetrates an oötype which is surrounded by a loosely organized, tubular Mehlis' gland composed of small, acinus-shaped cells. In this region the ovum and yolk are enclosed in a thin-walled, oval, operculate shell which is often provided with a comma-shaped posterior appendage. A very conspicuous uterus arises from the oötype anteriorly and proceeds as an irregularly coiled, intercaecal duct to the genital atrium. The uterus becomes larger in diameter anteriorly because of the massing of eggs. The egg gradually acquires a darker brown color and proceeds with cleavage until a fully formed miracidium is formed. Eggs pass to the outside through the genital pore and are carried by the bile to the host's small intestine; they are finally voided with the feces.

Excretory System. A clear, irregular excretory bladder is constricted posteriorly to form a narrow efferent canal which opens to the outside by a posterior excretory pore. The bladder proper is approximately the diameter of a gut caecum and can be followed anteriorly between the main bodies of the testes to the level of the seminal receptacle. The intertesticular course of the bladder is somewhat obscured because both testes have medial lobes which overlap the bladder.

Two lateral excretory vessels join the bladder slightly behind its blind anterior end. These vessels curve laterally and forward and follow an extracaecal course to the level of the ventral sucker, where they branch into anterior and posterior recurving tributary vessels. The continued branching of these major vessels results in increasingly finer vessels, the terminal units of which are the flame bulbs.

Nervous System. A bilobed brain with medial commissure and anterior and posterior nerves is located in the dorsal esophageal region. The extremely complex nervous system of a trematode is indicated by Fig. 113 of a preadult worm. The major portion of this system, if not all, persists unchanged into the adult stage but is

OPISTHORCHIS SINENSIS

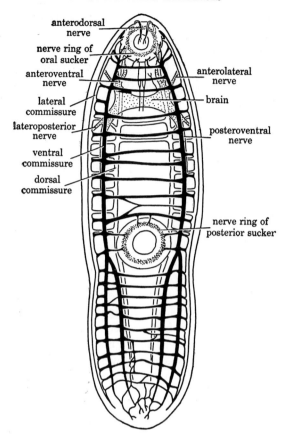

FIG. 113. *Opisthorchis felineus,* nervous system of the metacercaria. (Redrawn from Vogel.)

usually not visible in detail in ordinary stained whole mounts. Living worms treated with dilute methylene blue may show some parts of the system.

LARVAL STAGES

Miracidium and Sporocyst. Eggs when laid contain fully mature miracidia. Each ovoid miracidium bears long, motile cilia on an outer ectodermal sheath which completely covers the body wall except apically, where it is interrupted by an apical spine, by apical gland openings, and dorsally by a single excretory pore. An apical histolytic gland and a large, ventral, sausage-shaped secretory gland with a narrow duct opening beside the spine, together with about eight to twenty-five loosely organized germinal cells, fill the miracidial body space. Two dorsally placed flame bulbs and their ducts comprise the excretory system. A small, reniform nervous system is situated ventral to the apical gland. Miracidia do not ordinarily hatch free in nature; rather they are released from the eggs only after ingestion by appropriate snail hosts. The miracidium then actively penetrates the gut wall into the peri-intestinal blood sinus, where it becomes a thin-walled, sac-like sporocyst. This larva produces the redial generation which migrates to the hepatic blood sinuses.

Redia. A mature redia is elongated and has a rounded oral end and a somewhat pointed posterior end internally lined with a germinal epithelium. The thick, muscular integument lacks a birth pore and must burst to discharge mature cercariae. An anterior pharynx and a brownish sacculate gut extending about half way posteriorly are also present. Several germ balls develop synchronously into cercariae, so that six to eight of the latter mature and burst from the redia at the same time.

Cercaria. The long-tailed, oculated cercaria emerges from the snail host into the water and swims with a vigorous intermittent movement in search of its next host, a fresh-water fish. The propelling organ, the tail, is provided with muscle fibers and bears thin, lateral, transparent elytra. Two kidney-shaped eyespots are located near the anterior end of the body. An oral sucker, pharynx, and ventral sucker are present, but esophagus and gut caeca apparently are not yet formed. Seven pairs of unicellular penetration glands occupy the middle of the body; ducts lead away from these glands to open near the mouth. Dorsal and ventral peripheral mesenchyme cells are packed with cystogenous granules which are used in the construction of the metacercarial cyst. A large, triangular excretory bladder with anterolateral collecting ducts and a bifurcated posterior caudal excretory tubule comprise the cercarial excretory system.

Metacercaria and Young Adult. On coming into contact with fish, especially members of the Family Cyprinidae, the cercaria attaches to the surface, discards its tail, and creeps under a scale into the flesh. In the fish, the cystogenous material oozes out and hardens into a thin, tough, spherical or ovoidal membrane around the larva. Later, a capsule formed by the host is deposited about the true cyst. Organ development, e.g., the gut caeca, continues in the cyst. Mammalian hosts, chiefly man, become infected by eating fish containing infective metacercariae. The young worm, after liberation from its cyst in the duodenum, crawls along the mucosa to the common bile duct opening which it enters. Continuing its journey, it ascends the bile ducts to the distal biliary capillaries, where it establishes itself, grows to adulthood, and after approximately 25 days starts the cycle again by discharging ripe eggs.

Gorgodera amplicava
By C. G. GOODCHILD

Bladder flukes, of which *Gorgodera amplicava* (Fig. 114) is an example, may be obtained alive from the urinary bladders and excretory ducts of amphibians and fishes. The leopard frog, *Rana pipiens*, and the bullfrog, *Rana catesbeiana*, frequently harbor *G. amplicava* and species of *Gorgoderina*, another genus of bladder flukes.

In a pithed frog carefully excise the urinary bladder, which is a thin-walled sac lying against the posteroventral wall of the abdominal cavity, and pin it in an opened and stretched condition on the surface of a clean, waxed-bottom vessel containing amphibian Ringer's solution. Large flukes will be visible to the naked eye as somewhat opaque, worm-like organisms clinging tenaciously to the lining of the bladder. Young individuals are more difficult to see because of their small size and transparency; therefore, do not discard the preparation, in the event that a large fluke is not found, until it has been examined with a dissecting microscope or other optical aid. Often, bladder flukes may be recovered from the Wolffian ducts, and sexually mature specimens of *Gorgoderina* may occasionally be found in the tissue of the kidney. If time permits, search for flukes in these organs.

Before attempting to remove the fluke from the bladder, study its shape, motion, and its use of attaching organs, the oral and ventral suckers. The organism is usually firmly attached by the large, powerful ventral sucker. The anterior end is thin and mobile; it moves randomly from side to side and often is temporarily attached to the epithelium of the host. The flattened posterior end is moved rapidly and jerkily dorsoventrally in *Gorgodera*. In *Gorgoderina* the cylindrical posterior end is elongated and moved sluggishly from side to side. In locomotion the anterior end is stretched out and attached to the epithelium by the apical oral sucker; the ventral sucker is then released and, by an arching of the body, is moved forward leech-like and attached just behind the oral sucker. In the wall of the urinary bladder, notice small mounds and hemorrhage spots which are caused by the suckers. From these and other observations try to determine what possible harm is done to the host, why the ventral sucker is so

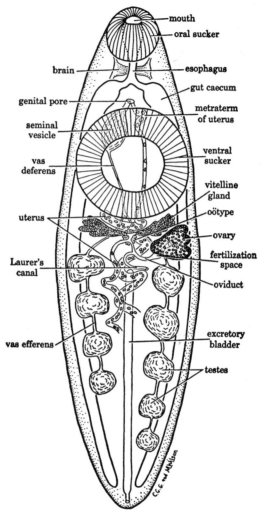

FIG. 114. *Gorgodera amplicava,* diagrammatic ventral view of adult.

large and powerful here as compared to most other trematodes, and what type of food is eaten by the parasite.

With needles or a blunt scalpel carefully transfer the fluke to a drop of spring water or non-chlorinated tap water. Carefully lower a coverglass on the worm and withdraw any excess water by touching a piece of absorbent paper to the edge of the coverglass. Observe the trematode with the aid of the microscope during this operation and quickly withdraw the paper at the instant the worm becomes immobile and well flattened. This process of flattening usually forces eggs from a midventral genital pore, located slightly anterior to the ventral sucker; the eggs may hatch at once and liberate ciliated larvae which will be described later. During the course of the study occasionally add sufficient water to the edge of the coverglass to replace that lost through evaporation. In addition to living worms, vitally stained ones, made by adding a small drop of dilute neutral red (1:5000) in distilled water to the preparation, as well as stained, permanent whole mounts and cross sections, should be available for supplementary study.

The Adult

Study the adult worm (Fig. 114) under low power and verify its dorsoventrally flattened, elongated body. Notice the narrow anterior end which apically bears the oral sucker. About one-third the body length from the anterior end is the very large ventral sucker. In fixed specimens this sucker, because of its size, is often displaced to one side. The broad posterior end tapers to a rounded point and is packed with reproductive organs, including an extensive, egg-filled uterus. Readily visible external openings include an anterior mouth, a posterior excretory pore, and the midventral genital pore. A fourth opening, the pore of Laurer's canal located dorsally, is visible in a cross section through the proper region but is usually difficult to see in whole mounts.

Integument and Muscular System. A thin, smooth cuticula, secreted by special hypodermal cells lying immediately beneath the surface, covers the body. Three rather loosely arranged muscle layers consisting of an outer circular layer, a middle diagonal layer, and an inner longitudinal layer underlie the hypodermis. In addition, dorsoventral muscle bundles, which provide for altering the thickness of the body, traverse the interior of the worm. These muscle groups are best seen in stained cross sections which should be examined, if available; however, they are resolvable in good whole mounts by careful focusing. The musculature of the suckers is chiefly radial. These fibers alter the size of the cavity and create the powerful suctions used in attachment and feeding.

Parenchyma. The space between the various organs in the interior of the body is filled with a dense, mesenchymatous tissue called parenchyma. Move the slide so that a transparent margin of the anterior body region is in the center of the field and then turn carefully to high power. Study a small section of the parenchyma. This tissue consists of irregular cells and large intercellular spaces containing tangled fibers. In life these spaces are filled with a fluid which carries on metabolic interchanges with all the organs and is functionally similar to the tissue fluid of higher forms.

Digestive System. The oral sucker bears, subapically on the ventral side, a transversely oval mouth opening. The opening leads into a funnel-shaped buccal cavity, whose wall is the oral sucker. This cavity, in turn, connects posteriorly to a narrow esophagus which proceeds posteriorly to a point midway between oral and ventral suckers, where it joins the two gut caeca. These anterior regions of the digestive tract are lined with inturned cuticula. Notice that a pharynx is lacking. The gut caeca are larger in diameter than the esophagus and are lined with a secretive and absorptive columnar epithelium. They extend caudad to a point near the posterior end where they end blindly.

Reproductive Systems. This worm is a true hermaphrodite with complete male and female reproductive systems in each worm and apparently capable of self-fertilization because single organisms produce fertile eggs. The reproductive systems will be described separately. In this study the stained whole mount, the cross sections, and the living worm should be examined.

Male Reproductive System. The chief taxonomic difference between *Gorgodera* and *Gorgoderina* is the number of testes. The former has nine testes distributed in two linear, lateral groups of five and four respectively, while the latter has only two simple testes located laterally but at different anteroposterior levels. The slightly lobed testes in *Gorgodera* are, in shape, roughly rectangular prisms and, in sexually mature individuals, are separated from each other by folds of the uterus. The testicular arrangement has been likened to a series of cigar boxes connected by a string through their centers. On each side, a vas efferens passes from the center of each testis to the testis ahead and continues forward from each testicular set to a point near the ovary where it joins a median vas deferens. The vas deferens near the genital pore unites with a sperm-filled, pyriform seminal vesicle. From the anterior margin of the vesicle a narrow ejaculatory duct passes ventrad to the common genital pore. The ejaculatory duct initially is surrounded with unicellular prostate glands. In the living worm examine carefully the testes and seminal vesicle and note the thread-like spermatozoa contained within them. In the cross section study a single testis and find peripheral clusters of developing spermatozoa and central groups of mature ones.

Female Reproductive System. Ova are produced in a three-lobed ovary located posterior to the ventral sucker and on the same side of the body as the set of five testes. In the cross sections it will be apparent that the smallest female germ cells are located peripherally and that as they increase in size they move dorsally and enter a dorsomedially directed oviduct. Initially, the oviduct is slightly expanded to form a sperm-filled fertilization space. In this chamber several spermatozoa surround an ovum and one succeeds in penetrating it. Slightly anterior to this region the oviduct receives the inner end of Laurer's canal. This canal, at first, curves dorsolaterally and then dorsomedially and finally opens to the dorsal surface near the posterior margin of the ovary. A prominent, dumbbell-shaped complex consisting of vitelline glands is located posterior to the ventral sucker. Each lateral group of glands is composed of nine to eleven follicles and gives off medianly a stout, transverse vitelline duct which meets the duct from the other side in a small, central vitelline reservoir. Located dorsally to the isthmus of the vitelline ducts is a compact, spherical oötype, best seen in cross section. The oötype externally is surrounded by chromophilic, unicellular glands of Mehlis, and internally is a zone of confluence of the oviduct and a common vitelline duct from the vitelline reservoir. In the oötype a thin, hyaline egg-shell is secreted around the fertilized ovum and cellular vitelline material which clusters about the ovum. The source of the eggshell is apparently vitelline material and secretions from Mehlis' gland. Leaving the oötype posteriorly is a narrow, thin-walled, egg-filled uterus which follows a rambling course in the posterior body region. In the mature worm the posterior body space is filled to capacity with eggs. Often the wall of the uterus is not visible, and the eggs appear to be free in the body. Near the level of the ovary the uterine wall becomes evident, and as the uterus continues forward from this region to the genital pore the eggs are forced into single file because of a narrowing of the lumen. The wall of the terminal portion of the uterus is muscular and, as the so-called metraterm of the uterus, functions as an ejector of the eggs.

As the egg moves in the uterus from the ovary to the genital pore, it slowly increases in size and completes embryogeny. In the host, eggs are passed from the genital pore into the bladder cavity and are eventually voided, unhatched, in the urine. When the urine is diluted with fresh water, hatching of the eggs occurs almost immediately and ciliated, free-swimming miracidia are liberated.

Excretory System. A posteromedian excretory bladder extends anteriad between the sets of testes from the excretory pore to a point near the ovary. Two main excretory canals join the bladder near its inner end and then proceed forward along the lateral margins of the body. Anterior to the ventral sucker each canal bifurcates; these smaller canals receive increasingly smaller tributaries arranged in a complicated pattern. The smallest canals end blindly in terminal ciliated bulbs, some of which will usually be visible in living worms. Giant, semilunar, terminal ciliated bulbs may be apparent in the largest worms.

Nervous System. In order to trace the finer ramification of the nervous system, specimens treated with methylene blue will be needed. Living flukes and stained whole mounts are satisfactory for viewing the nerve cords and the brain, a clear, bilobed mass located on the dorsal side of the body and with a cross commissure in the midesophageal region. Several nerve cords running anteriorly, and two large cords proceeding posteriorly, will be apparent in favorable specimens. Papillae, of perhaps a tactile function, may be visible in rows on the body surface, on the oral sucker, and on the rim of the ventral sucker.

Larval Stages and Life Cycle

Abundant, free-swimming miracidia can be obtained by teasing a sexually mature worm in a small dish containing spring water, well-aged aquarium water, or non-chlorinated tap water. In a few seconds the ripest eggs will begin hatching, and pear-shaped, free-swimming miracidia will be visible with the aid of a binocular microscope. Transfer several miracidia and unhatched eggs to a glass slide, add a small drop of dilute neutral red, and cover with a coverglass. The preparation should be studied with the oil-immersion objective. In the unhatched egg note the exceedingly thin shell and the fully formed miracidium inside. Now examine the miracidium, using either the hatched or unhatched specimen. The body is covered with cilia which begin to beat while the miracidium is still within the egg. Internally, several apical glands with coarse granular secretion, as well as two flickering terminal ciliated bulbs and several germinal cells, will usually be apparent.

The miracidium must penetrate a mollusk within about 24 hours or succumb. The next host in this cycle is *Musculium partumeium*, a bivalve mollusk belonging to the Family Sphaeriidae. Miracidia are drawn into the mollusk by the swift, incurrent feeding and respiratory stream of water and penetrate the gills with the aid of histolytic secretions. In the gills two generations of sac-like sporocysts occur, and eventually long-tailed, motionless cercariae are produced. Cercariae range in length from about 11 mm. to 18 mm. and emerge from the bivalve through the excurrent siphon into the water. Naturally infected mollusks may be found in streams and ponds harboring infected frogs. Examine the bottom and vegetation of such habitats and locate the bivalve hosts. These specimens should be isolated in small containers in pure water and the emergence of cercariae noted. After cercariae are obtained and studied, the bivalve should be sacrificed and its gills examined for sporocysts.

Snails and tadpoles ingest the cercariae; encysted metacercariae are formed in the body cavities of these vector hosts. When an infected snail or tadpole is eaten by an adult frog, excystment of metacercariae in the small intestine occurs and a posterior enteric migration brings the young worms to the cloacal openings of the excretory system. Invasion of the Wolffian ducts or urinary bladder then occurs, and in about one month eggs are produced to start the cycle again.

Fasciola hepatica
By C. G. Goodchild

The sheep liver fluke, *Fasciola hepatica*, the causative agent of a disease commonly referred to as "liver rot," was the first described trematode, being reported in 1379. Its life history, independently discovered by Leuckart and by Thomas, was also the first life cycle of a digenetic trematode to be traced. This large (up to 51 mm. in length), thin worm lives in the bile ducts of many mammalian hosts, including man. Because it is well known and adequately described in most textbooks, its morphology will be here given brief treatment only.

Stained whole mounts of the worm should be available for microscopical study. Examine the worm under very low power or with low-power projection apparatus and note the elongated, oval body with a small anterior conical projection and a large posterior region more rounded in front than behind. An oral sucker is situated apically, and a somewhat larger ventral sucker is located about one-seventh the body length from the anterior end. Between the suckers is a median genital pore.

Cuticula. Many sharp, posteriorly directed cuticular spines are present, especially anteriorly. These projections serve to anchor the worm in the bile duct and, incidentally, to irritate and inflame the epithelium of the duct.

Digestive System. A mouth and buccal cavity, surrounded by an oral sucker, are situated at the apex of the short, conical, anterior body projection. A muscular, ovoid pharynx lies behind the mouth. The next region is a very short esophagus which joins two gut caeca in the conical region. The caeca pass around the ventral sucker and follow a somewhat parallel, near-median course to the posterior end. Each caecum gives off many short, median diverticula and many long, lateral diverticula with primary, secondary, and tertiary branches which extend to the margins of the body. The liver fluke feeds apparently on blood and desquamated epithelium which are digested within the caecal lumen.

Reproductive Systems. The complexities of the well-developed hermaphroditic reproductive systems make this worm rather undesirable for study.

Male Reproductive System. Two extensively branched testes lie one behind the other, extending from about the middle of the body to near the posterior end. A fine vas efferens runs forward from the main anterior stem of each testis to the side of the ventral sucker; here it joins its mate at the base of an enlarged, conspicuous cirrus pouch. The latter contains three regions: an initial, distended seminal vesicle often packed with sperm cells; a middle, narrow *pars prostatica* surrounded by deeply stained prostate cells; and a terminal, thick-walled, protrusible cirrus which opens into a genital atrium. The cirrus, a sperm-transferring organ, is projected through the small, median genital pore normally only during copulation but is often permanently extruded during the convulsions of fixation.

Female Reproductive System. A dendritic ovary lies ahead of the anterior testis slightly to the right of the midline. Its club-shaped lobes, often mistaken for testicular ones, are seen, in favorable specimens, to be less tortuous and less branched. A short oviduct leaves the ovary medially and penetrates anterolaterally a centrally placed, spherical oötype, made conspicuous by surrounding chromophilous, unicellular Mehlis' glands. Posteriorly adjacent to the oötype is a triangular vitelline reservoir which receives two transverse vitelline ducts formed by a confluence of smaller ducts from many small, highly branched vitelline glands which lie in the lateral margins of the worm. A short, common vitelline duct, carrying yolk-filled vitelline cells, proceeds forward from the reservoir to join the oviduct in the center of the oötype. A Laurer's canal runs dorsad, from the junction of the oviduct and the common vitelline duct, to open at the surface. A large uterus, distended with developing eggs, follows a rambling course from the oötype to the common genital atrium. Eggs pass from the genital pore into the bile in which they are carried to the intestine. Further development occurs after they are evacuated to the outside.

Excretory and Nervous Systems. A median, posterior excretory bladder empties by a terminal excretory pore. Flame bulbs may be present in the adult, but neither they nor the excretory ducts are visible in most stained specimens. In fact, there are in the literature conflicting statements regarding the presence or absence of flame bulbs in this worm. If living worms are available, this problem may profitably be investigated. A bilobed brain and dorsal commissure are located in the esophageal region. A pair of lateral nerve cords with numerous branches runs to the posterior end of the worm.

Schistosoma haematobium
By C. G. GOODCHILD

Blood flukes apparently originated in the Nile and Yangtze river valleys and many parts of Africa and China are still heavily infected; in addition, South America, the adjacent Caribbean islands, Japan, and her adjacent islands, as well as the Philippines, have localized regions of heavy infection.

The medically important human blood fluke, *Schistosoma haematobium*, has been a scourge in Egypt since ancient times. It lives chiefly in the veins of the urinary bladder and intestine, where it causes grave pathological lesions. Two other important species of human blood flukes, *S. mansoni* (Fig. 115) and *S. japonicum*, have similar anatomies and life histories. These directions, with slight modifications, can be used for all three common species.

This worm, unlike most other trematodes, is dioecious. Larger, flattened males and attenuated, cylindrical females attach in approximately equal numbers to the lining of the abdominal blood vessels of the host. The female, during most of her sexual life, lives in a gynecophoral groove which is formed by a tube-like ventral infolding of the thin, postacetabular body margins of the male.

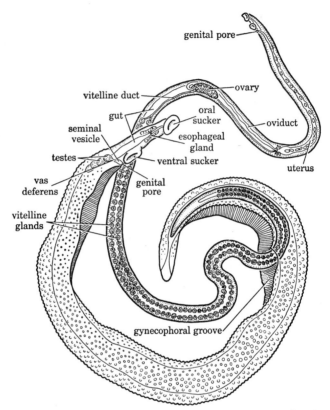

FIG. 115. *Schistosoma mansoni*, female in gynecophoral groove of male. (Modified after Fuhrmann, in Kükenthal and Krumbach.)

Prepared slides of male and female worms and a slide showing a pair with the male embracing the female in the gynecophoral groove should be available for study.

THE ADULTS

Integument and Muscular System. Small cuticular papillae cover the integument of the male but are confined to the anterior and posterior ends of the female. *Schistosoma mansoni* is similarly papillated, while *S. japonicum* lacks integumentary papillae. A small, weakly

SCHISTOSOMA HAEMATOBIUM

muscular oral and a somewhat larger, stronger, ventral sucker occur in the male, while both suckers are equal in size and weak in the female. The male gynecophoral canal is formed by lateral muscular body margins which have well-defined circular, oblique, and longitudinal muscle fiber layers; the female body muscles are ill defined.

Parenchyma. The spaces among organs of the body are filled with a typical mesenchyme having irregular cells and a fluid and fibrous intercellular matrix.

Digestive System. The digestive system is identical in both sexes and similar in the three species. A subapical mouth leads into a funnel-shaped buccal cavity surrounded by the oral sucker. This cavity connects directly to an esophagus without the interpolation of a pharynx. No doubt the absence of the pharynx is correlated with the abundant, easily ingested host's blood which is the food of these worms. Surrounding the esophagus is an hour-glass-shaped digestive gland. The esophagus reaches to the anterior margin of the ventral sucker where it joins two blood-filled gut caeca; these digestive tubes run posteriad and rejoin near the middle of the body to continue posteriad as a single, sinuous caecum which terminates blindly near the end of the body. The two caeca rejoin nearer the ventral sucker in *S. mansoni* and nearer the posterior end in *S. japonicum*.

Reproductive Systems. Basically, the arrangement of the blood-fluke reproductive system is like that of the corresponding system in hermaphroditic forms.

Male Reproductive System. Close behind the ventral sucker and located somewhat dorsally will be found four or five tightly clustered testes. Vasa efferentia from all the testes join a short vas deferens which at once enlarges to become a short, sperm-filled seminal vesicle. This vesicle opens directly to the ventral surface immediately behind the ventral sucker. Accessory copulatory devices are lacking, and insemination of the female, which, of course, is obligatory here, occurs by ventral apposition of the bodies with the genital pores in contact.

Female Reproductive System. An elongated ovary located near the middle of the body, immediately in front of the rejoined gut caeca, connects at its posterior end to a narrow oviduct which bends at once and proceeds forward together with a median common vitelline duct which it joins at an oötype. The posterior half of the body is filled with alternately lobed vitelline glands which empty through the common vitelline duct. Fertilization and shell formation occur in the oötype; the egg is then pushed into the uterus which may hold as many as twenty to thirty eggs in its anterior course to the ventral, postacetabular genital pore.

Schistosome eggs are used clinically to determine the infecting species. The egg of *S. haematobium* is oval and is provided with a terminal spine; the egg of *S. mansoni* is elongated to oval and has a long, sharp, lateral spine; while the egg of *S. japonicum* is oval to rounded with a short, blunt, lateral spine.

Nervous and Excretory Systems. A bilobed brain, in the dorsal midesophageal region, and paired posterior nerve trunks may be visible in well-stained specimens. The protonephridial excretory system has a posterior bladder, lateral collecting tubules with anterior and posterior branches, and terminal flame bulbs.

LARVAL STAGES

The female leaves the male at egg-laying time and moves into smaller veins of the bladder and intestine. The egg, by means of its spine, digs into the wall of the blood vessel and gradually is worked through the wall, the underlying tissue, and into the bladder cavity or intestinal lumen. The egg contains a fully mature miracidium when it is eliminated in the urine or feces. If the egg is passed into fresh water, hatching occurs at once, and a ciliated miracidium is liberated. This larva swims in search of a snail host. Active penetration of the snail occurs, and mother and daughter sporocyst generations occur within this host. In approximately six weeks fork-tailed cercariae are discharged from the daughter sporocysts into the water. The free-swimming cercaria has an elongated, oval body and a mobile tail, which consists of a proximal trunk and two distal, fork-like branches. On coming in contact with a human, the cercaria attacks and penetrates the skin by histolytic secretions, entering a vein or lymph vessel in which it is transported to its normal sites in the host. After a month, during which time the young worm feeds and grows to adulthood, ripe eggs are produced to start the cycle again.

Phyllobothrium laciniatum

By C. G. GOODCHILD

The adult of *Phyllobothrium laciniatum* (formerly known as *Crossobothrium laciniatum*) inhabits the small intestine of the sand shark,

Carcharias littoralis, while larval cestodes which infest the cystic duct of *Cynoscion regalis*, a teleost fish popularly known as the squeteague, have been suggested to be the plerocercoid stage of this worm.

The structure of the scolex (Fig. 116) may conveniently be studied by examining an adult or a living or preserved plerocercoid from the squeteague. The flared scolex possesses four elongated, concave attaching structures with somewhat thickened rims, called bothridia;

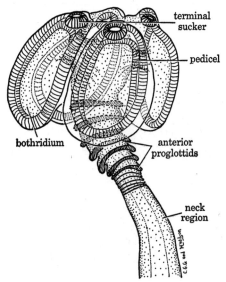

FIG. 116. *Phyllobothrium laciniatum*, scolex and anterior region of body.

the bothridia are borne on short, thick pedicels and each bothridium bears a single, accessory, terminal sucker. The muscular action of the bothridia and the accessory suckers should be observed in the living scolex, if available.

The body of the worm is profoundly emarginate and is composed of proglottids each narrower in front and characterized by four overhanging marginal lobes on the posterior border. In general, the proglottids increase in size from the scolex posteriorly, but there is not a graduated increase as in the taenioid cestodes. The proglottids are produced discontinuously from the neck region, and often several zones of maturing proglottids will occur along the length of the body. In some specimens an anterior zone of proglottids, contiguous to the scolex, is separated from the other proglottids by a smooth, narrower, proliferating neck region. The posterior proglottids are roughly triangular in surface view with a drooping, wavy, posterior margin consisting of the four lobes; moreover, they are tenuously joined and become easily separated from each other. Detached and independent proglottids may be recovered from the small intestinal chyme which is often rendered milk-white by their abundance.

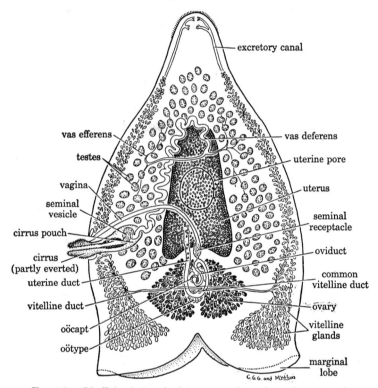

FIG. 117. *Phyllobothrium laciniatum*, surface view of proglottid.

Certain details of the anatomy, such as canals and flame bulbs of the protonephridial excretory system, are most easily seen in the living worm; other parts, such as the hermaphroditic reproductive systems, are easier to make out in well-flattened and stained specimens.

In the mature proglottid (Fig. 117) locate a heavy, secreted, outer cuticula, a mesenchymatous parenchyma, and muscle fibers which are responsible for the trematode-like locomotion of the detached, living proglottid. Clear, sinuous excretory canals may be visible, in favorable specimens, on the lateral borders, especially anteriorly. If they are not immediately apparent, do not continue searching for them or for their smaller tributaries and flame bulbs which are usually invisible. The proglottid lacks any trace of a digestive system. As is usual in flatworms, the reproductive systems are large and complex; they will be described separately.

Male Reproductive System. Many small spherical testes encircle a central uterine cavity. Examine a testicular follicle under high power and note the sperm cells in all stages of development. Leading away from the follicles are fine, efferent sperm ducts which unite slightly anterior to the center of the proglottid to form a wider, sperm-filled vas deferens. This extremely convoluted duct proceeds to the base of a seminal vesicle, which in turn joins a long, slender, marginal, eversible cirrus. The cirrus, when lying in the clear, bulbous cirrus pouch, is very much coiled, but is often seen protruded from the common, marginal genital pore. A demonstration of this protrusion will show the minute spines which cover the cirrus and an internal duct which opens apically and is an extension of the vas deferens.

Female Reproductive System. Immediately anterior to the marginal male gonopore is an opening of the female copulatory tube, the vagina. The vagina initially parallels the vas deferens but soon describes a smoothly arched course to a region near the posterior end where it dilates as an preovarian, cylindrical seminal receptacle. A deeply stained, ellipsoidal ovary, composed of polydendritic lobes, is located in the median plane slightly ahead of the posterior marginal notch. It surrounds a small, compact oötype and the posterior portion of the seminal receptacle. An oviduct leads away from the ovary and is surrounded initially by a sphincter, the oöcapt, which in whole mounts is usually visible as a circular structure in the center of the ovary. The oviduct runs posteriorly a short distance but then describes a U-shaped anterior bend and near the anterior edge of the ovary receives a constricted fertilization duct from the seminal receptacle. The oviduct makes a second abrupt curve, this time posteriorly, and passes into the spherical, lightly stained oötype lying immediately behind the oviducal sphincter. The oötype receives a common vitelline duct formed by left and right vitelline ducts running mediad from many lateral vitelline glands. The latter are grouped in the posterolateral margins and extend in a thin lateral band forward to a region near the anterior end. The fertilized eggs receive vitelline material in the oötype and are also provided with shells from vitelline materials, assisted perhaps by secretions from small encircling shell glands. When they are fashioned, the eggs pass into a short, anteriorly directed uterine duct which leads into the greatly expanded, sacculate uterus. In the gravid proglottid this cavity is filled with eggs and often shows a clear ruptured area, the uterine pore, through which eggs are normally discharged into sea water during characteristic swimming contortions of the eliminated ripe proglottid. The life cycle, as stated previously, is incompletely known.

Otobothrium crenacolle

By C. G. GOODCHILD

Adult tetrarhynchid cestodes usually occur in the small intestine of elasmobranchs, while their larvae are found in the tissues and body cavities of intermediate hosts such as holothurians, mollusks, crustaceans, and teleost fishes. The organ systems of mature tetrarhynchid proglottids are arranged like those in tetraphyllid proglottids, and their description, therefore, will not be repeated. The scolex, however, is so different in structure from all other cestode scoleces that it will be described in detail.

Abundant, living tetrarhynchid scoleces are available in the flesh of the butterfish, *Stromateus triacanthus*. A single fish may yield thousands of the cysticercoid larvae of *Otobothrium crenacolle* (Fig. 118) which occur in the muscles adjacent to the vertebral column. With a scalpel make a median sagittal incision in the fish and locate tiny, whitish, ovoid cysts in the flesh. Transfer several of these cysts to a drop of saline solution on a slide and remove the cyst membranes with well-sharpened needles. Notice a thin, brownish, external layer and a thinner, hyaline, internal one enclosing a mass of opaque granules which usually hides the small larva. If the worm does not spontaneously emerge, carefully tease it free. Observe the shape of the body and its method and speed of progression. In similar preparations test the effects of dilute hydrochloric acid upon larval activity and the effect of tap water upon the eversion of the proboscides.

With the aid of the microscope study a living animal imprisoned under a coverglass sealed at the edges with Vaseline; if living material is unavailable, then examine a stained whole mount. The scolex is elongated and has several anteroposterior regions: a head stalk is the anteromost portion and bears dorsal and ventral grooved bothridia; posterior to the head stalk is a neck region which is narrower than the head and is traversed by four canals or rhynchocoels enclosing spinous ribbons, the proboscides; the next zone, the bulbular region, is only slightly wider than the neck and contains four enlarged muscular bulbs continuous anteriorly with the rhynchocoels; terminally is a rounded postbulbular appendage which bears long cuticular bristles on the posterior half.

The scolex, externally, is completely invested with a thin cuticula beset with numerous short spines. Internally, it contains important organs (to be described below) in addition to mesenchyme cells, muscle fibers, many small clavate head glands opening at the surface, and numerous opaque calcareous bodies which will dissolve, with the evolution of a gas, when they are exposed to acid.

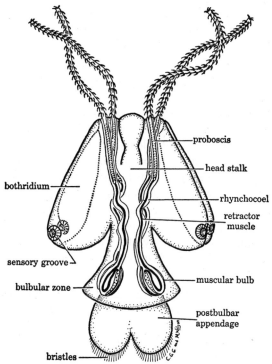

Fig. 118. *Otobothrium crenacolle,* cysticercoid larval scolex.

The two bothridia undoubtedly represent the fusion of former dorsal and ventral pairs of bothridia. The striking breast-stroke-like motions of the bothridia enable the newly excysted scolex to move rapidly through the thick intestinal chyme to the gut wall of the final host. Four sucker-like sensory grooves located on the posterolateral edges of the bothridia can, by radial and circular muscles, be everted to form small protuberances or inverted to form small suckers; moreover, their sensory surface is covered by long, thin spines resembling cilia.

The most conspicuous organs within the scolex are the proboscides. These eversible and retractile spinous ribbons are attached at the rim of an anterior pore and assist the bothridia in anchoring the worm to the enteric mucosa. Examine a partially everted proboscis at its anterior pore region. Notice as eversion occurs that the hollow proboscis turns inside out and that the hooks which are internal and point forward in the retracted condition become external and recurved in the everted state. Eversion is accomplished by muscular pressure, especially in the wall of the muscular bulb, on a corpusculated fluid contained in the rhynchocoel. Retraction is effected by a long, thin, undulous retractor muscle originating on the anterolateral wall of the bulb and inserting on the tip of the proboscis. The wall of the muscular bulb is continuous with the wall of the rhynchocoel; it is composed of several layers of muscle fibers and has, in addition, a longitudinal zone formed by a thicker layer of parallel, striated muscle fibers.

A concentration of nervous tissue, in the center of the scolex, with nerves distributed anteriorly, posteriorly, and to the proboscides may be apparent in favorable, methylene-blue-stained specimens.

Scattered flame bulbs and an anastomosis of clear excretory canals near the center will be apparent in living scoleces. Opaque granules, presumably identical to those embedded in the mesenchymatous parenchyma, often occur in the larger excretory canals. Moreover, similar granules pack the cyst in which the larva is held and seemingly are excretory substances produced by the worm.

Life Cycle. The final host of *Otobothrium crenacolle* is the hammerhead shark, *Sphyrna zygaena*, which feeds on butterfish. These encysted worms are absolutely harmless to human beings, and the host fish may be eaten with impunity. When swallowed by the shark, the cyst is digested and the scolex attaches to the wall of the spiral valve. Proglottid formation ensues, and sexual maturity is gradually attained. Finally, fertilized ova pass back into the sea to infect other intermediate hosts.

Diphyllobothrium latum
By C. G. Goodchild

The broad or fish tapeworm of man, *Diphyllobothrium latum*, a widely distributed, medically important parasite, is the largest human cestode, often exceeding eight meters in length.

A whole mount, a scolex, and stained mature and gravid proglottids should be available. Cross sections through a mature proglottid should also be examined. The anterior scolex is spoon-shaped and possesses dorsal and ventral longitudinal grooves, called bothria, with apposing lips which anchor the worm in the small intestine. A brain-like cerebral commissure with anterior nerves and large nerve trunks which extend posteriorly through the lateral margins of the body may be apparent in favorable specimens. The body is composed of these successive regions: an unsegmented neck, a region of immature proglottids, a zone of mature proglottids, and a region of gravid proglottids.

Mature Proglottid. The whole mount and the cross section of the mature proglottid should be studied for the following account. The

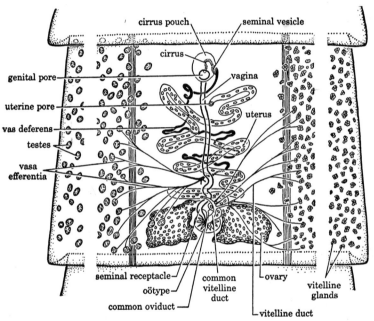

FIG. 119. *Diphyllobothrium latum*, surface view of central and peripheral zone of a mature proglottid. (Modified after Sommer and Landois.)

mature proglottid (Fig. 119) is broader than long. It is covered, as is the rest of the worm, with a thick, homogeneous cuticula secreted by a subcuticular layer of spindle-shaped cells. Longitudinal muscle fibers immediately under the surface and inner, transverse muscle fibers running from edge to edge are visible in the cross section and in favorable whole mounts. Tapeworms are usually well provided with muscles which assist the scolex in maintaining them in the intestine of their host by vigorous, forward-creeping movements. These motions oppose the posteriorly directed peristaltic waves which tend to eliminate the parasites. A typical, dense parenchyma, consisting of stellate cells surrounded by irregular fluid-filled spaces, fills most of the body. Some of the parenchymal cells are distended with calcium carbonate granules. A protonephridial excretory system consists of lateral longitudinal canals and transverse vessels which receive finer tributaries that originate in flame bulbs.

Reproductive Systems. The hermaphroditic reproductive organs are the most conspicuous structures in the proglottid.

Male Reproductive System. Many follicular testes, each approximately 50 μ in diameter, lie in a lateral band on each side of the proglottid. A thin vas efferens runs from each testis to the center of the proglottid, where it joins the other vasa efferentia in a small swelling which is also the base of a vas deferens. The sinuous vas deferens courses medioanteriad to near the anterior edge of the proglottid, where it expands to form a thick-walled seminal vesicle before proceeding into a cirrus pouch. The sperm duct narrows and coils in the cirrus pouch and then terminates as a strong, muscular copulatory cirrus. During copulation, the cirrus is protruded through the genital pore, located on the midventral line one-fifth of the proglottid length from the anterior edge, and transfers spermatozoa to the female aperture of an adjacent proglottid.

Female Reproductive System. A ventral, bilobed ovary, consisting of loose ovigerous tubes, lies in the medial, posterior third of the proglottid. A narrow oviduct leaves the ovarian isthmus and proceeds posteriad a short distance into an interovarian oötype. This latter egg-fashioning organ receives several additional female ducts: a narrow, sperm-transporting vagina runs directly posteriad from the genital pore to a point near the oötype, where it coils and distends, forming a seminal receptacle which in turn joins the oviduct. The combined vagina and oviduct immediately meet a common vitelline duct which is formed by left and right vitelline ducts running mediad from many spherical, dorsolateral vitelline glands. In the oötype, fertilization and the formation of an eggshell occur, with the latter process supposedly aided by secretions from encircling shell glands. A wide, egg-filled uterus leaves the oötype and follows a sinuous course anteriorly. The uterus terminates at the midventral uterine pore which is located slightly behind the genital pore. Immature, operculated, straw-colored eggs are passed intermittently from the uterine pore into the intestinal lumen of the host.

Gravid Proglottid. In the gravid proglottid the uterus is distended with eggs and has four to six compactly arranged uterine loops which form a central, rosette-shaped structure.

The posterior gravid proglottids, with their masses of eggs, are not ordinarily passed out of the host either singly or in strips but rather are sloughed off and release their eggs to mix with the intestinal contents. A single worm is capable of producing 1,000,000 eggs a day.

Larval Stages. Eggs deposited in fresh water develop for about 2 weeks and then hatch to liberate tiny, spherical hexacanth embryos, called coracidia, which swim by means of a ciliated outer envelope. These larvae are swallowed eagerly by certain species of copepods in the enteron of which the ciliated membranes are discarded and typical, naked hexacanth embryos appear. These larvae claw through the gut wall and establish themselves in the hemocoel where they become solid, procercoid larvae. Examine whole mounts of infected copepods, if available, and study the procercoids. Note an elongated oval body which has an apical, cup-shaped indentation, considered by some authorities to be the remnant of an ancestral mouth, and a caudal appendage, the cercomere, provided with the six larval hooks.

In about 2 weeks the procercoid larvae are infective for fresh-water fish. When an infected copepod is eaten by a plankton-feeding fish, the larval cestode discards the cercomere and bores through the stomach wall, eventually reaching the muscles of the body wall. Here it transforms into a non-encapsulated, worm-like, plerocercoid or "sparganum" larva which is glistening white in color and has a short, inverted groove on the anterior end which is the rudiment of the scolex. Examine, if available, museum mounts of preserved spargana and infected fishes. Note the occurrence of the flattened parasites in the myosepta of the host.

Mammals, including human beings, dogs, and bears, become infected by eating flesh containing living larvae. The anterior end of the plerocercoid everts and attaches to the small intestine; strobilization and maturation of the body follow rapidly. In approximately 20 days eggs are produced and are discharged in the feces to continue the cycle.

Taenia pisiformis
By C. G. GOODCHILD

The cyclophyllidean tapeworm, *Taenia pisiformis* (Fig. 120), from the small intestine of the dog, is selected as an example of the taenioid cestode because it is easy to obtain and is typical of the group. However, because the organs in this genus are so consistently arranged, other taenioid species can be studied with the following directions.

As this representative tapeworm is studied, compare it with the turbellarians and trematodes previously encountered. For example, digenetic trematodes may be found in almost any tissue or organ in the host's body, whereas adult cestodes, which have no digestive sys-

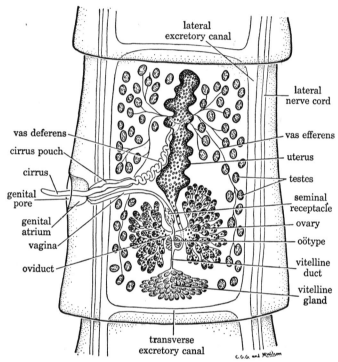

FIG. 120. *Taenia pisiformis,* surface view of mature proglottid.

tem, are found almost exclusively in the lumen of the alimentary canal. The quantity and quality of food ingested by the host are of vital concern to the cestode parasite, and rapid elimination of tapeworms often occurs in a starved host.

Examine macroscopically a whole specimen and note a small, anterior, rounded scolex and a thin, serrated, ribbon-like strobila or body which is narrower near the scolex than at the other end. The body consists of a thin, unsegmented neck region and a series of segments, called proglottids, which increase in size from their place of origin in the posterior neck region to their place of detachment at the posterior end.

Study microscopically the scolex and representative proglottids from prepared slides. Material for study of the proglottids should include an immature one from the region behind the neck, a mature one with functioning sexual organs from the middle zone, and a gravid proglottid packed with eggs and with degenerating sexual organs from the posterior zone.

Scolex and Neck. The attachment end of the tapeworm, the scolex, is somewhat square in front view and has one large, deep, well-defined sucker at each of the four corners.

Apically there is a retractile rostellum having a cup-shaped base and bearing two rows of alternating larger and smaller sharp, recurved hooklets. Study the hooklets under high power, and by careful focusing determine the approximate number. Select one hooklet near the edge and study it critically. Find the L-shaped body embedded in the tissue and the exposed claw-shaped extension of the longer arm. The scolex often uses the hooklets to claw its way beneath the host's mucosa, which makes medical elimination of the entire worm extremely difficult. Externally, the scolex, as well as the rest of the body, is covered with a thick homogeneous cuticula. The interior is filled with muscle fibers and parenchyma and, in addition, has a system of clear, interconnecting excretory canals which join lateral, longitudinal excretory canals which pass posteriad to the end of the body. The chief concentration of nervous tissue occurs in the scolex. Favorable whole mounts may show a pair of cephalic ganglia behind the suckers with nerve connections to postrostellar and postsucker nerve rings. The nerve rings in turn are connected by lateral, longitudinal nerves which continue posteriorly parallel to the excretory canals into the proglottids.

The unsegmented neck region is narrower than the scolex which it joins. Posteriorly it gives rise to the proglottids which, at first, are separated by shallow, incomplete grooves and are broader than long.

Immature Proglottid. The rudiment of the sexual organs makes its first appearance from the parenchyma of segments 25–30 as a slender, median cord of compact nuclei embedded in a mass of fibers. As succeeding proglottids are studied, other developmental stages are noted. Two secondary cell groups, the *anlagen* of vas deferens and vagina, grow peripherally from the first mass. The vas deferens ends blindly in the center but later receives the vasa efferentia, seldom visible in whole mounts, from many testes already starting to form as small isolated groups of cells laterally. The vagina grows posteriorly and medially and ends in an enlargement, the rudiment of a shell gland, oviduct, and proximal uterus. An ovary begins as a fan-shaped cell mass near the anterior edge of the enlargement, and a yolk gland forms behind the enlargement. From the proximal uterus a cord of uterine cells extends in the midline to the anterior edge of the proglottid.

Mature Proglottid. The mature proglottid (Fig. 120) possesses completely formed reproductive organs and should be studied in detail from whole mounts and cross sections. Observe in the whole mount the almost square form with overhanging posterior edges, and in the cross section the dorsoventrally flattened shape.

In the sectioned proglottid notice the thick investing cuticula, secreted by the hypodermis, and the underlying muscles arranged in several layers: there are outer and inner longitudinal layers between which is a layer of transverse fibers extending from edge to edge; dorsoventral bundles are also present. Tapeworms removed from their host often seem sluggish and non-muscular; in the host, however, they actually are capable of rather vigorous, forward-creeping movements. These movements nullify the efforts of peristalsis, which tend to eliminate all intestinal contents. The overhanging posterior margins of many cestode proglottids are probably also of importance in these sustaining muscular actions. A mesenchymatous parenchyma, certain cells of which in cestodes often produce opaque calcareous bodies, fills most of the body space between the transverse muscles. Reproductive organs and the lateral excretory canals and nerve cords will also show in the section.

Reproductive Systems. Examine a whole mount of one proglottid and note the extensive hermaphroditic reproductive systems, which you will find resemble those of the digenetic trematodes.

Male Reproductive System. Many small, peripherally located testes are connected by small, dendritic vasa efferentia to a vas deferens. This larger sperm duct begins near the center of the proglottid and proceeds as a convoluted tubule to the base of a marginal cirrus. An ovoid cirrus pouch containing prostate cells surrounds the cirrus which opens into a common genital atrium and to the outside through a marginal genital pore located in the center of a genital papilla. The genital pores alternate from one side to the other irregularly in the proglottid chain.

Female Reproductive System. Curving posteromediad from the genital atrium is a narrow, deeply stained vagina which enlarges to form a small seminal receptacle near its inner end. A two-lobed ovary, located in the posterior third of the body, is composed of loose, radiating cords of germinal cells. As ova mature, they move to a narrow, interconnecting, ovarian isthmus and into a short, posteriorly directed oviduct. The oviduct shortly joins the vagina, now functioning as a fertilizing duct from the seminal receptacle, and the combined fertilization duct continues posteriad into a small, interovarian oötype. The oötype is an important egg-fashioning organ which externally is provided with unicellular shell glands and internally receives additional female ducts. In the posterior midregion of the proglottid is a compact, transversely oblong vitelline gland which contributes granular vitelline material to the fertilized ova in the oötype via an anteriorly directed, common vitelline duct. Fully formed ova pass from the oötype into a median, club-shaped uterus which proceeds forward sinuously and ends blindly near the anterior margin of the proglottid. As additional eggs pass into the uterus, it becomes greatly distended and finally nearly fills the whole proglottid.

Clear marginal excretory canals and a transverse connecting canal at the posterior margin of each proglottid will usually be evident. Longitudinal nerves parallel the lateral excretory canals; they can usually be seen in the cross sections if not in the whole mounts.

Gravid Proglottid. The gravid proglottid is more than twice as long as it is broad. The egg-filled uterus usually develops ten to twenty lateral arms which subbranch and are likewise packed with eggs. The non-functioning gonads and accessory organs have usually degenerated and are not visible; however, the marginal genital pore and terminal male and female ducts persist but are likewise non-functional.

The most posterior gravid proglottids are detached from the worm and pass to the outside in the feces. They die through desiccation and liberate their resistant, infective eggs which contaminate the herbage on which rabbits and hares feed.

Larval Stages. The cestode egg hatches in the gut of the rodent and liberates a six-hooked embryo, the onchosphere, which claws through the mucosa into a lymph vessel and is transported to the liver or other abdominal viscera. A slow growth of the larva produces an ovoid, hollow vesicle, called a bladderworm or cysticercus, having a localized thickening of the wall from which develops an internal, inverted scolex.

Examine a whole and a sectioned cysticercus. Notice the thin bladder wall and the single inverted scolex invaginated into the fluid-filled bladder from a stalk-like neck. Understand that the suckers and hooks located on the inside in the bladder are everted in the small intestine of the dog, when swallowed, and anchor the scolex to the intestinal mucosa. The bladder is digested and a strobila is produced by the activity of the proliferous neck region.

Echinococcus granulosus
By C. G. Goodchild

The adult of this species is a small, benign, taenioid cestode living in dogs, in other species of Canidae, or in Felidae. The larval stage, the hydatid, is normally a parasite of sheep, but almost any mammal will serve as a host. It is potentially dangerous to human beings, in whom it may give rise to a fatal, metastasizing, cancer-like tumor.

Stained adult worms and sections of the hydatid stage should be available for study.

In the whole mount of the adult (Fig. 121), note a typical taenioid scolex with four deep, well-defined suckers and an anteriorly situated, retractile rostellum bearing a crown of approximately thirty to thirty-six hooklets. Examine a hooklet in side and face view, under high power, and note the embedded L-shaped body consisting of a flared shorter arm and a longer arm bearing an internal, anterior knob and an external recurved claw. The scolex possesses internal concentrations of parts of the nervous and excretory systems, as well as a muscular fiber network and a dense parenchyma all of which are also well represented in the body. Behind the scolex is a tapered, unsegmented, proliferous neck region. The body of the worm consists of three, or at the most four, proglottids, one immature, one mature, and one gravid. The immature segment is slightly wider than the neck region and has a cluster of primordial germinal cells in the center.

Mature Proglottid. The second proglottid is wider and about twice as long as the first one, and contains a full complement of taenioid-like male and female reproductive organs.

Male Reproductive System. In the central portion of the proglottid, encircling the other reproductive organs, are approximately twenty to thirty spherical testes each with a fine vas efferens not usually visible in whole mounts. In the postmedial portion of the proglottid the vasa efferentia join to the base of a vas deferens, an extremely convoluted, sperm-filled tube, which passes laterad to an enlarged, coiled, marginal cirrus lying in a bulbous cirrus pouch. The cirrus, during copulation, is protruded through a lateral genital atrium and genital pore.

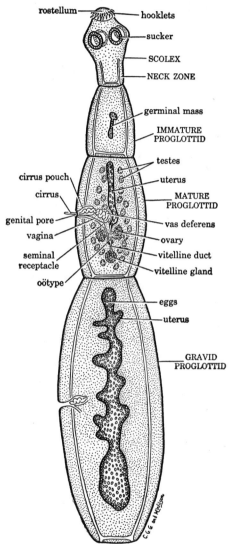

FIG. 121. *Echinococcus granulosus,* surface view of whole worm.

Female Reproductive System. Opening into the genital atrium closely behind the cirrus is a female copulatory tube, the vagina. The open end of the vagina is guarded by a muscular sphincter. Beyond the sphincter, the vagina passes medioposteriad and enlarges at its inner end to form a sperm-filled seminal receptacle. A third of the proglottid length from the posterior margin is a bilobed ovary with a narrow, central isthmus from which posteriorly is given off a short oviduct. The oviduct joins a duct from the seminal receptacle, and this combined fertilization duct continues posteriad into a median, compact oötype. The oötype is surrounded with unicellular shell glands and internally is a region of confluence of additional female ducts. A median, common vitelline duct penetrates the oötype and transports particulate vitelline material, to supply the ectolecithal eggs, from a posterior bilobed vitelline gland. The two lobes of the vitelline gland are oriented dorsoventrally, and in surface view this organ appears spherical or oblong because only the uppermost lobe is visible. After an eggshell is laid down, the egg passes from the oötype into a central, tube-shaped uterus which extends to a point near the anterior border of the proglottid.

Gravid Proglottid. The last proglottid, which is usually the third one, is gravid. It is wider and at least twice as long as the mature one. The egg-filled uterus usually develops lateral evaginations and assumes the appearance of a loosely twisted tube. The uterus eventually bursts to discharge the eggs in the small intestine of the carnivore host.

Larval Stage. The egg, upon being swallowed by a mammalian intermediate host, hatches in the duodenum as a typical, six-hooked onchosphere which attacks and penetrates the mucosa. The larva gains access to a portal vein and is transported passively to the liver. Implantation and growth occur here, or perhaps in the lung if the capillary filter system of the liver is traversed.

Larval cysts of several shapes may develop in this life cycle, but only the innocuous, so-called unilocular hydatid will be described. This hydatid is a spherical vesicle completely filled with clear fluid and having a wall of two layers. Examine the histology of the sectioned cyst wall (Fig. 122), and note an outer, amorphous, concentrically laminated ectocyst, which is surrounded by a tight, fibrous, cellular envelope, the pericyst, of host origin. The inner wall of the hydatid, the endocyst, is a thin syncytium from which brood capsules are budded off. Select a large, hollow brood capsule still attached by a narrow pedicel to the inner wall and examine it critically under high power. The pedicel, the inner wall of the hydatid, and the outer wall of the brood capsule consist of the continuous syncytium. Notice

ECHINOCOCCUS GRANULOSUS 207

that scoleces are formed from localized, evaginating thickenings on the inner surface of the wall of the brood capsule; they increase in size and for a time, at least, remain connected to the brood capsule by a narrow peduncle or stalk. Now study the scoleces in the brood capsule. Because they lie in all possible positions, end views and lateral

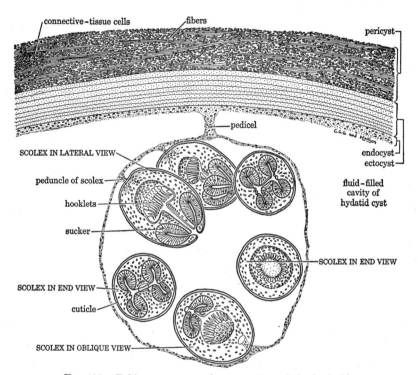

FIG. 122. *Echinococcus granulosus,* portion of the hydatid.

views can be obtained. Study one attached scolex in lateral view. Notice that the scolex is invaginated, i.e., hooklets and suckers are on the inside, and that, except for the posterior stalk, it is already covered with a thick, secreted cuticula. The anterior end is inturned to form a central canal into which the four suckers open and into the expanded base of which the hooklets project. The body of the scolex is filled with mesenchyme cells aggregated to form a subcuticular layer and to form a compact cup-shaped mass below the hooklets. The syncytium of the endocyst retains its embryonic potentialities, and new brood capsules may be produced indefinitely. Hydatids thus may expand to enormous size and contain nearly 50 liters of fluid as well as thousands of scoleces. The pressure so produced may severely damage adjacent host organs.

The infective brood capsule, when swallowed by the carnivorous host, is digested away, exposing the scoleces which quickly evaginate and attach to the duodenal mucosa. Proglottid formation and egg production ensue, and the cycle is begun again.

5.

Rhynchocoela and Acanthocephala

RHYNCHOCOELA

Amphiporus ochraceus

By C. G. Goodchild

The usually flattened and elongated proboscis worms, of which *Amphiporus* is an example, are free-living or parasitic. The free-living species are carnivorous worms living in the ocean, in fresh water, or on land. Many nemertine species are large and opaque and are not favorable for study because the internal organs are difficult to see. The smaller *Amphiporus* is selected for study because, in whole mounts, the internal structures are easy to observe, because it is commercially available, and because in gross morphology it resembles closely the fresh-water nemertine species, *Prostoma rubrum*, which can be rather easily cultured in inland laboratories (by feeding with small crustaceans and nematodes). However, these directions with slight, readily apparent modifications can be used to study other small nemertines often encountered, including *Tetrastemma* and *Zygonemertes*.

Living worms or preserved whole mounts, as well as cross sections through the midregion of the body, should be available for study. If time permits and living specimens are available, regeneration experiments should be attempted. Try cutting animals in two through the middle, and also cut off pieces anterior to the brain. Isolate the fragments in clean water which should be periodically changed.

The slender, yellowish *Amphiporus ochraceus* is dorsoventrally flattened and is 20 to 70 mm. long by 2 to 3 mm. wide (Fig. 123). The anterior end is rounded and is provided with six to fourteen conspicuous ocelli on each side which are often arranged in several short divergent rows and do not extend posteriorly beyond the brain. The posterior end of the worm tapers to a rounded point.

Body Wall and Parenchyma. The body wall (Fig. 124) consists of an outer layer of ciliated columnar epithelium and gland cells rest-

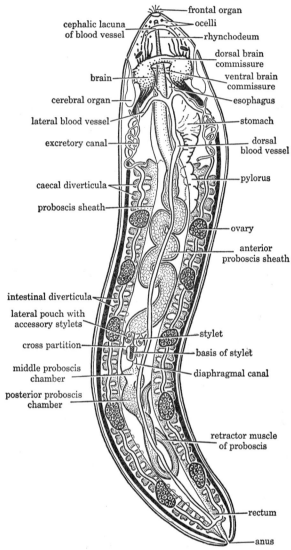

FIG. 123. *Amphiporus pulcher*, dorsal view of female. (Modified after Bürger.)

ing on a basement layer of hyaline connective tissue. Underlying the basement membrane are two layers of somatic muscles consisting of an outer circular and an inner longitudinal layer. Dorsoventral

AMPHIPORUS OCHRACEUS

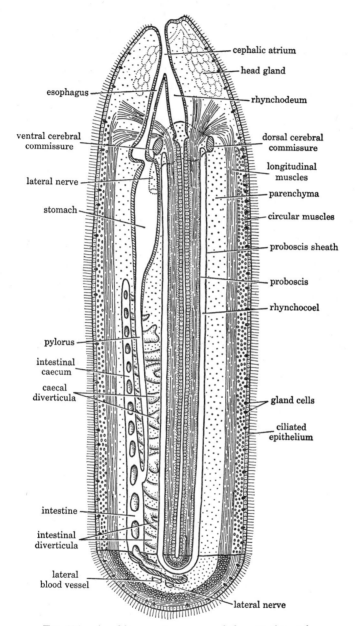

Fig. 124. *Amphiporus*, stereogram of the anterior end.

muscles may be apparent in the cross section. A mesenchymatous parenchyma, like that in flatworms, occupying the spaces surrounding the body organs should be identified and studied.

Proboscis and Proboscis Sheath. The proboscis is a long eversible organ of a formidable offensive and defensive nature contained, in its retracted state, in a proboscis sheath located dorsal to the gut. The proboscis is formed by an invagination of the anterior body wall, and the muscular layers, therefore, are continuous with those of the body-wall muscles. A retractor muscle originates on the posterolateral wall of the proboscis sheath and is inserted on the tip of the retracted proboscis. This retractor muscle aids the return of the everted proboscis into the sheath. The center of the proboscis is hollow and is lined with a glandular epithelium. In *Amphiporus* the proboscis is large in comparison with the body, is provided with sharp-pointed stylets, and has an anterior duct, the rhynchodeum, which opens anteriorly. Three distinct anteroposterior chambers occur in the proboscis. An anterior chamber is in the eversible region and terminates posteriorly at a cross partition which bears a sharp, median, conical stylet with a pear-shaped basis embedded in the tissue and which bears also two lateral pouches containing as many as four accessory stylets. The cross partition, or diaphragm, is perforated by a slender canal which connects the anterior and middle chambers and transports immobilizing or other types of secretions from a large posterior chamber to the base of the everted stylet.

The proboscis sheath is a closed tube with a wall of several muscular layers and a central cavity, the rhynchocoel, which houses the proboscis, and is filled with a corpusculated fluid. The sheath is nearly as long as the body and by muscular pressure on the fluid within causes a rapid eversion of the anterior portion of the proboscis. Only the region as far posteriorly as the diaphragm turns inside out as the proboscis is extended.

Digestive System. An anus is present in nemertines, and food materials such as minute organisms or body juices of larger prey pass from mouth to anus in an orderly fashion; accordingly, the enteron, which is located ventral to the proboscis sheath, is differentiated anteroposteriad into storage, digestive, absorptive, and eliminative regions. It is lined with ciliated columnar epithelium throughout.

The mouth opens ventrally into the rhynchodeum. A narrow esophagus connects to an expanded stomach, and the latter gives off posteriorly a tapered pylorus which parallels the intestine some distance dorsally before joining it. The blind anterior end of the intestine, the intestinal caecum, is provided with lateral, branched diverticula, as is the intestine proper. Posteriorly the intestine joins a tapered rectum which opens at the terminal anus.

Excretory and Vascular Systems. *Amphiporus* has a protonephridial excretory system with a pair of lateral collecting tubules and many terminal tributary ducts and flame bulbs. Parts of this system will be apparent in living organisms and may be seen in sections, but will be generally invisible in stained whole mounts.

A blood vascular system is present in nemertines. This system consists of a single dorsal vessel and paired lateral vessels which join immediately behind the brain before sending a narrow, cephalic, lacunar loop forward into the interocular region. In the intestinal region there are regularly arranged transverse vessels, alternating with the gut diverticula, which connect the dorsal and lateral vessels. A colorless, vascular fluid containing floating, nucleated corpuscles fills the blood vessels. This fluid ebbs and flows with bodily movements; there is no definite vascular circuit and a heart is lacking, although muscle fibers in the walls of the blood vessels occur generally over the whole system.

Nervous System and Sense Organs. In correlation with the active, predaceous life of *Amphiporus* the nervous system and sense organs are highly developed. The nervous system consists of a four-lobed brain in the cephalic region with fused dorsal and ventral paired ganglia and with commissures above and below the rhynchodeum uniting the ganglionic pairs. Five main longitudinal nerves proceed posteriorly from the brain. Large paired lateral trunks, from the ventral ganglia, will be visible subjacent to the longitudinal body muscles even in whole mounts; they are connected, in the rectal region, by a large dorsal commissure. A smaller dorsomedian nerve, from the dorsal commissure, and paired dorsolateral nerves, from the dorsal ganglia, will be visible in the cross section, if not in whole mounts. A network of nerve fibers connects these various trunks. In addition to these main nerve trunks other peripheral motor and sensory nerves emerge from the central nervous system. These peripheral nerves include general somatic nerves and, in addition: (1) cephalic nerves, from the dorsal cerebral ganglia, to the ocelli and other sense organs of the head; (2) esophageal nerves, from the ventral brain lobes, to the mouth and esophagus; (3) ten proboscidial nerves, from the ventral cerebral region, to the proboscis.

The outer surface of the body is provided with diffuse, generalized sensory cells, while the head bears ocelli, cerebral sense organs, and cephalic grooves. Each of the many ocelli is composed of a cup-like group of sensory cells surrounded by reddish brown pigment cells and connected to the cephalic nerves. The paired cerebral sense organs consist of ventrolateral surface pores on the head and small canals leading medioposteriad to expanded, highly specialized receptor organs lying in contact with the dorsal lobes of the brain. The cephalic grooves located on the dorsal surface of the head are V-shaped.

Reproductive Organs. Nemertine worms are usually dioecious and, in contrast to most Platyhelminthes, have simple reproductive organs. Gonads and germ cells develop from lateral mesenchyme between the gut diverticula. At maturity the testes are distended with many spermatozoa, and each ovary is distended with several large, greenish yellow ova. A narrow duct runs from each gonad to open on the dorsolateral body surface. Eggs and sperm are discharged into the water, where fertilization occurs. Marine species of nemertines often have a free-swimming larva, the pilidium. Varieties of *Prostoma rubrum* are hermaphroditic and possess well-developed ovotestes; in this fresh-water species the egg hatches and liberates a small worm which is a miniature of the adult.

ACANTHOCEPHALA

Neoechinorhynchus emydis

By C. G. GOODCHILD

The Acanthocephala, or spiny-headed worms, as adults are widespread, obligate parasites of vertebrates. As larvae they parasitize invertebrates, chiefly arthropods and occasionally mollusks. They are parasites without a free-living stage and are reduced morphologically to a reproductive sac devoid of a digestive system and of conspicuous receptor organs.

Neoechinorhynchus emydis, the form selected for study, is an intestinal parasite of various species of North American fresh-water turtles. Worms may often be recovered alive from these hosts even after they have been retained in the laboratory for as long as 6 months without food. Examine the small intestine of a recently killed turtle for acanthocephalans, and when a specimen is located note the flattened form of the body and the tenacious hold on the mucosa afforded by the proboscis. Remove the worm, being careful not to injure the proboscis, to a large drop of 0.85 per cent sodium chloride solution on a glass slide and cover with a coverglass which should be rim-sealed with Vaseline to prevent evaporation. Identify the organs mentioned in the following discussion in the wet-mount preparation or in well-stained permanent specimens. If *N. emydis* is not available, then other species of small acanthocephalans may be studied by making minor changes in the following account.

Adults. The elongated body consists of two regions, an anterior praesoma and a posterior, cylindrical trunk. The praesoma, in turn, is composed of an anterior, spinous proboscis and a neck. The proboscis in *N. emydis* is relatively small and globular, while the neck is unarmed and is a transitional boundary between the proboscis and trunk. The neck zone is marked by a cuticular infolding which isolates the subcuticular tissues of the trunk wall from those of the neck and proboscis. The female (Fig. 125) is 10–32 mm. long by approximately 0.7 mm. wide; the male (Fig. 126) is as wide but is somewhat shorter.

The retractile proboscis is provided with three transverse series of six hooks each. The hooks of the terminal series are not arranged in an exact ring; one hook on each lateral surface of the proboscis is attached at a level a little posterior to the other hooks of the terminal series. The six large hooks of this series are strongly recurved and are provided with flared, reflexed root processes. The hooks of the remaining two rows are more delicate and are non-flared basally. A proboscis receptacle into which the proboscis can be retracted is short and has a wall composed of a single layer of circular muscle fibers. Protrusion of the proboscis is assisted by the contraction of these fibers.

The smooth body is covered with a thin cuticula subjacent to which is a very much thicker hypodermis. The Acanthocephala are unique in possessing, in the hypodermis, a ramifying system of fluid-filled lacunae and, in *Neoechinorhynchus,* a fixed number of giant nuclei which often cause slight protuberances of the body wall. Invariably, in this species, five nuclei are located in the mid-dorsal line and one in the midventral line. Moreover, the hypodermis of the neck is confluent with two large organs, the dorsal and ventral lemnisci, which lie in the body cavity beside and behind the proboscis receptacle. The lemnisci likewise have a fixed number of giant nuclei; two nuclei occur in the dorsal lemniscus, but only one nucleus is found in the ventral lemniscus. These structures apparently function as reservoirs for lacunar fluid which assists the retraction and protrusion of the proboscis.

Digestive and excretory systems are lacking, although modified protonephridial organs do occur in some of the largest species of

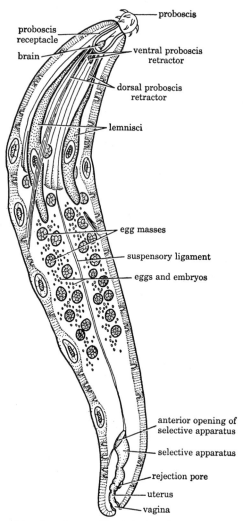

FIG. 125. *Neoechinorhynchus rutili*, adult female. (Modified after Meyer.)

Acanthocephala. Food materials and waste products diffuse through the body wall. The weak body-wall muscles, lying subjacent to the hypodermis, consist of outer transverse and inner longitudinal layers. Conspicuous muscle bundles, located anteriorly in association with the proboscis, should also be studied. Extending the length of the proboscis receptacle from the tip of the proboscis to the wall of the receptacle posteriorly are the invertors of the proboscis. Continuations of these same muscles, the retractors of the proboscis, pass through the

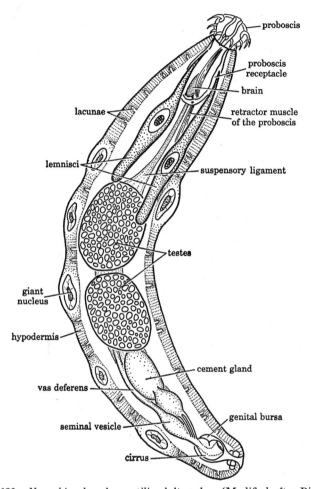

FIG. 126. *Neoechinorhynchus rutili*, adult male. (Modified after Bieler.)

receptacle wall, one dorsally and the other ventrally, to insert on the wall of the trunk.

The nervous system is reduced to an oval ganglionic brain in the wall of the receptacle of the proboscis and two nerve trunks, the retinacula, which emerge from the brain and run obliquely posteriad to the body wall.

As in all Acanthocephala, the sexes are invariably separate. Reproductive organs are located in a suspensory ligament, a cord-like structure, traversing the body cavity from the wall of the receptacle of the proboscis to the posterior region of the body. The reproductive systems will be described separately.

Male Reproductive System. Two ovoid testes are located tandem in the ligament slightly behind the middle of the body (Fig. 126). A vas deferens passes posteriad from the testes and proximally is paralleled by an elongated cement gland consisting of a syncytial mass with eight giant nuclei. A sac-like cement reservoir receives the secretion of the cement gland. The vas deferens passes posteriorly into an expanded sperm-filled seminal vesicle. An ejaculatory duct from the vesicle and paired ducts from the cement reservoir become confluent before penetrating a posterior, cone-shaped cirrus which is usually held in a retracted genital bursa. During copulation the genital bursa is extruded to form a bell-shaped holding structure.

Female Reproductive System. Most young acanthocephalan females possess one or two oval or elongated ovaries in the ligament; in this genus, however, only a single, elongated, postmedian ovary is present. The ovary is visible only in a young specimen because very early in life it fragments into many egg masses which henceforth float freely in the fluid of the body cavity where fertilization and embryogeny also occur (Fig. 125). Eggs and all stages of small embryos will be apparent in the body cavity of a mature female.

Fully formed embryos pass to the outside through a posterior female genital tract held in place by the ligament and consisting of three regions, a selective apparatus, a uterus, and a terminal vagina. The most anterior region of the selective apparatus is a funnel-like uterine bell which opens into a ventral division of the body cavity and into which eggs and embryos pass. Eggs and immature embryos are selectively returned to a dorsal division of the body cavity through a rejection pore at the posterior end of the selective apparatus. Mature embryos, however, are admitted into the thin-walled uterus and from here are passed into the thicker-walled vagina. The vaginal aperture is guarded by a sphincter and is located subterminally on the ventral surface.

Life Cycle. Within the egg membranes the early embryo develops into a larva, the acanthor, which is characterized by an apical crown of rostellar hooks. These hooks are strictly larval structures and have no continuity with the proboscis hooks of the adult. Functionally, at least, the hooks are similar to those of larval cestodes. The acanthor hatches only upon being ingested by an appropriate invertebrate host. Here a gradual metamorphosis occurs to form a second larval stage, the acanthella. By successive instars the acanthella becomes an infective juvenile with adult organs in rudimentary form. When the juvenile is eaten by the vertebrate host, attachment to the enteric mucosa and growth to the sexually mature worm take place.

The life cycle of *N. emydis* is incompletely known. First-stage larvae have been reported from a species of ostracod, while older stages, the juveniles, have been found in several genera of freshwater snails.

6.

Aschelminthes

Hydatina senta

By C. G. Goodchild

The rotifers, of which *Hydatina* is a classic example, are among the smallest of the metazoans; in fact, many rotifers are smaller than the larger protozoans. Rotifers are mainly free-living organisms encountered in abundance in marine and fresh-water habitats, even in temporary puddles. Other than being links in the food chain of fish, they are of little economic importance. They are of biological interest because of their ability to withstand desiccation, their use in genetical experiments, and the total absence of males in certain species.

Hydatina senta is a rather large fresh-water rotifer, occurring often in aquatic samples collected in the field. Living cultures or stained whole mounts may be purchased from commercial sources. Details of its anatomy are most easily understood from a study of the living organism; however, stained whole mounts are nearly as satisfactory.

Adult. The pear-shaped body of the adult (Fig. 127) consists of three regions: an anterior head, a middle trunk, and a posterior foot. The head consists chiefly of a flattened ciliated surface, the trochal disk. The trochal disk or corona is large and is bordered by a double ciliated ring, the velum. The anterior or preoral ciliary band is the trochus. The posterior or postoral band is the cingulum. Between the two zones is a groove which in *Hydatina* is evaginated to form five lappets bearing bristle-like setae or vibratile styles of perhaps a sensory function. The interlappet groove area is finely ciliated and leads into the mouth. The coronary cilia are used in locomotion and in creating vortical water currents which bring food particles within reach of the mouth. An optical illusion of two continuously turning wheels is often produced by the cilia of the velum. This illusion results from a greater visibility of the cilia during their slow recovery stroke than during their rapid effective stroke. A thin cuticula, with telescoping sections, covers a trunk region in which the chief visceral organs occur. The tapering foot contains a pair

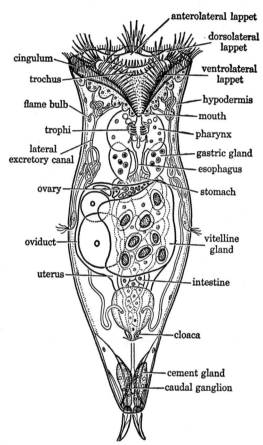

Fig. 127. *Hydatina senta*, ventral view of adult female. (Compiled and modified after Plate, de Beauchamp, and Remane.)

of conspicuous cement or pedal glands whose viscous secretion temporarily anchors the terminal, bifurcated, forceps-like toes to the substratum during certain feeding activities.

Integument and Body Wall. A thin cuticula is secreted by a subjacent syncytial hypodermis. Definite layers of muscles in the body wall are absent, although muscle bundles of several sorts traverse a fluid-filled pseudocoel.

Digestive System. An anterior mouth opens into a transversely elliptical, muscular pharynx or mastax which is an efficient chewing organ provided internally with hard jaws called trophi. The trophi (Fig. 128) include a midventral incus or anvil and two lateral mallei or hammers. The incus consists of a midventral piece, the fulcrum, from which run anterodorsally two pieces, the rami, which are hinged to the fulcrum. Each malleus likewise has two parts, a handle-like manubrium embedded in the muscular wall and a toothed claw, the uncus, extending into the cavity and coming in contact with the incus. Muscular movements cause a reciprocating masticating action between the clawed uncus and the median incus. Pulverized food is passed through an esophagus into a sac-like stomach which receives the secretions of paired anterolateral gastric glands. A straight intestine passes from the stomach to a cloaca which opens to the dorsal surface in the posterior trunk region. In the living animal actively beating cilia will be seen lining the digestive cavity; these cilia assist the movement of food through the tract.

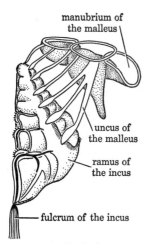

FIG. 128. *Hydatina senta,* bisected trophi in superior view. (Redrawn after de Beauchamp.)

Excretory System. This system is protonephridial in type and consists of terminal ciliated flame bulbs and paired efferent ducts opening to the surface. The paired convoluted excretory canals roughly parallel the lateral edges of the body and are formed of a succession of elongated, perforated cells. Opening into these ducts are irregularly placed, blindly ending, tag-like appendages having the form of flame bulbs. The lateral canals connect by an anterior renal commissure above the mouth and join a posterior, pulsatile bladder which periodically discharges its fluid contents to the outside through the cloaca. It has been estimated that some fresh-water rotifers may eliminate an amount of fluid equivalent to their own volume every 9 minutes.

Nervous System and Sense Organs. A suprapharyngeal ganglion constitutes the brain of the animal. From this brain, lateral and ventral pairs of nerves run the whole length of the body. An anterodorsal antenna and, near the middle of the body, paired lateral antennae, each consisting of a few sensory hairs borne on a slight prominence, as well as the vibratile styles of the lappets, are important tactile organs. Eyes, which are often present in rotifers, are completely absent in *Hydatina*.

Reproductive System. The male of *Hydatina* is well known, but because it is rarer than the female it will not be described.

The most conspicuous reproductive organ in the female is a sacculate and granular vitelline gland or vitellarium lying ventral to the stomach and possessing eight large nuclei. The ovary is a narrow cord of cells lying against the anterior edge of the vitelline gland. A dilated oviduct extends posteriorly along a lateral edge of the vitelline gland to open into a uterus. One or two eggs laden with yolk granules distend the oviduct, while smaller germinal cells extend to the anterior end of the ovary. Eggs are periodically discharged into the uterus and through the cloaca to the outside. Eggs are of three types: subitaneous, also called "summer eggs" or parthenogenetic eggs, which are thin-shelled and develop immediately into females; smaller, thin-shelled, parthenogenetic "male" eggs, which develop immediately into males; and thick-shelled "resting" eggs which are fertilized "male" eggs and develop, after a dormant period, into females only.

Chaetonotus brevispinosus

By C. G. Goodchild

Gastrotrichs are small fresh-water and marine organisms of the same general size and habits as rotifers, with which they are sometimes confused. Often gastrotrichs may be found creeping on the bottom in protozoan and rotifer cultures. With a small pipette transfer several living gastrotrichs to a drop of water on a slide and cover with a coverglass. In all probability you will not find the described species of gastrotrich; however, since the organ systems are fundamentally similar, these directions may be used for most common species.

The body of *Chaetonotus brevispinosus* (Fig. 129) is elongated and covered with short, curved, dorsal spines. The head is rounded and bears two zones of bristles. The body posteriorly is forked and provided with cement glands like those in rotifers. The animal glides along on two ventral ciliary tracts which extend from anterior to posterior ends.

ASCHELMINTHES

Integument and Muscles. The integument is a thin, syncytial hypodermis with an overlying, secreted, thin cuticula. Six pairs of unstriated longitudinal muscles traverse the body; circular and transverse muscles are lacking.

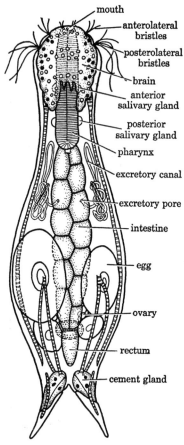

FIG. 129. *Chaetonotus*, dorsal view of adult female. (Modified after Remane, in Kükenthal and Krumbach.)

Digestive System. An anterior mouth leads into a straight muscular pharynx which has a cuticulate, triradiate lumen similar to that in nematodes. Posterior to the pharynx is an intestine having a wall composed of a single layer of cells. A few unicellular glands occur also in the wall, and their secretions digest algae and small animals sucked in by the pharynx. The intestine is separated by a sphincter or constriction from a posterior rectum which, in turn, opens by a small dorsal anus placed just in front of the posterior pedal process.

Excretory System. On each side of the midbody region is a greatly coiled excretory canal which ends internally in an elongated flame bulb and ends externally at a pore. The excretory pores are placed ventrally in the middle of the body.

Nervous System. A relatively large, saddle-shaped brain invests the anterodorsal part of the pharynx. From the brain a pair of dorsal nerve trunks extends posteriorly along the pharynx. Anteriorly directed nerve fibers innervate the cephalic bristles which are presumed to be tactile and gustatory. Eyes are lacking in this species.

Reproductive System. Parthenogenetic females only are known in the Chaetonotoidea. The midtrunk region of gravid females is distended with several large eggs. Eggs are produced in paired ovaries lying adjacent to the intestine posteriorly. The female gonopore is near the anus. Eggs are laid on the exuviae of crustacea and on weeds, to which they attach by a thick, ornamented shell. Embryonic development is direct, and young individuals hatch as miniatures of adults.

Turbatrix aceti
By C. G. Goodchild

Turbatrix aceti (formerly called *Anguillula aceti*), the vinegar eelworm, is a free-living nematode often occurring abundantly in vinegar. Examine, with the unaided eye and also with the binocular microscope, a thriving culture of the worms. Notice the vigorous, thrashing, swimming motions, and locate tangled groups of worms feeding on the "mother of vinegar," a conglomeration of bacteria and yeast.

Several large specimens should be selected, placed on a clean slide, and covered with a coverglass. If the swimming motions are too violent, kill the worms by holding the slide briefly over a small flame. Examine them unstained or after weak staining with dilute alcoholic Nile blue sulfate.

Males are about 1.4 mm. and females 2.0 mm. in length. The narrow, cylindrical body (Fig. 130) is blunt anteriorly, tapers to a fine point posteriorly, and, in killed specimens, is usually bent slightly ventrally behind the middle. The internal organs are somewhat obscured by fat droplets, but with care the structures can be resolved.

The body wall consists of a transparent, finely striated, external cuticula and an underlying syncytial hypodermis. The somatic muscles are relatively large, and their protoplasmic core is laden with fat droplets. The pseudocoel is filled with a clear body fluid in which are suspended many fat droplets, especially in the intestinal region.

A small mouth leads into a shallow buccal cavity which is lined with cuticula and bears a single large dorsal and a pair of smaller ventrolateral teeth. A pharynx, immediately behind the buccal cavity, is divided into three regions: an initial, cylindrical portion with radiating muscles surrounding a triradiate lumen; a middle, narrower isthmus; and a posterior, spherical muscular bulb containing a tripartite valvular apparatus. The pharynx joins an intestine which is a narrow tube leading to a thin-walled rectum opening at a subterminal anus. In the male the rectum opens into a cloaca which also opens subterminally.

A nerve ring is visible about midway in the pharyngeal isthmus region, and a ventral excretory pore may be visible in favorable specimens at the level of the posterior end of the isthmus.

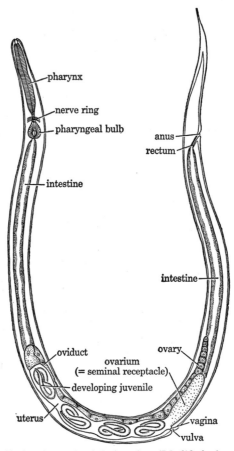

Fig. 130. *Turbatrix aceti*, adult female. (Modified after De Man.)

Female Reproductive System. A vulva is located slightly behind the middle of the body on the ventral side (Fig. 130). A single uterus, which contains approximately two tadpole-like and three vermiform embryos, runs forward from the vulva to the junction of anterior and middle body thirds, where the tube retroflexes and runs posteriorly a short distance as an oviduct and then continues as a filamentous ovary to a point just posterior to the vulva. Extending posteriad from the vulva is a short, distended, sac-like "ovarium" which probably functions as a seminal receptacle and is perhaps the homologue of a posterior ramus of the uterus and ovary. The young are born alive. Each female requires about 4 weeks to attain sexual maturity and will produce up to 45 embryos during a maximum life span of 10 months.

Male Reproductive System. The male reproductive system is usually seen in side view because the heat-killed male lies on its side because of the ventral flexure mentioned above. A thin, ventral testis begins about 0.3 mm. anterior to the cloaca, passes forward a very short distance, and then bends back upon itself. A sperm duct continues posteriorly and expands, slightly anterior to the cloaca, to form a sperm-filled seminal vesicle which tapers posteriorly and joins the cloaca (Fig. 131). Opening into the cloaca are two copulatory spicules with dorsally expanded head ends, curved bodies, and pointed ventral ends which may be protruded from the anus. Thin, ventrolateral lamina occur along the spicules and help to form a more or less closed tube for sperm transfer. Attached to the heads are protractor muscles running to the dorsal body wall. A keel-shaped accessory piece is present in the dorsoposterior wall of the cloaca; this plate probably functions to guide the spicules in protrusion. Four ventral pairs and one dorsal pair of small sensory papillae occur in the anal region.

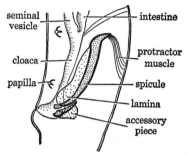

FIG. 131. *Turbatrix aceti*, cloacal region of adult male in lateral view. (Modified after De Man.)

Rhabditis maupasi
By C. G. Goodchild

The genus *Rhabditis* consists of small, rather simple nematodes living in the soil, in water, and in decaying organic material in the soil. *Rhabditis maupasi*, as a juvenile, is also found in the soil, but more commonly occurs in earthworms, *Lumbricus terrestris* and others.

A thriving culture of *R. maupasi* is easy to start and maintain in the laboratory. Infected earthworms are widespread and may easily be captured locally or purchased from commercial suppliers. To locate the nematodes slit the earthworm open by a dorsal longitudinal cut. Encysted juveniles are almost always to be found embedded in flattened, ovoid, brown bodies lying in the coelomic spaces of the posterior somites. Unencysted juveniles usually occur in the seminal vesicles and coelomic spaces and almost always are to be found near the nephridiopores in the expanded bladder regions of the nephridia. In heavily infected specimens nearly every nephridium is infected with as many as 12 or more active juveniles which are all approximately 0.5 mm. in length. Development beyond the juvenile stage is inhibited in the earthworm, because suitable food materials are apparently scarce. Furthermore, the juvenile appears incapable of lacerating the earthworm host to obtain nourishment and, therefore, is a benign parasite. Healthy earthworms may harbor hundreds of juveniles with no apparent impairment of function. The juvenile grows to the adult only upon the death of the host. The decaying flesh provides abundant nourishment, and growth to the sexually mature stage occurs in a few days. To establish a laboratory culture of the nematodes, place small pieces (5 mm. square) of the body wall of the earthworm on a layer of 2 per cent agar-agar in a covered dish and inoculate the meat with several brown bodies or active juveniles from the nephridia or seminal vesicles. The stomach-intestine of the earthworm should be carefully removed intact and discarded to prevent contamination with soil nematodes which may happen to be in the gut. Although cultures can be developed at laboratory temperatures, better results are obtained by keeping them at 7–20°C. Temperatures above 22°C. are detrimental to the worms and should be avoided. Subcultures should be started periodically by transferring eggs, juveniles, and adults to fresh pieces of beef or earthworm supported on 2 per cent agar-agar in petri dishes or in Syracuse watch glasses.

Observations of a flourishing culture should be made with the aid of a binocular microscope and intense reflected light. Notice the liquid character of the surface of the meat and find the nematodes crawling in this fluid and on the agar-agar. Study the character of the movement and observe the constant body length during the rather rapid locomotion. Small juveniles as well as intermediate and full-sized adults will probably be visible in the culture. Masses of eggs on the meat and in a zone surrounding it will also usually be apparent. Female worms (Fig. 132) can be identified by their large size, 2.0 mm. long, opaque eggs in the middle of the body, and a tapered thread-like posterior end. Smaller male worms (Fig. 133), 1.7 mm. in length,

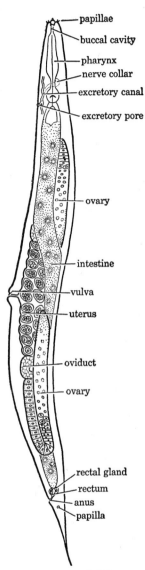

Fig. 132. *Rhabditis maupasi*, adult female. (Modified after Johnson.)

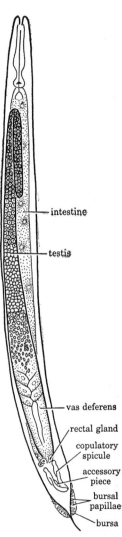

Fig. 133. *Rhabditis maupasi*, adult male. (Modified after Johnson.)

may often be formed in association with female worms, and such pairs should be examined with the greatest magnification available on the binocular microscope. Notice, in the male, the use in copulation of a truncated posterior end bearing a bursa, and notice the protrusion and retraction of amber-colored spicules from the center of the bursa. In old cultures dead worms will be present. They should be examined with the binocular or compound microscope, and in the carcass of the female masses of juveniles, which have hatched there and which consume the dead maternal tissues, should be seen.

With needles transfer several full-sized male and female adults, as well as juveniles and eggs, to a drop of fresh water on a slide. Before placing a coverglass on the preparation examine the worms under low power. In water the worms thrash about and are unable to accomplish effective locomotion. Notice again the body shapes, arrangements, and opacities of internal organs in uncompressed male and female specimens. Now place a coverglass on the preparation and remove any excess water by touching a piece of absorbent paper to the edge of the coverglass, after which the edges should be sealed with Vaseline.

THE ADULTS

The following anatomical study, except for the reproductive systems and associated organs, can be made with either a male or female worm, but a young, mature female is preferred. Before proceeding to the detailed study of the worm the following general features should be observed. Both sexes are elongated, somewhat cylindrical, pointed at each end, and provided with a very thin caudal termination. Almost invariably all mature worms will lie on their sides. The reason for this fact is apparent if the method of locomotion is correctly interpreted. The midbody region of the mature females is distended with ovoid eggs, and these serve as an easy means of distinguishing sexes. The male is more nearly cylindrical and can be easily identified by a clear midbody region and a flared posterior end comprising the bursa. The mouth is apical, while the anus is a subterminal, transverse slit on the ventral side. A vulva, the female reproductive opening, is located ventrally on a slight prominence near the midregion of the body.

Cuticula and Body Wall. A delicate, transparent cuticula covers the body. This outer layer is produced by a subjacent hypodermis. The body-wall muscles consist of a single layer of longitudinal cells lying between the hypodermis and a large, central pseudocoelic cavity.

Digestive System. The digestive tract is a simple tube lined with cuticula running from mouth to anus. The mouth is guarded by three lips, one dorsal and two ventral, each of which is subdivided by a shallow groove to form two lobes. Each lobe possesses a pair of small papillae of probably a receptor function. The buccal cavity is cylindrical and heavily cuticularized and is without teeth. Near its posterior end is a ridge-like constriction. The next region of the digestive tract is a conspicuous muscular pharynx which encloses the posterior portion of the buccal cavity and extends posteriorly approximately one-ninth the body length. Food is ingested by a pharyngeal sucking action caused by the contraction of lumen-dilating muscles. The small bore of the pharynx limits the type of food to microorganisms and fluids. The pharyngeal wall is a syncytium consisting of radial muscle fibers; the lumen is triradiate and is lined with cuticula. The pharynx has four anteroposterior regions: a cylindrial procorpus, an enlarged metacorpus, a narrow isthmus, and an enlarged pyriform bulb containing a well-developed, tripartite, rugate valve. Three glands are embedded in the wall of the bulbular region, one dorsal and two subventral. A duct from the dorsal gland runs forward and opens into the lumen of the procorpus near the buccal cavity. Ducts from the subventral glands open at the posterior end of the metacorpus. These details are usually easy to see in nematodes which have been placed in a dilute solution of neutral red for several hours. Posterior to the pharynx is an intestine whose wall is composed of a few large, uninucleate cells alternately arranged in two rows. The cytoplasm of these cells is usually packed with birefringent sphaerocrystals of rhabditin, a carbohydrate or "albumino-fatty" substance. These granules, as well as those in the pseudocoel, show strikingly well in worms viewed through a microscope equipped with a dark-field stop or dark-field condenser. In the specimens stained with neutral red the sphaerocrystals are colored orange or brilliant brick-red. The lumen of the anterior region of the intestine is wide; filamentous, branching fungi may be found growing in this zone. The wall of this anterior region is thin and practically devoid of sphaerocrystals. Posteriorly, the cells of the intestinal wall often contain the heaviest concentration of sphaerocrystals, and the lumen of this region often contains filamentous fungi also. A short rectum joins the center of the rounded posterior end of the intestine and passes posteroventrally to the anus. Initially the rectum is surrounded by three unicellular, rectal glands, one dorsal and two ventrolateral. In the male the rectum opens into a cloaca which also receives the sperm duct.

Excretory System. The system which has been called an excretory system consists of an H-shaped group of cells visible in living worms. Paired lateral canals extend from the region of the pharyngeal procorpus to a point near the posterior end of the worm where they end blindly. These canals are sinuous and are embedded in a cellular sheath which may represent the stretched outline of a median excretory sinus cell. Vacuoles are often present in the walls of the lateral canals. Posterior to the bulb of the pharynx each lateral canal joins a transverse canal running medioventrad. The transverse canal is embedded in the excretory sinus cell which lies against the wall of the pharyngeal bulb. The sinus cell is granular in appearance and possesses a conspicuous nucleus. Paired, granular, subventral glands connect to the posterior margin of the excretory sinus and extend posteriorly a short distance beside the intestine. A short, median cuticularized terminal duct passes posteroventrad from the sinus to a ventral excretory pore. A nucleus is said to occur along the course of the terminal duct.

Nervous System. Only the parts of the nervous system readily visible in the unstained, living worm will be described here. Surrounding the isthmus portion of the pharynx, slightly anterior to the pharyngeal bulb, is a nerve ring composed chiefly of nerve fibers. The nerve ring is obliquely placed with the dorsal edge slightly anterior to the ventral edge. Nerve cells may be seen laterally, lying in contact with the nerve ring, and major nerve cords may be seen joining the nerve ring dorsally and ventrally. The ventral cord is ganglionated ventral to the pharyngeal bulb and then fans out, ventral to the subventral excretory glands, and is lost from view in the ventral hypodermis. The dorsal cord is small and has fewer nerve-cell bodies in a dorsal ganglion adjacent to the pharyngeal bulb.

Reproductive Systems. The reproductive systems of *R. maupasi* are well developed, diagrammatically clear, and almost unsurpassed for study. The structure and function of all parts may be observed through the transparent body wall. Stages in spermatogenesis can be seen in the male, while maturation of the ova, fertilization, and early cleavage can be observed in the mature female. Several healthy, fully mature male and female specimens should be available for the following description.

Male Reproductive System. The male reproductive system is J-shaped (Fig. 133). The single, tubular testis is reflexed at its anterior end and is filled with spermatozoa in various stages of development. As spermatozoa pass posteriorly in the testis, they change in appearance. Initially they form a solid, conspicuously nucleated zone; in the midregion they assume a compact, honeycombed appearance with each cell having an increase in granular cytoplasm; near the posterior end of the testis the maturing spermatozoa become spherical and less compact. Posterior to the testis is a vas deferens which proceeds posteriorly and joins with the cloaca. The cloacal aperture is situated on a slight, median prominence within the concavity of the bursa.

Two stout, slightly curved spicules form a very evident V-shaped complex in the caudal region of the male. Each spicule is provided with an anterior, knobbed head and a slightly twisted shaft which ends in a point posteriorly. Protractor and retractor muscles connect to the enlarged head and shaft, while a V-shaped accessory piece embedded in the dorsal wall of the cloaca apparently guides the spicules during protrusion through the cloacal aperture.

The bursa is a flared, thin-walled extension of the posterolateral body margins. It is broadest in the middle and is emarginate posteriorly at the zone of junction with the fine caudal process. Nine pairs of bursal papillae, finger-like projections from the ventral surface, project from the margin of the body to a point near the bursal margin.

Female Reproductive System. The female reproductive system is a double J-shaped structure, consisting of anterior and posterior rami, converging to the vulva (Fig. 132). The blindly ending ovaries begin near the vulva and proceed anteriorly or posteriorly, in the two rami, to a point near the pharynx or rectum where they bend abruptly toward the vulva. Initially the ovaries contain germinal cells packed tightly together and rendered conspicuous by large, ovoid, grayish nuclei. Developing ova become compressed into single file near the vulvar ends of the ovaries, and abundant granular cytoplasm swells them into box-like shapes. Shortly beyond the point of ovarian retroflexion is a constricted oviduct, which is usually packed with sperm and which functions as a seminal receptacle. As the ovum passes through the oviduct, sperm penetration and fertilization occur. The fertilized eggs secrete thin, transparent shells and accumulate in the uteri which slowly are distended as the female matures. From a few to several hundred cleaving eggs may be present in the uterus. Midway between the uteri is a short, ventrally directed vagina which opens at the vulva. The vulva is situated slightly posterior to the midregion of the body and consists of a transverse slit bordered by two projecting lips.

The posterior end of the female tapers to a fine point and is not provided with a bursa. However, a pair of short, lateral papillae, approximately one-third of the distance backward from the anus, may be visible.

Eggs, Juveniles, and Life Cycle

Living eggs, 45 μ long, of *Rhabditis* are excellent for studying pronuclear fusion, spindle formation, spindle elongation, and cell division. The early fertilization phenonema are often visible *in situ* through the transparent body and uterine walls of the female. More precise control is possible by teasing open the body of a ripe female with needles and by placing the eggs thus liberated on a depression slide in fresh water. If the eggs are not crushed, deprived of oxygen, or allowed to dry, the entire sequence from pronuclear fusion through the second cleavage will occur, at room temperature, within a 2-hour period.

The juvenile resembles the adult in general body shape and morphology of the digestive system, especially the pharynx. The reproductive system is not yet formed. Several parasitic nematodes, e.g., *Necator americanus*, pass through, usually in the soil, a juvenile stage which resembles this one in the character of the pharynx; accordingly, they are called rhabditiform juveniles.

The life cycle of the worm is apparently relatively simple and direct. Juveniles grow to maturity in the decaying flesh of an earthworm. Adults reproduce bisexually, or hermaphroditically, and juveniles developed from eggs produced by these adults or later ones wander off in the soil. Those juveniles which chance upon a live earthworm penetrate it via the nephridiopores, dorsal pores, or reproductive apertures. The coelomic forms become covered by amoebocytes and encyst, while worms located in the other common sites remain unencysted. As previously mentioned, growth beyond the juvenile stage does not occur in the living earthworm.

Enterobius vermicularis
By C. G. Goodchild

Enterobius vermicularis, the pinworm or seatworm of man, is a rather small worm which frequents the caecum, the vermiform appendix, and adjacent regions of the digestive tract. Pinworm disease, enterobiasis, is a more common complaint in children than in adults and is often attended with loss of appetite, insomnia, and neuroses.

Only a single host is required in the life cycle. Because of the resistance to desiccation and the uniform distribution of the infective eggs in dust, most, if not all, persons in an infected household or institution finally become infected. The elimination, with drugs and enemas, of adult worms from the digestive tract and constant vigilance to prevent ripe eggs from again being swallowed are sanitary methods which will eradicate the disease.

Preserved male and female pinworms, as well as infective eggs, should be available for study.

Adults. The male worm, 5 mm. long, is about half the size of the female (Fig. 134). The posterior ends of the worms are different in the two sexes, that of the female being long and tapering, and that of the male being blunt, curved ventrally, and provided with copulatory organs to be described below.

A finely sculptured cuticula covers the body of the worm. Lateral cuticular wings, or alae, extend posteriorly a short distance from the anterior tip. In the young female lateral cuticular flanges occur in the midbody region. These flanges become inconspicuous as the uterus in this region is distended with eggs.

A slightly retractile mouth, surrounded by three lips, opens apically. Passing posteriorly from the mouth is a cylindrical, cuticulate pharynx. This structure presently dilates slightly, then narrows to an isthmus, and finally enlarges into an ovoid cardiac bulb containing a triradiate valve. Posteriorly the pharyngeal bulb connects to a tubular intes-

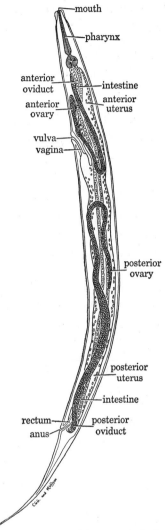

FIG. 134. *Enterobius vermicularis*, adult female.

tine, which proceeds posteriorly and terminates in a rectum which opens at an anus, located terminally in the male and subterminally in the female.

A nerve ring may be seen encircling the middle of the cylindrical pharynx. This ring possesses large ventral and lateral ganglia and smaller dorsal ganglia. Motor and sensory nerves radiate, in four longitudinal chords, anteriorly and posteriorly from the nerve ring. It is impractical to locate the finer distribution of nerves in whole mounts.

The system usually referred to as excretory is relatively simple. Four longitudinal canals join in an H-shaped fashion to a clear median excretory vesicle located in the esophageal region. The vesicle opens to the outside by a ventral cervical pore.

Reproductive Systems. The reproductive systems and the accessory genital organs will be described for each sex.

Male Reproductive System. The single, tubular testis begins near the middle of the body and after proceeding forward a short distance retroflexes and increases in diameter. Posteriorly this distended zone continues as a short, narrow, undulate vas deferens. This duct joins an enlarged sperm-storage organ, the seminal vesicle, which in turn connects posteriorly to an ejaculatory duct. The ejaculatory duct and the rectum open into a cloaca. A single, robust, sinuous, copulatory spicule with a ventrally curved tip lies in a sheath dorsal to the cloaca and the ejaculatory duct. The spicule is provided with protractor and retractor muscles and is used to facilitate sperm transfer during copulation. Subcuticular muscles which are visible in all parts of the body as fine wavy lines converge in definite bundles to inconspicuous caudal alae. These alae are supported by true, but reduced, papilla-like rays arranged as follows: a pedunculated preanal pair anteriorly; a larger caudal pair posteriorly; three additional pairs of sessile postanal papillae between the first two pairs; and lastly, a sessile pair lateral to the preanal pedunculated pair.

Female Reproductive System. Two sets of narrow, tubular ovaries and oviducts coil back and forth, one behind the other, in the middle and posterior thirds of the preanal body region (Fig. 134). The oviducts join the uteri, which during the life of the female slowly distend with eggs until she becomes a distended egg sac holding approximately 11,000 eggs. The uteri connect to a single vagina which originates at the middle of the body and proceeds forward to a vulva located at the zone of junction of the anterior and middle body thirds. Eggs are only infrequently laid by gravid females in the intestine of the host and accordingly are not often found in fecal samples.

Life Cycle. Gravid females migrate from the caecum to the anus from which they emerge usually at night. As they crawl over the perianal and perineal regions they oviposit. Moreover, their migrations cause an intense *pruritus ani*, often aggravated by scratching. Eggs lodged under the fingernails can be transferred to the mouth, from where they are swallowed to reinfect the host.

The eggs deposited perianally measure 50–60 μ by 20–30 μ and contain tadpole-shaped juveniles. Within a few hours, development to the rhabditoid juvenile occurs, and the egg is then infective. Examine an egg containing an infective juvenile and note the flattened ventral side and the transparent double-layered shell.

Eggs hatch in the small intestine, and rhabditoid juveniles begin a slow descent of the digestive tract. Several molts and mating occur before they attach to the caecal wall and grow to adulthood.

Necator americanus
By C. G. Goodchild

The American hookworm, *Necator americanus*, is not limited to the western hemisphere; rather, it also occurs commonly in Central and South Africa and in southern Asia and offshore islands. In the United States, the southern and particularly the southeastern states are now heavy endemic hookworm zones. It is believed that infected African Negroes brought into the United States established the hookworm in at least part of its present range.

The anatomy and life history of *N. americanus* are similar to those of *Ancylostoma duodenale*, the Old World hookworm. In both, infective, free-living juveniles, in fecally contaminated sandy soil or humus, gain access to human beings typically by boring through the skin; they migrate by the blood stream to the lungs, then ascend the air passages to the pharynx, from which point they are swallowed to become established as adult parasites in the small intestine.

Hookworm disease is a variable condition depending on several interacting factors. Healthy, well-fed individuals can harbor many worms with little or no ill effects. Other individuals, especially children, may be stunted in physical and mental growth because of hookworm-induced anemia. Mental retardation, lowered resistance to other diseases, and general loss of work efficiency are insidious effects of hookworm disease; moreover, this disease works inexorably to reduce the health and welfare of whole populations. Vigorous and

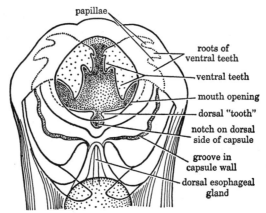

FIG. 135. *Ancyclostoma duodenale*, oral view of mouth capsule. (Redrawn from Chandler, after Looss.)

alert racial stocks have been rendered shiftless and apathetic by several generations of hookworm infestations.

With some alterations, chiefly in regard to the details of the lacerating teeth or cutting plates in the oral cavity (Fig. 135), slight dif-

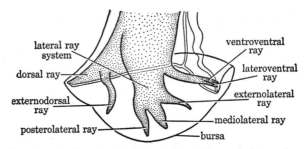

FIG. 136. *Ancyclostoma duodenale*, lateral view of bursa and bursal rays of adult male. (Redrawn from Chandler, after Looss.)

ferences in the positions of the vulvae, and some differences in the bursae of the males (Fig. 136), these directions can be used for studying either the American or the Old World hookworm or the dog hookworm, *Ancylostoma caninum*.

Adult male and female hookworms, as well as eggs and juveniles, and demonstrations of penetrating and migrating worms in the vertebrate host should be available for study.

ADULTS

Necator is a cylindrical, anteriorly tapered worm, with a cervical flexure which causes the mouth opening to be directed anterodorsad.

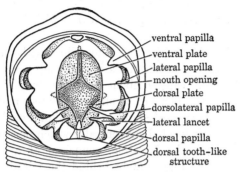

FIG. 137. *Necator americanus*, oral view of mouth capsule. (Redrawn from Chandler, after Ackert, Payne, and partly after Looss in Chandler.)

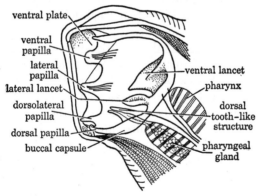

FIG. 138. *Necator americanus*, lateral view of mouth capsule. (Redrawn from Chandler, after Ackert, Payne, and partly after Looss in Chandler.)

This anterior hook-like bend probably is responsible for the common name, hookworm. The female is about 10–11 mm. in length, and the male, which can be immediately identified by its flared, posterior bursa, is about 7–8 mm. long. A cup-shaped buccal capsule (Figs. 137 and 138), lined with inturned cuticula, bears the mouth opening and has sharp cutting plates and lancets in its walls. The chief attaching organs in *N. americanus* are a pair of large ventral and a pair of smaller dorsal semilunar cutting plates. A duct of a dorsal pharyngeal gland opens at the apex of a conical dorsomedial tooth-like structure which lies between the dorsal plates. Extending from the lateral walls of the capsule toward the mouth opening are several pairs of finger-like oral papillae, and deep in the capsule are sharp, triangular, ventral and lateral paired lancets. These formidable, lacerating organs cause continuous bloodletting from the host's small-intestinal mucosa and thus produce the anemia so definitely associated with hookworm infestation.

A pharynx, which has a triradiate lumen lined with inturned cuticula and a wall composed of strong radiating muscle fibers, extends posteriorly from the buccal capsule about one-sixth the body length. The posterior region of the pharynx is slightly dilated, and embedded in its wall are three pharyngeal glands, one median dorsal, as previously mentioned, and a pair located subventrally which open into the lumen near the nerve ring. An intestine, composed of a single layer of columnar cells, begins immediately behind the triradiate pharyngeal valve and proceeds as a straight tube to a short rectum. In the intestine the host's blood and other tissues are digested and absorbed.

The somatic muscles consist of a single layer of cells underlying a hypodermis and arranged into four longitudinal groups. These four groups are interrupted by four longitudinal, quadrantal chords located on the dorsoventral and lateral axes. Because muscle cells are limited to two in each sectioned quadrant, the worms are called meromyarial. The almost complete lack of transverse muscle cells and the rigidity of the cuticula prevent any great change in transverse dimensions.

The organ complex usually designated as the excretory system is rather complicated. Flame bulbs are absent in nematodes, and homologies and functional comparisons with other invertebrate excretory systems have not been well worked out. The system here consists essentially of two canals embedded in the lateral chords and an isthmus in the pharyngeal region. Surrounding, or adjacent to the isthmus, are other excretory structures including an excretory sinus into which tributaries from the lateral canals open and from which ventrally is given off a short, cuticulate, terminal duct. The duct opens at a ventral excretory pore. Stretching posteriorly from the sinus are two large, subventral glands of uncertain function. Moreover, the initial portions of the tributaries of the lateral canals are embedded in these subventral glands. The tributary canals emerge from the subventral glands and penetrate the lateral chords where they join the lateral canals. These lateral canals extend from the region of the buccal capsule to the posterior end of the body where they end blindly. Encircling the pharynx immediately in front of the excretory pore is a nerve ring. From a large ganglion located on the ventral portion of the nerve ring, a large nerve runs in the ventral chord to the posterior end, and smaller nerves run in the lateral chords anteriorly and posteriorly. Four small nerves run forward from the ring to supply the sensory and motor apparatus of the anterior end. These main nerve trunks are connected in a complicated fashion by nerve commissures.

Reproductive Systems. The reproductive systems of the male and female will be described separately.

Female Reproductive System. The female reproductive aperture is a ventral, unpaired vulva located near the middle of the body. Eggs are produced in two extremely convoluted, tubular ovaries. One ovary is located anterior to the vulva, the other one behind it. From the ends of the ovaries nearest the vulva short, narrow oviducts carry the ova into wider, confluent seminal receptacles where fertilization occurs. From here eggs pass into the uteri where they accumulate. Later they pass into complicated, muscular ovejectors which expel the ova from the uteri into short cuticulate vaginae. The two vaginae join and then open to the surface at the vulva.

Male Reproductive System. The tubular testis occupies the middle third of the body. It begins at the junction of the middle and posterior body thirds and runs in a random fashion to the middle of the body, from which point forward it coils loosely about the intestine. Near the ends of the subventral glands the tube retroflexes and continues posteriad as the vas deferens to the middle of the body parallel to the transverse testicular coils. Here it joins a distended seminal vesicle which connects posteriorly to an ejaculatory duct leading directly to the cloaca. The ejaculatory duct, through most of its course, lies between two large, multicellular cement glands whose secretion apparently serves as an adhesive material holding the pair of worms together at copulation.

The posterior end of the male hookworm bears a wide bursa (Fig. 139), two copulatory spicules, and a gubernaculum. The bursa is a long, wide, bilaterally symmetrical structure, with supporting, paired, fleshy rays having a specific distribution of taxonomic importance. In *Necator*, each half of the bursa has a small dorsal ray with a short bifurcated tip; a slender unbranched externodorsal ray originates near the base of the dorsal ray; a robust lateral ray system is divided into a posterolateral, a mediolateral, and an externolateral ray. A ventrally directed ray springs from near the base of the lateral ray system; it bifurcates to form a lateroventral and ventroventral ray. Lying anterior to the other bursal rays on the ventral side is an inconspicuous prebursal ray. The two spicules are long and slender; they are protruded and retracted by special muscles and function to

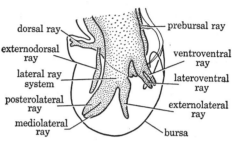

FIG. 139. *Necator americanus*, lateral view of bursa and bursal rays of adult male. (Redrawn from Chandler, after Looss.)

transport sperm to the vulva of the female. The gubernaculum, a hard piece on the dorsal wall of the cloaca, is said to be used to guide the spicules.

EGGS AND JUVENILES

A fecal smear containing the delicate-shelled, ellipsoidal eggs of *Necator* should be examined, and diagnostic features of the ova noted. The egg is usually in the eight-cell stage when it is eliminated from the host. During the first 24 hours, in favorable soil, it completes embryogeny and hatches as a free-living, first-stage, rhabditiform juvenile. This young worm has a mouth, pharynx, and simple intestine; it grows by feeding on bacteria and organic debris and within 48 hours undergoes the first molt. At about the fifth day it undergoes a second molt to become a filariform juvenile. In this infective, nonfeeding stage the mouth is closed and the pharynx is elongated; moreover, sharp buccal spears, of use in penetrating the human skin, appear. The third juvenile stage may remain alive in moist soil for nearly 4 months during which time it climbs up on objects, such as blades of grass, as far as a film of water extends. During the heat of the day it may retreat a short distance into the soil. If by chance the young worm should come into contact with human skin, penetration occurs.

Observe demonstrations of young worms in sections of skin, if these are available. Although this method of entry is most common, infection can also occur as a result of ingestion of infective juveniles in food or drink. Juveniles pass by way of the blood stream to the lungs; thence they ascend the trachea, in which they may be found 3 to 5 days or longer after penetration. From the glottis they are swallowed, and, finally, they become established in the small intestine. The third molt occurs soon after they reach the small intestine; the fourth, and final, molt occurs in about another week. Thereafter growth to sexual maturity with copulation and egg production ensues.

Ascaris lumbricoides
By C. G. Goodchild

Ascaris lumbricoides, a large nematode parasite of man, is world wide in distribution. It occurs more commonly in the small intestine of children than in adults because of the former's custom of playing in contaminated, egg-laden soil adjacent to dwellings. As many as 1000–5000 adult worms may be present in a single host; moreover, because of their strong, pointed, and cuticulate bodies and their habit of wandering, they may perforate the gut wall and cause peritonitis or they may enter the bile or pancreatic ducts or the vermiform appendix and interfere with normal digestive functions. Toxic products produced in the worms may cause disturbing symptoms in children, including convulsions and dulling of mental capacity.

Preserved male and female worms should be available for dissection. Ascarids obtained from the pig are morphologically indistinguishable from human ones and may be used for this study. Living pig ascarids may usually be obtained from local slaughter houses; moreover, preserved worms purchased from commercial suppliers are almost certain to be from the pig. Cross sections of both male and female worms through the regions of the gonads and pharynx, as well as demonstrations of eggs and migrating juveniles in blood or lung tissue, should be available for study.

Adults. Both sexes are elongated, cylindrical, and pointed at both ends. Males are 15–31 cm. long, while females are 20–35 cm. long. The males are further distinguished by a ventrally curved posterior end, from the slit-like anal opening of which often protrude two copulatory spicules. The cuticula is provided with transverse markings which give the worm a pseudosegmented appearance. Two broad, conspicuous, brownish, lateral chords and narrower, fainter, whitish, mid-dorsal and midventral chords run from anterior to posterior ends. The mouth opening is apical, while the anus is ventral and subterminal. Approximately 2 mm. from the anterior end is a small, ventral excretory pore visible only in favorable specimens. The female reproductive aperture, the vulva, is located at the junction of anterior and middle body thirds on the ventral side. This opening must be located because it is the reference point for a longitudinal incision to be made later. With suitable lighting, white, coiled structures, the female reproductive system, can be seen through the body wall in the posterior two-thirds of the worm; the vulva is located a few millimeters in front of the most anterior of these coils often in the center of a narrow, slightly constricted, girdle-like zone. Examine this general area carefully with the aid of a hand lens or binocular microscope and locate definitely the oval or slit-like vulvar opening placed at right angles to the long axis of the body. If the opening is still not visible, dry the cuticula of the worm and rub on to the area in question a small amount of ink, which should then be wiped off. Ink will fill in the vulva and make it visible. The excretory pore may be located similarly. The male reproductive system and the rectum join a cloaca which opens subterminally on the ventral surface.

Integument. The integument consists of a thick, resistant cuticula and an underlying, syncytial hypodermis. The cuticula in the intact specimen is a glistening, plicated layer. If a cross section is examined with the high power of the microscope, the layered condition of the cuticula may be noted. Seven layers are reported as occurring in the cuticula; of these, a central, homogeneous, matrix layer and an abutting, fibrillar layer are as thick as the other layers combined. The hypodermis consists of a thin, circularly streaked area with abundant nuclei but without cell walls.

Muscular System and Pseudocoel. The somatic muscles of nematodes usually consist of a single layer of contractile cells longitudinally disposed. The layer is quadrantally interrupted by the lateral and medial chords. From two (meromyarial species) to many (polymyarial species) muscle cells may occur in each quadrant; in the polymyarian *Ascaris*, approximately 150 muscle cells are present in each quadrant section. The morphology of these peculiar muscle cells may be understood by studying the muscular layer of the inner body wall in a dissected specimen and by examining the muscle cells visible

ASCARIS LUMBRICOIDES

in the stained cross section. In top or side view the entire muscle cell (Fig. 140) will be seen to consist of a hammock-shaped, peripheral, contractile sheath and a fibrous, greatly distended, and vacuolated, central, protoplasmic core. The protoplasmic core contains the nucleus and gives rise to a fibrous process which runs to a nerve trunk

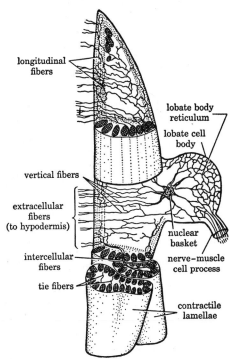

FIG. 140. *Ascaris lumbricoides,* stereogram of somatic muscle cells. (Redrawn from Mueller.)

in one of the medial, longitudinal chords. In the cross section study, with high power, a muscle cell complete with nucleus and identify as many of the parts as possible. The sectioned contractile sheath is composed of a U-shaped part with parallel muscle fibrils in its wall; if the contractile units adjacent to the nucleated one you have selected appear shorter, it is because they have been cut subcentrally.

Clear areas between the muscle cells and the body organs represent the pseudocoel, or false body cavity. This cavity is not lined with mesothelium and is not homologous to the coelom of higher invertebrates. In fact, the spaces are said to be large vacuoles in a few, enormous cells. A protein-rich fluid of characteristic, unpleasant odor fills the cavity. This fluid may produce some of the toxic and inflammatory symptoms of ascarid infections: it also may produce an allergic dermatitis in butchers and technicians who handle the worms.

Digestive System. The digestive system is a relatively simple tube extending from the mouth to the anus. To examine the three cuticulate lips which guard the mouth, sever, with a razor blade, the anterior 1 mm. of the worm and examine *en face* with the microscope. The single dorsal lip is broad and is provided with a forked fleshy core, from each lateral edge of which a pair of short papillae juts from the surface. The narrower subventral lips are paired, and each of their forked, fleshy cores gives rise ventromedianly to a single pair of short papillae. Finally, a very small papillary pair occurs on the lateral edge of each subventral lip. The inner edges of the lips are finely toothed and constitute rasping surfaces.

To expose the internal organs, cut through the body wall of the female slightly to the right of the mid-dorsal line and pin down, under water, the extended flaps. Since the body may be twisted, care is needed to maintain the proper position of the cut throughout the full length of the worm. The internal organs should be moved very gently as they are fragile and are easily broken.

Behind the lips in the anterior piece which was cut off is a small buccal vestibule. This vestibule connects posteriorly to a short, cylindrical, muscular pharynx. The latter is about 10 mm. long, is provided with a triradiate, cuticulate lumen, and has a wall composed of radially arranged, lumen-dilating muscles. A single, pinnately branched dorsal, and a pair of palmately branched subventral, uninucleate glands are embedded in the wall of the pharynx. The cross section through the pharynx should be examined for these details. The midgut, or intestine, is a dorsoventrally flattened, delicate tube extending posteriorly from the pharynx. Notice, in the cross section through the gonads, that the wall is composed of a single layer of tall columnar cells and an encircling, thick basement membrane. Each columnar cell possesses an ovoid nucleus near the basement membrane and several specialized zones at the lumen end. Jutting into the lumen is a non-mobile, cilia-like bacillary layer which arises from a compact, granular sub-bacillary layer. Digestion must be limited here because of the dearth of glands; no doubt predigested host chyme is ingested and absorbed without much further change. A short, flattened rectum, lined with cuticula, joins the intestine internally and opens

ASCARIS LUMBRICOIDES

ventrally at the anus. The latter is a transverse slit guarded by two folds and is situated about 2 mm. anterior to the posterior extremity.

Reproductive Systems. The reproductive organs are the most conspicuous internal structures. Locate the following parts in the dissection and, where possible, in the cross section (Fig. 141).

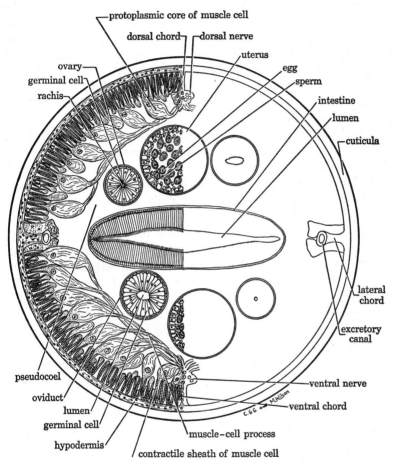

FIG. 141. *Ascaris lumbricoides,* cross section of the body through the region of the gonads.

Female Reproductive System. A short, median vagina opens at the ventral vulva, which previously has been located, and internally joins two large, mildly contorted uteri which run parallel to a region near the posterior end. At this point they are confluent with thinner oviducts which pass forward in a twisted fashion and grade into blindly ending ovaries. In the cross section, identify ovarian sections by the presence in each of a central rachis around which the young eggs cluster like spokes around a hub. The ovarian wall is composed of a single layer of greatly elongated epithelial cells, which are squamous or cuboidal in cross section, resting on a basement membrane. In the oviduct the rachis is missing but the wall is similar to that of the ovary. The uterine wall consists of elongated, tufted cells, which apparently can ingest excess spermatozoa, and muscle layers in addition to the basement membrane. The muscle layers consist of a thick, inner, circular layer and a thin, outer, oblique layer. The eggs in the uterus have heavy, secreted shells and often show early stages of cleavage.

If living females are available, tease the eggs from a small section of the uterus into a container of 2 per cent formalin. Examine a drop of the egg suspension with the microscope and identify the ovoid, thick-shelled, fertile eggs and the variously shaped, infertile ones. Set the container aside and record daily the development of the *Ascaris* juveniles within the eggs.

Male Reproductive System. In order to study the parts of the male reproductive system cut through the body wall beneath the right lateral chord and pin out the flaps. Notice an internal body cavity and digestive tract which are similar to those in the female. The reproductive system consists of a single, continuous, twisted tubule in the posterior body half. This tube has regional divisions and opens posteriorly with the digestive tract. The initial part is a long, narrow, greatly coiled testis. Developing amoeboid sperm cells are grouped about a central rachis as were the ova in the ovary of the female. The next region, a vas deferens, is about the same diameter as the testis, but is much shorter and lacks the rachis. This region joins an abruptly dilated, sperm-filled, seminal vesicle which occupies the posterior third of the body. The most posterior region of the male system is a short, muscular ejaculatory duct which joins the cloaca and opens ultimately at the subterminal anus. Two protractile, copulatory spicules lie in individual muscular sheaths located dorsal to the rectum. The spicules are used to align the reproductive apertures and to transfer sperm at copulation. Their function is aided by a chitinous guiding plate, the gubernaculum, in the dorsal cloacal wall.

Excretory System. The system usually designated as excretory is relatively simple and consists of only three huge cells. Carefully

ASCARIS LUMBRICOIDES 249

remove the pharynx from the dissected specimen and find a short, median, cuticularly lined, terminal duct running forward to the excretory pore from a distended, central, excretory sinus. From the posterolateral edges of the sinus two large excretory canals dip into the lateral chords and pass to the posterior end. Examine the cross section through the gonads (Fig. 141) and notice the fan-shaped expansion of the hypodermis into the heavily nucleated, lateral chord. The excretory canal, a thick-walled tube, is located at the medial end of an interstitial lamella which divides the lateral chord into dorsal and ventral halves.

Nervous System. The nervous system of *Ascaris* has been thoroughly studied, and its histology is amazingly complex. We shall con-

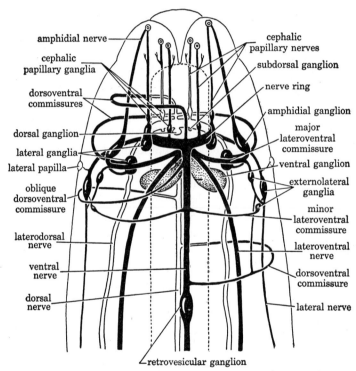

FIG. 142. *Ascaris lumbricoides,* diagram of the anterior portion of the nervous system. (Redrawn from Goldschmidt.)

sider here, however, only the gross morphology of the system. The central nervous system consists of a cluster of cephalic ganglia in close association with a pharyngeal nerve ring, and various nerve commissures and important longitudinal afferent and efferent nerves distributed to the body (Fig. 142).

Many of the structures mentioned in this paragraph are visible only in specimens stained in methylene blue solution. The cephalic ganglia consist of the following units: six small papillary ganglia are located slightly anterior to the nerve ring, and each gives off a nerve running forward to a labial papilla; two large ventral ganglia, two smaller subdorsal ganglia, and a single, small dorsal ganglion are located behind the nerve ring. From these cephalic ganglia six longitudinal nerves run to the posterior end of the body. These nerves include an unpaired, small nerve in the dorsal chord, a large ganglionated motor trunk in the ventral chord, and paired laterodorsal nerves and lateroventral nerves. Ventral to the rectum the ventral nerve ends by a preanal ganglion from which commissures pass to the other longitudinal nerves. A lateral sensory nerve, on each side, arises from large lateral ganglia which lie lateral to the pharyngeal ring. These lateral nerves, in close proximity to the excretory canal, pass posteriad where they innervate receptor organs including, in the male, the many preanal and postanal papillae.

Life History. The bile-stained, fertile egg passed in the feces will develop an infective juvenile within 9–13 days in favorable soil. If this egg is then swallowed by another human being, hatching occurs in the small intestine and a typical wandering period follows. The young juvenile penetrates the gut wall and is carried by the blood stream through the liver to the lungs where emergence into an alveolus occurs. After an ascent of the air passages to the pharynx the worm is again swallowed, and in about 2 months egg-laying begins. A rather constant feature in the life cycle of a nematode is a series of four molts before the attainment of sexual maturity. In *Ascaris* the first molt occurs in the eggshell. The second and third molts occur in the lungs; the fourth, and final, molt takes place after the juvenile has returned to the small intestine.

Hatching of eggs and the wandering phase of the ascarid life cycle will occur in non-human hosts even though the adult worm does not become established in the small intestine. Migrating juveniles can be obtained, by the following technique, from the infective eggs which have been isolated in 2 per cent formalin. Wash the eggs carefully and feed them to a guinea pig or rat, which should be sacrificed in about 1 day and the lungs and air ducts carefully searched for living ascarid juveniles. Study prepared slides showing young worms in the lungs. Notice the necrosis of tissue, the small areas of inflammatory hemorrhage, and the infiltration of leucocytes, especially eosinophils, around the juveniles.

Wuchereria bancrofti

By C. G. GOODCHILD

Wuchereria bancrofti, the most important human filarial parasite, is widely distributed in a tropical and subtropical circumglobar belt. In the United States the areas around Charleston, South Carolina, and perhaps other local areas along the southern and eastern sea coasts have formerly reported autochthonous infections.

The attenuated adult worms live in tangled masses in lymph nodes and vessels which they mechanically or inflammatorily block. Engorgment, and often rupture of the lymph vessels, may then ensue; untreated patients often develop elephantiasis, a hideous progressive enlargement of legs, scrotum, or mammae. The female worms give birth to swarms of small juveniles called microfilariae. These microfilariae appear in the peripheral blood with a diurnal periodicity as yet incompletely understood; they usually appear maximally in the hours between 10 P.M. and 2 A.M. and are scarce or absent during the daytime. Whatever is the basic cause for this cyclic swarming, it nevertheless brings the juveniles to the surface at the same time that mosquitoes, their other host, are most actively feeding.

Ingested microfilariae penetrate the wall of the mosquito's stomach. They lodge in the thoracic muscles and undergo a change in size and shape. After several weeks they migrate into the end of the insect's labium which they rupture to emerge on the skin of a human being during the next feeding. They actively invade the skin and grow to maturity in the lymph system within a year or less.

Stained, thick blood smears containing microfilariae of this species should be available. However, the common microfilaria, *Dirofilaria immitis*, from the blood of the dog may be substituted with only minor changes required in the description. Living microfilariae may be recovered from the blood of infected frogs. Some biological supply houses sell frogs infected with *Foleyella* and other genera of amphibian microfilariae.

Examine the stained smear of *W. bancrofti* with low power and notice the relative abundance of the worms and the graceful, non-kinked contour of the body. When you have found a good specimen, carefully study it under oil immersion.

The body (Fig. 143) is invested in a loose, faintly stained sheath which is very conspicuous at the extremities and is thought to represent a stretched membrane of the egg. The juvenile is long and cylindrical and is provided with a thin, transversely striated cuticula beneath which is a single layer of flattened subcuticular cells. The head end is blunt and possesses a clear cephalic space, the rudiment

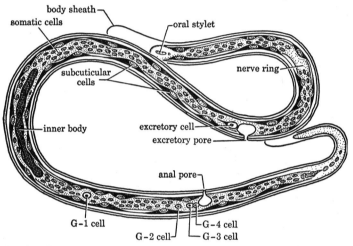

FIG. 143. *Wuchereria bancrofti*, microfilarial juvenile. (Redrawn from Fülleborn.)

of the adult buccal cavity, in which some workers report the presence of an oral stylet. The caudal end is tapering and enucleate.

Certain cellular and nuclear landmarks interrupt the central column of somatic cells. The relative positions of these structures are so definite that they are used to distinguish the various species of human microfilariae. Since absolute distance has been found to be less reliable than percentage distance, the various landmarks are located anteroposteriorly at fixed percentage distances of the total length.

A nerve ring (20 per cent) is a clear area in the neck region. A clear vesicular or diffuse blue-gray area in the neck region represents an excretory pore (30 per cent) and a large excretory cell (31 per cent) which lies immediately behind the pore. These two parts are rudiments of the excretory system. A rather large, darkly stained inner body (49 per cent) later forms the adult midintestine and genital organs. A G-1 cell (71 per cent) is situated posterior to the inner body. An anal pore (83 per cent), a posterior clear area, lies immediately behind a chain of G-2, G-3, and G-4 cells. The G-cells were formerly believed to form the adult genital organs but are now conceded to be rectal and anal *anlagen*.

Trichinella spiralis
By C. G. GOODCHILD

Trichinella spiralis, the trichina worm, a parasite of man, the pig, the rat, and other vertebrates, is the causative agent of the dreaded human disease, trichinosis. Any person eating undercooked, infected meat, which is usually pork, is a potential host of the worms; moreover, the cosmopolitan distribution of *Trichinella* renders it a constant threat to everyone. Europe and the United States are important zones of lethal and sublethal infection; general population infection rates of 15–90 per cent have been reported by reliable investigators in these, and adjacent, areas.

Male and female worms which are released by the digestion of viable cysts contained in ingested flesh are relatively innocuous parasites in the mucosa or lumen of the small intestine. Three days after ingestion, the young females are impregnated by the males which subsequently die and are eliminated. During the next 2 days the females burrow deep into the intestinal wall and on the fifth day begin to produce small juvenile worms. Each ovoviviparous female, during a maximum period of 3 months in the intestinal wall, produces at least 1500 juveniles which are distributed by the host's lymphatic and blood streams. The young worms invade chiefly the ends of active skeletal muscles, such as the diaphragm, intercostals, and those of the tongue, throat, and upper trunk. Mass encystment in the muscle cells with attendant necrosis is the critical phase of the disease. The victim may succumb at once to a combination of toxemia, heart inflammation, pneumonia, and kidney failure. In mild cases, prognosis is usually good; however, the enteric phase of a mild infection is often confused with "intestinal flu" or other gastrointestinal disturbances, while the phase involving migration and penetration of the juveniles may be confused with rheumatism or other muscular pains. After several months a slow calcification of the cysts begins, and although this process is usually completed in about 18 months, the encysted worms may remain alive for several years longer.

254   ASCHELMINTHES

Whole mounts of adult males and females as well as living encysted and unencysted juveniles should be available for study. Living rats or rat carcasses which are already heavily infected with encysted *Trichinella* juveniles are available from commercial supply houses. Small fragments of muscle from an infected rat can be compressed between glass slides to see the living worms or can be digested in a proteolytic solution to excyst them. By feeding bits of infected muscle to small rodents, adult worms and a second generation of juveniles may be recovered. The small intestinal mucosa of the host should be examined, with the aid of a dissecting microscope, 3 days after feeding to recover sexually mature male and female worms. In 3 weeks or slightly longer, a heavy juvenile infection will undoubtedly be found in the muscles of the host. Stained permanent whole mounts of juveniles as well as sectioned infected muscle should be examined to note details of the encysted worm and of a capsule wall of host origin. Furthermore, typical changes in pericapsular host tissue, as are mentioned below, should also be noted.

Morphology of the Adults. Males, which are approximately 1.5 mm. long by 0.05 mm. wide, and females (Fig. 144), which are approximately 3–4 mm. long by 0.08 mm. wide, have pointed anterior ends and somewhat fleshy posterior ends which in the male bear two large, conical papillae. The adults have a smooth cuticula and an underlying, thin, scattered hypodermis. A simple, apical mouth opening leads into a short, muscular pharynx which in turn connects to a long, capillary esophagus extending posteriorly nearly half the length of the body. The esophagus is partially enclosed in a column of large granular cells, the stichosome, which is thought to be glandular. A simple intestine runs from the esophagus to a terminal anus. Muscular and nervous systems are difficult to observe in whole mounts and will not be described. An excretory system is apparently absent. The reproductive systems will be described separately.

Male Reproductive System. A relatively large testis begins near the posterior end and passes forward to the level of the esophagus where it recurves sharply and merges into a vas deferens. This sperm duct runs posteriorly to an ejaculatory duct, which opens to the outside through a terminal cloaca. The cloaca is eversible and functions as a sperm-emissive organ. It is guarded by the two fleshy, subterminal papillae which are undoubtedly sensory and assist in copulation.

Female Reproductive System. The female reproductive system passes anteriorly from a cylindrical, posterior ovary. In front of the ovary there is a short, narrow oviduct which joins a distended uterus, the first part of which serves as a seminal receptacle; a narrow vagina joins the uterus and in the midesophageal region opens to the outside through a vulva. The thin-shelled eggs develop as they move

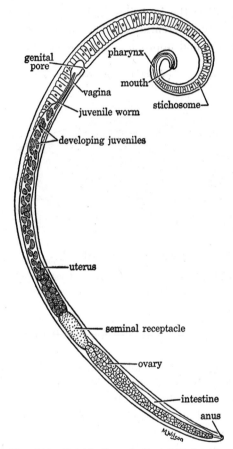

FIG. 144. *Trichinella spiralis*, adult female.

forward, and by the time they reach the anterior end of the uterus they have hatched into small "microfilaria-like" juveniles which are deposited by the female in the enteric tissue.

Morphology of the Juvenile. The newly born juveniles are approximately 100 μ in length by 6 μ in diameter and begin to reach the skeletal muscles 9 days after the ingestion of viable cysts. Locate living worms in small pieces of carefully compressed rat muscle and notice their general lack of internal organization. The reproductive system is not yet formed, and except for the presence of an oral stylet, which apparently is used together with extracellular, histolytic enzymes to lacerate and invade the muscle cell, the digestive system is rudimentary.

In the muscle fiber the juveniles rapidly grow to 1 mm. in length and become sexually differentiated. They assume characteristic spiral shapes in the fibers and approximately 2 weeks later are enclosed in

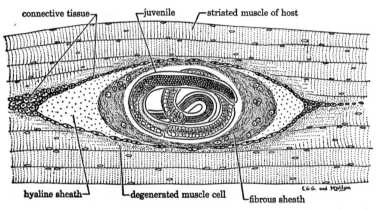

FIG. 145. *Trichinella spiralis*, encysted juvenile. Cyst is shown in section; juvenile is shown as an intact specimen.

thin, adventitious cysts of host origin. The lemon-shaped cyst is fully formed in 9 weeks, at which time it is 0.5 mm. in length by 0.2 mm. in diameter.

Longitudinal sections and whole mounts of well-formed cysts should be available for study at this point. The cyst (Fig. 145) lies parallel to the muscle fibers and has a wall of two layers, an inner, nucleated and fibrous zone and an outer, chromophobic, hyaline layer. Usually one, often two, and as many as seven juveniles may be contained in one cyst. Each worm is nearly mature and possesses all the organ systems of the adult. This precocity results in rapid sexual maturity of the excysted juvenile. Various types of connective tissue cells of the host may be seen surrounding the cyst, and degeneration of adjacent muscle fibers is usually apparent. Study the pericystic zone under high power or oil immersion. Notice the disappearance of the typical structure of the striated muscle cell. Notice also the abundance of nuclei surrounding the cyst, and the thickening of the sarcolemma. Fatty degeneration may also be apparent.

Trichuris trichiura

By C. G. Goodchild

Trichuris trichiura, the whipworm, is a human parasite frequenting the caecum and adjacent regions of the digestive tract. It occurs commonly in children in warm, moist regions of the world, and although world wide in distribution, it is rather locally restricted in the United States. Several species of whipworms are common in many mammals, including sheep, goats, hares, rabbits, mice, dogs, and foxes. However, *T. trichiura* is generally restricted to man, with certain other primates and the pig harboring perhaps the same species.

The life cycle of the whipworm is completed in a single host. The egg, which undergoes cleavage and juvenile development outside the host, requires more or less continuous moisture over a 10–14 day period for successful embryogeny. Infection occurs when embryonated eggs gain entrance to the body, often in contaminated food or drink. Juveniles which hatch from eggs in the small intestine temporarily penetrate, or attach to, the intestinal wall to obtain nourishment. A slow, posterior migration, during the next 10 days, finally brings them to the region of the caecum where they permanently attach. Three months later they attain sexual maturity and begin to produce eggs.

Clinical symptoms are usually absent in persons with light infections. This worm can produce grave complications, however, because of its habit of penetrating the mucosa, and often the wall, of the digestive tract. Ulceration of the digestive tract or peritonitis may result; moreover, the smaller lacerations may provide pathways of entrance for secondary bacterial invaders such as staphylococci or streptococci. Whipworm-induced inflammation of the appendix may result in an attack of acute appendicitis.

Preserved specimens of adult male and female worms as well as eggs are desirable for a complete study of this species.

Morphology of the Adults. This dioecious worm is 30–50 mm. long and possesses a unique and characteristic body shape (Fig. 146). The anterior three-fifths of the body is a narrow tubule, while the posterior two-fifths is thick and fleshy. The common name, whipworm, was given because the anterior region resembles the lash of a whip, while the thick posterior part resembles the handle. A secreted, finely sculptured cuticula covers the body.

A simple digestive tract proceeds from a small mouth at the apex of the narrow tubule to a posterior anus. A layer of mucus usually encloses the anterior end; the worm undoubtedly aspirates liquefied

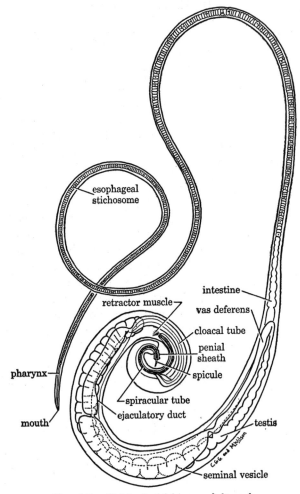

Fig. 146. *Trichuris trichiura*, adult male.

mucosal tissue and blood as food. Liquefaction of the host's tissue is accomplished by extracorporal enzymes which emerge from the mouth; blood is obtained by laceration of the host's blood vessels with a hard, sharp, piercing oral stylet. The digestive tube anteriorly is differentiated into a short, muscular, sucking pharynx followed by a long glandular stichosome, or esophageal gland, composed of a single column of granular stichocyte cells partially enclosing a small-bore, capillary esophagus. The esophagus joins an intestine which passes posteriorly to a terminal anus in the female; in the male, the intestine opens subterminally into a cloaca which will be described in the account of the male reproductive system.

An excretory system is absent, and the muscular and nervous systems are difficult to study from whole mounts; they will not be described. The reproductive systems are the prominent organs, and they will be described in detail.

Male Reproductive System. The male reproductive system forms a tight, elongated, U-shaped complex of organs in the fleshy part of the worm (Fig. 146). A sinuous, slightly distended testis originates near the posterior end and passes forward to the anterior level of the fleshy portion. At this point a constricted vas deferens joins the testis, and immediately it turns abruptly posteriad and parallels the testis a short distance. The next region is a sperm-filled seminal vesicle which connects anteriorly to the vas deferens and posteriorly to a muscular-walled, narrow-lumened ejaculatory duct (cirrus organ) which opens dorsolaterally into the cloaca.

The posterior end of the male worm is coiled dorsally in a flat 360° spiral and is provided with a conspicuous ensheathed copulatory spicule. Examine the end of the male worm and observe the following parts and determine their arrangement. The cloacal tube proceeds posteriorly from its junction with the ejaculatory duct and the intestine to within 1 mm. of the end, where it joins a spiracular tube. This latter tube contains a single, lanceolate spicule which is about 2.5 mm. long and is enclosed in a retractile penial sheath beset with many sharp, stout, recurved spines. The spicule may function like a plunger to transfer sperm in a copulating pair held firmly together by the spines of the penial sheath. Retractor muscles are inserted on the inner ends of the spicule and sheath. Digestive and reproductive systems open to the surface at the terminal cloacal pore.

Female Reproductive System. The reproductive organs in the female are located also in the fleshy part of the worm, which in this sex is straighter than it is in the male. A sinuous, distended ovary originates near the posterior end and runs to a point near the narrow tubule region. At this point, it joins a narrower oviduct which immediately bends posteriad and passes to the posterior tip. Here it recurves anteriad and joins a greatly expanded uterus filled with eggs. The uterus proceeds to the anterior ovarian level, where it joins a narrow, muscular, serpentine vagina which opens to the outside at a vulva. The reproductive pore is located on a slight prominence at the junction of the narrow tubular and the fleshy body regions; it is a transverse slit with a radially striated cuticula.

Several thousand fertilized eggs emerge from the reproductive opening each day. The lemon-shaped egg measures about 50 μ in length by 22 μ in diameter. It is provided with a double shell, the outer one of which is slightly thicker and becomes bile stained in the enteron. Each end of the egg is provided with an unstained, hyaline plug. The ovum contained in the egg is covered with a vitelline membrane and is undivided at the time it is discharged in the feces.

7.

Entoprocta and Ectoprocta

ENTOPROCTA

Barentsia

By M. D. Rogick

Barentsia occurs as soft, tufted, brownish or gray colonies ranging up to half an inch in height, and is usually found on various submerged objects in sea water. This organism can be obtained very conveniently for study by submerging slides in sea water for 2 or 3 weeks. By mounting some of the bryozoan material on a slide and studying with a compound microscope most anatomical detail can be discerned.

The individuals or zooids consist of a calyx or head, a stalk, and a swollen base or musclium. The individuals of a colony are connected with each other basally by stolons growing over the substratum. Incomplete septa separate the stalks from the musclia and the musclia from the stolons.

The calyx (Fig. 147) is from 0.22 to 0.78 mm. in height and from 0.15 to 0.52 mm. in width. It is gibbose posteriorly and flattened anteriorly and dorsally. The transparent, thin-walled calyx bears 13–23 slender, active tentacles. It contains a digestive tract, protonephridia, ganglia, nerves, reproductive organs, and a brood chamber for embryos. A decapitated stalk can regenerate a new calyx.

The slender ciliated tentacles can be curved inward into a ciliated atrium. When the tentacles are turned inwards, the calyx rim or velum may contract over them. The mouth and the anus, as well as the openings of the reproductive and protonephridial ducts, lie within the tentacular circle or lophophore. The two latter sets of openings are very difficult to see.

The most conspicuous organs in the calyx, exclusive of the embryos, belong to the digestive system. They are the mouth, the wide pharynx, the narrowing tubular esophagus which is vertically placed along the

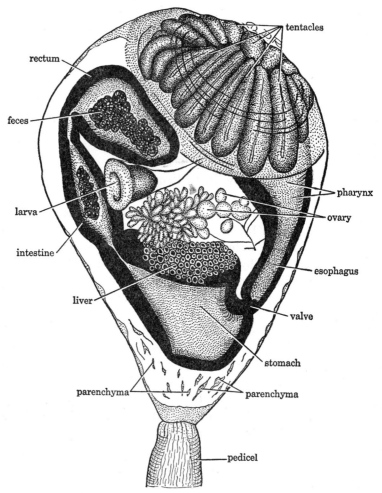

Fig. 147. Calyx of *Barentsia laxa*.

flattened side of the calyx, and the large sac-like horizontal stomach whose upper wall is yellowish and thickened into a digestive gland or "liver" and whose lower wall follows the ventral contours of the calyx closely. A tubular intestine lies vertically against the gibbose side of the calyx, and a similarly shaped rectum is bent horizontally forward over both the gonads and brood chamber and below the floor of the atrium. The epithelium of the entire gut is ciliated; the cilia are longer and more prominent in the stomach, near the esophageal valve.

Barentsia feeds on protozoans, bacteria, and other small organisms and organic detritus. These food particles are swept towards and into the mouth by currents created by the cilia on the tentacles. The food mass revolves within the stomach continuously, as does the material within the intestine and rectum; the material revolves in one direction for a number of turns, then in the other.

The reproductive organs are absent from some calyces, presumably immature ones. When present, they occur as large masses above the liver, in front of the intestine and beneath the rectum. In the male calyx the testes appear as two very large, lateral, elliptical, opaque masses. Within the testes are numerous, fine, thread-like, densely packed spermatozoa which are in constant motion as time for release approaches. In the female calyx the ovaries are in the same position relatively as are the testes. Some species of *Barentsia* are hermaphroditic. The ovaries contain oval or pear-shaped developing eggs. The same zooid may also be carrying embryos in various stages of development in the brood chamber, just posterior to the ovary. Embryos most advanced in development push upward toward the atrial floor to be released. Sometimes three or more of these large opaque larvae may be lined up in the reproductive passages before release. Released larvae are about 0.09 mm. wide and 0.15 mm. long and are partially ciliated, somewhat translucent, and provided with a digestive tract which is already functional. Their digestive tracts resemble very closely those of the adult zooids. The larvae swim very actively for a time after their release and before they become permanently attached.

Ganglia of the simple nervous system are located between the ovaries. These ganglia and the nerves proceeding from them are difficult to see.

The protonephridia are anterodorsal to the ovaries and posterior to the pharynx and esophagus. They are slender, short, tubular processes with internal, intermittently flickering flagella which can be discerned only when the flagella are active.

Muscle fibers occur in the body wall, tentacles, wall of the digestive tract, stalk, and the enlarged muscular base. A diffuse mesenchyme tissue or parenchyma occurs in the calyx.

The cylindrical stalks bearing the calyces are transparent and of varying degrees of flexibility, depending upon the age of the zooid and the species. They range from 0.37 mm. to 9 mm. in length. The part of the stalk near the calyx is more flexible and is called the pedicel; the rest of the stalk is the peduncle and the musclium in which the muscle fibers are clearly visible.

Occasionally, in addition to the basal musclium, there may be one or two other musclia inserted along the peduncle. These may sometimes have stolons arising from their sides. Each basal musclium is connected with the musclium of another zooid by a slender, transparent, cylindrical stolon of varying dimensions (0.07 to 1 mm. in length and 0.03 to 0.06 mm. in width). From one to four stolons may arise from the base of the basal musclium.

In many colonies which are observed there are live stalks which have lost their calyces. Sometimes new heads may form under degenerating heads before the degenerated ones drop off. At other times, the calyx may be broken off before a new one has begun to develop at the pedicel tip. In the latter case, just as in the former, a new head is regenerated if conditions are suitable.

ECTOPROCTA

Crisia

By M. D. Rogick

Crisia consists of dense, bushy, white, dendritic masses on seaweeds, hydroids, and other submerged objects. The colonies or zooaria may attain a height of about 20 mm. The animals feed on suspended organic particles which they bring to their digestive tract by means of their ciliated tentacles.

The zooarium is usually attached to seaweed by a small rounded calcareous primary disk which in some instances is separated from the erect primary zooid by a constricted chitinous joint or basal tubulus. In addition to the basal disk there are sometimes slender calcareous rootlets helping to attach the colony to the substratum.

The zooarium consists of several kinds of zooids: autozooids, gonozooids, and rhizooids. The autozooids are by far the most numerous and common. The other two types are few in number and hard to find but are taxonomically important.

The primary zooid gives rise by budding to autozooids which by repeated budding produce the branching tufted colony. The branches of the dendritic zooarium consist of groups (internodes) of zooids, each group or internode being separated from the ones below, above, or to the side by a yellow or brown narrowed chitinous joint or node. The number of zooids in an internode varies from one to over thirty, depending upon the species, and other factors such as location in the zooarium.

The autozooids possess long, narrow, cylindrical, calcified tubes whose apertures are terminal and whose walls are marked by numerous widely spaced, but small, pseudopores where the calcareous layer of the body wall is interrupted. The large terminal aperture is closed by a soft delicate terminal membrane which forms part of the soft polypide occurring inside the calcareous tube. If the living colony is undisturbed for a time, a tentacular crown, or lophophore, with its eight slender tentacles may be seen to protrude through the terminal aperture.

The gonozooids or oöecia are large, swollen, vase-like zooids which serve as brood chambers for developing embryos. Their walls are marked by numerous pseudopores which are placed very close together. The proximal part of the gonozooid is a narrow, tapering tube wedged in between neighboring autozooids and sometimes partly hidden by them; the middle part, or brood chamber proper, is very greatly dilated but tapers gracefully into the proximal part; and the distal part, or oöeciostome, is a narrow, constricted tube, usually very short, through which the larvae leave the gonozooid.

Bowerbankia

By M. D. ROGICK

Bowerbankia is a soft colonial ctenostome, occurring as a dense gray or brownish fuzz only 1 or 2 mm. high on seaweeds, hydroids, rocks, and other submerged objects. These animals are ciliary feeders, bringing suspended organic particles into their digestive tract by means of their ciliated tentacles. Mount a small fragment of the colony on a slide. Add some sea water and a coverslip and study the preparation with the aid of a compound microscope. Slides which have been immersed in the sea for 2 weeks or so may have young colonies containing very few individuals; these are ideal for study.

The zooarium consists of numerous tall cylindrical zooids closely packed together. These arise by budding from a basal stolon which is easily discernible in young colonies but more difficult to see in older ones because of zooecial crowding. The zooids range from 0.42 to 1.5 mm. in height and from 0.16 to 0.24 mm. in width, while the diameter of the stolon is from 0.07 to 0.11 mm.

The thin-walled, cylindrical zooids are flexible, transparent, and frequently have square distal tips. They arise usually from the sides of the stolon singly, in pairs, or sometimes even in groups paralleling one another. Their basal ends may or may not be drawn out into a narrowed caudal process.

The zooid has eight to ten ciliated retractile tentacles which are borne on a circular lophophore. When protruded they are surrounded by a short delicate setigerous collar which is almost imperceptible because of its transparency. The tentacles can be retracted completely into the body cavity by the action of retractor muscles. The body wall, directly below the tentacles, is also pulled inward, to form a sheath for the tentacles.

Above the retracted tentacular sheath is a narrow channel, the vestibule, also formed as a result of the partial inversion of the body wall. Toward its lower end are several groups of muscle fibers, the parieto-vaginals, leading from the outer body wall to its wall. There are several other bands of muscle fibers, including the parietals which are arranged hoop-like, horizontally along the body wall proper.

The digestive tract consists of a mouth, pharynx, esophagus, gizzard or proventriculus, stomach, intestine, and anus. The mouth opens within the tentacular circle; the anus opens outside the circle. The pharynx is directly beneath the tentacular crown and tapers into a long narrow esophagus which ends in an enlarged, bulbous gizzard, lined with teeth. The brownish or yellowish gizzard has a firm muscular wall. The stomach is a large, wedge-shaped, elongate, soft sac extending downward to the base of the zooid. A slight constriction, the pylorus, separates the stomach from the long narrow intestine which parallels the esophagus within the tentacular sheath. When the tentacles are retracted, the digestive tract is thrown slightly out of position; the esophagus folds into a loop and the stomach and intestine are also slightly bent.

The stolon is slender, transparent, and occasionally septate, giving off buds which develop into zooids. These buds vary in size and stage of development. Some are simple transparent sacs; others have very short young tentacles and a tiny digestive tract.

Electra

By M. D. ROGICK

Electra appears in nature as a fragile, white, flatly encrusting calcareous patch on seaweeds, stones, and other submerged objects. Young colonies are fan-shaped. Older colonies are more circular but may sometimes be quite irregular peripherally and consist of hundreds of zooids. The zooids bud in rows, somewhat radially.

Each zooid consists of two parts: the zooecium and the polypide. The zooecium constitutes the peripheral body wall and the firm outer, partially calcified case; the soft inner polypide includes the tentacular crown, digestive tract, and associated musculature.

Each zooecium has the following surfaces: (1) a surface for attachment of the zooid to the substratum, (2) an upper free or frontal surface, (3) two lateral surfaces, (4) a proximal surface nearest the parent zooid, (5) a distal surface farthest from the parent zooid. The lateral surfaces, the side walls of each zooid, are perforated by a few pores. The upper surface consists of two parts: a calcified proximal perforate part, the gymnocyst, and a distal elliptical membranous area, the aperture. The latter large area is covered by a transparent, lightly chitinized membrane, whose distal part has a hemispherical flap or operculum through which the lophophore, bearing the ciliated tentacles used in feeding, may be protruded. The crescent-shaped, distal border of the operculum is more heavily chitinized. There is a circle of several short, curved spines bordering the frontal area and curving over it. In some zooids these may be broken off. In addition to these spines there is another and larger spine, the basal spine or flagellum, which is placed proximally below the frontal area. The gymnocyst is perforated by a number of large or small pores, depending on the species, being large in *E. pilosa* and small in *E. hastingsae*.

Bugula

By M. D. ROGICK

Bugula appears as soft, yellowish dendritic colonies on seaweeds and other submerged objects. Some species have a spiral growth habit (*B. turrita*), others appear fan-shaped (*B. flabellata*), and still others have an irregular bushy appearance.

The colony or zooarium consists of fronds of numerous box-like or roughly tubular zooids which arise by repeated budding. These zooids are arranged in two or more rows in each frond, depending upon the species. The zooarium is attached to the substratum by a firm holdfast of irregular shape. Slides which have been submerged in the sea for 1 or 2 weeks may be an excellent source of young *Bugula* colonies containing only a few individuals and showing the holdfasts well.

The yellowish cuticula forming the outer wall of each zooid is so transparent that the internal organs can be readily studied. There is a large U-shaped area, the aperture, occupying the upper half or two-thirds of the frontal wall of each zooecium. It is covered over by a thin membrane which is a part of the body wall of the animal. The tentaculated lophophore can be extruded through the aperture. A specialized individual shaped like a bird's head, the avicularium, consists of head, beak, and stalk and is attached alongside the aperture. The head part contains fan-shaped muscles whose fibers converge toward the beak or mandible. The avicularia in living specimens nod constantly and snap their jaws, but their function is unknown.

Small, shallow hemispherical or beret-shaped ovicells overhang the distal end of some of the zooecia. These ovicells or oöecia are brood pouches for developing embryos. If embryos are still in them, the ovicells look opaque or reddish. If the embryos have left, the ovicells are transparent.

There are about fourteen tentacles. They are borne on a circular, fleshy ridge, the lophophore, which surrounds the mouth. The tentaculated lophophore can be withdrawn, pulling in after it a portion of the peripheral body wall, the introvert, which forms a sheath for it. This retraction is accomplished by two fairly thick bands of retractor muscles which insert on the lophophore and originate on the body wall.

The tentacles are ciliated and create currents which bring suspended food material to the mouth. The digestive tract is V-shaped and consists of a mouth, pharynx, esophagus, stomach, caecum, intestine, and anus. The pharynx is just below the circle of tentacles. The caecum is that portion of the stomach which forms the bottom of the V and is anchored to the body wall by a double strand, the funiculus. The anus is near the mouth and pharynx but lies outside the tentacular circle.

Many of the zooids contain ovoid brownish structures called brown bodies. These are degenerating polypides; new polypides normally regenerate from the zooecium.

Both male and female reproductive organs may occur in the same zooid. The testes are soft, fluffy growths on the funiculus, consisting of germinal tissue and actively moving spermatozoa. The ovary is attached to the body wall more distally than the testes and near the midregion of the body cavity. The egg is fertilized by a spermatozoan, presumably from the same zooid; the zygote then passes into the ovicell where it develops further.

A small ganglion is present just slightly beneath the tentacular crown, between the upper part of the pharynx and the intestine.

Plumatella

By M. D. ROGICK

Plumatella constitutes firm-walled, brown, moss-like growths on stones and other submerged objects in fresh-water habitats. The colonies may be extensive, sometimes over a foot in area. In some cases the colony may adhere closely to the substrate, while in other cases the zooids may crowd upward into dense masses 1 or 2 inches high. The colony consists of long, cylindrical, slender zooids (Fig. 148*D*) each of which buds to produce one, or sometimes more, new zooids.

The zooid body wall consists of two major layers: ectocyst and endocyst. The ectocyst in some species is keeled, opaque, and encrusted with debris and sand grains. In other species it is quite translucent. The endocyst is the soft inner membrane of the body wall. It is generally continuous from zooid to zooid.

The tentacles, numbering from thirty to fifty, are borne on a horseshoe-shaped lophophore which is retractable into an introvert, or inverted tentacular sheath. The digestive tract, retractor muscles, ganglion, and mode of polypide retraction are somewhat similar to those of *Bugula*. A funiculus is present.

Plumatella reproduces sexually and asexually. The latter method, involving statoblasts, is peculiar to fresh-water species and will be described briefly. Statoblasts are produced in large numbers throughout the summer. With the onset of winter most of the colonies die, sometimes leaving the zooecial tubes filled completely with statoblasts;

in other instances the statoblasts are released earlier. The statoblasts possess hard brown bivalve cases. They are elliptical and flattened and about 0.3–0.5 mm. in length. They contain living germinative

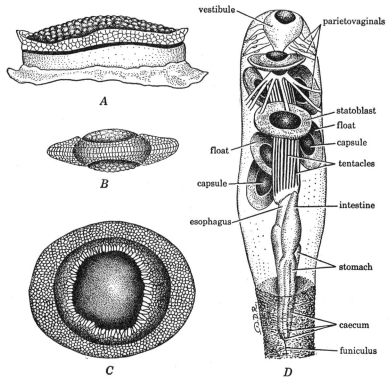

Fig. 148. *Plumatella*. A. Sessoblast, side view. B. Floatoblast, side view. C. Floatoblast, dorsal view. D. Zooid tip containing seven mature floatoblasts and polypide.

tissue from which a new colony may arise when environmental conditions are favorable for germination. The floatoblasts (Fig. 148B, C) have a central, rounded, brown capsule surrounded by a light-colored float of "air cells." The sessoblasts (Fig. 148A) possess a very large capsule with a vestigial chitinous float or annulus. Statoblasts are of taxonomic importance in distinguishing various species of fresh-water bryozoans.

8.

Annelida and Sipunculoidea

ANNELIDA

Neanthes virens

By F. A. BROWN, JR.

Neanthes virens, commonly called the clamworm, is one of the largest and most common of the marine annelids of the North Atlantic coast of North America. This worm was formerly known as *Nereis virens*. Large individuals range up to 40–45 cm. in length and to nearly 1.25 cm. in width. Their coloration ranges from steel blue through greenish, with tinges of orange and red especially on the parapodia. In general the tendency towards development of an orange or reddish coloration is greatest in females during the breeding season.

Neanthes normally lives in tubes which it forms in the sand or mud. These excavations are made by activity of the protrusible pharynx and are lined with mucus. The worms remain within the tubes most of the time, occasionally protruding the anterior end of their bodies in search of food. They are reported to be omnivorous.

Only during the breeding period do these worms commonly leave their burrows and become free-swimming. This phase of their life cycle is often associated with the differentiation of two distinctly different regions of the body, an anterior atoke and a posterior epitoke. The latter portion bears the gametes; both its parapodia and chaetae are larger and more leaf-like, an adaptation to the swimming habit. This special reproductive individual is called a heteronereid.

If living specimens are available, the patterns and rhythms of locomotory activity may be studied. Locomotory patterns appear to be of three types. (1) Slow ambulation. In this, only parapodial action is observed. Each parapodium shows during slow (and also rapid) ambulation a characteristic type of circular movement involving an effective and a recovery stroke. In the effective phase the parapodium is protruded and moved posteriorly and ventrally in contact with the substrate. In the recovery phase it is retracted, lifted upwards, and moved anteriorly. Observe the character of each phase. Note that, as a quiescent worm becomes active, the sequence of initiation of activity passes rapidly from anterior to posterior end. The waves of rhythmical activity of an active worm, however, are readily seen to pass in an anterior direction, with the activity on the two sides of the body in opposite phase. (2) Rapid ambulation. This involves lateral undulations of the body of small amplitude in addition to the parapodial activity discussed under slow ambulation. (3) Swimming. This involves lateral undulatory waves of large amplitude in addition to the usual parapodial activity. Determine for the lateral undulations of the last two types of locomotion the direction taken by the wave of initiation of the pattern and also determine the direction of movement of the activity wave itself.

Place a glass tube of a bore slightly greater than the diameter of the worm in the container with the worm. Observe the readiness with which the worm enters the tube and its reluctance to leave it. Study the intermittent undulatory activity which pumps water effectively through the tube. By means of a little added carmine suspension determine the direction the water takes through the tube. Observe the pattern of activity of the respiratory movements. Unlike for the locomotory rhythms, the undulations are dorsoventral and the waves of activity pass posteriorly.

With forceps hold a small piece of mollusk flesh or other natural food at the tube entrance. Observe the manner of feeding in *Neanthes*.

External Anatomy. For this study, a specimen freshly killed by immersion in 7 per cent alcohol is best, though a preserved specimen is quite satisfactory.

The body is composed of an anterior prostomium followed by a variable number of more or less similar segments ranging sometimes to as many as 200 or more (Fig. 149). Segments are increased in number during the growth of the worm by production of new ones just anterior to the caudal one. With the exception of the first segment or peristomium, and the caudal one, these segments possess lateral appendages, the parapodia.

The prostomium bears a pair of small tentacles at its anterior tip. Laterally it bears a pair of stout, fleshy, and somewhat contractile palps. Dorsally are located two pairs of pigmented eyes.

The peristomium bears four pairs of tentacular cirri, two dorsal and two ventral. The two longer dorsal cirri may extend backwards to the region of the fifth to ninth segments.

The remaining segments of the body bear parapodia (Fig. 150). These increase in size over the first few segments, are large in the

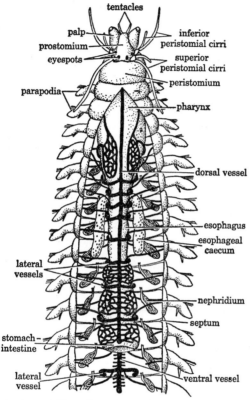

FIG. 149. Dorsal view of *Neanthes virens,* showing the general arrangement of internal organs.

midregion of the body, and decrease in size towards the posterior end. Study carefully with the aid of your microscope a segment cut from the midregion of a preserved specimen. The parapodia are biramous, possessing a dorsal division or notopodium and a ventral one or neuropodium. Each division contains a large, black, tapering, supporting rod or chaeta, called an aciculum, largely embedded within it. At its proximal end each aciculum has attached musculature whose activity results in protrusion and retraction of the parapodium. Both rami of a parapodium typically possess, in addition, a bundle of chaetae projecting from chaetigerous sacs also possessing associated muscle fibers capable of bringing about movements of these chaetae. The chaetae are jointed structures, having a blade and a shaft. Two types of chaetae are present; one possesses a short, the other a long, distal blade. The first and second pairs of parapodia lack the notopodial chaetae.

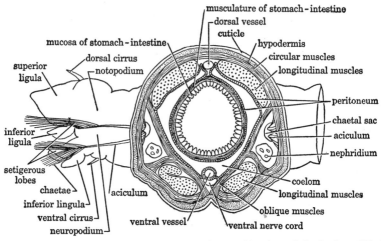

FIG. 150. Cross section of *Neanthes virens* from midregion of the body. (Modified after Turnbull.)

Each notopodium has two flattened, richly vascularized, and ciliated lobes, the superior and inferior ligulae. Dorsally, at the base of the superior one is the dorsal cirrus. The neuropodium has a single ligula, narrower than the dorsal ones, and a ventral cirrus. Other lobes, at the bases of the chaetal sacs, are called the chaetigerous lobes.

Nephridiopores may be seen upon the ventral surface of each segment except the first and the last, close to the base of the ventral cirrus.

The last segment, bearing the terminal anus, has no parapodia but has a single pair of elongated ventral cirri, the anal cirri.

Coelomic Cavity and Its Fluid. Make a short, longitudinal incision immediately to the right or left of the dorsal midline near the midregion of the body. Place a small drop of the coelomic fluid on a slide and examine it with the compound microscope. The fluid contains amoeboid corpuscles and, especially during the breeding season, numerous reproductive cells in various stages of their development. If reproductive cells are present, determine the sex of your specimen.

Now continue the dorsal incision anteriorly to the end of the worm. Tease carefully the internal septa and pin out the body wall. The specimen should be submerged during the dissection and study.

The coelom in *Neanthes* is an extensive perivisceral cavity. It is divided incompletely into compartments by segmental septa which pass from the body wall to the visceral organs. Perforations in the septa allow passage of the fluid from one region to another. A visceral peritoneum encloses the digestive tract which is suspended within the coelom. Segmentally arranged through much of the length of the worm there are areas of ciliated peritoneum covering the dorsal longitudinal muscle tracts. These patches are known as the dorsal ciliary organs, which are believed to be vestiges of the primitive coelomoducts. Search for these if your specimen has been freshly killed.

Body Wall and Musculature. Study the structure of the body wall (Fig. 150). The body is covered by a thin cuticle whose iridescence is due to the possession of numerous striations. There are also numerous pores through which hypodermal gland cells discharge their products to the exterior. The hypodermis is a single-celled layer, thicker on the ventral surface where its contained glands are larger and more numerous. A portion of the dorsal hypodermis is well vascularized by a plexus of blood vessels, thereby appearing to assist the ligulae in respiratory gaseous exchange. Internal to the hypodermis is a thin layer of circular muscle. Internal to the latter are four longitudinal muscle tracts, two dorsolateral and two ventrolateral. Oblique muscle fibers originate in the median ventral line and pass dorsolaterally into the lateral body wall. A parietal peritoneum covers the musculature on its inner surface, lining the coelom.

Alimentary Canal. The mouth is located ventral to the prostomium and is bordered by the peristomium which forms the buccal ring. The mouth opens into a buccal cavity. The pharynx lies just posterior to the latter and has a highly muscular wall. The posterior portion of the pharynx, the dental region, possesses special muscle tracts in which are embedded laterally the hollow bases of a pair of large, powerful jaws. All the foregoing regions of the gut, constituting the foregut, are lined with a cuticle continuous with that of the exterior. Numerous regularly arranged denticles are also on this cuticle of the dental region. The foregut may be completely everted as a proboscis by means of protractor muscle bands connecting the pharynx with the peristomial wall and by elevated internal hydrostatic pressure (Fig. 151). As the dental region is everted, the two large jaws open widely. As the proboscis is retracted by retractor muscle bands which extend to the body wall from the posterior dental region, the jaws are brought together and cross one another, effectively holding anything which comes between them. Food organisms can be killed and torn apart by their action. The pharynx opens posteriorly into a narrower esophagus (Fig. 149). Two large caeca open into it. After passing through a few segments, the esophagus enters the stomach-intestine which extends the remainder of the length of the body with little change. A sphincter muscle lies at the junction of the esophagus and stomach-intestine. The stomach-intestine has both longitudinal and

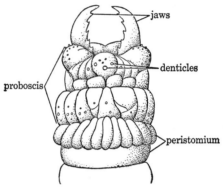

FIG. 151. Everted proboscis of *Neanthes virens*, showing jaws. (Redrawn from Turnbull.)

circular muscle fibers, the former lying nearer the peritoneum, the latter nearer the gastric mucosa. The mucosa is a columnar epithelium. The stomach-intestine is segmentally constricted by septa. The stomach-intestine passes in the last body segment into the rectum, and it, like the pharynx, has a cuticular lining.

Circulatory System. The closed circulatory system of *Neanthes* comprises a highly differentiated system of vessels containing blood which is red because of hemoglobin in solution in its plasma. The blood is propelled through the system by peristaltic waves passing over the vessels. The principal vessels of the system are longitudinal and in the median line (Figs. 149, 150). One is dorsal and the other ventral to the alimentary tract. The blood is propelled anteriorly within the dorsal vessel and posteriorly within the ventral one. The peristaltic waves of the dorsal vessel are often clearly evident in an intact living specimen. This vessel receives a pair of lateral vessels in every segment except the first four or five. The lateral vessels bring blood dorsally from the ventral vessel. In addition each lateral vessel receives, shortly before entering the dorsal vessel, a vessel bearing aerated blood from the notopodium and the vascularized dorsal body wall. In each segment also, the dorsal vessel has vessels connected with an anastomosing network of vessels upon the stomach-intestinal wall. The dorsal vessel at the anterior end of the worm bifurcates to produce vessels passing lateroventrally to a vascular reticulum upon the esophageal wall and then continuing as lateral vessels to join the ventral vessel in the fifth segment.

The ventral vessel, like the dorsal, possesses connections in each segment with the stomach-intestinal network of vessels. From each circumintestinal segmental vessel, shortly after it leaves the ventral vessel, a branch passes into the neuropodium to provide blood to the respiratory capillary system of this appendage. At the posterior extremity of the body the ventral vessel is continuous with the dorsal one through a simple circumrectal vascular ring.

Metanephridia. In every segment of the body except the first and the last is a pair of metanephridia. Each metanephridium consists of a somewhat convoluted, ciliated tube embedded in a syncytial mass of protoplasm. The internal opening is in a ciliated funnel, the nephrostome, which is located in the segment just anterior to that containing the bulk of the tubule. The external opening, or nephridiopore, is located on the ventral surface of each parapodium, close to the point of origin of the ventral cirrus.

Reproductive System. *Neanthes* is dioecious. The gonads are simply seasonal outgrowths of the ventral coelomic lining of many segments. They become enlarged during the breeding season. At other times they are not evident. The gonads possess no ducts but simply discharge their products into the coelom, which may at times become quite filled with them. While floating in the coelomic fluid, the reproductive cells undergo development. The ripe gametes are discharged to the exterior by rupture of the body wall, and fertilization occurs externally in the sea water. There is often a modification to the heteronereid form during the breeding season, as described earlier.

Nervous System and Sense Organs. Carefully remove the gut forward as far as the pharynx to expose the ventral nerve cord.

The ventral nerve cord (Fig. 152) lies along the median floor of the body cavity immediately ventral to the ventral blood vessel. From each of the first three ganglia two pairs of lateral nerves arise; the nerves forming the anterior pair branch to supply the anterior septum of the segment and the next anterior segment, and the nerves constituting the posterior pair pass within the same segment to the parapodium and to the skin. In each of the remaining segments three pairs of nerves arise; the anterior pair innervates the segment next anterior, and the remaining two are intrasegmental.

Anterior to the ganglia of the first segment two circumpharyngeal connectives pass upwards to the brain located dorsally in the prostomium. The lobed brain sends nerves anteriorly to the tentacles and palps and four short optic nerves dorsally to the four eyes.

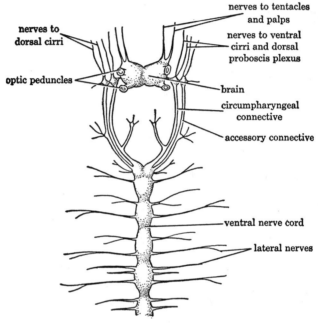

Fig. 152. Dorsal aspect of the central nervous system of *Neanthes virens*. (Redrawn from Turnbull.)

The circumpharyngeal connectives near their point of junction with the brain bear small ganglia from which arise nerves supplying the ventral pairs of peristomial cirri. Also from these small ganglia arise nerves passing to a system of ganglia over the dorsal surface of the proboscis. A pair of nerves arising from the connectives nearer their ventral ends supply a similar but ventral ganglionated plexus of the proboscis. These are ordinarily visible only after special staining such as with methylene blue. Arising from the first ventral ganglion and paralleling the connectives through much of their course is a pair of accessory connectives. Near the brain these give rise to nerves supplying the dorsal pairs of peristomial cirri.

The prostomium bears upon its dorsal surface over the brain two pairs of conspicuous eyes and ciliated nuchal pits. The eyes are pigmented vesicles containing a transparent refractive body. The wall of the vesicles comprises the light-receptive cells which are continuous with the optic nerve fibers. The nuchal pits are considered to be chemoreceptive in function. It is highly probable that the tentacles, palps, and cirri also possess important sensory functions.

Arenicola cristata

By F. A. BROWN, JR.

The lugworm, *Arenicola cristata*, is found along the Atlantic coast of the United States from Florida to Cape Cod. It is the most common species of the genus in regions south of Cape Cod. North of the Cape a smaller species, *A. marina*, replaces it. The latter rarely occurs south of the Cape and then is typically in deeper water.

Arenicola cristata dwells in regions of muddy sand near or slightly below the low-tide mark. It occupies a tube which it forms by burrowing. Part of the tube possesses a lining of mucus. The tube is usually U-shaped, 1 or 2 feet deep, with one opening to the surface possessing castings about it, the other having a funnel-like form. The worm pumps aerated water through its tube by rhythmic peristaltic waves which pass anteriorly over the body. Its delicate gills are partially protected by the notopodial chaetae which project over them.

Arenicola both burrows and feeds by forcefully everting its proboscis into the sand and then retracting it with its load of sand. The burrow is enlarged by firm muscular peristaltic waves which also pass anteriorly. Organic material in the ingested sand is digested and absorbed as the material passes through the gut.

During the breeding season large numbers of eggs in elongated gelatinous masses are found at the mouth of the burrow.

Arenicola is best studied immediately after being killed by immersion for an hour or two in 7 per cent ethyl alcohol in sea water. Preserved specimens are, however, quite satisfactory for study of most aspects of their morphology.

External Anatomy. Study the general form and organization of the worm. The body comprises four general regions (Fig. 153). (1) An anterior head region comprises the prostomium, peristomium, and first achaetous segment. Posterior to this zone is a region (2) possessing six segments bearing parapodia. (3) A third region possesses eleven segments bearing both parapodia and gills. Finally, the most posterior region (4), the tail, has a variable number of segments and bears neither parapodia nor gills.

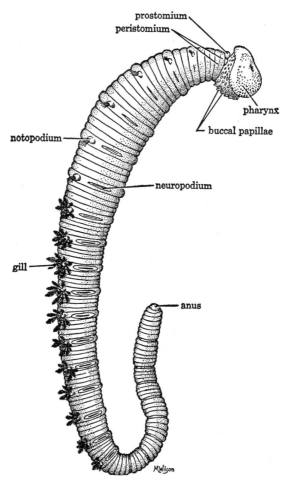

Fig. 153. Lateral view of *Arenicola cristata*.

The prostomium is small and is located anteriorly in the mid-dorsal line. It is trilobate and is bordered laterally and posteriorly by a deep groove, the nuchal groove. The prostomium bears no appendages. The peristomium possesses a single annulus which may be subdivided in two. The mouth is a transverse slit ventral to the prostomium and largely bounded by the anteroventral edge of the peristomium. It is bordered by papillae. The anterior region of the gut may be everted as a proboscis.

Following upon the peristomium are two annuli, the anterior one of which may be subdivided into two. These represent a first body segment which has lost its chaetae early in development.

The first three chaetigerous segments possess respectively two, three, and four annuli; the remaining segments characteristically have five annuli per segment, with parapodia located, when present, in the next to the last annulus. Therefore, in general, the parapodia are located quite far posteriorly in their segments.

The annulation of the tail region is quite variable. The number of annuli per segment increases from the anterior region backwards. Near the posterior end of each segment in the tail is seen a larger and more heavily pigmented annulus. This annulus appears to correspond to the chaetigerous annuli of the more anterior body regions.

The anus is terminal in the last tail segment.

Each of the chaetigerous body segments possesses a pair of reduced biramous parapodia. A dorsolaterally located notopodium is conical in form, and a small bundle of needle-like chaetae projects from a chaetal-sac opening at its tip. The chaetal sac is controlled by retractor and protractor muscles to be studied later. The notopodial chaetae are very sharply pointed; their distal portions bear a serrated edge and often also are blade-like.

The neuropodia are ventrolaterally located and comprise dorsoventrally elongated muscular ridges. Each neuropodium has a longitudinal slit-like opening into a chaetal sac. The sac contains large hook-like chaetae, or crochets. These chaetae are formed at the ventral end of the sac, older and more fully formed ones being found in the dorsal portion.

The parapodia of *Arenicola* have neither acicula nor cirri.

Nephridiopores occur only in the 5th to 10th chaetigerous segments, and may be readily seen. They are oval slits lying slightly dorsal and posterior to the dorsal ends of the neuropodia.

Eleven pairs of gills are intimately associated with the notopodia in chaetigerous segments 7 to 17 inclusive. They are red in color because of their contained blood. Each gill is an extensively branched outgrowth of the body wall and contains an extension of the coelomic cavity. It is thin-walled and consists of thin epidermal and peritoneal layers between which lie fine muscle strands and fine interconnecting branches of the afferent and efferent branchial vessels. In the living worm the gills may be seen to undergo a rhythmic contraction.

ANNELIDA AND SIPUNCULOIDEA

Coelomic Cavity and Its Fluid. Make a small longitudinal incision in the dorsal line near the midregion of the body. Collect some of the coelomic fluid, place it upon a microscope slide, and examine

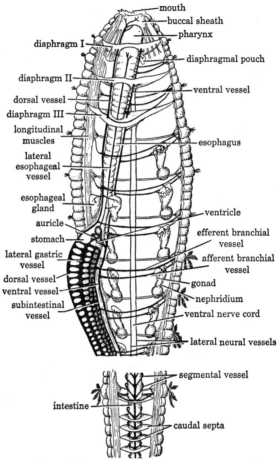

FIG. 154. Dorsal view of *Arenicola*, showing arrangement of major internal organs. (Modified after Ashworth.)

it with a compound microscope. Determine the sex of the specimen from the character of the reproductive cells which are almost always present in this fluid. In females oöcytes are observed in varying stages of growth. Their diameters range from about 20 μ to over 150 μ. The coelomic fluid of males contains clusters of spermatogonia each possessing anywhere from eight to myriads of cells. Ripe flagellated sperms occur in very characteristic discoidal masses. The abundance of reproductive cells in the coelomic fluid varies seasonally, with huge concentrations present during the breeding season.

Coelomic corpuscles, of two types, are also present. These are large fusiform cells from 40 to 50 μ in length and smaller amoeboid cells. The latter often contain yellowish or brownish granules. These amoeboid cells are phagocytic.

Now carefully continue the mid-dorsal incision anteriorly to within 3 or 4 mm. of the anterior end and posteriorly a short distance into the tail region. In the last the gut will be entered unless extreme precaution is exercised.

The coelom is very extensive and extends throughout the length of the worm. Anteriorly it is incompletely divided into compartments by three fenestrated septa known as the diaphragms (Fig. 154). The first of these marks the anterior boundary of the first chaetigerous segment; the second, the posterior boundary of the second chaetigerous segment; and the third, the posterior boundary of the third chaetigerous segment. All three diaphragms contain muscle fibers. The fenestrations are of such size as to permit free passage of the coelomic fluid and its smallest cells. The first diaphragm bears two finger-like pouches projecting backwards but opening into the anteriormost coelomic space. The pouches are contractile and richly vascularized. Short dorsal and ventral mesenteries containing the dorsal and ventral blood vessels are present in the region of the first and second diaphragms.

The coelom is spacious and undivided by septa between the third diaphragm and the tail. Vestiges of septa appear to be present as strands of connective tissue following the branchial and nephridial vessels, especially the posteriormost ones.

The tail region possesses transverse septa which are incomplete ventrally in the region dorsal and dorsolateral to the nerve cord. Dorsal and ventral mesenteries bearing the dorsal and ventral vessels connect the gut to the dorsal and ventral body wall in the midline.

Body Wall and Musculature. Study these in your specimen. The body is covered by a distinct cuticle secreted by the underlying hypodermis. The latter possesses numerous differentiated areas in which the cells are more deeply columnar and include abundant mucous and pigment cells. These areas are responsible for the macroscopically observed pattern of elevated pigmented patches in the skin of the worm. Underlying the hypodermis is a thin layer of connective tissue followed by a layer of circular muscle. Internal to the last is a layer of longitudinal muscle bands. They are separated from the coelomic space by the coelomic peritoneum.

Between the third diaphragm and the tail another group of muscles is very conspicuous. These muscles, the oblique, arise at the sides of the ventral nerve cord and pass dorsolaterally into the body wall at the level of the notopodia.

Associated with each notopodial sac is its musculature. This includes a number of protractor fibers and a single retractor one.

Digestive System. This tract extends from mouth to anus supported by the septa and mesenteries in the anterior and posterior regions of the body, but lying moderately freely through most of the midregion. In the latter region this freedom is increased by the great lengths of the branchial and nephridial blood vessels. At the anterior end of the stomach, however, is found a pair of fine connective tissue strands following the nephridial vessel and firmly attaching the stomach to the body wall.

The mouth opens into a buccal cavity whose surface bears conspicuous sensory papillae. The buccal cavity is followed by the pharynx. The pharynx is encased in a sheath of musculature, the buccal sheath. From this sheath fibers extend posteriorly, penetrate the first diaphragm, and become attached to the body wall between the first and second diaphragms. These fibers serve as retractor muscles for the buccal mass and pharynx which, when everted, constitute the proboscis. Protrusion of the proboscis is probably in part effected by increased hydrostatic pressure of the body fluids and in part by contraction of longitudinal muscle fibers of the body wall. The buccal cavity and pharynx possess a cuticular lining.

Posterior to the pharynx is a long tubular esophagus apparently capable of considerable extension, required during proboscis eversion. The esophagus passes through the three diaphragms. The posterior portion of the esophagus has a glandular and ciliated columnar epithelial lining. Opening into the posterior region of the esophagus is a pair of esophageal glands. These are hollow, dorsolateral diverticula lined with a secretory epithelium which is plicated, thus providing extensive internal surface. The esophagus possesses layers of both longitudinal and circular muscle fibers.

The stomach follows the esophagus. It is an elongated portion of the gut whose outer surface is extensively provided with yellowish chloragogue tissue separated into patches by a regular network of blood sinuses. This chloragogue tissue is reported to be excretory in function. Musculature is poorly developed in the stomach. The mucosa is richly glandular, particularly in its anterior region. A ciliated groove begins in the midregion in the ventral midline and continues posteriorly to the anus. Lateral ciliated-groove tributaries join with the ventral groove. The cilia maintain a current in the direction of the anus.

The intestine continues from the stomach to the anus with little or no differentiation other than possessing the ventral ciliated groove. Glandular cells are not abundant in its lining.

In view of the weak musculature of the gut proper, it appears reasonable to presume that activity of the somatic musculature is largely responsible for the passage of ingested material through the gut.

Circulatory System. By carefully stretching the digestive tract to right or left in the dissected specimen nearly all the major vessels of the circulatory system may be traced and their interconnections determined. The vessels are red because of the blood plasma being richly provided with hemoglobin. The blood contains a low concentration of colorless nucleated corpuscles 5–10 μ in diameter.

The dorsal blood vessel, carrying blood anteriorly, extends from the posterior to the anterior tip of the worm. Peristaltic waves rhythmically pass anteriorly over this vessel. In the tail it receives segmental vessels originating in the ventral vessel. Passing over the intestine anterior to the tail it receives branchial efferents from the last seven pairs of gills and several intestinal vessels per segment. Passing over the stomach, the vessel joins with numerous lateral vessels from the gastric plexus. Over the esophagus it gives rise to lateral vessels to the esophageal glands, to the first three nephridia and body wall, to the second diaphragm and adjacent notopodial sac region, and to the body wall in the region of the first chaetal sac. The dorsal vessel finally breaks up into a capillary network in the buccal region.

The ventral blood vessel originates in the anterior tip of the worm and conveys blood posteriorly to the posterior extremity. It collects blood from the buccal region, and as it passes posteriorly it receives branches from the first diaphragm and body wall, the esophagus, the second diaphragm, and the lateral neural vessels. The ventral vessel supplies the afferent branchial and nephridial vessels. Branches of the last also supply the body wall. At the level of the hearts (discussed below) the vessel receives branches from the ventricles. In the tail the ventral vessel gives rise to the segmental vessels which connect directly with the dorsal vessel in this region.

The gastric plexus is differentiated into certain longitudinal tracts. A gastric lateral vessel is evident on the anterior portion of the stomach. This passes anteriorly, opening into a thin-walled distention, the heart auricle, and then continues directly anteriorly as the lateral esophageal vessel supplying the esophageal glands and the esophagus, finally terminating immediately posterior to the first diaphragm.

Subintestinal, longitudinal vessels are found in the plexus. These, at the level of the stomach, receive the anterior four branchial and nephridial efferents.

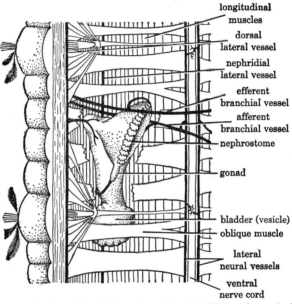

FIG. 155. Enlarged view of the region of a mixonephridium of *Arenicola*. (Modified after Ashworth.)

The ventricles of the heart are highly muscular bodies capable of considerable distention. They propel blood from the auricles into the ventral vessel.

Some longitudinal vessels are found in the body wall. A dorsal lateral vessel lies just dorsal to the row of notopodial sacs (Fig. 155). A nephridial lateral vessel lies just ventral to the level of the nephridiopores. Lateral neural vessels are located dorsal to each edge of the ventral nerve cord. These peripheral longitudinal vessels are interconnected by commissural vessels in those segmental annuli bearing the chaetae.

Mixonephridia and Gonads. There are six pairs of large mixonephridia, located in chaetigerous segments 5 to 10 inclusive. Study these carefully *in situ* in your specimen, both before and after removing some of the oblique musculature which partially obscures them. Observe carefully their blood supply, supporting attachments, and relation to the gonads. Finally, remove one of the organs and study it with the aid of your compound microscope.

Each mixonephridium comprises three distinct regions: nephrostome, excretory tube, and bladder (Fig. 155). The nephrostome is located anteriorly and has an opening directed anteromedially between dorsal and ventral lips. The nephrostome is richly vascularized over its whole surface, hence has a pinkish cast. The fimbriated dorsal lip of the nephrostome is especially well supplied. The lips and internal surface of the nephrostome possess a ciliated epithelium which maintains a current from the coelom into the organ.

The blood vessels of the nephrostome unite to produce the gonadal vessel passing posteriorly from the posterior end of the dorsal lip. On this vessel immediately posterior to the nephrostome of each mixonephridium except the first is a minute gonad, usually only a fraction of a millimeter in length.

The nephrostome opens into the more posterior and laterally located excretory portion of the mixonephridium. The lining of this portion is a ciliated epithelium, the cells of which are charged with yellowish and black granules.

The excretory portion opens into the bladder. This portion, unlike the preceding portions, has muscle fibers in its walls. It is capable of considerable distention. The bladder opens to the exterior by way of the nephridiopores.

The mixonephridia are supported in the coelom by attachment of the nephrostome both to the ventral body wall and to the overlying oblique musculature.

The minute gonads slough off reproductive cells in a very early stage in their development. The development of the gametes occurs largely while the cells float freely in the coelomic fluid.

The oöcytes, at the time they break away from the ovaries, are only 15–20 μ in diameter. At the completion of their development they are large discoidal bodies approximately 90–150 μ in diameter.

The spermatogonia first freed into the coelomic cavity comprise groups of nine or more cells all attached to a protoplasmic mass, the blastophore. The spermatogonia multiply and undergo transformation into fully differentiated sperm cells while still held together in these groups. The result is the production of discoidal masses of flagellated sperm.

The gametes at the height of the breeding season may largely fill the coelom. They escape to the exterior by way of the mixonephridia, whose bladders may at times be observed to be enormously distended by these products.

Nervous System. Remove the digestive system as far anteriorly as the pharynx and expose the whole central nervous system. Study its organization.

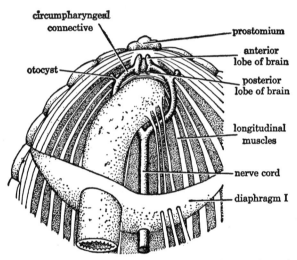

Fig. 156. Dorsal view of *Arenicola*, showing the anterior portion of the central nervous system. (Modified after Ashworth.)

The ventral nerve cord lies longitudinally in the ventral median line. It occupies a position just interior to the circular muscle layer. The cord is dorsoventrally flattened. It is non-ganglionated and appears superficially to be single. An examination of sections shows, however, that as in other polychaetes it contains two longitudinal tracts. Ganglion cells are moderately uniformly distributed along the cord on its ventral and ventrolateral surfaces. The cord sends off small lateral nerves at the level of each interannular groove. At the level of each parapodium a pair of larger nerves leaves the cord.

Near the anterior end of the worm the cord bifurcates, giving rise to the circumpharyngeal connectives (Fig. 156). Along their course these connectives give off numerous nerves to the stomatogastric system of the buccal and pharyngeal complexes. A nerve supplies the conspicuous otocysts. These latter organs are easily found and lie near the connective at the place where it crosses the interannular ring of the peristomium. Other nerves pass to the peristomial epidermis and its underlying musculature. Remove an otocyst and examine with the microscope first it and then its inclusions.

The connectives continue dorsally and enter the anterior lobes of the brain, which are situated in the prostomium. Posterior lobes of the brain extend posteroventrally to a position ventral to the nuchal groove. The brain is a small organ, usually less than a millimeter in diameter. Nerves from the brain pass anteriorly into the eversible region of the buccal mass.

Eyes are usually present, closely associated with the dorsal surface of the anterior brain lobes. These eyes are usually not visible externally because of their small size and the pigmentation of the overlying epidermis.

Amphitrite ornata
By F. A. Brown, Jr.

Amphitrite ornata is common along the eastern coast of the United States between North Carolina and Cape Cod. It dwells near the low-tide level in tubes buried in sand or mud. The tubes are the burrows of the worm, which have been lined with mucus. Another common species of the genus, with somewhat overlapping distribution, is *A. brunnea*, whose distribution is largely north of Cape Cod. South of the Cape it is usually found in deeper water than is *A. ornata*.

Very commonly and symbiotically sharing the tube of *A. ornata* is one of the scale worms of the Family Aphroditidae, *Lepidometria commensalis*.

External Anatomy. The body of *Amphitrite* (Fig. 157), usually about 20–30 cm. long, can be seen to comprise three distinct regions: (1) a head, (2) a "thorax," and (3) an "abdomen." The head is formed by a fusion of the prostomium and peristomium. The thorax comprises that thicker portion of the body lying posterior to the head, all of whose segments except those close to the head bear notopodia containing bristles. The abdomen is a long, many-segmented region without bristles which in specimens out of their tubes is usually strongly coiled. Determine the number of segments in the thorax and abdomen.

The head bears numerous long, greatly extensible tentacles on its anterolateral surfaces. These are outgrowths from regions of two

290 ANNELIDA AND SIPUNCULOIDEA

lateral lobes of the prostomium and are believed homologous with the palps of more generalized polychaetes. The mouth is located on the anterior surface of the head amongst the tentacles and possesses thick dorsal and ventral lips of peristomial origin. The tentacles are feeding organs. They are hollow, containing an extension of the coelom filled

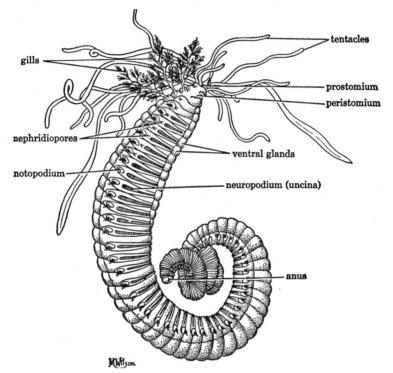

FIG. 157. Lateral view of *Amphitrite ornata*.

with the coelomic fluid and coelomic corpuscles. They are poorly vascularized and hence yellowish buff in color. The tentacles are slightly flattened orally-adorally and carry a ciliated groove on their oral surface. Food particles are swept to the mouth along these ciliated grooves. Mucous cells are abundant at the groove. The tentacles have a well-developed musculature; a circular muscle layer underlies the epidermal layer. Internal to the circular is a layer of longitudinal muscle. Three longitudinal nerve tracts lie in the adoral wall. The tentacles are capable of very considerable extension and rapid contraction, and show writhing activity for long periods after isolation from the worm.

The first three segments of the thorax bear dorsolaterally a pair of gills. Each gill is attached to the body by a single large stalk, but branches repeatedly and profusely away from the body to produce bushy organs. The gills are believed homologous with dorsal cirri. The finer branches are usually much coiled. The gills are hollow outgrowths of the body, containing, like the tentacles, extensions of the coelom, and are filled with coelomic corpuscles. The epithelial walls contain numerous mucous glands. Internal to the epidermis is a layer of longitudinal muscle fibers. On opposite sides of each gill filament are branches of the afferent and efferent blood vessels which unite at the gill tips. The gills normally exhibit a rhythmical contraction, with change of color from deep red to lighter red as blood is rhythmically forced out of the organs.

Note on which segments the bristled notopodia and the elongated uncini-bearing neuropodia commence and upon what segments they become discontinued. With fine forceps remove a few chaetae from organs in the midregion of the thorax. The notopodial chaetae possess blade-like, serrated terminations. Observe carefully the form of the stout hook-like uncini of the neuropodia. These are oriented in the neuropodia in such ways as to facilitate the movement of the worm in its tube.

On the ventral surface of the first few thoracic segments are the ventral shield glands. They are supplied with mucus for the lining of the burrows by the yellow ventral glands, conspicuously seen projecting internally into the coelomic cavity in the worm. This will be conspicuously evident when the worm is opened later.

The anus is found terminally in the last abdominal segment as a dorsoventrally elongated aperture.

Coelom and Coelomic Fluid. Make a longitudinal incision through the dorsal body wall immediately to the right of the midline. Place a small amount of the coelomic fluid upon a microscope slide and examine the preparation for the three contained types of coelomic corpuscles. The largest and most abundant type is the pinkish eleocyte. These cells are approximately spherical and charged with granules. A second type is a smaller amoebocyte. The third type is a fusiform cell darkly granular except for clear ends. Look for sex cells which may be present in the coelomic fluid. If they are present, determine whether they are oöcytes or developing sperm cells.

Now continue the dorsal incision, anteriorly close to the anterior tip of the worm and posteriorly for some distance into the abdomen. Stretch out and pin down the body wall. Note in so doing that the gut is attached to the dorsal median body wall by thin suspensory muscles.

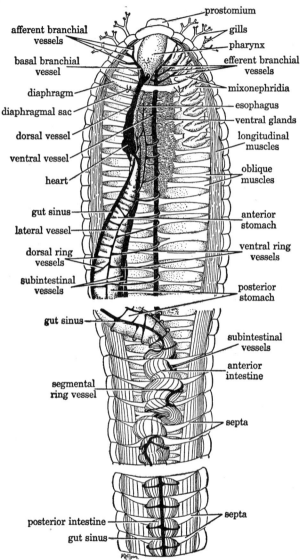

Fig. 158. Dorsal view of *Amphitrite ornata*, showing general internal anatomy.

Note that transverse septa are almost completely absent from the anterior portion of the worm. Only at about the level of the fourth or fifth segment is there a thick diaphragm separating the coelom into an anterior and a posterior chamber (Fig. 158). Find the muscular sacs projecting posteriorly from the diaphragm and opening into the anterior chamber.

The posterior portion of the abdominal region is divided completely by numerous intersegmental septa.

Body Wall and Musculature. The body is covered with a thin cuticle secreted by an underlying hypodermis in which are numerous unicellular mucous glands. Immediately internal to the latter layer is the circular musculature. Internal to the last is the longitudinal musculature, which throughout most of the length of the body is clearly differentiated into four columns of fibers, two dorsolateral and two ventrolateral. Both sets of tracts terminate in the head at the level of the brain.

Oblique musculature, comprising fibers extending from the midventral line to the lateral body wall just ventral to the dorsal longitudinal tracts, is found through most of the length of the worm, dividing the coelom rather effectively into dorsal and ventral portions. Dorsal and ventral suspensory muscle fibers anchor the gut in the dorsal and ventral median lines, respectively.

Each of the chaetal sacs of the notopodia is provided with a number of protractor fibers and a single retractor.

Later, when the nervous system is studied, a median ventral longitudinal muscle will be observed lying immediately dorsal to the nerve cord and extending the full length of the body.

A somatic peritoneum covers all the foregoing musculature of the body wall.

Digestive System. The digestive tract is differentiated into six regions. The mouth opens into a dorsoventrally compressed chamber, the pharynx. The line of division between this and the narrower tubular esophagus is the diaphragm. Both these regions are ciliated internally and contain musculature. There are numerous mucous cells in their lining epithelium. Food coated with mucus is swept posteriorly by peristaltic and ciliary movement to the third region, the anterior stomach. This is a thinner-walled, less muscular region. Its internal layer of cells contains numerous secretory ones, apparently sources of digestive enzymes. This region is covered externally with a yellowish chloragogue tissue reportedly of excretory function. The posterior stomach possesses a thick muscular layer and internally a thick secreted membrane. This region appears to act as a powerful churn, and, with the cooperation of sand grains, as an effective gizzard. The posterior stomach opens into a coiled anterior intestine. The anterior intestine is covered with chloragogue tissue. Both intestinal portions are relatively non-muscular, possess a ciliated ventral groove, and probably function primarily in absorption.

Circulatory System. The blood of *Amphitrite* is red; it contains hemoglobin.

A large amount of blood is found in an extensive gut sinus which occurs over the stomach and intestine between the gut epithelium and musculature. The blood slowly moves anteriorly in this sinus. Ventral to the gut in these regions is a pair of large subintestinal ventral vessels having numerous short vessels connecting it with the gut sinus. Blood also flows anteriorly in these vessels. At the anterior end of the stomach the subintestinal vessels curve dorsally over the gut and join to continue as a large median dorsal vessel over the esophagus. This short dorsal vessel has projecting into it a large mass of cells comprising the heart-body, of unknown function. Speculations have attributed a valvular action and a hemoglobin destructive function to this body. Immediately anterior to the diaphragm the dorsal vessel divides into right and left basal branchial vessels, each of which divides into three afferent branchial vessels supplying the three gills of that side of the body.

The three pairs of efferent branchial vessels unite to form a large median ventral vessel conducting blood posteriorly. In the stomach region the ventral vessel sends numerous pairs of segmental ventral ring vessels to a pair of longitudinal lateral vessels of this region. The latter have numerous dorsal ring vessels connecting them with the gut sinus. In the region of the intestine, to which the lateral vessels do not extend, blood is carried by segmental ring vessels directly to the gut sinus.

In the intestinal region the ventral vessel supplies, by segmental connectives, a small supraneural vessel.

The esophagus and pharynx are vascularized by a plexus and by vessels arising directly from the anterior end of the gut sinus.

Mixonephridial and Reproductive Systems. Carefully cut away the oblique muscles in the anterior thoracic region to expose the numerous pairs of mixonephridia. Note that mixonephridia lie both anterior and posterior to the diaphragm. Eight pairs of mixonephridia are present.

Each mixonephridium opens into the coelom by a nephrostome which is attached to the body wall near the midlateral line. The nephrostome possesses ciliated and fimbriated dorsal and ventral lips. It leads into an excretory sac which bends sharply back upon itself. The cells of the excretory sac include many granule-containing and many ciliated ones. The lack of vascularization of the mixonephridia indicates that these organs must, in their functioning, regulate the composition of the coelomic fluid. Nephridiopores open to the exterior between the notopodium and neuropodium.

The gonads are on the ventral surface of the anterior thoracic cavity just posterior to the diaphragm. These organs are numerous and interspersed among the yellow ventral glands.

The growth and development of the gametes occur while the cells are floating freely in the coelomic fluid. The gametes are conveyed to the exterior by way of the mixonephridia.

Nervous System. The nervous system occupies a primitive position, in contact with the epidermis. The whole system is quite difficult to study in dissections. In fact, it is usually futile to attempt to study the suprapharyngeal ganglia located in the prostomium and the circumpharyngeal connectives by any method other than in histological sections. The supraesophageal ganglia send tentacular nerves to each tentacle and nerves to the dorsal and ventral lips.

The ventral nerve cord begins anteriorly in the midventral peristomium where it receives the connectives. It then extends posteriorly as a pair of cords with segmental commissures. This portion of the nervous system can often best be studied by pinning out the specimen ventral side up and teasing away the epidermis to expose the double cord lying directly beneath it.

The ventral cord is not distinctly ganglionated. Lateral nerves are given off at brief intervals and pass between the circular muscle fibers to innervate the body wall.

Lumbricus terrestris

By F. A. Brown, Jr.

The common earthworm, *Lumbricus*, is locally very abundant in moist earth in Europe and North America. It remains wholly in its burrow during daytime in response to light and humidity. It may come out of its burrow at night, especially after rain. During such periods these oligochaetes may be easily collected in large numbers with the aid of a flashlight.

Study the patterns of locomotory activity and the mechanics of locomotion in an active, living specimen. Observe the worm carefully as it creeps forward upon moist filter paper. Note the posteriorly progressing peristaltic locomotory waves. The regions of the worm in which the longitudinal musculature is contracted at any given moment comprise the "feet." They are provided with good traction by the chaetae, which will be studied later. Observe the responses of a forward-crawling worm (1) to touch of its anterior end, (2) to touch of its posterior end, and (3) to having the anterior end come to project off the edge of a table.

Determine the locomotory pattern of a worm moving backwards.

External Anatomy. The body is subcylindrical; there are numerous annuli which correspond exactly with the true segmentation of the body. The number of segments is usually well over 100, and variable in number posterior to the clitellum.

The dorsal and ventral surfaces are usually clearly distinguishable; the dorsal surface is more darkly pigmented.

The prostomium (Fig. 159) is the anteriormost portion of the body and is not considered to be a true metamere. Ventral to the prostomium, and bounded largely by the anterior edge of the peristomium or first body segment, is the mouth. Neither the prostomium nor the peristomium bears any appendages. The remainder of the metameres of the body are quite similar, except for five or six segments beginning with number 32 or 33, whose walls are swollen because of large hypodermal glands responsible for the formation of the cocoon. These swollen segments constitute the clitellum. Terminally located in the last segment is the anus.

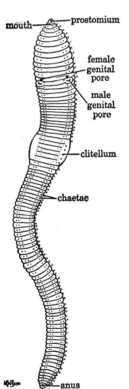

FIG. 159. Ventral aspect of *Lumbricus terrestris*.

Four pairs of chaetae are found in each segment. Two pairs are ventral and the other two pairs are nearly lateral in position. The chaetae are short, stout bodies, each located in a chaetal sac. The direction in which the chaetal tips are pointed can be detected by rubbing a finger first forwards and then backwards along the chaetal rows.

Some of the chaetal sacs near the region of the reproductive organs of the body contain smaller chaetae and are quite glandular. These secrete an albuminous fluid into the cocoon at the time of reproduction. These specialized chaetal sacs are termed the capsulogenous glands.

A pair of external openings of the metanephridia, the nephridiopores, may be observed upon the ventral surface of each segment except the first three or four and the last. These pores lie slightly lateral and anterior to the outer chaeta of each ventral pair.

Ventrally in segment 15 the male genital apertures are a pair of small transverse slits lying between two swollen lips. The female genital apertures occupy a corresponding position on segment 14, but are less conspicuous (Fig. 159).

Small openings into the seminal receptacles may be found on the lateral surface in the grooves between segments 9 and 10, and 10 and 11. Sperm enter the seminal receptacles through these pores during the mutual exchange of sperm by two copulating individuals.

Dorsal pores are found in the midline in the intersegmental grooves from the 10th or 12th segment, posteriorly. These pores open directly into the coelomic cavity. Coelomic fluid is exuded from these pores onto the surface of the worm when the worm is subjected to desiccation or to certain irritants as, for example, acetic acid.

Coelomic Cavity and Its Fluid. Make a short longitudinal incision in the dorsal midline of a freshly killed (by 7 per cent alcohol) specimen. A preserved specimen may be used but is less preferable. Draw out a small amount of the coelomic fluid and examine it for cells with a compound microscope. Examine the types of cells present. If possible, examine the coelomic fluid from a region of the body into which a carmine suspension was injected on the preceding day and note the evidences of phagocytosis.

Now, with the worm submerged in water, continue the incision forward to the anterior end, and backward, teasing away the septa and pinning out the body wall. Take care not to penetrate the gut. Note that the coelomic cavity is divided by septa into segmental compartments. Each septum consists of a thin muscular layer sandwiched between two layers of peritoneum.

Body Wall and Its Musculature. These are studied to best advantage in a prepared cross section. The body wall is bounded externally by a well-defined albuminoid, iridescent cuticle which is secreted by the underlying hypodermis (Fig. 160). The hypodermis contains many unicellular mucous glands which liberate their product to the exterior by numerous fine pores in the striated cuticle.

Immediately beneath the hypodermis is a thin connective-tissue layer. Then, in order, come a layer of circular muscle fibers, **a layer** of longitudinal fibers, and the somatic peritoneum. The longitudinal muscle is divided into seven tracts, one dorsolateral pair, two ventrolateral pairs, and a single median ventral tract. The circular muscle layer is intersegmentally discontinuous. Small muscles are associated with the chaetae.

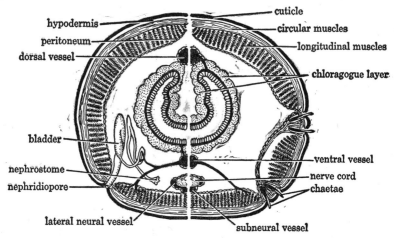

Fig. 160. Cross section of *Lumbricus terrestris* at region of stomach-intestine. (Modified after Marshall and Hurst.)

Digestive System. Study the digestive system, taking care not to damage the circulatory and reproductive systems.

The gut (Fig. 161) is a straight tube extending from mouth to anus. It is lined with a cuticule secreted by an underlying epithelial layer. Many of the latter cells are ciliated. The wall of the gut possesses an inner circular and an outer longitudinal muscle layer. These layers show different degrees of development in the various regions.

The mouth opens into a buccal cavity which in turn leads into a large, highly muscular pharynx containing, in addition to well-developed circular and longitudinal muscular layers, a rich mass of radial fibers passing to the body wall. These last are able to dilate the pharynx very effectively and play an extremely important role in the feeding activity of the worm. Posterior to the pharynx a narrower and thin-walled esophagus continues through segments 6 to 12 inclusive. Laterally in segment 10 is a pair of small diverticula, the esophageal glands. Laterally in segments 11 and 12 are small diverticula filled with particulate calcium carbonate. These last are the calciferous glands which excrete excess calcium into the gut.

The esophagus becomes enlarged at its posterior extremity to form the crop. Lying just posterior to the crop is a highly muscular region, the gizzard. The cuticular lining of the gut is much thickened in this organ.

Continuing posteriorly from the gizzard is a long stomach-intestine, extending practically the remainder of the length of the body. That portion of the stomach-intestine located in the last metamere is often

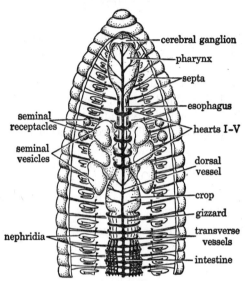

FIG. 161. Dorsal view of *Lumbricus terrestris*, showing general arrangement of internal organs.

called the rectum. The stomach-intestine has a poorly developed musculature. It is segmentally constricted by the septa, effectively thereby increasing its internal surface. Further increasing its surface for digestive and absorptive functioning is a mid-dorsal infolding of the stomach-intestinal wall, the typhlosole. The intestine is covered with a layer of yellowish cells called the chloragogue tissue.

The number of secretory cells in the stomach-intestinal epithelium indicates that not only absorption but also digestion takes place largely in this region of the gut.

Circulatory System. The circulatory system contains blood which is red because of hemoglobin in solution. The blood corpuscles themselves are not pigmented.

A closed system of blood vessels is well developed in *Lumbricus*. The major vessels of the system are (1) a dorsal longitudinal vessel passing above the digestive tract, and (2) a ventral longitudinal vessel below the gut. These two vessels are connected with one another segmentally by paired tranverse vessels passing around the gut. Five of these transverse vessels, those lying in segments 7 to 11 inclusive, are larger and more muscular than the remainder and comprise the "hearts."

In a freshly anesthetized worm, it can readily be observed that peristaltic waves force blood anteriorly in the dorsal vessel, ventrally within the "hearts," and posteriorly within the ventral vessel. Valves in the dorsal vessel and the hearts aid the peristaltic contraction to produce an efficient circulation of the blood.

In *Lumbricus*, most of the other vessels are difficult or impossible to study other than in sectioned material. Other major vessels comprise three more longitudinal vessels, namely a pair of lateral neural vessels and a single subneural vessel all intimately associated spatially with the ventral nerve cord. They are best seen in study of a cross section.

Branches of the ventral vessel supply the metanephridia. The dermis of the skin is also richly supplied with capillaries, rendering the general surface of the body an effective respiratory structure.

The Metanephridia. Each segment of the body except the first three or four and the last contains a pair of metanephridia. They lie to right and left of the digestive tract. Each metanephridium macroscopically appears as a feathery mass of tissue.

The metanephridium is a complexly coiled tubule opening to the exterior, in the segment in which it is principally located, by way of the nephridiopore. The tubule terminates internally in a ciliated and funnel-shaped opening into the coelom, the nephrostome. The end of the tubule bearing the nephrostome projects through the anterior septum of the segment and opens into the segment next anterior.

Carefully remove a metanephridium, including that portion of the septum through which the nephrostome projects, and study this structure with a compound microscope. The lumen of the tubule is coiled and intracellular and embedded in a chain of glandular cells. The end of the tubule nearest the nephridiopore is intercellular and dilated to produce a bladder or terminal vesicle. Muscle fibers are found in the walls of this vesicle.

Obtain if possible a preparation of a metanephridium from a freshly killed specimen. In this preparation the ciliary activity of the nephrostome and also often of a portion of the tubule can readily be studied.

Male and Female Reproductive Systems. Obtain a clearer view of these systems by transecting the stomach-intestine near the clitellum and carefully lifting up and teasing free the gut as far anteriorly as the pharynx. Observe in segments 10 and 11 the large whitish sperm reservoirs lying across the median line (Fig. 162). They possess three pairs of lateral seminal vesicles, the anterior reservoir having two pairs, an anterior and posterior pair of vesicles, and the posterior reservoir only a large posteriorly directed pair.

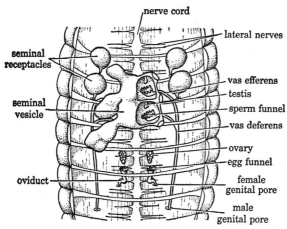

Fig. 162. Reproductive system of *Lumbricus terrestris*. (Modified from Vogt and Yung.)

Dissect open the anterior sperm reservoir at its dorsal midregion. Place a little of the contents upon a slide and examine it with a compound microscope. Spermatogonia and spermatocytes in various stages of development may be found. These cells are split off from the testes very early in their development. These preparations commonly also contain *Monocystis* (see p. 28), which parasitizes a large fraction of the earthworm population.

Now carefully pick off much of the dorsal walls of the sperm reservoirs, and with the aid of a pipette wash out the contents of the reservoirs. Look for the large ciliated and fimbriated funnels which lead into the vasa efferentia and thence into the vasa deferentia. These funnels are located on either side of the midline attached to the posterior face of the reservoir. On the opposite face, anterior to each funnel, is a minute testis. These testes are often difficult to locate with certainty since their color and texture are not much different from the reservoir contents. A similar set of reproductive organs may also be found in the other sperm reservoir.

The two minute vasa efferentia which may sometimes be seen in dissections unite on each side to produce the vas deferens which leads posteriorly to the male genital pore. The ovaries are very minute. They are located on the anterior septum in segment 13 just to the right and left of the ventral midline. Examine an ovary under the compound microscope and look for oöcytes.

On the posterior septal wall of segment 13, opposite the ovaries, are minute ciliated funnels leading into the oviducts which open to the exterior at the female genital pore of segment 14. The oviducts and their funnels can usually be observed only with difficulty in dissections.

Two pairs of globular, whitish, seminal receptacles are attached by short ducts to the lateral wall at the intersegmental junctions between segments 9 and 10, and 10 and 11.

In reproduction two individuals attach themselves to one another by their ventral surfaces, their anterior ends pointing in opposite directions. The attachment is made by chaetae and by a sticky secretion from the clitellum and from the capsulogenous glands. Sperm cells mutually pass from one specimen into the seminal receptacles of the other along longitudinal seminal grooves which facilitate this transfer.

Mature eggs, and sperm from the receptacles, are later liberated, together with albuminous fluid, into a cocoon which is slipped off over the anterior end of the body. Fertilization and development occur within the cocoon.

Nervous System. The central nervous system comprises principally a suprapharyngeal pair of ganglia or "brain," a pair of short circumpharyngeal connectives leading ventrally around the pharynx to join the ventral nerve cord at its anterior end, and the ganglionated ventral cord.

The two suprapharyngeal ganglia lie in segment 3, dorsal to the pharynx and to right and left of the midline. A commissure connects the two ganglia. Nerves from these ganglia innervate the prostomium whose ganglia they primitively are. Stout connectives pass posteroventrally, joining with the subpharyngeal ganglion. This ganglion, like all the remaining ones, is actually composed of two ganglia pressed against one another in the midline. This ganglion innervates the peristomium and the following two segments, hence would appear primitively to be a fusion product of three embryonic ganglia.

There is a ganglionic enlargement of the ventral cord in each segment of the body. In each segment are three pairs of lateral nerves, two close together near the posterior portion of the segment and one in the anterior region. These nerves lead to the gut and body wall.

Crossing longitudinally within the ventral nerve cord near its dorsal surface are three giant fibers, a large median and a pair of slightly smaller lateral ones. The giant fibers are readily evident in cross sections. They are important in shock responses of the worm as a whole.

In addition to the foregoing, but not evident in dissections, is a complex of stomatogastric nerve fibers in the walls of the anterior gut. This system is connected with the central nervous system proper by nerve fibers originating in the circumesophageal connectives.

Hirudo medicinalis

By F. A. Brown, Jr.

The medicinal leech, *Hirudo medicinalis,* is a European species which has been introduced into certain ponds and streams of the eastern portion of the United States. It is a relatively large leech often growing to 10 or more centimeters in length. It feeds upon blood of a vertebrate to which it periodically attaches itself.

This leech may be readily obtained at biological supply houses and in larger cities may be obtained alive through large wholesale druggists who still sell it for blood-letting.

General Form and Locomotion. Examine carefully a living specimen if one is available. Observe its dorsoventrally flattened, leaf-shaped body. Note that the broadest region is located toward the posterior end. Directed ventrally at the anterior end is an anterior sucker, and ventrally directed at the posterior end is a larger ventral sucker. Observe the firmness of the grip of these suckers to the substrate by attempting to pull them free.

The leech has two general types of locomotion, creeping and swimming. The character of the swimming is usually readily seen by dropping an individual freely into a deep container of water. Notice in a free-swimming leech the greater dorsoventral flattening and the undulatory waves which pass longitudinally over the body. Determine whether these waves progress anteriorly or posteriorly, i.e., in the direction of locomotion or in the opposite direction. Study the creeping type of locomotion which involves alternate fixation of the two suckers and contraction of the body upon fixation of the anterior sucker and extension upon fixation of the posterior. This is a type of activity often called "leech-like" locomotion.

The leech may now be killed for the following study. Killing in a relaxed condition may be accomplished by submerging the animal in a 7 or 8 per cent solution of alcohol or a weak solution of chloroform or ether. If the specimen is to be preserved for study at a later time, this may be done in 4 per cent formalin.

External Anatomy. The leech *Hirudo*, like other members of the class but unlike other annelids, possesses primitively a constant number of segments: thirty-two. Externally, in *Hirudo*, one is able to differentiate the anteriormost twenty-six of these. The remaining six are believed, from a study of central nervous organization, to have contributed to the development of the large posterior sucker.

Study the distribution of papillae in the midregion of the worm. There are transverse rows of these passing over both dorsal and ventral surfaces of the body. There is one transverse row of papillae in each true segment; in the greater portion of the body midregion they are located upon the middle, or third, annulus of the five annuli occupying each true segment. Trace the rows of papillae anteriorly, and notice on the dorsal surface near the anterior end five pairs of pigmented eyespots which occupy positions corresponding entirely to the segmental papillae of the more posterior region.

Now confirm the following external characters of the segments of your specimens:

SEGMENTS	NUMBER OF ANNULI	ANNULUS BEARING EYESPOTS, PAPILLAE, OR BOTH
1	1	1st
2	1	1st
3	2	1st
4	2	1st
5	3	2nd
6	3	2nd
7	4	2nd
8–23 incl.	5	3rd
24	2	1st
25	2	1st
26	2	1st

HIRUDO MEDICINALIS

Having now determined by external means the segmentation of the body of the leech, observe any other characters of the body which appear to indicate the true segments of the worm.

Study the anterior sucker. The mouth is located within it. Study the posterior sucker. Dorsally in the midline, at the point of junction of the posterior sucker with the 26th segment, is the small anus.

A pair of nephridiopores is located on the ventral surface of each of the segments 7 to 23 inclusive. The pores are located at the posterior edge of the annulus immediately anterior to that bearing the papillae.

In the ventral median line, in the annulus posterior to the papillae-bearing one of the 11th segment, is a single aperture, the male genital pore. In the corresponding locus of the 12th segment is the female genital pore.

Segments 9, 10, and 11 comprise a functional clitellum; abundant secretory cells in the hypodermis of this area are responsible for the formation of the cocoon in which the young normally develop.

Body Wall and Musculature. Submerge the specimen in water and cut through the body wall longitudinally in the dorsal midline. Take great care that you do not cut into the underlying organs to which the body wall is attached by connective tissue and radiating muscle fibers. Lift the body wall upwards and slowly stretch it laterally, while carefully cutting and teasing away the attaching tissues. Pin out the body wall in the dissecting pan.

The body wall is seen to be bounded externally by a thin cuticle which is secreted by the underlying hypodermal layer. A well-developed circular muscle layer lies directly beneath the hypodermis. Internal to the circular muscle is a thick longitudinal-muscle layer. Also within the body wall are numerous radial muscle fibers passing from the external layers of the wall to the viscera; segmentally arranged through the trunk are well-developed dorsoventral tracts of muscle whose contraction results in greater dorsoventral flattening of the body.

That portion which would perhaps correspond with the position of the major part of the coelomic cavity in most annelids is, in the leech, filled with a richly vascularized mass of pigmented cells, the botryoidal tissue.

Gland cells associated with the hypodermis, especially in the clitellum, have their cell bodies proper deeply located in the body wall, and are connected with the exterior only by narrow stalks.

Digestive System. Most of the principal portions of the digestive system should now be visible in your dissected specimen.

ANNELIDA AND SIPUNCULOIDEA

The mouth, as mentioned earlier, is located ventrally, within the anterior sucker. The mouth contains three tooth-covered muscular elevations, the jaws. One is dorsal, and the other two are ventrolateral. These paws are used by the leech to produce the characteristic triradiate incision through which it draws blood from its host. Do not make any incisions in the region of the mouth at this time, since they would be likely to damage the nervous system which will be studied later.

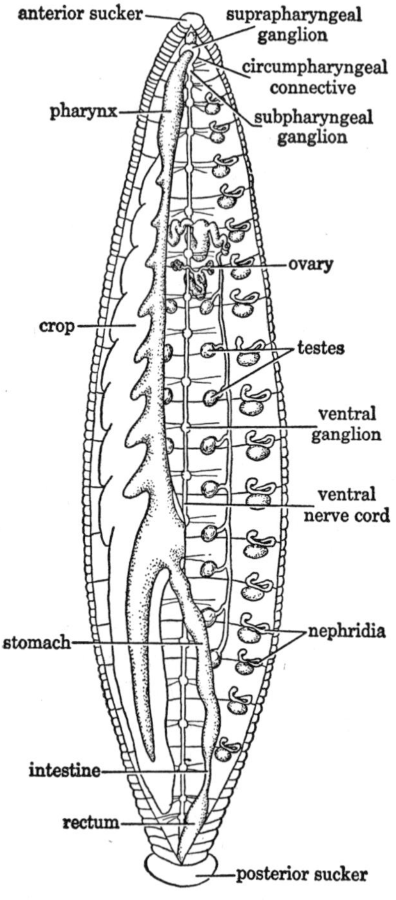

The mouth opens into a thick-walled muscular pharynx located in segments 4 through 8 (Fig. 163). The musculature of this region includes both longitudinal and circular fibers as well as a rich supply of radial fibers passing to the body wall. The last group obviously dilates the pharynx in producing its pump-like, blood-sucking activity. Unicellular glands of the pharynx secrete an anticoagulant, hirudin, into the blood which is drawn in as food. Hirudin also prevents blood-clotting in the wound of the host.

The pharynx opens into an enormous crop which occupies much of the space in the trunk region. It possesses eleven pairs of segmentally arranged, lateral diverticula, the most posterior of which extend posteriorly through most of the remaining segments of the body. The crop is capable of becoming considerably distended by ingestion of blood; it can contain enough to supply the nutritional requirements of the inactive leech for several months.

FIG. 163. Dorsal view of *Hirudo medicinalis*, showing general arrangement of internal organs. (Modified from Parker and Haswell.)

The crop opens into a small stomach at segment 19. Its inner surface bears a spiral elevation, thereby greatly increasing its surface. Digestion of the blood occurs in the stomach.

The stomach opens posteriorly into a narrower intestine which, in turn, opens into an enlarged chamber, the rectum. The rectum opens dorsally to the exterior by way of the anus.

Circulatory System. *Hirudo* appears to possess no true blood-vascular system. The entire circulatory system is considered to represent a very much reduced coelom. It consists of a well-developed system of vessels containing a red coelomic fluid. The latter contains hemoglobin in solution. The corpuscles are colorless. The system is difficult to study in dissections with any degree of completeness, but parts of it will become evident during the study of other organs.

There is a pair of longitudinal lateral vessels lying immediately lateral to the metanephridia. These possess muscular walls. Branches of these lateral vessels pass dorsally and ventrally, finally breaking up into anastomosing vascular beds in most of the tissues of the body. The two lateral vessels unite with one another at both the anterior and posterior ends of the animal.

A longitudinal sinus lies in the midline just dorsal to the gut; a second one is ventrally located and contains within it the ventral nerve cord. These two longitudinal sinuses unite with one another at the posterior end of the animal. These sinuses have numerous branches which are connected with sinuses associated with the testes and the metanephridial nephrostomes, and also with the vascular bed of the branches of the lateral vessels described earlier.

Be continuously on the alert for the various portions of these circulatory vessels in your dissection and study of the remaining organ systems of the leech.

Metanephridial System. Carefully remove the digestive system except at the anterior end of the worm. Lateroventrally in each of the segments 7 to 23 inclusive is a pair of metanephridia.

The main body of each metanephridium comprises a loop-like lobe (Fig. 164). This is made up of a large main lobe and a small apical lobe; the two are connected by the recurrent lobe. Extending medially from the main body is the testis lobe which terminates as a ciliated nephrostome within the testis sinus. Another branch, directed posteriorly, opens into a metanephridial vesicle which in turn opens to the exterior through the nephridiopore.

The lobes of the metanephridium of *Hirudo* comprise a glandular mass of cells within which is an anastomosing system of intracellular ductules which finally collect into a larger intercellular duct. The duct leads to the vesicle. In most of the metanephridia, the lumen of the nephrostome does not connect directly with the system of ductules within the testis lobe.

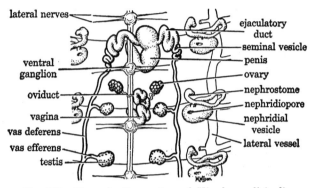

Fig. 164. Reproductive system of *Hirudo medicinalis*.

Reproductive System. *Hirudo* is an hermaphrodite, containing well-differentiated male and female reproductive systems. The gonads of each sex possess a lumen which may be considered a constricted-off portion of an ancestral coelomic cavity, and the gonadal ducts may be interpreted to be differentiated coelomoducts. This peculiar situation, compared with annelids in general, of having gonads directly continuous with their ducts appears to be an evolutionary outcome of the great restriction of the coelom within this class of annelids.

The male system has usually ten pairs of small globular testes located ventrally to right and left of the median line in segments 12 to 21, inclusive. Small vasa efferentia connect the testes with the more laterally located, longitudinal vas deferens. The vas deferens passes anteriorly to become much-coiled seminal vesicles in segment 10 and then, slightly more anteriorly, the somewhat more muscular ejaculatory ducts. The latter open into the base of a curved, muscular penis. Also associated with the base of the penis are numerous secretory cells comprising the prostate gland. With the aid of the secretion from the prostate the spermatozoa are aggregated into bodies known as spermatophores. The male genital pore opens to the exterior in segment 11.

In the female system there is a single pair of globular ovarian sacs in segment 11. These each have short, posteriorly directed oviducts which unite to form a single, median oviduct. The single oviduct is richly provided with albumen-secreting cells. The oviduct opens posteriorly in a vagina which bends anteriorly upon itself and opens to the exterior by the female genital pore of segment 12.

Cross fertilization is the rule in *Hirudo*. At copulation there is mutual transfer of sperm from the male system of one to the female system of the other. Upon separation of the individuals, eggs and sperm, together with albumen from the oviduct, are liberated into a cocoon produced by the clitellum. The cocoon is then passed off over the head of each individual. Development of young proceeds within the cocoon. When they hatch they look like minute adults.

Nervous System. Much of the ventral nerve cord should now be clearly evident in your preparation. Clear away as far as possible obscuring tissues. Carefully dissect anteriorly to expose the subesophogeal ganglion, the short circumpharyngeal connectives, and the suprapharyngeal ganglion.

Study carefully the organization of the nervous system. The suprapharyngeal ganglion is small and is located immediately posterior to the dorsal, median jaw. The subpharyngeal ganglion is a fused ganglion made up of five embryonic ones. Except for the last one, the remaining ganglia of the ventral cord are simple segmental ones, connected with those more anterior and posterior by double connectives. The last ganglion is a product of fusion of the embryonic ganglia of the last seven segments of the embryonic worm, six of which appear to have participated in the formation of the posterior sucker.

Lateral nerves leave the ventral cord at the ganglia and innervate the body wall and other organs.

SIPUNCULOIDEA

Phascolosoma gouldii

By F. A. Brown, Jr.

Phascolosoma gouldii is found in Massachusetts Bay and southward into Long Island Sound. Other similar species are found in most parts of the world. It is most abundant in muddy sand just below the low-tide mark. In some highly localized patches it is so abundant that a dozen or more may be brought up in a single shovelful of sand. It burrows vertically, obliquely, and horizontally in an irregular fashion, often going as deep as about 2 feet.

The general manner of burrowing is readily observed if the worms are placed in a jar containing sea water and an appropriate sand and mud mixture. The anterior end of the animal is capable of being progressively rolled outwards or inwards. In burrowing, the everting anterior end of the worm is forced into the sand where it is anchored by a localized dilation of the anterior tip of the body. The body is then pulled toward the new excavation by contraction of the general longitudinal musculature. This process is repeated over and over as the worm progresses.

Preparation of Specimens for Study. *Phascolosoma* is a relatively hardy animal that may be readily procured alive from biological supply houses. It is best studied in the fresh condition, though formalin-preserved specimens are quite satisfactory. It is easily shipped and can be kept alive in containers of sea water in the laboratory. If the animals are to be kept for some length of time before use, they can be maintained alive satisfactorily by placing the containers in a refrigerator.

The extraordinary contractility of these worms makes them difficult specimens to study and dissect while fresh unless they are thoroughly anesthetized. One good way of anesthetizing is to place them in 7 per cent ethyl alcohol in sea water, or in sea water to which is added gradually a saturated solution of chloroform in sea water. In response to this treatment the animals become well extended and the general musculature relaxed. The internal organs including the cilia remain in a living state for several hours. The dissection should be performed in sea water. It is desirable to have available a source of intense illumination, and, if possible, both dissecting binocular and compound microscopes.

External Anatomy. Study carefully the external form of your specimen. *Phascolosoma gouldii* is typically whitish or light gray in color. Adult specimens, fully extended, measure about 25–30 cm. in length and 0.6–0.8 cm. in greatest diameter. The body is divisible into two general regions: (1) an anterior introvert, which is capable of being rolled into (2) the posterior portion. The body wall of the introvert is less muscular than the remainder and extends posteriorly almost as far as the pair of conspicuous lateroventral nephridiopores of the single pair of giant nephridia, and the dorsal anus of the essentially U-shaped digestive tract. The anus is located in the mid-dorsal line. (In a fully contracted living specimen the body is characteristically curved, with the maximum convexity along the dorsal midline.) At the anterior tip, the mouth is terminal and surrounded by an oral disk possessing tentacles.

Just posterior to the tentacular crown and on the median dorsal surface is the *cephalic lobe*. It comprises a ciliated elevation divided into four portions by three longitudinal grooves. This organ lies dorsal and, in part, posterior to the brain. At its anterior end the longitudinal ridges overhang a horizontal transverse ciliated pit or groove. At the lateral edges of this groove tubes of cells, terminating blindly, pass downward into the brain. In the latter region the cells of the tube wall appear to possess the form of typical ganglion cells. The function of the pits is unknown. The ciliated cells of the dorsal ridges appear richly innervated, and hence there is suggestive evidence of a sensory function.

The Coelomic Cavity and Fluid. With fine-pointed scissors make a short longitudinal incision through the body wall just posterior to the introvert and slightly to the right of the midline. With a pipette remove some of the contained body fluid and examine it with a compound microscope for coelomic corpuscles and reproductive cells. The coelomic fluid contains several types of cells and has a pinkish cast. The color is due to hemerythrin, located in abundant, disk-shaped, nucleated corpuscles. The latter vary considerably in size, ranging in diameter from 6 to 24 μ. These cells, singly, when fresh appear optically homogeneous and slightly yellowish. Other, and less abundant, corpuscles are small and large granular amoebocytes showing clear pseudopodia and some giant (up to 123 μ) multinucleate disk-shaped bodies. Freshly drawn coelomic fluid shows the capacity to form a clot rather rapidly. The amoeboid cells appear to participate in this process, in part through agglutination of their pseudopodial processes.

Reproductive cells are also usually abundant in the coelomic fluid. In the female they vary from small immature eggs about 25 μ in diameter, having large germinal vesicles, to large mature eggs (to 200 μ) possessing a thick gelatinous capsule and with nucleus obscured by yolk granules. In the male, sperm cells in various stages of development may be found. These cells may occasionally be so abundant as to give the coelomic fluid a milky appearance. They usually become motile only after being put in sea water.

Now continue the longitudinal incision to the posterior end of the worm, and also nearly to the anterior tip, and determine the extent of the coelom. Pin out the body wall. The coelom is of an extensive perivisceral nature undivided by septa. It is lined by peritoneum which, especially over the gut, is covered by a thin, unicellular layer of chloragogue cells frequently stated to be excretory. The latter are easily broken away from it. These chloragogue cells contain yellowish granules. Over most of the peritoneum the nuclear region of each peritoneal cell possesses a tuft of long cilia which produces a circulation of the coelomic fluid.

Body Wall and Musculature. The animal is encased in a thick cuticle possessing pores to the exterior scattered abundantly but irreg-

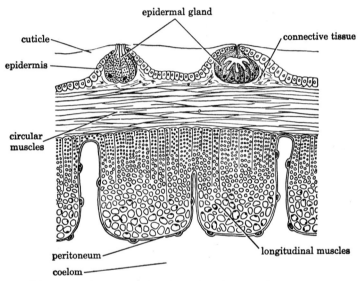

FIG. 165. Diagrammatic cross section through body wall of *Phascolosoma gouldii*. (Redrawn from Andrews.)

ularly over the surface (Fig. 165). The pores are associated with thickened regions of the underlying hypodermis, which have differentiated into spheroid or ovoid cell masses possessing histological evidences of secretory activity. The cuticle over these organs is usually considerably thinner than over the remainder. Near the anterior and posterior extremities of the body are integumentary thickenings of the structure suggesting a sensory function. Immediately internal to the hypodermis is a thin connective-tissue layer separating it from a thick, continuous layer of circular-muscle fibers. Internal to the latter are the longitudinal-muscle fibers. These are separated into thirty or forty longitudinal bundles, possessing many interconnecting bundles, by longitudinal grooves which penetrate partially or completely outwards to the circular-muscle layer. A peritoneal membrane covers the muscle layers completely on their inner surface, bounding the extensive coelomic cavity.

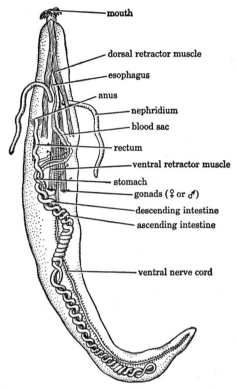

FIG. 166. Dorsal view of *Phascolosoma gouldii*, showing arrangement of internal organs. (Modified from Andrews.)

The retractor muscles (Fig. 166) comprise four dorsoventrally flattened band-like muscles with origins in the longitudinal muscles of the body wall and insertions near the anterior tip of the introvert. At this latter region the four muscles are fused to form a nearly continuous sheath about the pharynx. It is completed dorsally and ventrally by connective tissue. There are a number of connective-tissue fibers also connecting the anterior end of the esophagus with this muscular sheath. The retractor muscles are covered with a peritoneal layer essentially like that of the body wall with which they are continuous.

Vascular System. Very little of this system can be worked out by the ordinary techniques of dissection. A closed system of blood vessels is restricted to the anterior end of the animal. This comprises an oral-disk plexus of vessels which connects with a larger circular vessel surrounding the pharynx. There are three vessels in each tentacle. Blood actively circulates through the tentacles by action of ciliated tufts on the lining epithelium of the vessels. The circumpharyngeal vessel opens into a larger dorsal vessel immediately ventral to the brain. This dorsal vessel, clearly evident in dissected specimens, continues posteriorly along the dorsal surface of the pharynx and esophagus, finally ending blindly over the posterior portion of the latter. This longitudinal vessel, called the blood sac, possesses contractile walls in addition to internal ciliated tufts. The fluid contained within the above system of vessels is far richer in red corpuscles than the coelomic fluid, and hence is distinctly more reddish to the eye. The fluid also contains amoebocytes substantially like those of the coelomic fluid and shows a similar clotting ability.

In view of the organization and activities of the vascular system, together with a reported observation that in oxygen-poor sand the tentaculated crown may be extended out of the sand, it appears that the tentacles may serve as functional gills.

Digestive System. Study the alimentary tract, carefully tracing it from mouth to anus. The mouth is anterior and terminal and is located within the tentaculated oral disk. The oral disk and the median concave faces of the tentacles are ciliated, as is also the lining of the pharynx. The tentaculated crown is withdrawn into the pharynx when the animal retracts. The remainder of the alimentary canal comprises the esophagus, stomach, descending and ascending limbs of the intestine, and the rectum. The *esophagus* is long. Its anterior portion is surrounded by the four retractor muscles to which it is attached by connective-tissue fibers. The remainder lies free except for a single connective-tissue thread of attachment. Internally its epithelial lining is thrown into a number of longitudinal folds. The *stomach* is short and has four longitudinal pink bands which are due to four corresponding longitudinal internal ridges containing radial muscle fibers. The stomach lies at about the level of origin of the ventral retractor muscles. The *descending intestine* extends from the stomach to the posterior tip of the animal. The longitudinal bands are here red but less conspicuous. Cilia are numerous within this portion of the tract. Ciliated pits occur internally on the longitudinal ridges of this region. These pits become deeper in the more posterior portion, even to the extent often of forming numerous minute externally visible diverticula in this area. The *ascending intestine* possesses a ciliated groove. This is visible externally as a light line between two pinkish longitudinal bands. The bands are a concentration of radial muscle fibers on each side of the groove. Their contraction appears to alter the depth of the groove. Cilia in the groove maintain a current of fluid moving in the direction of the rectum. The ciliated groove terminates at the rectal end of the intestine as a small, externally evident, diverticulum—the rectal caecum. The ascending intestine has associated with it a longitudinal mesentery containing muscle fibers. This is known as the spindle muscle. Its contraction results in tighter coiling and hence shortening of the ascending intestine, and simultaneously of the descending intestine which has numerous connective-tissue points of attachment with the former. This shortening is possible because the gut hangs quite freely in the coelomic cavity. The only supports of the digestive tract other than those already mentioned are three elastic threads attaching the ascending intestine to the body wall near its rectal end, and the anal connective-tissue fibers.

Numerous microscopic ciliated crescentic funnels known as "pseudostomata" occur on the external surface of the ascending intestine. They are believed to have a role in removing particulate matter from the coelomic fluid, inasmuch as carmine particles injected into the coelom become concentrated in the chloragogue cells found within these ciliated funnels.

The *rectum* possesses a lining of ciliated epithelium thrown up into numerous longitudinal ridges. There are no pseudostomata, nor is there a ciliated groove in the region.

Nephridia. The single pair of giant nephridia hangs freely into the coelom from their attachment at the point of the external opening. The free end is usually directed posteriorly and is closed. The internal opening, or nephrostome, is very close to and just anterior to the external opening. Thus the tubule is essentially U-shaped. The nephrostome is a slit lying between a semicircular flap of the anterior arm of the tubule and body wall. Both sides of the nephrostome are ciliated. The tubule walls contain well-developed circular- and longitudinal-muscle fibers lying between an internal secretory layer of cells and the covering peritoneum. After studying a nephridium *in situ*, remove it and examine it for action of the cilia on the nephrostome.

Reproductive Organs. The sexes are separate. The gonads are minute ridges of tissue lying transversely on the posterior faces of the ventral retractor muscles near their point of origin. These ridges are continuous with one another across the median line beneath the ventral nerve cord. The gonads comprise small masses of reproductive cells supported in connective tissue. The largest cells of the gonads, those about to be liberated into the coelomic space, are nearest the free tips of the fimbriated organs. The sexes can be distinguished in them by the fact that the largest cells in an ovary (24 μ) are nearly twice the size of the largest cells of a testis. The breeding season lasts from June through August. Sperm cells are forcefully ejected from males by way of the nephridia. The presence of the discharged male products induces the female to eject her eggs. Egg-laying normally occurs at night between the hours of 8 P.M. and 5 A.M. A few hours before liberation of eggs or sperm the nephridia become enormously distended with these sexual products.

Nervous System. The *ventral nerve cord* lies free in the coelom except for its attachment to the body wall by its lateral nerves. The cord is more or less ovoid in cross section in the posterior portion of the body but is wider and more dorsoventrally compressed in the region of the introvert. It terminates posteriorly in one or two posteriorly projecting nerve fibers; anteriorly in the region of the pharynx it branches to form the circumpharyngeal connectives which pass dorsally and enter the suprapharyngeal ganglion or brain. The ventral cord and lateral nerves are loosely invested in a peritoneal sheath. Between the sheath and the cord proper is connective tissue. In the posterior region of the body a single median muscle tract lies within the sheath dorsal to the cord. As the cord enters the introvert, the muscle tract bifurcates, the two branches going to right and left sides of the cord. These arrangements of the muscle tracts of the cord in the two regions account for the differences in cross-sectional form. Contraction of these muscle fibers results in shortening of the cord.

The *lateral nerves* pass from the dorsolateral aspects of the cord. They are numerous but appear to come off quite irregularly (Fig. 167). The cord shows no evidences of segmentation. The lateral fibers pass to a position between the two muscle layers of the body wall where they produce "ring nerves" passing around the circumference of the body. The "rings" send off numerous branches. Some branches of the "ring nerves" pass to produce a series of longitudinal nerve fibers in the connective tissue underlying the hypodermis. There is considerable fiber anastomosis here also. The subhypodermal, longitudinal fibers send branches to the hypodermal sensory organs and to the circular muscle; the ring nerves innervate the longitudinal-muscle fibers. These two major nerve plexuses are continuous with the plexuses of the gut and of the retractor musculature.

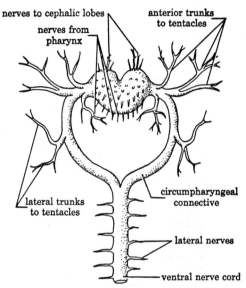

FIG. 167. Anterior region of the central nervous system of *Phascolosoma gouldii*. (Redrawn from Andrews.)

The *brain* lies dorsal to the pharynx with its overlying "blood sac," and between the dorsal retractor muscles of the introvert. It is a relatively large structure, being about 2 mm. in diameter. The dorsal aspect of the brain is separated from the coelom only by the peritoneum and a thin layer of connective tissue. Carefully expose the brain and the circumpharyngeal connectives. The brain has numerous fibers passing to the cephalic lobes anteriorly and to the pharynx and dorsal retractors posteriorly.

A number of large nerves originating in the circumpharyngeal connectives innervate the tentacles, the pharynx, and probably also the ventral retractor muscles.

9.

Mollusca and Brachiopoda

MOLLUSCA

Chaetopleura apiculata

By MADELENE E. PIERCE

Chaetopleura apiculata is a small chiton about 2 cm. in length. It lives in the shallow waters of the Atlantic coast from Cape Cod to Florida. By means of its broad flat foot, this small mollusk can cling fast to the rocks or empty shells upon which it lives, or creep slowly over their smooth surface in search of diatoms and algae. Chitons are easily collected either by examining the small stones and shells at the low-tide level, or by dredging in water up to 8–10 meters over a substrate which is composed largely of old shells. The dorsal surface of *Chaetopleura* is covered by a shell which is composed of eight calcareous, transverse plates overlapping posteriorly. These overlapping plates allow the animal to bend slightly during locomotion or to curl up into a ball if pried loose from the substrate. They are surrounded and kept in place by a muscular integumental fold called the girdle (Fig. 168A).

A living specimen can best be studied when it is attached to a glass plate. In order to prevent the animal from curling into a ball, transfer it quickly from its original stone so that it will attach to the glass in a flat position. Slight pressure on the shell from above may induce the animal to remain flat. Later, study the natural behavior of the chiton when it is allowed to move its plates freely.

On the ventral surface, the most conspicuous organ is the flattened, elliptical foot which is adapted for creeping and clinging. Locomotion is accomplished by waves of muscular contraction and ciliary action. Adherence to the substrate is maintained by suction and by the aid of mucus secreted by numerous glands. Projecting over the sides of the foot is a fold of the mantle called the girdle. This fold encloses a deep furrow, the mantle cavity, in which the gills are located. They are composed of a series of twenty-four short transverse ridges. The head fold, containing the mouth, occupies the anterior portion of the ventral surface. If the mouth is enlarged by cutting, the radula, a chitinous ribbon bearing rows of teeth, can be seen. This ribbon should be removed and mounted on a slide for further study. (The radula mechanism will be explained in detail under the study of the gastropod, *Busycon*.) The anus is located at the posterior end of the ventral side (Fig. 168*B*).

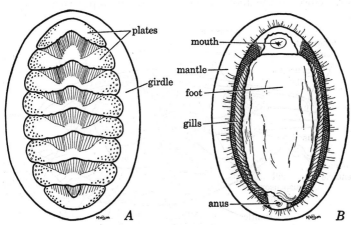

FIG. 168. *Chaetopleura apiculata*. *A*. Dorsal aspect. *B*. Ventral aspect, with foot slightly reflected to show gills and anus.

The breeding habits of *Chaetopleura* make it a satisfactory form for embryological study. At Woods Hole, spawning may occur from early June to late September, but the sex products are most abundant from July 10 to August 20. If the specimens are placed in finger bowls of clean sea water, they will usually shed in the evening from 8 to 11 o'clock. Adult males and females are indistinguishable from one another. However, since no evidence for interdependent stimulation exists, isolated individuals spawn satisfactorily. Eggs can then be fertilized and studied. The entire embryological development is completed in 6–10 days.

Yoldia limatula

By MADELENE E. PIERCE

Yoldia limatula belongs to the most primitive order of the bivalves. Geographically this small clam has a wide distribution, for it is found

320   MOLLUSCA AND BRACHIOPODA

not only on the Atlantic coast from Labrador to North Carolina, but also on the Pacific coast and in Europe. It lives in the sandy mud of shallow coves and inlets and may be found in water up to 9 meters in depth. It is most commonly found in the intertidal zone near low-tide level, where it can burrow rapidly by means of its large and powerful foot.

Place a lively specimen in a dish of sea water, and study its behavior. Observe the foot which is thrust out repeatedly. Notice its

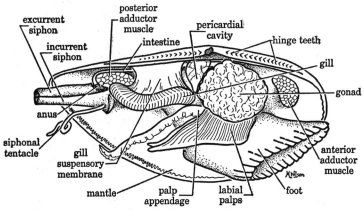

FIG. 169. Right valve and mantle removed to show internal anatomy of *Yoldia limatula*.

general shape, but particularly the contour of its ventral surface. The temporary broadening at the distal end, together with the temporary concavity of the sole, results in rapid and efficient burrowing. The mantle is open along the greater part of its margin. If the specimen is allowed to remain quiet for a while, the elongated, united siphons may be extended. If the animal can be induced to feed, the long feeding appendage may be protruded. This structure is an extension of the outer palp and should not be confused with the siphons. During the process of feeding, the appendage is thrust out and inserted into the mud some distance from the shell. Cilia within the groove carry mud mixed with microorganisms to the mouth.

Remove one valve to study the internal anatomy (Fig. 169). The gills, the most primitive type, are small thick organs composed of a double series of parallel plates, suspended from the body wall by a thin muscular membrane (Fig. 170). The motion of the gills of *Yoldia* is characteristic and can be observed through the shell of young specimens. As blood is forced into the gill plates, they are pressed ventrally, so that water passes through the gills. Contraction of the suspensory membrane then pulls back the gill plates to their dorsal position. This results in a discharge of water through the dorsal excurrent siphon and an inflow of water through the ventral incurrent siphon to the mantle cavity. This respiratory action is rhythmical.

The nervous system is composed of the three pairs of ganglia typical of the bivalves. In a living *Yoldia*, however, they are particularly easy to identify because of their brick-red color. Even the connectives are slightly tinged, so that it is easy to trace both the cerebrovisceral and the cerebropedal connectives.

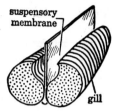

FIG. 170. Diagram of gill of *Yoldia* to show suspensory membrane.

Along the hinge line of each valve, numerous tiny teeth are present. This condition is considered a primitive characteristic.

Pecten irradians

By MADELENE E. PIERCE

Pecten irradians is the small edible bay scallop with the fluted shell. Its range extends from Maine to the Gulf of Mexico. *Pecten* inhabits the shallow waters of protected bays, where the bottom is usually a mixture of sand and mud. Although it is often found swimming about in the eelgrass where water may be only a foot deep, it may be dredged from depths up to 20 meters. Unlike most bivalves, *Pecten* is often an active swimming form. Usually it swims with a fairly regular motion which is the result of two jets of water sent out through temporary openings formed by the mantle edges near the hinge line. Since these jets are sent dorsoanteriorly and dorsoposteriorly, the scallop progresses ventrally (forward). Occasionally, if startled, the scallop swims by a series of quick, violent jerks, the result of a rapid closing and opening of the valves. This motion is possible because of the striated muscle fibers within the single, highly modified and enlarged adductor muscle. In this case, water within the mantle cavity is ejected ventrally, and the animal therefore moves dorsally (backward). The scallop is an early example of jet propulsion.

Normally the scallop rests on its right valve. On each side of the beak are projections of the shell called wings. The posterior wing has a broad connection with the valve; the anterior wing possesses a notch or indentation, with a resulting narrow connection to the valve. If the specimen is allowed to relax, the valves will gape slightly. Along the edges of the mantle, now slightly separated, notice the

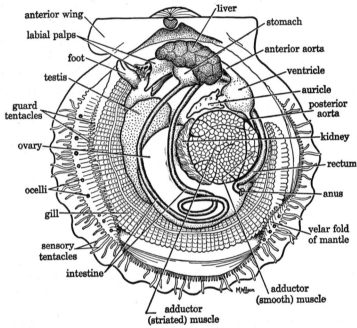

FIG. 171. *Pecten irradians.* Left valve, mantle, and gills have been removed to show internal anatomy, and part of the outer wall of the visceral mass dissected to make visible the digestive tract.

velar fold, the pallial eyes, and the tentacles. The velar fold, which guards the tentacles, appears to be sensitive to certain irritating chemical stimuli. To test this sensitivity, place a few drops of extract of starfish near the velum. The tentacles are particularly sensitive to mechanical stimulation. Test this statement by touching one or two with a needle. The eyes are sensitive to a decrease in light, such as a shadow passing over the animal. However, only the eyes directly in line with the light are affected, and stimulation of at least two eyes is necessary to elicit a response. Structurally, the eye shows considerable differentiation, for it is composed of a cornea, lens, retina, and optic nerve. These are not homologous to the structures of the same name in the vertebrate eye. The retina is composed of two parts, a distal and a proximal layer. In the distal retina, the sensory cells are inverted so that light must pass first through the nerve fibers before reaching the sensory elements.

Pry open the valves and cut the huge adductor muscle which is composed of two sections, a large anterior striated section used in rapid swimming, and a small posterior unstriated section used for holding the valves closed over a long period of time. The gills are crescentic structures located on the anteroventral margin of the vis-

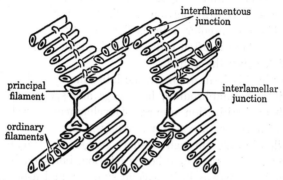

FIG. 172. Diagram of *Pecten* gill to show relation of ordinary and principal filaments within the lamellae. The lamellae are shown in their folded state.

ceral mass (Fig. 171). Study their muscular movement which allows them to be drawn together during swimming. This contraction results in the formation of frequent vertical folds which cause temporary elevations and depressions along the surface of the lamellae. The gills are composed of a series of parallel filaments loosely attached to each other. There are two types of filaments, the larger principal filaments and the smaller common filaments (Fig. 172). The former usually lie at the bottom of each depression. Examine a mounted portion of each gill. Between the principal filaments, the interlamellar junctions are present. Distinguish in addition the interfilamentous junctions. Near the periphery of the gills they are simple clumps of modified cilia, but near the interlamellar junctions they become much larger and more prominent. There are no siphons. Water flows freely through the mantle cavity. At the anterior dorsal edge of the visceral mass lie two small organs of similar size, the degenerate foot, and the mouth surrounded by a fringed palp. A dissection of the entire digestive tract is not always practicable. However, the free posterior end of the intestine, posterior and dorsal to the adductor muscle, is easily seen. The heart lies in the mid-dorsal line between the dark mass of liver and the adductor muscle. The ventricle, which is a large smooth-walled chamber, surrounds the intestine. The paired triangular auricles, which are smaller and have rugose walls, unite with the ventricle at its ventrolateral margin.

In fresh specimens the kidney is easily identified as a rose-colored organ, somewhat crescentic in shape, which lies ventral to the auricles and closely follows the contour of the adductor muscle.

Reproductive System. The hermaphroditic gonad is a conspicuous organ lying anterior and ventral to the adductor. The ovary occupies the anterior and ventral portion and is bright pink when the eggs are ripe. The cream-colored testis lies slightly dorsal to the ovary. In the scallop there is often a dense black pigmentation in the membranes which cover the gonad.

Spawning in this form occurs in the late summer and early autumn. Although it is hermaphroditic, both eggs and sperms are shed into the sea where fertilization occurs. Development is rapid with the formation of a veliger larva in a few days. Like many bivalves, the young scallop passes through a short period of attachment by its byssus thread to eelgrass or other vegetation, but soon becomes free. It grows rapidly, spawns in its second summer, and usually dies near the end of its second year.

Venus mercenaria

By MADELENE E. PIERCE

The clams, mussels, and scallops of this group are often referred to as the lamellibranchs because of their flat, plate-like gills. The general directions for the dissection of a typical lamellibranch will be given for *Venus*, a common clam of this area. However, these directions may be used profitably in the study of other lamellibranchs for which there are brief directions limited to the structures particularly characteristic of each animal.

Venus mercenaria is known by a variety of common names, such as quahog, hard-shelled clam, and littleneck clam. Its entire range includes the shallow, protected waters of the Atlantic coast from the Gulf of St. Lawrence to Texas, but it is most abundant from Cape Cod southward. By means of its powerful foot, this form burrows about over muddy or sandy flats near the lower portion of the intertidal zone. Although it lies buried a few inches, it keeps its siphonal end above the surface of the mud in order to maintain a current of sea water through its body. This current is essential for respiration and feeding.

Shell. The shell of *Venus* is composed of two symmetrical halves called valves. Because of this arrangement, the mollusks possessing two valves are called bivalves, as distinguished from snails, which are called univalves. At the dorsal margin the valves are joined by a brown, external hinge ligament. On either side of this ligament there is a swelling in each valve called the umbo. This ends in a point, the beak, which is directed anteriorly. Growth has proceeded from the beak as an origin and has resulted in concentric lines of growth. The lunule is a heart-shaped configuration which lies anterior and ventral to the beak. This is a useful character for field identification (lunule absent in fresh-water mussel). Before opening the valves, determine the correct orientation, right, left, dorsal, ventral, anterior, and posterior. Obtain, if possible, a young *Venus* for comparative shell study.

Pry the two valves apart and try to insert your knife between the shell and the mantle which lines it. Detach the mantle carefully, and, at the anterior and posterior ends, cut the adductor muscles. Since the contraction of the adductors closes the valves, they will now gape and be readily separated. The gaping is due to the action of the ligament which opposes the adductor muscles.

On the inside of the shell are various markings, indicating the former attachment of organs. Near each end of a valve is the large, round scar of the adductor muscle. Closely associated with these are the two smaller scars of the retractor muscles of the foot. The small impression of the anterior retractor is separated from the larger one of the anterior adductor and lies dorsal and a little posterior to it. The impression of the posterior retractor is usually merged with that of the corresponding adductor. (In the fresh-water clam the scar of the anterior retractor is similar in position; that of the posterior retractor similar but more distinct; and in addition the scar of an anterior protractor lies ventral and slightly posterior to the anterior adductor.) Running from one of these large scars to the other is a distinct line called the pallial line. This line, parallel to the margin of the shell, marks the attachment of the pallium, or mantle, to the shell. At its posterior end, the line forms a triangular sinus, the pallial sinus (absent in the fresh-water clam). This indicates the position of the retractor muscle of the siphons.

Along the free ventral edge notice the tiny teeth. This arrangement is the result of secretion by a fringed edge of the mantle. The crenulated margin ensures a firmer closing of the valves. Along the dorsal edge notice the larger teeth which are of two kinds. Anterior to the ligament are the prominent cardinal teeth; below the ligament are the less obvious lateral teeth which extend posteriorly for some distance. Notice that the teeth fit tightly and form an efficient interlocking mechanism. In *Venus*, the thin periostracum, or outer layer of the shell, is partly worn away by constant friction with the sand. The innermost layer, the laminated nacreous or pearly layer, is very thick but does not appear as shining and lustrous as in many mollusks. The middle, or prismatic, layer is present but not easily visible.

Mantle. With one valve removed, place the specimen in a dish of sea water. The mantle, now exposed, forms a thin covering for the soft parts. The mantle consists of two lobes of tissue, one of which is applied to the inner surface of each valve. These two lobes are continuous dorsally. The free border of each lobe is thickened and contains muscles which attach along the pallial line. These muscles serve the function of a washer. The posterior edges of the mantle lobes are also thickened and fused at two points, one above the other, to form a double tube, the siphons. (In the fresh-water clam the ventral siphon is not a perfect tube. It is only by the contact of the two ventral posterior edges of the mantle that the "tube" is maintained.) In *Venus*, these siphons are very short but are easily identified by their dark pigmentation. Water enters through the ventral, incurrent siphon, and after circulating through the branchial chambers, leaves by the dorsal, excurrent siphon. A few grains of powdered carmine placed near the siphons of an undissected, living specimen will demonstrate these currents clearly.

Visceral Mass and Foot. Carefully cut the mantle and reflect its parts. The ventral portion of the median mass now visible is the foot, which is somewhat hatchet-shaped. Identify the small foot muscles, the attachments of which have already been seen on the shell. These muscles may be seen at the extreme anterior and posterior margins of the foot, where they extend dorsally toward their attachment on the valves. The dorsal portion of the median mass is called the visceral mass. As its name suggests, it contains the viscera which will be studied in sequence (Fig. 173).

Respiratory System. With the mantle removed, the gills are the most conspicuous organs. They consist of two pairs of thin membranous folds, ridged in appearance, which lie on each side of the visceral mass. They are attached to the body on each side along a line extending from the wall which separates the two siphons to a point even with the beaks on the exterior. Only the dorsal edge is attached; the ventral edge hangs freely. The outer gills are attached to the mantle lobe near its dorsal region; the inner gills are attached to the visceral mass. The entire space enclosed between the right

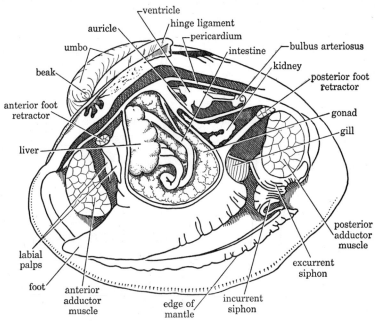

FIG. 173. *Venus mercenaria,* left valve, mantle and gills removed to show internal anatomy. The wall of the visceral mass has been partly removed, and the pericardium has been cut away. (Modified from Woodruff.)

and left lobes of the mantle is called the mantle cavity. Since the gills are attached to the mantle, to each other, and to the visceral mass, their attachments form a continuous horizontal partition separating a small cavity, the suprabranchial or cloacal chamber, from a much larger ventral cavity, the branchial chamber. Into the latter the incurrent siphon leads, and the gills hang freely. The suprabranchial chamber is not visible at this point. It is a median, dorsal tube, extending anteroposteriorly, and connecting with the excurrent siphon (Fig. $174A_3$).

Each gill is a double fold, the inner and outer surface being composed of parallel ridges or filaments. Each surface is referred to as a lamella. Each gill, therefore, consists of two lamellae, an inner and an outer one (Fig. 174A3). The inner and outer lamellae of a single

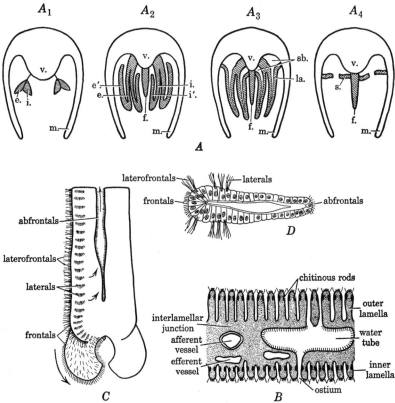

Fig. 174. A. Diagrams of gill structure of the Pelecypoda. Types of gills found in the different orders of Pelecypoda. A_1. *Taxodonta*. A_2. *Anisomyaria*. A_3. *Eulamellibranchiata*. A_4. *Septibranchiata*. v.: visceral mass; m.: mantle; e.: external row of filaments; e'.: external row of filaments turned back; i.: internal row of filaments; i'.: internal row of filaments turned back; s.: septum; f.: foot; sb.: suprabranchial chamber; la.: lamella. (Modified from Lang.) B. Cross section of a mussel gill. (Modified from Borradaile and Potts, courtesy of The Macmillan Company.) C. Portion of gill filament, near free edge, to show typical sets of cilia present and direction of beat. (Modified from Orton.) D. Cross section of one limb of a filament, to show same sets of cilia. (Modified from Orton.)

gill are separated by a thin space, which at regular intervals is divided vertically into a series of narrow tubes, the water tubes. These tubes connect dorsally with the suprabranchial chamber. The

VENUS MERCENARIA

partitions forming the walls between the tubes are called interlamellar junctions, and they extend dorsoventrally between the lamellae. They contain blood vessels. From one gill, cut a transverse section 1 mm. wide and stand the section on edge. If this is examined under a dissecting microscope, many of the above-mentioned structures can be seen (Fig. 174B).

Separate one lamella from the other. This can best be done by working with the free dorsal border of the inner lamella of the inner

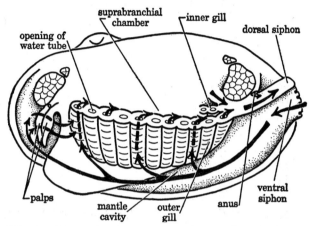

FIG. 175. Diagram illustrating the circulation of water through the mantle cavity and gills of a fresh-water mussel. The suprabranchial chambers are shown as if cut open from above. (Modified from Wolcott.)

gill. Mount a small piece on a slide and study it under the low power of the microscope. Try to identify several structures already seen grossly, as well as the following. The filaments extend the width of the gill, and between them are connecting bridges, the interfilamentar junctions. The inhalant ostia are small but frequent openings bounded by filaments and their junctions. It is through these ostia that water enters the gills, the surface of which is abundantly ciliated (Fig. 174C, 174D). A few grains of powdered carmine, placed at the ventral scalloped edge of the gill, will demonstrate the current produced by the prominent cilia present. Because of the constant beating of the cilia, a continuous current of water is pumped through the incurrent siphon into the branchial chamber, through the ostia, and into the water tubes. From here the water enters the suprabranchial chamber and is carried out by the excurrent siphon (Fig. 175).

Not only oxygen but also food particles are transported by these currents through the branchial cavity. The strength of these currents is markedly influenced by change in temperature of the surrounding water. At 22°C., *Venus* actively pumps for 96 per cent of the day, but with a drop to 5°C. ciliary activity decreases, and below 5°C. hibernation of the individual occurs.

Circulatory System. Near the dorsal midline and just anterior to the posterior adductor muscle, there is a thin-walled triangular chamber, the pericardium, within which lies the heart. Carefully open the pericardium. If the heart is still beating, perhaps you can determine its rate. The heart is composed of a single median ventricle, and lateral paired auricles. The thick-walled ventricle is somewhat pyramidal in shape with its apex pointing anteriorly. It surrounds the intestine. Two aortae leave the ventricle; the anterior one runs dorsally to the intestine, the posterior one ventrally to the intestine. Posterior to the heart, but within the pericardial cavity, the posterior aorta forms a conspicuous swelling, the bulbus arteriosus (not present in fresh-water mussel). The paired auricles, triangular in shape, are very thin-walled and can be distended to a considerable extent.

The circulatory system of *Venus* is an open system. This means that arteries communicate with veins by way of large spaces called sinuses or lacunae. Blood is pumped by the ventricle of the heart through the two aortae to the viscera and neighboring sinuses. From here, by a network of veins, blood is returned through the kidneys to the gills and finally brought back to the auricles of the heart.

The pelecypod heart is an excellent subject for physiological study. It can be perfused with various solutions, and the effect upon rate of beat observed. If heart levers and kymographs are available, a record of the experiment can be made.

Carefully remove one valve and open the pericardial cavity. Place the animal in a finger bowl containing molluscan Ringer's solution, and determine the rate of heart beat. Remove the animal to a second bowl, and observe the effect of dropping $M/2$ KCl on the heart. Add $M/2$ CaCl$_2$, and note the effect. Return the animal to Ringer's solution for recovery. How does the effect of these salts on the pelecypod heart compare with their effect on the vertebrate heart?

The experiment can be repeated, using different drugs such as nicotine, adrenalin, acetylcholine, and atropine in varying dilutions.

Excretory System. Ventral to the pericardium and between it and the posterior adductor muscle lies a pair of dark glandular organs, the nephridia. Each communicates with the pericardium by a minute opening which is difficult to find, and also with the branchial cavity by another opening more easily seen. Reflect both gills dorsally. This second opening is located on a small papilla which will be described under the genital system. The nephridia are regionally differentiated in their composition, and consist of a broad U-shaped tube. The ventral anterior portion communicating with the heart is glandular; the dorsal posterior part is thin-walled and functions as a bladder. (These organs are much more satisfactorily seen on the fresh-water clam.)

The coelom of the mollusks is reduced to three small cavities, the pericardial, the gonadial, and the nephridial. The extensive spaces are the blood sinuses, which have enlarged to such an extent that they appear to have replaced the coelom.

Digestive System. On either side of the anterodorsal edge of the visceral mass lie two triangular flaps, the labial palps. (They are more nearly oval in the fresh-water clam.) Both the inner and the outer flap unite with the corresponding structure of the other side, above the mouth, which lies slightly posterior to the dorsal border of the anterior adductor muscle. The palps are abundantly ciliated and cause streams of water to pass into the mouth. They therefore continue the transfer of food particles already brought in by the currents from the gills.

For dissection of the alimentary canal, pick off carefully the tissue from one side of the visceral mass and foot. (As you dissect, consult the directions for the nervous system so that you will not destroy the cerebral ganglia.) The mouth leads directly into the esophagus. There is no radula or buccal mass in this form. The esophagus soon widens into a thick-walled stomach, lying dorsal to a large brown gland, the liver. The liver connects with the stomach at its anterior end. The stomach continues posteriorly and ventrally, and gradually narrows into the intestine. Trace the loops of the intestine through the visceral mass, until the tube turns dorsally and joins with the portion already identified in connection with the pericardium. From the posterior wall of the pericardium, trace the intestine dorsal to the adductor muscle and finally to its end, the anus, which lies at the base of the dorsal siphon. Return to the pyloric region of the stomach where there is a diverticulum, the caecum, which contains the crystalline style. This style is a solid gelatinous rod which is very conspicuous in some forms (*Mya*, *Mactra*), but is not always seen in

Venus. It is concerned with the liberation of certain enzymes, chiefly for digestion of carbohydrates. Its presence appears to be correlated with an abundant supply of oxygen.

Venus is a typical filter feeder. This means that food consisting of diatoms, Protozoa, and organic detritus, brought in by the respiratory currents, is passed anteriorly along the edge of the gills and becomes entangled in strings of mucus. Larger food particles are dropped to the edge of the mantle and discarded. As the mucous strings approach the mouth, the cilia on the palps continue their transportation into the mouth. Digestion is chiefly intracellular by phagocytes lining the digestive tract and by wandering leucocytes. Digestive juices from the liver are known to occur, and in addition the crystalline style liberates an enzyme aiding in the breakdown of carbohydrates.

Nervous System. Although possessing no well-marked head region, the bivalves display a central nervous system similar to that of other

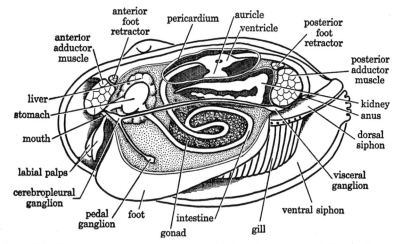

FIG. 176. Diagrammatic representation of internal anatomy of the fresh-water mussel after left valve, mantle, and gills are removed. (Modified from Wolcott.)

mollusks. It consists of three pairs of ganglia and their connectives (Fig. 176). Carefully dissect the visceral mass away from the sides of the esophagus. The cerebral ganglion will be seen lying posterior to and about 3 mm. from the dorsal edge of the anterior adductor muscle. It is a round, yellowish organ, about the size of the head of a pin. It is connected with the corresponding ganglion of the other side by a commissure which passes anterior to the esophagus. Two conspicuous connectives pass from each ganglion. One of these courses posteriorly to unite with the visceral ganglion of the same side. The other passes into the foot to unite with the pedal ganglion. The cerebral ganglia also furnish the anterior pallial nerves. With a razor, make a median sagittal section through the foot and into the visceral mass. This should expose the pedal ganglia which lie embedded in the foot dorsal to the muscular portion. To find the third pair of ganglia, cut the united lamellae of the gills ventral to the posterior adductor muscle. On the anterior surface of this muscle, the visceral ganglia will be seen. They are slightly pear-shaped bodies lying so close together that they appear to be single. Trace all connectives as far as practicable.

Reproductive System. The genital glands consist of a pair of simple gonads, either testes or ovaries, which occupy varying amounts of space in the visceral mass. During the breeding season they are large and conspicuous and fill the spaces around the loops of the alimentary canal. It is difficult to distinguish the paired nature of the reproductive glands. Reflect the gills and follow dorsally along the posterior edge of the foot. Under the attachment of the inner gills lies a small projection, the genital papilla, which contains the opening of the genital ducts. A minute opening, the nephridiopore, is also located on the papilla but is difficult to see. It is, of course, the external opening for the kidney.

The spawning of *Venus* occurs in the Cape Cod region from late June to early August, when the temperature of the water has risen to a critical level, about 75°F. Eggs and sperm, numbering into the millions, are freely shed into the sea where fertilization takes place. Spawning in this species is not necessarily completed by one discharge, but may occur several times during the summer. The early embryology of this form is typical for the bivalves. The trochophore and veliger stages are rapidly completed, and during the late summer the young quahog becomes temporarily attached to a substrate by a byssus thread. It soon regains a free existence. For a period of 2 years the young clam lies relatively dormant during the winter but grows rapidly during the warm months. By the third summer, it has matured and is ready to spawn.

Mactra solidissima, a close relative of *Venus,* possesses mature eggs and sperm in the late summer. The sex products can be easily obtained from the ovary (usually pinkish) and from the testes (white) of mature clams. The eggs and sperm should be placed in separate finger bowls of sea water. If the eggs are then fertilized with a dilute sperm suspension, development will proceed. Within 10 minutes of insemination, the large germinal vesicle reacts; within 30 minutes the polar bodies begin to form; and after about an hour first cleavage takes place. A swimming form is produced in 5 hours. With careful attention, these embryos can be carried over to later stages.

Fresh-water Clams or Mussels

By MADELENE E. PIERCE

Since all lamellibranchs are essentially similar, the directions previously given for *Venus* can be used satisfactorily for the following forms. Specific additional suggestions for the study of several fresh-water forms are given below (Fig. 176).

The Family Unionidae is nearly world wide in distribution and includes practically all the large fresh-water mussels. The Genus *Unio* formerly included many bivalves which have since been separated into several genera, such as *Elliptio, Quadrula, Lampsilis, Lasmigona, Amblema, Cyclonaias,* and *Ligumia*. The Genus *Anodonta* has long been, and still is, recognized as a genus entirely distinct from *Unio* or any of the genera recently separated from *Unio*. In the United States at least 500 species have been described. These mussels are found abundantly in streams and lakes, where they lie buried in the sand or mud, generally with only the posterior tips of their valves exposed. A large active foot enables them to burrow. Commercially a few species are used in the manufacture of buttons.

If possible, collect and examine several species for a comparative study. Notice the great variety of size, shape, color, and thickness of the shells. The general outline of the shell may vary from a slender ellipse, as *Ligumia nasuta*, to a nearly perfect circle, as *Cyclonaias tuberculata*. In some species of *Anodonta* the valves may bulge to such an extent that the name "floater" has been given to that form. In certain species of *Elliptio*, on the contrary, the valves show only a slight swelling which results in a very short lateral dimension. The color of the periostracum varies from the bright greenish yellow of many *Anodontas* to the dull dark brown of *Elliptio*. The periostracum of some mussels is often decorated with radiating lines of darker hue, upon a lighter background, as in *Lampsilis*. The outer surface of the valves may be smooth or marked with corrugations. In the region of the umbo the surface is often deeply corroded because of the action of acids in the water. The thin, fragile shell of *Anodonta* contrasts markedly with the thick, heavy one of *Quadrula*.

The inner pearly layer is often lustrous and colorful, shading from a silvery white through salmon and pink to lavender. The nacreous layer, which may be very thick in these forms, often composes the greater part of the shell. The hinge teeth are almost as variable as the number of genera. At one extreme is *Anodonta*, which virtually lacks teeth; at the other *Elliptio* and *Quadrula* with strong, well-developed cardinal and lateral teeth.

The fresh-water clams are dioecious. The paired gonads which surround the coils of the intestine are brightly colored when the animal is mature. Sperm is shed into the dorsal siphon, carried out to the surrounding water, and brought by way of the ventral siphon of the female into her gills. The eggs, which are shed into the mantle cavity, are carried through the ostia into the water tubes of the gills, where fertilization and development occur. Usually only the outer gill serves as the marsupium or brood pouch, which is easily recognized by its swollen appearance.

The cleavage is complete but unequal and results in the formation of a peculiar parasitic larva called the glochidium. This is a tiny bivalve creature, possessing a large adductor muscle, and often a conspicuous hook at the ventral point of each of its triangular valves. After the glochidia are discharged into the water, many die for lack of the proper host. However, some come in contact with the proper species of fish, upon the skin or gills of which they encyst. For several weeks they lead a parasitic life. During this period they undergo metamorphosis into the adult. Such a life cycle ensures a wide dispersal of the species.

Two types of breeding are recognized among the fresh-water mussels. In the case of the short-term breeders, the eggs are fertilized in June and July and are usually discharged by September. The glochidia attach for a relatively short period to the fish host. Representatives of this type of breeding are *Amblema, Cyclonaias, Elliptio,* and *Quadrula*. The eggs of the long-term breeders, fertilized in August, remain in the marsupium until the following spring when they are discharged. The glochidia of this group spend from 3 to 12 weeks in the parasitic stage. Examples of long-term breeders are *Anodonta, Lampsilis, Lasmigona,* and *Ligumia*.

Busycon canaliculatum

By Madelene E. Pierce

Busycon, formerly known also as *Fulgur*, *Sycopsis*, and *Sycotypus*, is a large whelk or winkle of the east coast of the United States. It occurs commonly in shallow waters from the south shore of Cape Cod to the Gulf of Mexico. By means of its tremendous foot, it can plow through the sand in search of bivalves upon which it feeds.

Observe the behavior of a living specimen. Notice the large, tough, muscular foot on which the animal glides along the walls of the aquarium. If the head is extended, watch the motion of the tentacles and the proboscis. Touch the specimen and study the position and function of the horny plate, the operculum, which fits into the aperture of the shell as the animal is withdrawn.

The Shell. In your study examine the entire shell and one which has been sawed along its axis. The shell is a structure somewhat like a spiral staircase. The apex lies at the peak of the spire which is composed of several whorls. The number is constant for the species, but a young specimen may not possess the complete number. Notice the increase in size of each whorl from the apical region to the body whorl which is both the largest and most recently formed. The large opening of the body whorl is called the aperture. The axis around which the whorls are wound is the columella. This can be studied better from a hemisected shell. The siphonal canal is the elongated end opposite the apex. On the external surface, notice the lines of growth which represent successive periods of growth. In one half, identify the three layers characteristic of molluscan shells: the outer, rough periostracum largely worn off; the smooth, inner nacreous layer lining the shell; and the thick, middle prismatic layer.

To determine whether your specimen is wound dextrally or sinistrally hold the shell with the apex above, the siphonal canal below, and the aperture toward you. If this aperture is on your right, the animal is considered dextral.

Surface Anatomy. Compare a specimen which has been removed from its shell with the shell just studied, and try to understand their relationship. On the right side of the animal, notice the columellar muscle which attached the animal to its shell and thus enabled it to withdraw. This muscle continues into the large, flat, ventral foot, which is composed almost wholly of muscle fibers. The foot is used primarily for locomotion and attachment, and its flat, pliable surface is eminently modified for the purpose. By a series of rhythmic contractions of muscle, the foot is lifted off the substrate, moved forward, and set down again. These pedal waves can be observed in a living specimen. As an organ of attachment, the foot adheres firmly to the rocks by means of suction and the aid of copious mucus secreted by the cells of the pedal gland. The opening of this gland, the pedal groove, lies near the anterior end of the sole of the foot. Near the

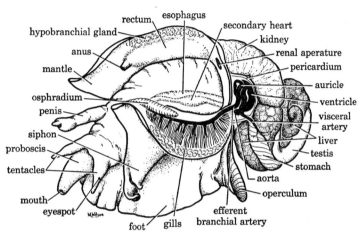

FIG. 177. *Busycon.* Removed from shell; mantle cut and reflected to show interior of mantle cavity.

posterior surface of the foot and above the sole is the operculum, a thin, horny disk, which serves as a trap door to lock the animal securely within its shell. By means of it the animal is protected not only from mechanical injury and chemical irritants, but also from excessive dryness.

The head, an anterior projection above the foot, bears a pair of triangular tentacles, on the lateral surface of which are the pigmented eyespots. From beneath the tentacles the proboscis may protrude, but this will be true only in a relaxed specimen (Fig. 177). At the tip of the extended proboscis, the triangular mouth opening can be easily identified. Frequently the teeth of the radula can be seen through this opening. If the specimen is a male, the large penis will project to the right of the right tentacle.

Above these structures lies the visceral mass which in the living animal occupies the spire of the shell. This soft area is covered by a thin, transparent, closely applied membrane, the mantle. Near the base of the visceral mass the mantle becomes thicker and forms a collar which extends around the body along a line corresponding to the aperture of the shell. It is at the free edge of the mantle that the most active secretion of the shell occurs. The space enclosed within the mantle is called the mantle cavity. The elongated portion of the mantle is the siphon and forms the lining of the siphonal canal. It is through this canal that water is drawn into the mantle cavity for aeration of the gill.

Through the mantle many of the structures of the visceral mass can be identified. Beginning at the apex, notice the grayish brown liver, which occupies the major part of the first two whorls. Its right lobe fills the apex; its left lobe lies below in the next whorl. Between the two lobes is the gonad, usually conspicuous because of its orange color. On the outer surface of the left liver lobe, and partly covered by it, lies the stomach. This is neatly curved to fit the contour of the whorl. Notice the large brown kidney which lies along the dorsal surface of the mantle. It is somewhat rectangular in shape and is composed of two regions, a small tubuliferous region lying to the left of a larger acinous region. These two parts can be seen more clearly after the mantle cavity has been opened. To the left of the anterior end of the kidney lies the pericardial cavity. The heart is often visible through the thin dorsal wall. Anterior to both the pericardial cavity and the kidney lies the large obvious gill which is oblong in shape and usually brownish in color. If the specimen is injected, the left portion will be red, the right, blue.

Open the mantle cavity by cutting the mantle along the medial edge of the gill. Continue cutting until the pericardium is reached. Take care to leave the heart undisturbed at this time. Turn aside the mantle and re-examine the gill. Between the anterior end of it and the base of the siphon lies an elongated elliptical organ, the osphradium, over which all water passes before reaching the gill. The osphradium is composed of 90–100 leaflets covered with epithelium containing sensory, glandular, and ciliated regions. It has an abundant nerve supply. Although the function has not been satisfactorily determined, the organ appears to test the water entering the mantle cavity. Along the median side of the gill lies a modified glandular area of the mantle called the hypobranchial gland. This gland secretes mucus rapidly and copiously and thus protects the gills by removing dirt and other foreign particles. If the specimen is a female, the large yellow nidamental gland will easily be seen on the right side of the mantle cavity. The opening of the gland, which is the genital aperture, is located on a slight elevation near the edge of the mantle. Cut open the nidamental gland by making an incision parallel to the oviduct within. From this point trace the oviduct into the ovary. Posterior and to the left of the genital opening lies a conspicuous papilla which encloses the anus. Probe through the anus into the broad rectum.

Circulatory System. An injected specimen is preferable for this study. Open the mantle cavity if this has not been done, and carefully cut the pericardial wall to expose the heart lying in its chamber. The heart is composed of an anterior, triangular, thin-walled auricle, and a posterior, round, thick-walled ventricle. The auricle receives blood from two large vessels: the nephridiocardiac vein returns blood from the tubuliferous portion of the kidney; the efferent branchial artery returns blood from the gill. The latter artery can be seen along the lateral margin of the gill to its posterior end. Here the artery turns sharply medially along the border of the pericardial cavity where it enters the auricle. The blood then enters the ventricle and emerges from it by a large single aorta. Near its origin notice a branch, the visceral artery, which soon divides into several smaller arteries supplying the visceral hump. Trace these as far as practicable. The main aorta bends sharply anteriorly and continues parallel with the esophagus below the floor of the mantle cavity. Soon both aorta and esophagus dip deeply into the head region. Here the aorta enlarges to form the secondary heart. Trace the smaller branches into the head and foot. The heart, a typical molluscan one, consists of smooth muscle fibers. The blood is carried by arteries to all parts of the body. The venous blood is returned by less well-defined vessels, the large lacunae or sinuses. From them, blood passes into the kidney, through the gill for aeration, and finally reaches the auricle. Since blood also enters the auricle directly from the kidney, it is obvious that the arterial and venous blood of *Busycon* is mixed.

Excretory System. The two parts of the kidney have already been studied. However, observe at this point that the tubuliferous part appears to be composed of a series of parallel structures. They are the tubules. The thicker walls of the acinous part are composed of lobules. Near the anterior end of the kidney, the mantle is pierced by a conspicuous slit, the renal aperture.

Digestive System. The proboscis is an extensible, retractable organ, the walls of which are composed of a superficial integument and two layers of muscle, an outer longitudinal and inner circular one. Slit the head fold between the tentacles, thereby freeing the proboscis as far as possible without removing it. At its posterior end are numerous strong muscles, the retractors, which attach it to the integument. The proboscis functions as follows. Shortening of the proboscis is caused by a contraction of the longitudinal muscles within the wall; retraction of the proboscis itself within the animal by the retractor muscles already mentioned. Elongation is the result of simultaneous relaxation of the longitudinal muscles with contraction of the circular ones. Protraction is not satisfactorily understood. It is thought that, when blood flows rapidly into the sinus of the head fold, the contraction of these muscles forces the enclosed proboscis out of the animal.

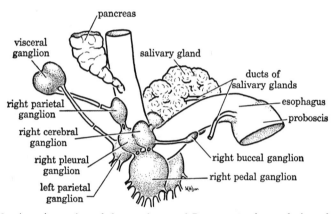

FIG. 178. Anterior region of the esophagus of *Busycon*, to show relation of ganglia to each other and to the digestive tract and its glands.

The digestive tract continues posteriorly from the proboscis as a narrow tube, the esophagus. Lying on either side of it are the large paired yellow salivary glands. Trace their long slender ducts forward to their entrance into the extreme anterior lateral wall of the esophagus (Fig. 178). More posteriorly, on the right side of the esophagus, lies a small gland, the pancreas. Its duct enters just posterior to the abrupt bend of the esophagus where several ganglia are located. Complete the dissection of the digestive tract by tracing the esophagus to the stomach and identifying the intestine, rectum, and anus.

Turn to the study of the nervous system before completing the internal dissection of the proboscis.

There are two satisfactory procedures for dissection of the proboscis.

1. Cut the proboscis along the midventral line, reflect the skin, and pin it flat on a dissecting tray. This will expose the buccal mass. Superficially this mass comprises several sets of muscles which conceal three important organs, the esophagus, the odontophore, and the

radula. The esophagus lies flat along the inner median dorsal wall of the proboscis. The odontophore is a forked cartilage, the ends of which can be seen near the posterior end of the buccal mass. It is the support for the ribbon-like radula which bears many rows of teeth used in procuring food.

2. Cut the proboscis along the dorsal midline, and pin it onto the foot. With this method it is easier to keep the orientation. However, there is the danger that the esophagus may be injured unless extreme care is taken in dissection.

Study the various sets of muscles which control the action of the odontophore and radula (Fig. 179A, B, C). The odontophore protractors, seven or eight slender muscles on each side, run forward from the odontophore to the lateral walls of the proboscis. The odontophore retractors, two broad flat muscles which are attached to the posterior horns of the odontophore, run posteriorly and dorsally to the base of the proboscis. The radula protractors, three pairs of slender muscles on the ventral surface of the buccal mass, attach anteriorly to the radula sac. Posteriorly the lateral protractors attach to the horns of the odontophore; the medial protractors attach to the base of the proboscis. The radula retractors, a strong set of muscles, attach the buccal mass to the posterior wall of the proboscis. These muscles can be seen most clearly from the dorsal aspect. Since the effective stroke of the radula is the back stroke, these muscles are particularly strong and bulky. Carefully remove the muscles and examine the odontophore, radula, and radula sac. The odontophore is the cartilaginous support over which the radula is pulled back and forth during the feeding process. The radula is a long, flat ribbon-like structure bearing a series of short rows of teeth. On the ventral side of the odontophore the radula lies flat. As it is pulled dorsally over the tip of the odontophore, the greatest width is exposed and comes in contact with the food. As the ribbon passes dorsally and posteriorly, the lateral edges fold in upon themselves, and therefore all teeth are completely enclosed in the loose membranous radular sac. Since the teeth point posteriorly, their orientation is another indication of the effectiveness of the back stroke (retraction). As the anterior teeth are worn away, new rows which are constantly being formed by odontoblasts at the dorsoposterior end, move forward to replace them. At first the young teeth are chitinous, but later become hardened by deposition of mineral salts.

As the number and arrangement of teeth on the radula are used as criteria in the classification of many Gastropoda, a comparative study of radulae is worth while. *Littorina, Urosalpinx,* and any nudibranch offer characteristic differences. The directions for preparation of the radula are simple. Kill the animal by immersion in hot water for a

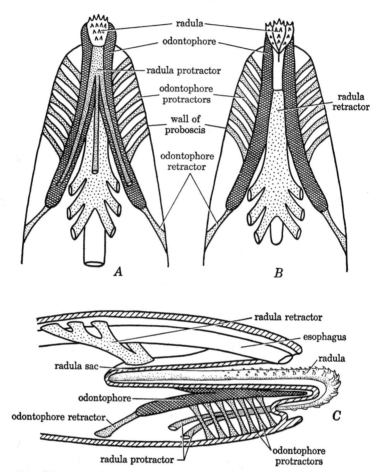

Fig. 179. Diagrams of the proboscis of *Busycon* to show relation of muscles to odontophore and radula. *A.* Ventral aspect, with wall of proboscis removed. *B.* Dorsal aspect. *C.* Lateral aspect.

minute or two. Remove the radula sac and boil it gently in 10 per cent potassium hydroxide for several minutes, or until the radula sac is dissolved. The radula should then be mounted in a drop of glycerine. As you examine the teeth on the radula, keep in mind the habitat, type of food, and feeding activity of the animal under observation.

In a living specimen, the rasping action of the radula can be demonstrated by placing a few drops of macerated clam at the tip of the proboscis. In a living, but narcotized, specimen, the same reaction can be elicited by alternate stimulation of the retractor and protractor muscles.

Nervous System. The nervous system of this form shows a high degree of cephalization. With the single exception of the visceral pair, the ganglia are concentrated in close proximity to each other around the anterior part of the esophagus (Fig. 178). They form an irregular brownish yellow mass of tissue which completes a ring around the esophagus. With a hand lens, notice on the ventral side of the esophagus the small paired buccal ganglia, which are located most anteriorly. They connect with each other and with the cerebral ganglia. Posterior to this pair lie the much larger pedal ganglia, which are fused yet distinctly paired in outline. The pedal ganglia possess numerous large nerves which supply the foot muscles as well as smaller connectives to the cerebral and pleural ganglia. On the dorsal surface of the esophagus is a conspicuous commissure which unites the paired cerebral ganglia. These ganglia are the most centralized, for they connect directly with the pleural, pedal, and buccal ganglia, and indirectly with the parietal and visceral. On the left side, posterior and ventral to the cerebral, lies a smaller ganglion, the left pleural. It extends nearly to the ventral surface of the esophagus. On the right side, four ganglia lie close together. The right cerebral and right pleural are fused, but retain a distinct constriction. Posterior and dorsal to the right pleural lies the right parietal. Short connectives unite these two ganglia. The remaining ganglion, the left parietal, lies ventral and slightly anterior to the right parietal. It nearly touches the right pleural and right pedal. The visceral ganglion, probably the fusion of a pair, lies below the external opening of the kidney. The elongated visceral connectives unite the visceral ganglion with the parietal ganglia.

Reproductive System. The gonad, ducts, and external openings have already been described. The sexes are separate, and fertilization is attained internally by means of the large penis. In the Cape Cod area, eggs are laid during August, but in the Carolinas they may be deposited as early as May or June. Like many Prosobranch mollusks, *Busycon* protects its eggs by surrounding them with a tough case and attaching the cases to a stable object. As the eggs pass out of the oviduct and through a temporary fold in the foot, they are covered by a secretion from the nidamental gland. This secretion forms not only the disk-shaped egg cases, but also a stem or cord, along which the disks are attached in a series. One end of the cord is fastened to a buried object. Often as many as 70 disks, each containing 20 or more eggs, are found on a single cord. The rate of formation may be as fast as 1 disk every 2 hours when animals are in aquaria.

The eggs of *Busycon* are very large, but they contain enormous quantities of yolk, cleave slowly, and in general are not particularly satisfactory for embryological study in the classroom. Development is completed within the egg case. The young mollusks emerge with body form of miniature adults.

The eggs of another gastropod, *Crepidula*, are classic forms for the study of molluscan embryology, and fortunately *Crepidula* can be obtained throughout the summer. By pulling apart the individuals from the groups into which they are usually aggregated, the yellow egg masses may be found deposited on the shells of the underlying animal. Various stages in development may be discovered by examining several different masses of eggs. In the veliger larva of this or any other mollusk, try to identify the following structures: (1) velum, with large powerful cilia around the margin; (2) transparent shell; (3) head vesicle and eyes (dorsal); (4) foot (ventral); (5) mouth, just above the foot; (6) heart, usually beating (dorsal).

Among the fresh-water forms, the common pond snail, *Physa*, will live in a laboratory aquarium and often deposit its transparent gelatinous egg cases upon the walls of the aquarium. These eggs are excellent material for embryological study.

Eolis

By Madelene E. Pierce

Eolis is often referred to as a nudibranch, for it is a member of that large and varied group of mollusks which have, as the name so aptly suggests, no shell to cover the gills. It is also sometimes called a sea slug. *Eolis* is a small nudibranch, 1 or 2 cm. in length, which avoids the light whenever possible. It is often seen crawling about on the underside of seaweeds or, more frequently, creeping about in masses of hydroids upon which it feeds. It inhabits the shallow waters of the Atlantic coast from Cape Cod northward and also the northern coasts of Europe.

EOLIS

In studying the specimen, allow it to crawl freely, but in a somewhat limited space like a watch glass. Small nudibranchs often crawl easily on the underside of the surface film. This is a particularly profitable position from which to study one of them. Later, if the specimen is small, you will find it advantageous to mount it on a slide and depress it slightly with a coverglass. If it is desired to keep live nudi-

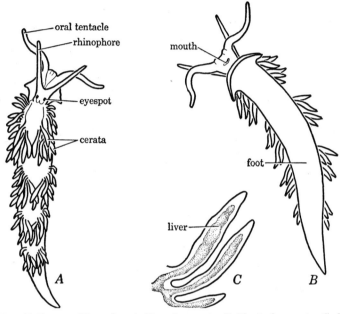

FIG. 180. *Eolis,* a nudibranch. *A.* Dorsal aspect. *B.* Ventral aspect. *C.* Several cerata, showing extensions of liver into these adaptive gills. (Modified from Alder and Hancock.)

branchs for several days, they should be kept in cool running sea water or in the ice box.

The soft body, which is without a shell, mantle, or true gills, is covered with a transparent integument. The cerata, called adaptive gills, are the respiratory organs. They are cylindrical extensions of the integument, containing hollow extensions of the liver and nematocysts derived from ingested hydroids (Fig. 180C). The cerata have a characteristic arrangement. They may occur in many transverse rows or in a few longitudinal rows. The color also varies, shading from brown through salmon to red. Often the ends are tipped with white. Mount one of the cerata on a slide and study it.

The head bears cylindrical tentacles, usually two pairs, one of which may be retractile. These pairs may be of different lengths. If so, the anterior or oral pair is the longer. The posterior or dorsal pair, often called rhinophores, bear simple sense organs which appear to test the surrounding water. At the base of each rhinophore is an eyespot. The mouth is prominent and can be most easily seen from the ventral view. On the right side of the head is the rather conspicuous genital opening, and just posterior and dorsal lies the small anus (Fig. 180A and B).

The foot is a muscular organ on the ventral side of the animal. Its proportions vary with the species. Locomotion is accomplished by waves of muscular contraction.

The extent to which the internal organs can be observed varies with different individuals and species. The following are usually visible in most specimens.

The dorsal systemic heart pulsates actively, between 50–100 beats per minute. Blood is sent, somewhat indirectly, to the cerata and to the skin generally for aeration. Upon its return to the heart it is mixed with blood from the viscera, so that only partially aerated blood enters the auricle.

The most conspicuous organ of the digestive tract is the mouth. It is guarded by fleshy lips, which cover both the radula and the horny lateral jaws within the buccal cavity. The powerful jaws cut the soft hydroids into smaller pieces which the radula then carries back into the buccal cavity and towards the esophagus. Although the esophagus, the stomach, and the intestine are well defined, they are not easily seen through the transparent skin. The liver, already identified, has numerous delicate ducts from the cerata which form trunk channels, eventually uniting to enter the digestive tract.

The nudibranch is hermaphroditic, and the common gonad is protandric. In the posterior part of the abdominal cavity the numerous eggs of the large ovary are usually visible. Copulation in these mollusks results in an exchange of sex products, and spawning takes place a day or two afterward. Nudibranchs are most prolific. The eggs are extruded in a long, ruffled string of sticky gelatinous substance. Because of its consistency, this jelly adheres firmly to the substrate upon which the eggs are deposited, and therefore the egg masses are saved from undue buffeting. Nudibranch egg masses are commonly found upon the hydroids on which the mollusks have been living. In the laboratory, nudibranchs will often deposit eggs upon the sides of the dishes in which they are kept.

If several groups of eggs from different individuals are examined, a representative series of embryological stages can be obtained. The veliger larvae, especially, are excellent. Because of their transparency, many structures are visible. (See more detailed directions for study of the veliger under *Busycon,* Reproductive System.)

Loligo pealeii

By MADELENE E. PIERCE

Loligo pealeii, commonly known as the squid, is a littoral form inhabiting the east coast of the United States from Maine to South Carolina. Although the eggs may be found in water as deep as 46–92 meters, the adult squid is usually found as an actively swimming form of shallow waters. The winter habits are little known, but in the spring and early summer the squid appear forming in schools of 10–100 or more and breed near shore.

Study a living specimen in the aquarium or in a large finger bowl at your desk. Notice its position as it is resting, and watch the characteristic rhythmic, gentle movements of the lateral fins. These rhythmic contractions of muscles ensure a continuous supply of water for the gills which are within the mantle cavity. Touch the squid so that it swims actively. Try to observe the funnel through which water is ejected in spurts. When the funnel is ejecting water backward, the animal moves forward; when it is ejecting water forward, the animal moves backward. The squid was an early perfector of jet propulsion. Observe the movements of the arms which help in steering the animal.

Notice the rapidity of the color change in the integument. This is due to the simultaneous expansion or contraction of many chromatophores. Their activity can best be studied on a newly hatched specimen placed under the microscope. The chromatophores in the squid are more like small organs than the individual cells of the vertebrates. Each is composed of a small sac of fluid containing pigment and surrounded by a highly elastic membrane to which 10–30 muscles are attached. Innervation is from the mantle nerves. Contraction of the muscle cells results in the dispersion of the pigment within the chromatophore and the consequent blush of color over the animal.

If the animal becomes irritated, he may eject the fluid contents of his ink sac. This fluid owes its dark color to the presence of a pigment, sepia.

Allow your fingers to come in contact with the suckers of the arms. Place a few small fish, 1–2 inches, in the aquarium with the squid and observe his reactions.

Orientation. If possible, dissect a fresh specimen. It is important to orient the squid in terms not only of its morphology but also of its activity, since these are not identical in this animal. Hold the animal with its pointed end uppermost, its head down. The apex of the long, cone-shaped body is the most dorsal point of the animal. The arms, tentacles, and funnel form the ventral surface. The mouth indicates the anterior surface, the funnel the posterior. When the animal swims, it moves forward and backward, with the anterodorsal side uppermost. Functionally, therefore, the morphological anterior side is dorsal; the posterior side is ventral; the ventral surface is anterior; and the dorsal surface is posterior. The functional meaning of the terms will be used in the following directions.

External Anatomy. At the end of the head is the modified foot, which is drawn out into five pairs of arms surrounding the mouth. These arms are of two types, four pairs of true arms which are non-retractile and of about equal length, and one pair of elongated retractile tentacles. Each arm bears two rows of suckers which decrease in size from the base to the tip of the arm. Each sucker is composed of a cup which is attached to the arm by a pedicle. Remove a sucker and study it under the microscope. Notice the toothed chitinous ring which supports the edge of the cup, and the piston which forms the central basal portion of it. Compare the elongated peduncle and terminal club of the tentacle with an arm. If the specimen is an adult male, the lower left arm (4th left) will show a modification of the suckers called hectocotyly. This results in a decrease in the size of the suckers with a simultaneous increase in the size of the pedicles. It is thought that the hectocotylized arm is modified for the transference of sperm to the horseshoe organ on the buccal membrane of the female.

Turn back the arms and tentacles and observe the muscular membrane extending from their bases toward the mouth. This membrane is divided into an outer part, the buccal membrane, and an inner part, the peristomial membrane. The buccal membrane consists of seven projections, each with suckers on the inner surface. The peristomial membrane surrounds the mouth opening through which the chitinous beaks are usually visible. If the specimen is a female, notice the horseshoe organ or sperm receptacle which lies on the buccal membrane in the median line below the mouth.

On the head is a pair of large well-developed eyes, in which the cornea, iris, pupil, and lens can easily be identified. Behind the eye is a fold of tissue, the free edge of which is covered by the mantle. This is the olfactory crest. Within its concavity is a special sense organ to which has been attributed the function of olfaction, but experimental evidence in proof of this is lacking. In front of the eye is the small aquiferous pore which communicates by the aquiferous canal with the outer chamber of the eye, thus regulating the pressure within this chamber. The activity of the pore is regulated by a sphincter muscle. On the underside, projecting from beneath the mantle, lies the funnel. Although the funnel of the squid and the siphon of the clam both eject water from the mantle cavity and are therefore similar in function, the two organs cannot be homologized. The funnel is a modification of the foot, while the siphon is an extension of the mantle.

The mantle completely envelops the remaining part of the body. The free anterior edge of the mantle is called the collar. Notice the three scallops into which the collar is divided. The projections on either side of the funnel mark the position of the pallial cartilages. The single projection on the dorsal midline indicates the anterior end of the pen, or internal shell, which will be described under the skeletal system. Posteriorly the mantle extends as a long cone, flanked at the apex by two lateral fins. They consist of three sets of muscles, longitudinal, transverse, and vertical, which result in strong yet flexible fins. Not only are they able to make slow, undulating movements which propel the squid forward or backward, but they can also make swift, powerful strokes which instantly reverse the direction of progression.

Internal Anatomy. Open the mantle cavity by cutting the mantle a little to one side of the midventral line from the edge of the collar below the funnel to the apex of the cone. The thick wall of the mantle is composed almost entirely of a sheet of circular muscles covered both inside and outside by integument. The funnel contains both circular and longitudinal fibers. The latter are continuations of the large prominent pair of retractors of the funnel. These are attached dorsally to the pen. There is also a pair of small protractors which attach the funnel to the cephalic cartilage of the head. Cut open the funnel in order to find the muscular valve on the dorsal wall near the external opening. The valve prevents water from entering the funnel from the exterior. Notice the valve-like extension of the muscular side walls of the funnel into the mantle cavity. These prevent the escape of water from the mantle cavity by way of the space between the collar and the funnel. Medial and dorsal to the retractors of the funnel are the cephalic retractors which extend from the head cartilage to the pen.

The contraction of muscles in the mantle controls the passing of water into and out of the mantle cavity. When the circular muscles relax and the longitudinal muscles contract, the mantle cavity enlarges, and consequently water is drawn in between the collar and the head. When the circular muscles contract and the longitudinals relax, the cavity decreases in size, the collar is drawn tightly around the head, and water is expelled through the funnel.

The visceral mass occupies most of the space in the mantle cavity. For ease in further study identify the following conspicuous organs contained therein, although they may seem to have little relation to each other at this time. Extending for a short distance into the funnel is the white rectum, the opening of which lies between two small flaps of tissue, the rectal papillae. Dorsal to the rectum is the silvery ink sac. Avoid breaking it. The paired gills lie along the lateral border of the visceral mass to which their proximal ends are attached. The distal ends extend anteriorly and are attached to the lining of the mantle. At the base of each gill is a round organ, the branchial heart.

If the specimen is a female, a pair of large, white, elliptical nidamental glands will cover a considerable portion of the anterior viscera. Each gland has a small opening at its anterior end, in front of which lies a small accessory nidamental gland. The accessory glands, which are often bright red before egg laying, have a small opening on the ventral surface. The function of the nidamental gland is to secrete the outer capsules for the egg masses, but the function of the accessory glands is still unknown. Lift the left nidamental gland to expose the large, flaring opening of the oviduct. Slightly posterior to the opening, the wall is swollen to form the oviducal gland, the function of which is to produce the spherical capsules of the individual eggs. The conical portion of the viscera usually contains a large mass of eggs, which are considered to be lying within the ovary and proximal parts of the oviduct.

If the specimen is a male, the slender penis lies at the left of the rectum and slightly posterior to it. Between the bases of the gills are located the paired, somewhat triangular kidneys.

Respiratory and Circulatory Systems. An injected specimen is desirable for this study. The paired branchial hearts have already been identified. They receive venous blood from the large veins which drain the body. From the head, a single large vein, the anterior vena cava, proceeds in the median line to the kidneys. As it meets the kidneys, it divides into a right and a left precavae, which enter and pass through the nephridial sacs. As the precavae emerge from the kidney, they diverge and enter the corresponding branchial hearts. Before entering the branchial heart, the right precava receives three branches: the right mantle vein returning blood from the anterior

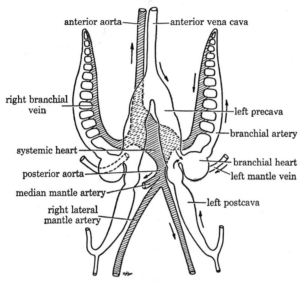

Fig. 181. Diagram of the circulatory system of *Loligo*, to show the main arteries and veins in their relation to the hearts and the gills.

portion of the mantle, the right postcava collecting blood from the posterior portion of the mantle, and the small genital vein. The left precava receives two branches, the left mantle vein and the left postcava. The branchial artery carries blood from each branchial heart to the gills. It courses along the dorsal edge of the filaments. The branchial vein returns blood from each gill to the systemic heart, an asymmetrical organ lying in the midline. The opening of these veins into the heart is protected by a pair of semilunar valves. Two aortae leave the heart. Each is protected by a single semilunar valve. The anterior aorta, the larger one, leaves the heart at its anterior projection and runs parallel with the esophagus to the head. The posterior aorta divides immediately into three mantle arteries: the median one which runs anteriorly along the ventral surface of the mantle cavity, and the paired lateral arteries which run obliquely and posteriorly to supply the dorsal surface of the mantle cavity. From the anterior border of the heart a small genital artery runs forward a short distance and then bends dorsally (Fig. 181).

Excretory System. If the specimen is a female, remove the nidamental glands to expose the underlying paired kidneys. They are usually white or pale, except in injected specimens. The intimate connection of the kidney with the venae cavae results in the assumption by the kidney of the color of the injection fluid. The kidneys appear to be simple, triangular structures with the longest extension anterior. The papilla, enclosing the nephridiopore, lies at the tip of each extension and on either side of the rectum. However, the kidney is more complex than it appears, for it is composed of a large bilobed sac which encloses and hides the urinary gland.

Digestive System. Remove the funnel and its muscles and dissect the head along its median line. The buccal mass or bulb is a hard, round, muscular organ, on the anterior surface of which are the horny beaks. Pry them open, and note their interlocking arrangement as well as the odontophore and radula enclosed within. Posterior to the buccal bulb and partly embedded in its muscular wall are the paired buccal salivary glands. The duct of each gland opens into the buccal cavity, near its posterior limit. The esophagus, a thin-walled, narrow tube, leads out of the bulb and into the stomach. It enters the liver, passes dorsally, later ventrally, and joins the stomach near the middle of the visceral mass. In the adult squid, the liver is a single, median, cone-shaped organ, with its broad base near the collar. In this anterior portion, a small, ovoid, median salivary gland lies embedded. It lies dorsal to the esophagus and between two diverging nerves, the visceral and pallial. The pancreas is a U-shaped organ lying anterior to the stomach. The ducts of the pancreas and liver unite and enter the caecum as a single hepatopancreatic duct. The stomach is a thick-walled muscular organ, which communicates directly with the caecum, a long sac which may extend to the apex of the visceral mass. It is into this caecum that the single hepatopancreatic duct leads. Close to the entrance of the esophagus into the stomach is the opening of the intestine, a short tube which runs forward between the lobes of the pancreas; a constriction marks the beginning of the rectum which terminates in the anal opening.

The squid is a carnivorous mollusk, preying upon small fish which it catches easily. The suckers of the arms hold fast the prey once it is caught, and the radula and jaws devour it.

Nervous System. On the inner dorsal surface of the mantle at the level of the tip of the gills lie the large stellate ganglia. From these, several nerves radiate into the mantle, and one, the visceral nerve, connects with the visceral ganglion in the head. With the exception of the stellate, the paired ganglia of the central nervous system are concentrated in the head. Four pairs lie clustered around the esophagus a short distance posterior to the buccal bulb. With a razor cut a median sagittal section through the head. Above the esophagus lies

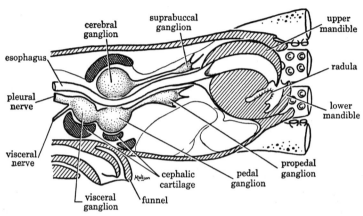

FIG. 182. Diagram of head region of *Loligo* to show cephalic cartilages and ganglia.

the supraesophageal ganglion, a single, rounded mass, which represents the fused cerebral ganglia. Below lie two ganglia, which, although close to one another, are readily distinguishable. The anterior one is the pedal, the posterior the visceral. Completing an esophageal ring by forming its side are the paired pleural ganglia which cannot be seen from the median section. Fibers from the pedal ganglion extend forward a short distance to the propedal ganglion. Those from the cerebral ganglion pass anteriorly to connect with the small suprabuccal ganglion (Fig. 182).

The nerves which supply the mantle are composed in part of large fibers called giant nerve fibers. These fibers have their origin in the cephalic ganglia, pass through the stellate ganglia, and terminate in the circular muscles of the mantle. From extensive experimental study, it is known that the speed of conduction varies with the diameter of the nerve fiber; the larger the diameter, the higher the rate. Upon dissection it has been found that the larger fibers of the mantle nerve innervate the areas farthest from the ganglion; the smaller fibers the nearer regions. The result is a simultaneous contraction of the complete mantle, enabling it to act as a unit. This integrated activity is of great value in the swimming of the squid.

The eye of the squid is a highly modified one, similar to but not homologous with that of the vertebrate. Each eye is closely applied to the cephalic cartilages which partly surround it and thus form an incomplete orbit. The cornea, iris, pupil, and lens have been previously identified. Carefully remove the eye, and as you do so observe the large optic ganglion to which the eye is connected by a short optic nerve. Float the eye in a dish of water. The eye consists of an anterior and posterior chamber which are separated from each other by the lens. The outer wall of the anterior chamber is formed by the broad, transparent cornea. The dark iris, just anterior to the lens, may surround either a circular or reniform pupil. When the eye is exposed to light, a fold of the iris projects across the circular opening, resulting in the reniform or crescentic shape of the pupil. The lens is attached to the ciliary body located behind it. The posterior chamber is filled by a semifluid, transparent mass, the vitreous humor. Forming the posterior and lateral walls of this chamber is the retina which extends forward as far as the ciliary body. The inner layer of the retina contains the rods which are directed anteriorly. The outer layer is composed of ganglionic cells which converge to form the optic nerve. Change of focus is accomplished in the following manner. The muscles of the ciliary body, by changing the pressure on the vitreous humor, indirectly move the lens forward or backward and thus increase or decrease the distance between the lens and the retina. External to the retina is a tough, cartilaginous layer, perhaps comparable to the sclerotic coat of the vertebrate eye.

The olfactory crests have already been noted.

Two statocysts, probably organs of equilibration, lie embedded in the cephalic cartilage below the visceral ganglion. Each is composed of a cavity lined with epithelium, a portion of which is ciliated. Upon this ciliated area the statolith lies.

Skeletal System. The most conspicuous part of the skeleton is the pen, a transparent chitinous structure which is the internal shell. The pen lies in the mid-dorsal line and extends from the edge of the collar to the apex of the mantle. It articulates at its anterior end with the nuchal cartilage which is embedded in the muscle lying between the visceral mass and the pen. On either side of the base of the funnel lie the infundibular cartilages which articulate with the pallial cartilages embedded in the mantle. In the head is the cephalic cartilage which serves as a protection to the ganglia concentrated there. The cephalic cartilage consists of very irregular plates perforated by foramina for esophagus, blood vessels, and nerves (Fig. 182).

Reproductive System. Remove the left gill and branchial heart. In the male, at the apical end of the visceral mass, lies the single large white testis, which opens directly into the coelom by a slit at its anterior end. Near this opening is located a ciliated funnel, the entrance to the vas deferens. The vas deferens is a very slender coiled tube which appears opaque white because of the tightly packed sperm contained within it. The tube continues forward between two organs into which the vas deferens eventually leads. On the right is the transparent spermatophoric sac containing packets of sperm called spermatophores. On the left is the somewhat convoluted, thick-walled sac, the spermatophoric organ which receives the sperm brought to it by the vas deferens. Here in a series of chambers the sperm receive appropriate secretions and membranes and are packed into bundles, the spermatophores, which are then transferred to the spermatophoric sac. In this chamber the spermatophores are temporarily stored before being ejected by the muscular penis.

Remove several spermatophores from their sac and place them in a dish of sea water. If the animal is a fresh specimen, the spermatophores may ejaculate as they are being transferred. Place a few in a solution of ¼ saturated magnesium chloride for 10 minutes. This treatment will slow down the ejaculation process so that it may be watched.

Each spermatophore consists of an outer transparent tunic which has a cap and a cap thread at the smaller oral end. Within the tunic are three important structures. At the oral end is the ejaculatory organ, which has the appearance of a coiled spring. Beyond it is the small cement gland, followed by the large sperm mass (Fig. 183). When the cap is broken, the ejaculatory organ turns wrong side out and pulls the cement gland and sperm mass with it. The sperm mass is thus automatically cemented to the object upon which it is thrown. Under usual conditions, this object is the sperm receptacle or horseshoe organ on the buccal membrane of the female.

If the specimen is a female, the apical region of the visceral mass will be occupied by the ovary. Through a ciliated funnel which is not readily seen, the eggs pass into the oviduct. In this region the oviduct is a transparent tube tightly packed with eggs and scarcely distinguishable from the ovary. Trace the oviduct as it runs first

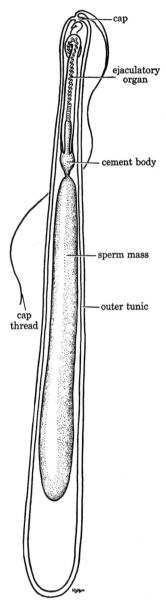

Fig. 183. A spermatophore of *Loligo*.

anteriorly, then posteriorly, and finally anteriorly again to become the thick-walled glandular portion, the oviducal gland. The terminal portion is the short, flared opening of the duct. The horseshoe organ, the nidamental glands, and the accessory nidamental glands have already been described.

During the breeding season the male squid becomes very excited, and displays a courtship behavior which culminates in the deposition of spermatophores within the female. He may thrust them into the mantle cavity of the female, or he may place them within the seminal receptacle on her buccal membrane. The eggs may therefore be fertilized either within the cavity, before their extrusion through the funnel, or after their expulsion as they are held within the arms of the female squid before transference to the place of attachment.

At the time of extrusion the eggs are enclosed in two cases, the single capsule about each egg, and the gelatinous matrix surrounding the capsules. The female holds the fluid, gelatinous mass within her arms, until she finds a suitable spot for attachment. Then she deposits 10–20 elongated egg masses at one central point. These slender masses of semifluid substance stiffen in contact with the sea water, so that they soon have the consistency of a stiff jelly and the appearance of a long finger. Squid egg masses are usually found attached to rocks below the low-tide mark. The eggs, which are relatively large, contain an abundance of yolk, and cleavage is therefore meroblastic. A rudimentary trochophore larva, but no veliger. occurs during early development. The young squid hatch within 2–3 weeks. It is thought that 2 years is the normal life span of the squid, although possibly some live through the third summer.

Squid eggs develop readily in the laboratory if kept in plenty of cool running sea water. They can be examined from time to time, since their development is slow, and successive stages studied. It is very difficult to raise the young squid beyond the point of hatching.

BRACHIOPODA *

Terebratulina septentrionales or *Terebratella*

By MADELENE E. PIERCE

Terebratulina septentrionales, although found most abundantly on the Maine coast, occurs from Norway to New York, and has been taken in depths ranging from low-tide mark to 3800 meters. *Terebratella* is a Pacific coast genus. Throughout life, these forms are attached by a strong muscular pedicle to rock. They feed upon the plankton which is carried into the mouth by cilia located on a complicated organ, the lophophore.

Because of anomalies of orientation the valves of the shell are known as the pedicle and brachial rather than the ventral and dorsal. In the specimen to be studied, the pedicle valve is larger than the brachial valve. Along the hinge line the pedicle valve is prolonged into a conspicuous posterior beak, which is perforated by a muscular stalk, the pedicle. Pry open the valves and cut along the hinge line. Most of the space between the valves is filled by a conspicuous looped structure, the lophophore, which is attached to the brachial valve. Each half of the lophophore is called an arm or brachium. A vertical membranous partition or mesentery divides the interior of the shell into a posterior coelomic and anterior brachial cavity. The mouth, a slit-like opening, is located between the two arms of the lophophore, at the point where they diverge from the vertical membrane. Examine an arm of the lophophore. Along the axis a double row of small tentacles is borne. Cut along this line to expose an internal, calcareous, ribbon-like loop, which serves as the support of the lophophore. Cilia are present on the surface of the mantle as well as on the tentacles.

* Thanks are due Professor P. E. Cloud, Jr., of the Museum of Comparative Zoology, Cambridge, Mass., for helpful suggestions and criticisms.

They set up two incurrent streams of water along the arms, and one excurrent stream between the arms.

Within the brachial valve, notice the paired adductor muscles which converge to a single bundle of fibers as they pass between the prominent teeth to reach their origin on the pedicle valve. Careful examina-

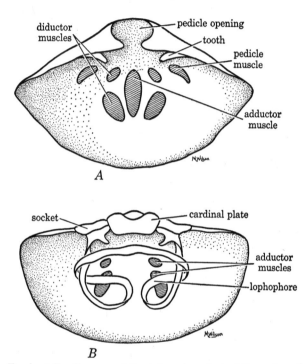

FIG. 184. *Terebratella*, diagram of the internal surface of the valves to show the muscle attachments. A. Pedicle valve. B. Brachial valve.

tion of the insertion of the adductor muscles reveals four insertions and therefore four scars on the brachial valve. These muscles close the valves. In the pedicle valve, and from four origins, the fibers of the diductor muscles converge to a single insertion which is located on the cardinal plate of the brachial valve. The surface of muscle attachment, or myophore surface, is a small, ridged, cup-like depression in the middle of the hinge line. The diductor muscles open the valves. Lateral to the origin of the diductor muscle lies another fan-shaped muscle, with insertion on the pedicle. (There is a similar muscle from the brachial valve.) This muscle is called the pedicle muscle and permits rotation on the pedicle.

Clean the valves. On the pedicle valve, notice the pair of teeth along the hinge line, and identify as many of the muscle scars as possible. On the brachial valve, notice the sockets into which the teeth fit, the myophore surface on the cardinal plate into which the diductor muscle inserts, and any other muscle scars which are visible (Fig. 184A, B).

10.

Arthropoda

Xiphosura polyphemus
By J. H. LOCHHEAD

Xiphosura polyphemus, long known to zoologists as *Limulus*, is commonly called the king crab or horse-shoe crab. It is, however, not really a crab at all, but is a member of an isolated order of the Chelicerata, bearing the same name, Xiphosura, as has been given to the genus. It is abundant on the eastern coast of North America, from Maine to Mexico. Attempts to establish it in Europe and on the California coast have succeeded only temporarily. Two closely related genera, embracing four species, occur in Asiatic waters, from Japan to India.

The bizarre appearance, large size, and isolated taxonomic position of *Xiphosura* have long excited the interest of zoologists. The closest relatives of the order appear to be found in the extinct Eurypterida (= Gigantostraca), with which the Xiphosura often are allied in the Class Merostomata. This is an exclusively aquatic class, though a theory was once proposed that it had its origin in terrestrial Arachnida. Anatomically the closest living relatives of the Xiphosura probably are the scorpions. At one time the view was vigorously upheld that *Xiphosura* lies close to the ancestral tree of the vertebrates, this view being advanced in two quite conflicting theories by Gaskell and by Patten. Patten always used the terms haemal and neural in place of dorsal and ventral, because of his belief that the functionally ventral surface in *Xiphosura* is represented by the functionally dorsal surface in a vertebrate. The larva of *Xiphosura* has been said to resemble the extinct trilobites, but this resemblance would appear to be purely superficial. However, certain features of the embryos do seem to suggest a trilobite relationship.

The adults of *Xiphosura* live for the most part in water a few fathoms deep, where they plow through the mud, probably feeding mostly on worms. Laboratory observations indicate that they are most active at night. Commonly they harbor a number of ectocommensals, notably the turbellarian, *Bdelloura*. In the spring and early summer they briefly visit the shore for breeding, coming in on a high tide. The eggs are laid in a hollow scooped out in the sand, usually at or just below the average high-water mark, and the accompanying male then deposits sperm on top of them. The mass of eggs is then quickly covered over with sand. In the late summer tiny specimens of *Xiphosura* sometimes are abundant on sand or mud flats exposed at low tide.

It has been calculated from measurements of a large number of specimens that males of the Japanese species become mature at the thirteenth postlarval instar, the females one instar later. The mature females of all species are considerably larger than the males. There is evidence that at least the males do not molt after reaching maturity. Eggs, larvae, young, and adults all are remarkably hardy and can be shipped alive for long distances. The adults can survive for weeks out of water, even without food, provided that their gills are kept slightly moist.

Body Plan. The body is much depressed and is covered by a tough exoskeleton. The following divisions of the body are distinguished: a prosoma or cephalothorax, bearing the appendages around the mouth; a mesosoma or preabdomen, bearing the operculum and the gill flaps; a metasoma or postabdomen, which is without appendages; and a telson, which is long and spine-like. The mesosoma and metasoma are fused together into an opisthosoma. Six movable spines on each side mark the lateral margins of the somites of the mesosoma. The opisthosoma is jointed both with the prosoma and with the telson. Powerful muscles can flex or extend the body at each of these joints.

The body is concave below, so that when the animal is resting on a flat surface the limbs of the prosoma are entirely concealed by the overhanging carapace.

The mouth is midventral in the prosoma, between the bases of the walking legs. The anus is located ventrally, on the proximal end of the telson.

Paired Appendages, Operculum, and Gill Flaps. A pair of small chelate appendages, termed chelicerae, is located in front of the mouth. Each has three segments. The appendages of the next pair are much larger and are termed pedipalps. Each has six segments and is chelate except in mature males. A masticatory process, or gnathobase, beset with movable spines, is present on the basal segment, pointing towards the mouth. In structure and in number of segments these two pairs of appendages are closely comparable to the corresponding pairs in the scorpion. The appendages of the following three pairs are closely like the pedipalps (and so differ somewhat from the condition in the scorpion). The appendages of the sixth pair are constructed like the others, but differ in two particulars. Firstly, on the outside of the basal segment of each of these appendages there is a movable spatulate process, perhaps representing an exopodite, and said to be used in cleaning the gills. Secondly, the appendages are not chelate, but instead each has four large, blunt, flattened spines, movably articulated with the distal end of the penultimate segment, besides a small movable spine on the preceding segment and two small spines at the tip of the terminal segment (Hansen interprets one of these two spines as a special terminal segment, lacking on the other appendages). Just behind and partly between the gnathobases of the sixth pair of appendages, a pair of blunt, movable processes is present, bearing a medial row of short spines. These are called chilaria, and it has been shown that they are the appendages of a rudimentary pregenital somite. Autotomy of the prosomal appendages does not occur.

The second through fifth pairs of appendages are used for walking, and their chelae pass food up to the gnathobases. The gnathobases work in an alternating fashion, which "cards" the food into fragments and gradually pushes it into the mouth. The chelicerae also are used in manipulating the food, and they and the chilaria behind probably help to keep fragments of food from escaping. The sixth pair of appendages is used to push against sand or mud, or to push this out from under the carapace. Each time one of these appendages is extended, the four flattened, movable spines automatically are spread out by the resistance of the sand or mud, so that either some sand or mud is pushed back, or the animal gets pushed forward. In plowing forward through the surface layer of the substratum, *Xiphosura* also uses the telson and an alternate arching and straightening of the back. The back is arched, the sixth legs drawn forwards, and the telson thrust into the substratum. Then the body is straightened, while the sixth legs push backwards against the firm purchase provided by the telson.

At the molt preceding maturity, the pedipalps of the male change from a chelate to a subchelate condition. The immovable finger becomes much reduced, while the distal segment becomes thick and hook-like. When the animals migrate inshore for breeding, the male clings to the back of the female, using these claspers to hold on to the edge of her opisthosoma, with a tenacious grip. In the Asiatic genera the third pair of appendages also becomes modified as claspers, but in one of these genera the claspers are relatively little modified, retaining a chelate condition.

On the opisthosoma six flap-like structures are present. The first of these does not bear gills and is termed the genital operculum. It acts as a protective cover to the gills when they are laid back against the body. Paired genital openings are present on its posterior surface, close on either side of the midline, slightly less than half way towards the distal margin. Each of the succeeding five flaps bears on each side a so-called book gill, containing a large number of gill lamellae, stacked one on top of the other. There is some resemblance between these external book gills and the internal book lungs of scorpions and other arachnids. According to most authorities, the flaps on the opisthosoma of *Xiphosura* represent paired appendages which have fused together in the midline, and each of which has an endopodite and an exopodite reminiscent of the Crustacea. In an extension of this theory it has been suggested that the book lungs of terrestrial arachnids have been derived by an insinking of book gills. A less widely accepted theory holds that the Xiphosura are descended from terrestrial arachnids, the book gills thus having been derived from book lungs, and the flaps of the opisthosoma representing much modified sternal plates which now hang down free from the body. This theory, however, as also the one deriving book lungs from book gills, fails to take into account some important differences in the embryology of the two kinds of respiratory lamellae.

In the living animal all six of the opisthosomal flaps beat to and fro most of the time. Usually the beat is rather leisurely and serves only to aerate the gills. At times, however, the beat may become powerful, and it then is a means by which the animal can swim. This the smaller animals do quite frequently, and the larger ones do when coming inshore to breed. It is reported that when swimming the animals usually lie on the back. The beating of the opisthosomal flaps probably also is of use during burrowing, in creating a current carrying out mud or sand posteriorly.

Each time that the flaps move forward the gill lamellae fill with blood, while each time the flaps move backward the blood empties from these lamellae.

Exoskeleton and Hypodermis. The exoskeleton is uncalcified, but it is made leathery by a high content of scleroprotein. It contains a brown pigment, but the underlying epidermis or hypodermis does not contain chromatophores. One worker claims to have found hypodermal mucous glands which pour out mucus onto the surface of the exoskeleton when the animal is strongly stimulated.

In molting, the animal escapes from its skeleton through a slit which appears nearly all around the anterior and lateral margins of the prosoma.

For internal dissection the dorsal exoskeleton has to be removed, from either a large or a small specimen, depending on which is selected. If a large specimen is chosen, it may as well be as large as possible, which means that a female will be preferred. On the other hand, there are some advantages to be gained from using a small specimen, say about 8–10 cm. long, exclusive of the telson. Observations on the activities of the living animal are readily made with such specimens, which do not require an especially large aquarium. Injection of a suitable color mass into the heart of a small specimen is a simple matter, the needle being inserted through the ligament at the posterior border of the prosoma. Many of the arteries then show up through the semitransparent exoskeleton. Removal of the dorsal exoskeleton can be accomplished easily, with ordinary scalpel and scissors, in contrast to the procedure necessary with a large specimen. Among the internal organs it should be possible to locate most of the major structures without too much difficulty, especially if a spotlight and a binocular microscope are used. But, if a detailed dissection is desired, particularly of the nervous, muscular, or reproductive systems, a large specimen is to be preferred.

Removal of the dorsal exoskeleton from a large specimen is a rather laborious task. The following method is suggested. First of all the animal is bled, by cutting off the legs with tin-snips or with bone scissors or by slitting the thin membranes at the points where the legs join the body. As much blood as possible should thus be drained out, since coagulated blood internally will hamper the dissection. The loss of blood also rapidly kills the animal. Next a median dorsal cut should be made through the exoskeleton of the prosoma, from the anterior to the posterior margin. This can be done with a hack saw, bone scissors, or even an old-fashioned can opener. Care has to be taken not to cut deeply at the posterior end, where part of the heart is located. Next a horizontal cut should be made all around the edge of the prosoma, about 1 cm. above the free margin. For this the can opener is good, or the hack saw can be used. Transverse dorsal cuts, from the lateral margins inwards as far as the ligamentous joint between prosoma and opisthosoma, are the next to be made. Then two cuts running the full length of the opisthosoma must be made, each one parallel to and about 1 cm. outside of one of the rows of six depressions visible on the dorsal surface. The ligament at the joint between the opisthosoma and the prosoma next has to be cut with a sharp scalpel, restricting the depth of the cut to about 5 mm. on a large specimen. The dorsal membrane at the base of the telson also has to be cut. If it is desired, cuts can be made around each of the lateral eyes and around the median eyes (located just in front of the most anterior median spine on the prosoma), so that the arteries and nerves supplying these sense organs will not be broken.

Portions of the dorsal exoskeleton on the opisthosoma and on the prosoma can now be pried up, starting at the posterior margin in each case. Numerous muscle attachments will be met, particularly in the prosoma, and they should be cut with a scalpel as close to the exoskeleton as possible. A sharp scalpel or small bone scissors should be slipped under the integument of the opisthosoma to cut the strong apodemes, or entapophyses, one of which projects downwards from each of the twelve depressions already noted. In doing this, care has to be taken not to damage the heart, which here lies not far below the surface. At the posterior margin of the prosoma, two especially strong entapophyses, in line with those in the opisthosoma, have to be cut with a saw or with bone scissors.

Muscular System. The origins of a considerable number of muscles will be seen as soon as the dorsal integument has been removed. The more ventral insertions of most of these, as well as some other deeper-placed muscles, will not be met with until later in the dissection. Furthermore, in describing the muscles it will be necessary to refer to other structures, particularly of the circulatory and endoskeletal systems. Despite these considerations a description will be given here of all those muscles most likely to be seen, or which have special functional interest. Actual dissection of the muscles should be deferred until later, and indeed most students probably will not wish to do more than identify some of the muscles that attract their attention during the course of the rest of the dissection.

Two pairs of extensors for the telson will readily be found, as well as one pair of flexors.

The internal extensors of the opisthosoma arise from the dorsal integument of the prosoma, above the heart (the location and appearance of which are described under the circulatory system). Paired external extensors arise from the large entapophyses at the posterior margin of the prosoma (already mentioned under the directions for removing the dorsal integument). Both pairs insert medianly on the anterodorsal border of the abdominal integument.

There are three large pairs of flexors for the opisthosoma, two of which arise dorsally. These two will be seen just outside the pericardial sinus, opposite the first and second ostia of the heart respectively. They run together on each side and insert on a series of six tendinous apodemes that project inwards from the so-called stigmata, which can be seen as transverse slits on the proximal posterior surfaces of the abdominal appendages. Besides flexing the abdomen these muscles make an important contribution to the movement of the gill flaps. The two members of the third pair of abdominal flexors arise from the posterior dorsal side of the endocranium (described under the endoskeleton) and pass backwards close to the midline, to insert on the ventral integument behind the last abdominal appendage. These muscles also insert, in the opisthosoma, on the six entapophyses of each side and on the six median endochondrites (likewise described under the endoskeleton). They also receive on each side four small slips of muscle, arising ventrally at the level of each of the first four gill flaps.

The endocranium, which lies dorsal to the ventral nerve mass, between the bases of the prosomal appendages, is suspended by a number of muscles from the dorsal integument. Those that will be seen on one side are as follows. One especially large muscle arises lateral to the crop, about in line with the two dorsally attached abdominal flexors of that side. Just medioposterior to this muscle a much smaller muscle arises. Another small muscle arises beside the anterior suspensory ligament of the heart. A fourth, slightly larger muscle, arises between and medial to the two abdominal flexors, just outside the pericardial sinus. Fifth and sixth muscles in this series arise from the large entapophysis near the posterior border of the prosoma. The function of these various muscles is to counteract the pull on the endocranium of various other muscles, such as the ventral abdominal flexors already described, and some of the numerous muscles running to the prosomal appendages. It also has been suggested that by compressing the prosoma these muscles may aid in the expulsion of genital products. In this function they may be aided by numerous small strands of muscle which pass between the dorsal and ventral integuments in the lateral parts of the prosoma.

In the opisthosoma, seven small dorsoventral muscles are present on each side. The first arises from the anterior border of the abdominal carapace, while each of the remaining six arises just anteromedial to one of the entapophyses. The first five pass downwards between the pericardial sinus and the collateral artery, while the last two pass outside this artery. All seven of these muscles run ventromedially, each of the first six inserting on the outer border of one of the endochondrites, and the last one inserting on the ventral integument behind the last gill flap. One function of these muscles seems to be to compress the opisthosoma, and the first six also serve to oppose the pull exerted on the endochondrites by small internal branchial muscles, which arise from the outer borders of the endochondrites and run ventrolaterally into the corresponding mesosomal appendages.

Muscles running to the base of each of the prosomal appendages are divided into two sets, those that have their origin on the ventrolateral margin of the endocranium, and those that originate on the dorsal integument. The chelicerae form an exception in not receiving any muscles from the endocranium; each is supplied by four small muscles arising on the dorsal integument. Each of the remaining five prosomal appendages receives four muscles from the endocranium (five in the case of the sixth appendage) and five muscles from the dorsal integument (four in the cases of the second and sixth appendages). Each chilarium receives four muscles from the endocranium and one from the dorsal integument, plus another which inserts just lateral to the chilarium. The origins of the muscles running from the dorsal integument to the third through sixth prosomal appendages lie lateral to the other body muscles, forming rather conspicuous large blocks. In an ordinary dissection the division into a number of separate muscles running to each appendage will not be distinguished.

Each of the abdominal appendages receives on each side the small internal branchial muscle already mentioned and two much larger muscles which respectively pull the appendage forwards, or abduct it, and pull it back towards the body, or adduct it. All these muscles actually run into the appendage, in contrast to the proximal muscles of the prosomal appendages. The first abductor (for the operculum) arises from the large entapophysis of the prosoma and from the dorsal integument in front of this. The second abductor (for the first gill flap) also arises from the entapophysis of the prosoma. The third through sixth abductors arise from the first four abdominal entapophyses. The first adductor arises near the anterior border of the dorsal abdominal integument, slightly lateral to the line of entapophyses. The second through sixth adductors arise just lateral to the first five abdominal entapophyses.

A pair of flattened dorsoventral muscles opens the anus. A slender pair of ventral longitudinal muscles draws the anus forward, and by elongating the slit probably aids the constricting action of the sphincter muscle which surrounds the hindgut.

Strong circular muscles surround the crop and gizzard. Muscles run from the ventral integument to the crop and esophagus, and from the endocranium to the esophagus near the mouth.

Of functional interest, and of interest, too, because similar muscles are found in the scorpion, is a series of eight slender, rather transparent muscles on each side, which run ventrally from the floor of the pericardial sinus. Presumably they pull the pericardial floor downward.

Reproductive System. In a mature female, the ovary consists of an intricate reticulum of tubes, with small follicles on the walls of many of them. These tubes extend into nearly all parts of the body not occupied by other organs; however, they do not invade the appendages, telson, and more lateral parts of the opisthosoma. The hepatopancreas is equally extensive, and its tubules, together with some loose connective tissue, may seem to be inextricably mingled with the branches of the ovary. If the specimen has not laid its eggs, details of the ovarian tubes will be much obscured by an enormous number of ova, which tend to clump together in small groups, greatly distending the very thin, transparent walls of the tubes. Earlier stages of oögenesis occur particularly in the lateral parts of the prosoma, outside the row of limb muscles. The oviducts are large muscular tubes, each of which starts medial and posterior to the block of muscles supplying the sixth prosomal appendage, and passes posteroventrally to its opening on the genital operculum.

In the mature male the testis is very similar in structure and extent to the ovary of the female. With a hand lens it may be possible to see the globular sperm sacs in which the earlier stages of spermatogenesis occur, attached to the ducts in grape-like clusters. The vasa deferentia are located in the same position as are the oviducts of the female. Each spermatozoan is provided with a long, vibratile tail.

The male genital openings on the operculum are round, and each is located on a well-defined papilla. In the female each opening is a transverse slit and there is no definite papilla. These differences can be seen even in the smallest specimens and afford a simple means of distinguishing the sexes before the appearance of the male claspers.

The spawning habits of *Xiphosura* and the use of the male claspers have been described above. Reports that certain of the Asiatic species carry the eggs on the abdominal appendages, instead of laying them in the sand, have been shown to be erroneous.

Circulatory System. About one-half of the elongated, dorsal heart lies in the prosoma, and the other half in the opisthosoma. The whole organ is surrounded by a thin-walled pericardial sinus. This is invaded by a certain amount of areolar connective tissue, and its dorsal wall is indistinguishable from the epidermis. In the opisthosoma, five branchiocardiac canals enter the pericardial sinus on each side, coming up dorsomediad from the gills, just in front of each of the first five abdominal entapophyses. The first one receives a canal from the operculum just as it passes behind the dorsal part of the opercular adductor muscle.

The heart is suspended from the walls of the pericardial sinus by nine pairs of elastic ligaments. The members of the first pair arise near the front of the heart and are exceptional in that they pass lateroanterodorsally through the pericardial wall to attach to the dorsal integument. The remaining eight pairs occur just opposite the eight pairs of ostia and pass laterally out to the pericardial wall, the last five on each side attaching inside the upper ends of the branchiocardiac canals. The ostia are paired, transverse slits on the dorsal surface of the heart, through which blood enters the heart. Each ostium is guarded internally by a bipartite valve. Three longitudinal nerves on the dorsal surface of the heart will be more fully described under the nervous system.

The arteries which leave the heart are four pairs of lateral arteries, one pair of aortae, and a median frontal artery. The lateral arteries are placed opposite the first four pairs of ostia, passing outwards below the suspensory ligaments. Two semilunar valves occur inside the proximal portion of each. Just outside the pericardial wall, the lateral arteries of each side run into a longitudinal collateral artery. The two large aortae and the much smaller frontal artery leave from the anterior end of the heart. Backflow of blood from these three arteries is prevented by one large semilunar valve on the dorsal wall of the heart. Just in front of this valve, at the level of the anterior suspensory ligaments, a pair of rudimentary ostia can be detected on the inner surface of the heart, but their slits do not penetrate the heart wall.

The collateral arteries run the full length of the heart, on either side, each giving off about fourteen lateral branches which supply adjacent muscles and other tissues. Anteriorly each collateral ends in a few small branches just lateral to the beginning of the aorta. Posteriorly the two collaterals come together, so close behind the heart that the mistake is easily made of supposing that an actual connection with the heart exists at this point. From the point of junction a large superior abdominal artery passes backwards to enter the telson. Medially each collateral artery gives off a series of branches which pass ventromedially to the stomach-intestine. The more anterior of these intestinal arteries are hidden under the four lateral arteries. More posteriorly, four intestinals are visible on each side, under the last four suspensory ligaments. It requires close inspection to see that these are not additional lateral arteries coming from the heart, but that instead they curve ventral to the heart and supply the stomach-intestine.

Nearly opposite the second lateral artery, the collateral of each side gives off an especially large lateral branch. This passes outwards and backwards to near the posterolateral angle of the prosoma, where it curves forward as the anterior marginal artery. Before this curve is reached, about halfway out from the collateral artery, two important branches, anterior and posterior, are given off. The posterior branch runs back into the opisthosoma, where from its position it is known as the posterior marginal. This anastomoses with a branch which leaves the superior abdominal near the posterior end of the opisthosoma. The anterior branch passes forward lateral to the prosomal limb muscles and is known as the hepatic artery. Near the lateral eye this anastomoses with the lateral eye artery, which comes up from below at a point lateral to the muscles of the second prosomal appendage.

The frontal artery is sometimes absent and sometimes arises from one of the aortae instead of between the two. When present it passes anteriorly above the gizzard and crop, to divide farther forward into left and right anterior marginal arteries, which anastomose with the arteries of the same name already described.

All the arteries thus far described can be seen in dorsal view. Further dissection of the vascular system should be delayed until after the digestive and endoskeletal systems have been studied, but the description of the vascular system will be completed here.

The two aortae curve downward and then backward, behind the crop. At the most anterior part of the curvature each gives off a branch which supplies the gizzard, crop, and esophagus. Ventrally the aortae pass back on either side of the esophagus and so disappear below the endocranium. After part of the endocranium has been removed, it will be seen that the two aortae come together immediately behind the esophagus where this curves downward to the mouth. Just in front of the junction, three or four transverse arterial commissures cross the posterior or dorsal surface of the esophagus, between the two aortae. If the esophagus is lifted up, it will be seen that the junction of the aortae is part of a vascular ring, which encircles the short vertical part of the esophagus just above the mouth. Numerous arteries radiate out from this ring in all directions, and a large ventral artery passes backwards from it in the midline.

The vascular ring and associated arteries of *Xiphosura* are very remarkable in that enclosed within them are the principal ganglia and nerves. In an uninjected specimen the impression is given that it is only the nerves that are seen, since the nerves are white and somewhat opaque, whereas the arterial walls are thin and transparent. In an injected specimen it will be seen that the dorsal nerves, which radiate out to supply more especially the lateral parts of the prosoma, are not enclosed by arterial walls except just where they come out from the vascular ring. Practically all the other nerves, however, including the ventral nerve cord and its branches, are enclosed within arteries. The arteries thus can be given the same names as the nerves, and a description of either one will serve for the other.

Most conspicuous are the large arteries which radiate out from the ventral edge of the vascular ring, to supply the six pairs of prosomal appendages. A much smaller opercular artery, and a still smaller chilarial artery, leave the ring ventroposteriorly and run back for some distance parallel with the large ventral artery. Each of the second through sixth prosomal appendages also receives two or three smaller arteries which arise just dorsal to the main artery.

The anterior part of the vascular ring forms a rounded swelling, enclosing the cerebral ganglia. From the dorsal surface of this swelling, a relatively large, lateral eye artery, with a bulbous, swollen base, arises on each side and passes forward medial to the cheliceral artery. Farther forward this artery curves outward, upwards, and finally backwards, passing under the lateral eye and then anastomosing with the hepatic artery as already described. A much smaller, stomodeal artery, also with a bulbous swelling at the base, arises on each side of the cerebral part of the vascular ring, from its dorsolateroposterior border, in contact with the esophagus. This artery passes forward along the ventrolateral surface of the esophagus, supplying the esophagus, crop, and gizzard, and then anastomoses with the aorta where this starts to bend upwards behind the crop. From the anterior border of the cerebral part of the vascular ring, three small arteries can be seen running forwards, medial to the lateral eye arteries. Two of these arise from the ring slightly ventrally and supply the so-called olfactory organ, located midventrally, in front of the chelicerae. The median one of these three arteries arises slightly dorsally, a little off center, runs forward below the esophagus, curves upwards in front of and slightly to the right of the crop, and finally reaches the median eyes on the dorsal surface.

The large ventral artery gives off branches to the five gill flaps, the abdominal muscles, and the stomach-intestine. Near the posterior end of the opisthosoma a branch passes dorsally on each side to anastomose with the superior abdominal artery. Nerves for the telson emerge from branches of the ventral artery just behind the anus, the arteries not continuing any farther.

All the arteries finally discharge into hemocoelic spaces, as in other arthropods. But in two different features concerning these spaces *Xiphosura* is perhaps unique. In the first place the hemocoel develops by a coalescence of the primary body cavity with several coelomic pouches. The cavity so formed has been termed a myxocoel. Second, there are present a number of minutely branched collecting channels which drain the blood from the lateral parts of the prosoma. The system of collecting channels is not likely to be seen unless specially injected, but will be described here to complete the functional picture. Ventrally one main collecting channel runs nearly the full length of the body on each side. In the prosoma it lies medial to the outer row of large limb muscles, and for part of its course is close to the stolon of the coxal gland (described under the excretory system). In the opisthosoma it lies just lateral to the ventral abdominal flexor. Three large branches enter it laterally in the prosoma, each in turn greatly branched, so that the drainage of the lateral parts of the prosoma is perhaps as complete as that provided by the lymphatics in a vertebrate. Drainage of other parts of the body and of the appendages is through wider and less clearly defined hemocoelic spaces, which gain access to the main collecting channel on each side through ventral and lateral openings. Opposite each of the gill flaps, an efferent branchial channel leaves each of the main channels ventrally and passes to the inner border of the gills. From the lateral side of the gills the blood is returned to the pericardial sinus by a branchiocardiac canal, as has already been seen. The operculum, also, seems to have on each side an afferent branchial channel and a branchiocardiac canal.

One investigator states that the openings from the main collecting channel into the afferent branchials are associated with the tendinous apodemes on which the dorsal abdominal flexors insert. Contraction of these muscles, he states, results in closing the openings, while dilation of the openings is brought about by contraction of the small lateral slips of muscle which join the ventral abdominal flexors.

The blood of *Xiphosura* contains hemocyanin and a single type of white blood cell. When observed in one of the gill lamellae placed under a microscope, these cells are seen to be spindle-shaped, large, and without amoeboid activity. As soon, however, as they come in contact with any foreign body, as when blood is withdrawn from the animal, they become amoeboid, agglutinate with one another, partially disintegrate, and may throw out long fibrous processes. The agglutinated mass of blood cells forms a soft, gelatinous clot, around which there may later occur a slight secondary clotting of the surrounding fluid.

Digestive System. The different parts of the digestive system usually are not dissected in a functional sequence, but a brief description of them in the functional order will first be given. The esophagus passes upwards from the mouth, through the vascular ring, then runs forwards to the more swollen crop. This curves upwards and backwards, merging into the gizzard without any noticeable change in external appearance. A marked constriction indicates the boundary between the gizzard and the stomach-intestine, which is a wide tube running to near the posterior end of the opisthosoma. A short rectum leads to the anus. On each side of the stomach-intestine two ducts from the hepatopancreas enter, one usually in front of and the other slightly behind the level of the mouth. The very much branched hepatopancreas is coextensive with the gonad, as has already been described.

Dissection of the digestive system usually starts with the removal of the heart to expose the stomach-intestine. However, if a study of the cardiac nerves is planned, the dissection should be made from one side, leaving the heart in place. To expose the full length of the esophagus a wide median strip must be cut from the endocranium (described under the endoskeleton). This is most easily done if the stomach-intestine is first lifted out of the way after cutting it across near its anterior end.

If the digestive tract is slit open longitudinally, all of it except the stomach-intestine will be seen to have a cuticular lining. The lining of the esophagus is longitudinally folded. That of the crop is wrinkled, permitting wide distention of this organ. The gizzard is much less distensible, and its lining is thrown into longitudinal ridges studded with somewhat flattened teeth. From the gizzard a conical valve projects into the stomach-intestine. The cuticular lining of this valve is thrown into about thirteen longitudinal ridges which extend freely at their posterior ends. The inner wall of the stomach-intestine is considerably folded, and in the part overlapped by the conical valve, bears numerous closely set papillae.

When studied histologically the lining of the stomach-intestine is found to consist of columnar, glandular, epithelial cells, having a striated border. In the hepatopancreas there are absorptive cells with a striated border, and glandular cells which secrete the digestive enzymes. All parts of the gut, including the tubules of the hepatopancreas, are provided with both circular- and longitudinal-muscle fibers.

Food is ground up in the gizzard and admitted to the stomach-intestine through the conical valve. Coarse fragments are ejected through the mouth. Digestion occurs in the stomach-intestine, by means of enzymes from the hepatopancreas, except for the splitting of dipeptides, which occurs intracellularly in the hepatopancreas after absorption has occurred. A second set of enzymes, acting in acid medium in the foregut, has been reported, but where these enzymes would come from is not clear. The stomach-intestine secretes a mucilaginous coating for the fecal pellets, which may be up to 10 cm. long. Fat, glycogen, and protein are stored in the connective-tissue cells between the tubules of the hepatopancreas, rather than in the cells of the hepatopancreas itself. The absorptive cells also have an excretory function, as is illustrated by the fact that at times they pass considerable quantities of calcium phosphate into the lumen of the hepatopancreas, whence crystals of this material are conveyed to join the feces in the stomach-intestine.

Endoskeleton. A full dissection of the endoskeleton is not possible without damage to other structures yet to be described. Only the endocranium and portions of the entapophyses should be exposed to view at this stage of the dissection. The other parts can be located by feeling with a probe.

Three distinct classes of structures make up the endoskeleton, differing in their chemical composition, histology, and embryology.

First of all there are a number of infoldings of the exoskeleton. Most noticeable of these are the six pairs of dorsal entapophyses in the opisthosoma, plus the single large pair at the posterior margin of the prosoma, and ventrally the six pairs of tendinous apodemes, which project inwards from the stigmata visible on the proximal posterior surfaces of the operculum and gill flaps. The cuticular linings of the fore- and hindgut also develop as infoldings of the exoskeleton.

Second, there are the six median endochondrites and the endocranium. The endocranium is a structure of complex shape, mainly in the form of a horizontal shelf, located below the anterior third of the stomach-intestine, dorsal to the esophagus and the vascular ring. A posterior portion of the endocranium encircles the ventral artery near its anterior end. The endochondrites are small transverse bars, located below the ventral artery, one at the base of each of the abdominal flaps. Both the endocranium and the endochondrites develop from the ends of embryonic muscles. Histologically and chemically they have been found to show some resemblance to vertebrate fibrocartilage.

Lastly there are seven pairs of so-called branchial cartilages, and on each side of the body a series of longitudinal cartilaginous bars connecting the ventral ends of the abdominal entapophyses. Each member of the first pair of branchial cartilages runs from the posterior margin of the endocranium into one of the chilaria. The remaining six pairs run from the ventral ends of the abdominal entapophyses into the operculum and the gill flaps. Those that enter the gill flaps can be found just medial to the gills. Embryologically these cartilages develop from mesoderm of the somite walls. Histologically and perhaps chemically, they resemble the hyaline cartilage of some of the lower vertebrates.

The entapophyses, tendinous apodemes, endochondrites, endocranium, and branchial cartilages all serve for the attachment of muscles, some of which were mentioned earlier.

Nervous System. When dissecting for the esophagus, a median strip probably was removed from the endocranium. If necessary this strip can be widened so as to expose more fully the nerve ring and associated nerves. Care should be taken not to proceed too far laterally, since parts of the excretory organs lie just lateral to the endocranium. If an uninjected specimen is used for the dissection of the nervous system, application of Schaudinn's fluid will be found to whiten and harden the nerves. The time usually allowed for the dissection probably will not permit more than the identification of the nerve ring, ventral cord, and the proximal parts of the larger nerves. Tracing out the more distal parts of the nerves requires much more time and care. In any event, after the nervous system has been studied *in situ* from the dorsal surface, as much as possible of the nerve ring and attached nerves should be removed to a petri dish for more detailed study, more especially of the nerve roots and of the ventral surface.

Those nerves which run inside arteries need not be described here, since the student can refer to the descriptions already given for the corresponding arteries. This is true for the neural ring, with its three or four postoral commissures and anterior cerebral swelling, the nerves to the operculum and chilaria, the several nerves to each of the prosomal appendages, the stomodeal nerves, the olfactory nerves, and the nerves to the median and lateral eyes. The nerve ring is made up of fused ganglia, but details of its structure are not easily made out in an ordinary dissection.

Some of the nerves not adequately covered in the description of the circulatory system are the following. Emerging through the dorsal surface of the vascular ring there are eight dorsal nerves on each side, which do not run within arteries. The first of these is a very small nerve, which arises just lateroposterior to the root of the lateral eye nerve. Although small, this nerve has a lengthy course. It starts by following the lateral eye nerve, to which it at first is lateral, becoming medial as the two curve posteriorly. Passing medioventrally to the lateral eye, it continues back in the ventral epidermis, curves posteromedially around the base of the sixth prosomal appendage, then curves slightly laterally again near the adductor muscle for the first gill flap, and finally ends in the anterior part of the metasoma. It appears to supply ventral sense organs in the opisthosoma. The largest dorsal nerve is the second one, which is about the same size as the lateral eye nerve. It arises dorsal to the roots of the nerves to the pedipalp, close to the stomodeal nerve, then passes forward just lateral to the first dorsal nerve. At the point where this and the lateral eye nerve curve laterally it divides into three main branches, which continue forwards, one to the dorsal and two to the ventral epidermis. Each of the third, fourth, fifth, and sixth dorsal nerves arises dorsal to the nerves for the corresponding appendage and runs out in front of that appendage, to supply the lateral parts of the epidermis, usually with three main branches, one dorsal and two ventral. The fifth and sixth of these nerves at first run back above the nerves to their corresponding appendages, then turn sharply laterally. The seventh and eighth dorsal nerves arise close together, between the origin of the sixth nerve and the start of the ventral nerve cord. They branch like the more anterior dorsal nerves, to supply the lateral epidermis in the posterior part of the prosoma and anterior part of the opisthosoma respectively.

On the ventral surface of the nerve ring, a preoral transverse commissure will be found, at the posterior border of the cerebral ganglia. From it three tiny nerves pass to the anterior lip of the mouth.

The ventral nerve cord is actually double, as in all arthropods, though this can be seen only after removal of the arterial sheath. Opposite each of the first four gill flaps it is enlarged as a ganglion, from each side of which an anterior dorsal nerve passes outwards to the lateral epidermis and into one of the marginal spines, and a posterior ventral nerve passes into the corresponding gill flap. In addition two small nerves arise on either side from each ganglion close to the dorsal nerve, and pass dorsolaterally to supply the ventral abdominal muscles and the stomach-intestine. Intestinal branches also arise from the sixth, seventh, and eighth dorsal nerves, near the points where these curve laterally. A network of anastomoses connects together the various intestinal nerves and also the nerves to the ventral abdominal muscles.

Behind the fourth abdominal ganglion, the ventral nerve cord appears to fray out on each side into about eight nerves, which supply the last gill flap, the stomach-intestine, the rectum, the more posterior abdominal musculature, the epidermis and the last two marginal spines, the telson, and the posterior parts of the heart and pericardium.

On the dorsal surface of the heart three longitudinal nerves probably have already been noticed. There is a median ganglionated cord, from which a number of minute nerves pass out to a plexus, with some of the strands on each side forming a more definite lateral nerve. Both the two lateral nerves and the median ganglionated cord run nearly the full length of the heart.

Dissection of the remaining cardiac nerves and of certain associated nerves requires more time than is usually available. However, a

378 ARTHROPODA

description will be given here to complete the functional picture and to aid those who may wish to attempt the dissection.

A series of segmental cardiac nerves arises from the sixth to the thirteenth dorsal nerves, inclusive, just lateral to the ventral abdominal flexor. From the base of each of these, excepting the first two, or from the adjacent part of the corresponding dorsal nerve, connection is made with a lateral longitudinal nerve, which runs forward onto the ventral surfaces of the two dorsal abdominal flexors. This latter nerve receives a further connection from the fourteenth dorsal nerve. In the opisthosoma the lateral longitudinal nerve lies just lateral to the tendinous apodemes and medial to the branchial cartilages.

The segmental cardiac nerve arising from the sixth dorsal nerve passes up to the wall of the pericardial sinus, on which it branches at a level midway between the first two pairs of ostia. No connection has been found from this nerve to the heart, so that it may not be a functional cardiac nerve, although it is so named because it is in series with the others. The next two of the segmental cardiac nerves join together to form a somewhat larger nerve, which reaches the pericardial sinus at a level between the second and third pairs of ostia. This nerve and each of the succeeding five segmental cardiac nerves send branches to the dorsal epidermis, connect with a longitudinal nerve running in the wall of the pericardial sinus, and also connect with the median ganglionated cord on the surface of the heart.

The beat of the heart is maintained by a rhythmic discharge of impulses from the ganglionated cord. Unless shifted by experimental means, the pacemaker for the beat occurs in this cord at the level of the fifth and sixth pairs of ostia, which is where the greatest number of nerve cell bodies are located. The beat is subject to regulation by impulses arriving through the segmental cardiac nerves. Of these, the one which takes its origin from the seventh and eighth dorsal nerves is an inhibitor nerve. Those which originate from the ninth, tenth, and eleventh dorsal nerves, and which join the cardiac ganglion at about the level of the fourth, fifth, and sixth pairs of ostia, are augmenter nerves. Presumably the last two segmental cardiac nerves also are augmenters, but this point appears not to have been determined.

After removal from the body of the animal, the *Xiphosura* heart will continue to beat for hours, if merely moistened with sea water, provided that some nervous connection is left between the dorsal ganglionated cord and the heart muscle.

Sense Organs. The most complex sense organs are the large lateral eyes. The exoskeletal covering over each of them is divided into numerous, relatively large facets, corresponding with an internal organization into separate receptor units. Each unit has about ten to fifteen receptor cells, and a conical lens formed by an ingrowth of the exoskeleton. There is no curtain of surrounding pigment cells and no crystalline cone, and there are no tapetal cells. The structure of these receptor units is so different from that of the ommatidia in the compound eyes of insects and Crustacea that their homology has been doubted. The differences, however, may well be the result of degeneration, associated with the burrowing and nocturnal habits of *Xiphosura*.

The median eyes are located just in front of the most anterior of a row of small spines on the dorsal surface of the prosoma. Normally there are two, close together on either side of the median line. Each is a hemispherical cup, with a nearly spherical lens formed from an ingrowth of the exoskeleton, beneath which are a columnar epidermis and a layer of receptor cells which tend to be arranged in groups. The two nerves from these cups shortly unite, and in the angle thus formed there is a mass of tissue, which when examined histologically proves to be a vestige of two medianly fused eyes. Occasionally adult specimens are found with this posterior double eye in a functional condition, provided with a lens like the pair of eyes just in front.

On the ventral surface, a median, wart-like elevation of the exoskeleton, about 5-8 mm. wide, can be seen just in front of the "elbows" of the chelicerae. A distinctive oval area on each side of the wart will be noted. This is the structure supplied by the two so-called olfactory nerves. Its internal structure changes remarkably during development. In the larva there is a median, pigmented "frontal organ," of doubtful significance, on either side of which is a fully functional ventral eye, supplied by one of the so-called olfactory nerves. Actually the fibers of these nerves originate, even in the adult, from the same ganglionic centers within the cerebral ganglia as do the fibers of the lateral eye nerves. Each of the ventral eyes has an exoskeletal lens, a cup of receptor cells, and a backing of pigment cells around this cup. During further development the frontal organ largely breaks down, the pigment cells become arranged in scattered groups, and apparently the photoreceptor cells become transformed into bipolar and multipolar sensory neurons, the distal fibers of which extend into tiny sensory setae on the surface of the organ! There are also numerous vertical pores through the exoskeleton, smaller than those leading to the setae and quite different from the pores found in other parts of the exoskeleton. Possibly these pores serve for the escape of a glandular secretion, but a recent worker has been unable to find gland cells. Whether this organ in the adult condition is actually olfactory, as has been surmised, has not been determined. The possession of ventral eyes by the larva probably can be correlated with the larval habit of swimming on the back, contrasting with the more usual burrowing habits of the adult.

The movable spines on the gnathobases of the second through fifth prosomal appendages appear to contain taste receptors. When they are touched with appropriate food, reflex chewing movements of the limb, which is thus stimulated, and bending and snapping movements of the chelicerae are initiated. Below each of the spines is a group of bipolar sensory neurons, the distal fibers of which extend into the cavity of the spine. Taste receptors of some sort may also be present in the chelae of the first five pairs of appendages, which have been reported to open when touched with food.

The external spatulate process of the sixth prosomal appendage receives an unusually large nerve, the fibers of which have been reported to be connected with an enormous number of receptor cells located underneath the anterior surface of the process. Since water going to the gills passes over this anterior surface, it has been surmised that the function of these sense receptors is to test some quality of the water in the respiratory current.

Relatively large nerves supply the telson and the six movable spines on each side of the mesosoma, suggesting that these structures, too, are sensory.

On the jointed distal processes on either side of the midline of the gill flaps a small number of blister-like elevations are present. The exoskeleton of these has been reported to be perforated by peculiar pores, with underlying receptor cells, but what sense, if any, is here involved has not been determined. On other parts of the gill flaps sensory setae have been reported, likewise of unknown function.

Xiphosura has been claimed to be remarkably sensitive to slight increases of temperature. The margins of the gill chamber, the margins of the carapace, and the gill flaps are said to be the most sensitive areas, with other areas of the body only slightly less sensitive. The presumed temperature receptors have not been found.

Excretory System. In the ventrolateral parts of the prosoma a pair of complicated coxal glands is present, presumed from their structure to be excretory. The glandular portion of each of these organs consists of four brick-red lobes, located near the bases of the second through fifth prosomal appendages, and each containing a maze of fine, partly intracellular ducts. Each elongated lobe runs laterally from a point near the nerve ring into the basal part of the corresponding appendage, just dorsal to the smaller nerves supplying that part of the appendage, and between anterior and posterior muscles running from the endocranium to the appendage. The first lobe often is considerably smaller than the others. All four lobes receive a copious arterial blood supply. The ducts of the first three lobes open into a labyrinth of slightly larger ducts in a longitudinal stolon, which connects the inner ends of the lobes. The stolon runs at right angles to the lobes, at a level dorsal to the nerves of the appendages and ventral to the dorsal nerves which pass outwards between the appendages. Posteriorly the stolon widens into a coelomosac, into the cavity of which open the ducts of the stolon and of the fourth lobe. From the coelomosac a relatively wide, convoluted tubule leads to the outside, following a tortuous course. First it passes above the fourth lobe to a point posterodorsal to the outer third of this lobe. Here the tubule coils a few times, then runs forward with many coils, dorsal to the outer thirds of the third and second lobes. Anterior to the second lobe, or dorsal to the first lobe, the tubule widens considerably, passes mediad, and then runs posteriorly above the middle portions of the second and third lobes, close to the lateral border of the endocranium. Above the fourth lobe it turns ventromediad and passes through part of the lobe on its way to the external aperture. This aperture is located on the articular membrane of the fifth prosomal appendage, proximal and posterior to the basal segment. When the folds in this membrane are smoothed out by stretching, the position of the opening will be seen to be indicated by a small light area surrounded by a ring of darker gray, located in the membrane near the midpoint of the posterior proximal border of the appendage. In young specimens the position of the opening may be very hard to see.

Of the internal parts of the coxal gland only the lobes and the stolon are at all likely to be seen in an ordinary dissection. The tubule is very transparent, and probably will be seen only if a successful injection is achieved through the external aperture. Except for a short length of the excretory tube near the external aperture, all parts of the coxal gland are mesodermal in origin.

The excretory activity of the hepatopancreas has already been mentioned.

Argiope aurantia*

By J. B. Buck and M. L. Keister

The familiar golden garden spider (sometimes called *Miranda*) is abundant in gardens, weed patches, and vacant fields in late summer and fall. The female, which is the sex ordinarily seen, sometimes attains a leg spread of 5 cm. and a body length of 2½ cm. The thorax is covered with a fine felt of silvery hairs, and the abdomen is conspicuously marked in a mottled pattern of black and yellow (white in preserved specimens). The four pairs of legs are black except for basal brown regions on the last three pairs. The third pair of legs is markedly shorter than the others.

The male of *Argiope* has the same general coloration as the female, but is only a quarter of her length and has a cylindrical, rather than a bulbous, abdomen. He is said to spin a small companion web near the female's, but is rarely seen.

The banded garden spider (*Argiope trifasciata*) is similar to *A. aurantia* in general appearance and distribution, but can be distinguished because of its smaller size, more cylindrical abdomen, and cross-banded color pattern. Both species are found hanging head downward in the centers of orb webs spun vertically between tall weeds or grass. The first two legs on each side are held close together, and the last two likewise, so that the legs present a characteristic X-shaped arrangement. Both species are entirely harmless to man.

Argiope spins well in captivity, and if confined in a large glass crock covered with cheesecloth will demonstrate the whole process of

---

* The descriptions of the spider, millipede, and roach were written from liquid-covered specimens dissected under a stereoscopic microscope at 10 to 20 diameters magnification, using a bright spotlight for illumination. Watchmaker's forceps and insect micropins (held in pin vises) served as the principal dissecting instruments, and micropins were used, sparingly and with care, to anchor the specimens. Under such conditions, dissections are possible which are as complete as can be achieved on mammals with the unaided eye; without them, no serious study of the internal anatomy of such small animals is possible. Commercially preserved material was used except where otherwise noted. Our descriptions include more details than a student can be expected to find in a single dissection.

We are happy to acknowledge helpful suggestions from H. S. Barber, J. Franklin Yeager, S. C. Bishop, Dietrich Bodenstein, Leigh Chadwick, Helen Trembley, and particularly R. E. Snodgrass.

weaving the beautifully symmetrical and intricate orb web. The spider ordinarily first walks about, paying out a silken dragline from the spinnerets, and dabbing the substratum with attachment disks for anchoring the main peripheral supporting framework of the web. These attachment disks are made by spreading and bringing together the tips of the spinnerets, and if examined with the dissecting microscope will be seen to consist of ragged patches formed of hundreds of short, delicate filaments. After the supporting framework is attached, the radial threads, or spokes, are put in, then a temporary spiral, beginning at the outside, and finally the permanent spiral filament, beginning near the center. The web is generally finished by the installation of a heavy, vertical, zig-zag reinforcing band across the center. Careful examination will show that the various types of structural element of the web are made of different kinds of silk, some being multiple stranded, some beaded, and so on.

Argiope feeds well in captivity, though she seems unable to capture prey except by means of the web. If a fly becomes entangled, the spider rushes to it by way of the radial lines, seizes it with fangs and palps, and bites it repeatedly, presumably injecting poison. Then, hanging head upward by the two fore pairs of legs, and manipulating or rotating the prey with the palps and the third pair of legs, she quickly wraps it in sheet-like swathing bands of silk which she draws from the spinnerets, using her two hindmost legs alternately. (Ewing gives a remarkable photograph of the swathing band.) The pinioned prey, usually still alive, is then carried or hauled to the center of the web, anchored there, and consumed at leisure. As in most spiders, digestion is initiated outside the body by powerful enzymes in the digestive juice which the spider exudes on to the prey. *Argiope* apparently crushes and tears her victims, in addition to sucking in the liquefied soft parts, and can reduce a large blowfly to a compact pellet the size of the head of a pin in a half hour or less. This pellet, after the swathing band has been thriftily consumed, is cut loose and discarded.

Captive spiders will demonstrate a number of other instructive behavior patterns, such as drinking, "alarm" web-shaking, web-repairing, and, in gravid females, egg-laying. The eggs are laid in a hanging cocoon about the size and shape of a thimble. Its outer jacket is marvelously constructed of numerous felted lamellae of at least two kinds of silk and encloses an inner, continuous, exceedingly thin, transparent membrane.

Since male spiders are rarely available as dissection material, and since the anatomy of those organs which distinguish spiders from other arthropods is more easily worked out in the larger female, the present descriptions will be confined to that sex. If males are available, however, the remarkable copulatory apparatus in the terminal segment of the pedipalp may be examined microscopically in an uncovered glycerine mount, after preliminary boiling in 20 per cent potassium hydroxide.

External Anatomy. The body of *Argiope* consists of two principal divisions, the cephalothorax or prosoma and the abdomen or opisthosoma, neither showing any clear external sign of segmentation in the adult, aside from the appendages. Viewed from above, the cephalothorax is roughly rectangular in shape and shows radial furrows on its dorsal surface which mark the sites of internal attachment of the leg muscles. On the anterior dorsal extremity of the cephalothorax are four pairs of simple eyes of different sizes, mounted so as to command a view forward, laterally, and upward (Fig. 185). Anteriorly, below the eyes, are a pair of vertically oriented chelicerae, making up most of the "face" of the spider as seen from the front. Each consists of a robust basal segment and a dark brown, conical fang hinged to its distal extremity. The chelicerae are believed to be homologous with the crustacean second antennae, rather than with the mandibles. When not in use, the fangs are folded against the medial surface of the basal segment, where they fit between two rows of tooth-like tubercles. In biting, the fangs stand out laterally at right angles to the basal segments of the chelicerae and move horizontally as the basal segments move laterally. Internal (posterior) to the chelicerae is the rostrum or labrum, forming a sort of "upper lip" to the mouth. The inner and ventral surfaces of the labrum are densely covered with light brown bristles.

On the ventral surface of the cephalothorax is a large, flat, central sternum surrounded by the articulations of the paired pedipalps and the four pairs of walking legs. The pedipalp is a six-segmented appendage with an enlarged basal segment or coxa bearing bristles which assist in feeding. The succeeding segments of the pedipalp are a short, cylindrical trochanter, a long femur, a tiny, triangular patella, a tibia, and an unjointed tarsus bearing a claw at the tip. The pedipalps are ordinarily carried protruding in front of the chelicerae and function, in the female, primarily as tactile organs. Between the coxae of the pedipalps, at the anterior end of the sternum, is a fleshy protuberance, the sternum of the pedipalp segment, serving as the posterior wall of the mouth cavity. It is provided with feeding bristles.

The walking legs are similar to one another. Each consists of seven more or less cylindrical segments, a stout coxa, a very short trochanter, a long, tubular femur, a wedge-shaped patella, a long tibia, and a long bipartite tarsus bearing one straight and two curved claws. (The distal part of the tarsus is often called the metatarsus, in reverse of the order used with vertebrates.) All the segments of the legs are supplied with bristles or setae, some reported to be sensory.

A slender, armored waist, the pedicel, connects the cephalothorax to the much larger abdomen, which is oval and bulbous, with blunt anterolateral projections. The abdomen is covered with a leathery exoskeleton having a pattern of alternating darkly pigmented and transparent spots. Patches of yellow or chalky material beneath the transparent areas contribute to the characteristic coloration of the species.

At the anterior end of the ventral surface of the abdomen occur two lateral, brownish, triangular spots close under the integument, marking the position of the book lungs (Fig. 186). Just posterior to each spot is a long, transverse slit, the opening or spiracle of the book lung. By stripping forward the integument covering the lung, it is possible to see the numerous thin, white, respiratory leaves or lamellae, arranged vertically in the anteroposterior axis. Each leaf is supported on its free edge by a brown, hardened rib, and the leaves are held apart by numerous microscopic trabeculae. Between the spiracles is a triangular, scoop-like, backward-pointing process, the epigynum, which conceals the female genital openings.

At the posterior end of the ventral surface of the abdomen is an eminence consisting of three pairs of spinnerets and the anal papilla (Fig. 187). The anterior and posterior spinnerets are large, conical, two segmented, and readily movable in life. They approach one another rather closely in the midline, partially concealing the small, triangular, unjointed middle pair. The distal surfaces of all the spinnerets are studded with a variety of types of minute silk spouts, each of which is associated with a particular type of silk secreted by a specific internal silk gland. The tiniest, hair-like type of spout, the spool, can be picked off in bunches with a forceps and will repay examination with the compound microscope. Even at low magnification, silk threads can be seen issuing from the largest type of spout, the short, stout, sparsely distributed spigot.

ARTHROPODA

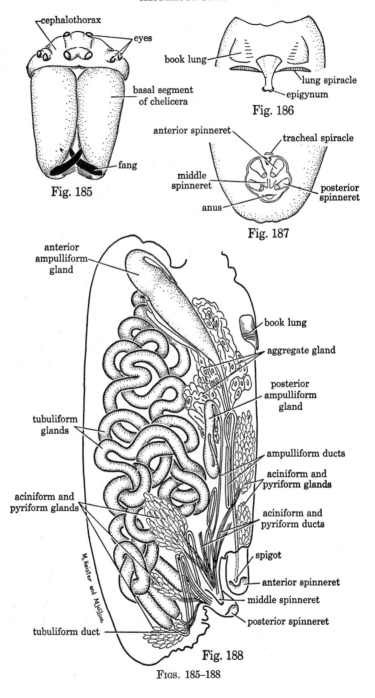

Figs. 185–188

Behind the spinnerets, on the anterior surface of the anal papilla, is a transverse slit, the anus. Just anterior to the group of spinnerets is another transverse slit, the tracheal spiracle, which is often difficult to see.

INTERNAL ANATOMY

Before dissection is begun, the legs should all be removed. The specimen should then be laid ventral side up in an alcohol-filled, paraffin-bottomed dish, and carefully sawed into mirror halves with a sharp razor blade. The more complete half should be selected for study, and the other kept in reserve.

The viscera will be described from one sagittal half of the body, and it should be kept in mind that, except for the gut, heart, endoskeleton, and nerve mass, each structure is duplicated in the other half of the body.

Digestive and Nervous Systems. The chelicera, mouth, pedipalp, and labium may now be re-examined to gain a clearer picture of their structure and relationships. From the mouth the slit-like pharynx runs dorsally to about the middle of the cephalothorax, where it joins the slender, tubular esophagus approximately at right angles. Iridescent sclerotized plates form the anterior and posterior walls of the pharynx (epipharynx and hypopharynx). The esophagus has thickened dorsal and lateral walls. It runs posteriorly to about the level of the second pair of legs, where it empties into the so-called stomach, the wall of which is also strengthened by plates. The stomach is connected to the dorsal wall of the cephalothorax by powerful muscles, and similar muscles anchor it ventrally to a horizontal endoskeletal plate, the endosternite. When one considers the arrangement of these muscles, and the fact that the spider feeds on liquefied food, the true function of the stomach should become apparent.

The stomach narrows at its posterior end and leads into a somewhat wider midgut, which narrows again in passing through the pedicel

---

FIG. 185. *Argiope.* Anterior view of cephalothorax of female. ×9.
FIG. 186. *Argiope.* Ventral surface of epigastric region of abdomen of female.
×3.5.
FIG. 187. *Argiope.* Ventral surface of posterior region of abdomen of female.
×3.5.
FIG. 188. *Argiope.* Silk glands of left half of body of female, as seen from the median plane. Posterior aggregate gland omitted, and other glands simplified.
×7.

into the abdomen. Near its beginning the midgut gives off a large, lateral caecum which will be traced later. Lying close to the dorsal surface of the midgut, the slender, transparent aorta runs forward. It bifurcates just behind the stomach. (If not seen, it may appear in the other half of the specimen.)

In the abdomen, the midgut enlarges greatly and abruptly turns dorsally (Fig. 189). The wall of this dilated portion is pierced by several holes, some quite large. They are the openings of much-branched diverticula which ramify obscurely through the fatty or glandular-appearing tissue which fills the spaces among the viscera. Dorsally the midgut narrows again and runs directly back to the short rectum, which ends on the anal papilla. Dorsal to the rectum, the gut is expanded into a very large storage space, the stercoral pocket.

In the anterodorsal corner of the cephalothorax lies a large, glistening white, sausage-shaped poison gland, with a spirally muscled wall. It is regarded as a modified salivary gland. Ventrally the poison gland narrows abruptly into a duct which runs through the basal segment of the chelicera into the fang, where it terminates in a minute opening near the tip, on the outer side. The poison is primarily a neurotoxin.

The nervous system of the spider is concentrated in a large ganglionic mass encircling the esophagus just anterior to the stomach. Nerves are given off from the dorsal part of this mass to the eyes and mouth parts, and from the ventral part to the legs. Nerves, but not a nerve cord, run back through the pedicel into the abdomen, and supply, among other structures, muscles and spinnerets.

If the stomach, stomach muscles, and the ganglia are carefully removed, the brownish, pebbled tube of the intestinal caecum can be traced laterally and anteriorly from its point of origin on the midgut (Fig. 189). In its course it gives off a large branch which runs anteromedially beneath the poison gland, and a branch toward the base of each leg. The caeca, together with the abdominal intestinal diverticula, presumably form the main digestive and absorptive areas of the alimentary canal and also provide space for sufficient food storage to carry the spider through extended fasts.

Excretory System. Lying just ventral to the most anterior leg branch of the caecum, and in close contact with it, is a small, elongated, gauzy sac, the coxal gland. This gland is best dissected from the dorsal surface, after removing the poison gland and the head muscles behind (external to) it, the chelicera, and the dorsolateral exoskeleton of the cephalothorax. This approach will also be found

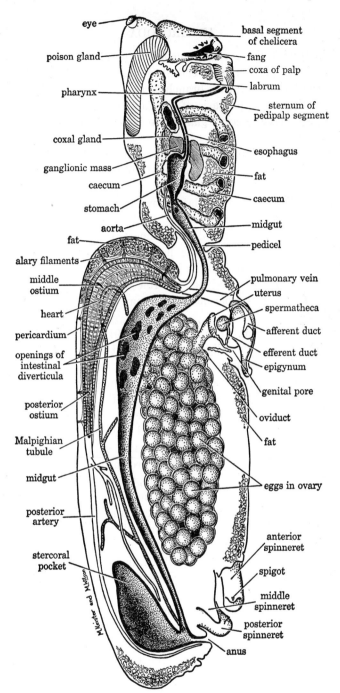

FIG. 189. *Argiope.* Viscera of left half of body of female, as seen from the median plane. Fat which fills most of space between viscera indicated diagrammatically in a few regions only. ×8.

helpful in tracing the caecal branches. The gland rests ventrally on the anterodorsal portion of the endosternite and sends its duct ventrally along the anterior caecal leg branch to an opening medial to the base of the first leg. On the basis of the fact that they pick up injected carmine particles, the coxal glands are believed to have an excretory function.

Lateral to the region where the stercoral pocket overlies the dorsal wall of the intestine, a slender, white Malpighian tubule may be found by displacing the intestine. It can be traced forward from the point where it empties through the lateral wall of the intestine just as the latter enters the stercoral pocket. It first receives a large, dorsal branch near the posterior end of the heart, then runs anteriorly and far laterally to the level of the heart, then curves back and ends blindly in the fat ventral to the heart (Fig. 189). As in insects, the Malpighian tubules are assigned the role of extracting soluble wastes from the body fluids, concentrating them, and excreting them into the hindgut.

Circulatory System. The heart (Fig. 189) is a large, thick-walled tube which extends dorsally from the point where it gives off the median anterior aorta near the pedicel, then curves posteriorly, close beneath the dorsal integument, to a point almost midway to the posterior end of the abdomen. The lining of the heart, when exposed by clearing away the clotted blood where necessary, is seen to consist of slender, parallel strands of longitudinal muscle, overlying a compact layer of circular muscle. The heart lies in a pericardial space, the wall of which adheres to the spongy, brownish mass of fatty tissue which permeates the abdomen. Dorsally, the heart is attached through the pericardial wall to the integument in nine places by tufts of fine alary filaments. Under the first of these tufts, which is close to the pedicel, is the first slit-like ostium, guarded posteriorly by a valvular flap. In about the same region, and lateral to the heart, a large, thin-walled, sinus-like pulmonary vein (filled, in preserved specimens, with coagulated blood) empties into the pericardial space. It carries aerated blood from one of the book lungs.

A second ostium is situated dorsolaterally under a tuft of filaments, at about the middle of the length of the heart. Like the first ostium, it is provided with a valve which would be closed by blood flow in the anterior direction. Ventral to the ostium is the opening of an artery, the diverticular artery, which runs laterally into the abdomen. The third and final ostium is near the posterior end of the heart, close to the origin of a large artery running posteroventrally. Just beyond this point the heart ends, giving off a slender, median, transparent posterior artery which runs along the dorsal wall of the stercoral pocket, and another which runs posterolaterally.

The circulatory system is of the open type. The blood, which is driven forward and backward by the contractions of the heart, runs out of the arteries into tissue spaces, or lacunae. Here it percolates past the tissues to ventral sinuses which route it through the book lungs. It is finally returned to the pericardial space by way of the pulmonary veins and passes thence through the ostia into the heart.

Reproductive System. In the immature female the ovary is a white, flaccid sac with small wart-like follicles on its surface. In the mature female the egg-packed ovary occupies most of the space between the intestine and the ventral body wall. The large yellow eggs, of which there are about 250 in each ovary, can be dissected away to disclose a large, white, flattened oviduct, which runs anteroventrally around a large longitudinal muscle and terminates in a short, flat, median uterus, close above the epigynum (Fig. 189). The uterus opens posteriorly by a horizontal slit-like passage into a deep furrow in the (ventral) surface of the abdomen dorsal to the epigynum. Lateral and dorsal to the base of the epigynum, and immediately beneath the uterus, is a tiny, bright-red, hard sphere, the spermatheca, where the sperm are stored from the time of copulation to that of egg-laying. The spermatheca receives the sperm by way of a slender, dark-brown duct, which curves within the anterior half of the cavity of the epigynum (Fig. 189). The spermatheca in turn sends a minute sclerotized duct to the uterus.

In male spiders the testes are long, slender, convoluted tubules, in the analogous position to the ovaries, under the intestine. They run forward and, as vasa deferentia, unite with each other and terminate at the genital opening. Before copulation sperm are transferred from the genital opening to the specialized tips of the pedipalps of the male. In copulation the pedipalps are inserted into the epigynum and the sperm are expelled into the funnel-like openings of the paired afferent spermathecal ducts.

Tracheal System. The tracheae are difficult to find in *Argiope*, but of such unusual structure as to repay the effort well. If possible, the dissection should be made on a freshly killed spider, so that the tracheae will be air-filled. In any case, an intact specimen is preferable to a sagittal half, since a ventral dissection is required. A 3-mm. transverse cut through the integument is made about halfway between the tracheal spiracle and the epigynum. From the ends of this incision, two longitudinal slits are teased back to the level of the tracheal spiracle. The resulting rectangular flap is very carefully teased away from the underlying tissues, beginning at its anterior extremity. The tracheae appear on the surface of the tissues thus exposed as two pairs of very slender white threads. One of them runs straight forward from the spiracle, the other anterolaterally. They cannot usually be traced far forward. If a length of one of the main trunks is mounted in water and examined under the microscope, it will be seen to consist of a rough, unbranched tube with a very much pitted and ridged wall. The wall itself can be seen in optical section to be double, the two layers being connected by numerous fine, irregular trabeculae. No typical spiral thickenings (taenidia) are present in the main trunks, but with the oil-immersion lens, particularly in preparations made from a distal section of the trunk, a few very tiny and delicate tracheae with spirally thickened walls can usually be seen.

Silk Glands. To expose the silk glands lateral and ventral to the ovary, the gut, heart, and eggs should be very carefully removed. In doing this, three large dorsoventral muscle bundles will be encountered, two of which run from the dorsal body wall through the ovary to skeletal supports below, while the other connects the pedicel region with the dorsal body wall. Other large muscles run longitudinally under the ovaries from pedicel to spinnerets. Just posterior to the epigynum, these are interconnected with a fan of smaller muscles leading to the ventral body wall, and with the middle dorsoventral muscle by a large fascia. A similar fascial interconnection occurs between the posterior dorsoventral muscle, the longitudinal muscles, and the large muscles inserted in the spinnerets.

There are five types of silk glands in *Argiope* (Fig. 188). (1) Ventral to the position of the ovary and about midway in the abdomen is the smaller of two ampulliform glands. It is a slender, white sac, about 2 mm. long, containing, like most of the silk glands, an amber-colored core of coagulated silk. It is embedded in fat and a confusing tangle of ducts from other glands, which must be freed and cleared with great care. Anteriorly it narrows into a thread-like duct which bends sharply backward, runs back almost to the spinnerets (where it is overlain by the small, bead-like lobes of the pyriform glands, which should be laid aside with care), curves anteriorly again for about 1 mm., then turns posteriorly again, and terminates in a spigot on the median surface of the middle spinneret. The larger ampulliform gland

ARGIOPE AURANTIA 393

lies quite far anterior in the abdomen. It is a spindle-shaped body about 5 mm. long and 1 mm. in diameter, with a straw-colored core. Anteriorly, it has a narrow, blind, recurved end. Posteriorly, its slender duct takes the same kind of course as the duct of the small ampulliform gland, and eventually ends on a spigot on the anterior spinneret. The ampulliform glands secrete the draglines and dry threads. (2) The anterior of the two aggregate glands appears as a snarl of cream-colored, warty gland tissue and large, convoluted, thin-walled ducts in the neighborhood of the ampulliform glands. It empties on the posterior spinneret. The posterior aggregate gland will be seen later. These glands are thought to form the sticky drops which festoon the cords of the web. (3 and 4) Anterior, lateral, and dorsal to the spinneret region are a number of clusters of tiny bead-like glands, each individual of which gives off a microscopic duct. The glands are actually of two types, bulb-shaped (aciniform) and pear-shaped (pyriform), but will be considered together. Their ducts, in which extraordinarily fine threads of silk occur, are gathered into loose bundles or cables and terminate in individual spools on the various spinnerets. Without attempting to describe the glands in detail, it may be said that the ducts from the anteroventral cluster of glands open on the anterior spinneret; those from the several clusters surrounding the spinneret muscles terminate on the middle spinneret; and the two posterodorsal clusters supply the posterior spinneret. The duct cables must be dissected with care, since they are mixed up with the ducts of all the other glands, and also are easily confused with the spinneret muscles. The pyriform glands form the silk for the attachment disks of the web, and the aciniform glands the swathing bands in which the prey is enveloped. Lateral to the posterodorsal groups of pyriform glands is a mass of large, pimply, cream-colored or pinkish tubes, the posterior aggregate gland. It opens on the posterior spinneret. (5) The fifth and final type of spinning gland is the tubuliform, of which there are three in each longitudinal half of the abdomen. The glands themselves can be found as very much contorted tubes, ranging from black or purple to light yellow in color, lateral to the ovarian wall, and occupying most of the lateral part of the abdomen. Toward the posterior end of the abdomen, lateral to the aciniform glands, they terminate in three parallel nozzles, which give off stiff, slender ducts, just visible to the naked eye, which curve briefly forward, then ventrally to spigots, one on the middle and two on the posterior spinneret. The tubuliform glands are be-

394   ARTHROPODA

lieved to form the egg cocoon (they are much reduced in immature females). However, as already described, the cocoon contains at least two types of silk in addition to the relatively coarse brown threads which appear to be the product of the tubuliform glands.

Artemia

By J. H. LOCHHEAD

Artemia, the brine shrimp, is not a true shrimp, but belongs to a quite different and primitive order of crustaceans. It is able to live in a notably wide range of salinities, from a salt concentration not much above that of fresh water up to one over three times that of sea water. Normally, however, *Artemia* is found only in highly saline lakes and pools, probably because it falls an easy prey to the predators present in other environments. The distribution is world wide, mostly in the warmer and more arid regions. Dried eggs, from which laboratory cultures can be started, are on sale by biological supply companies. The closely related fresh-water "fairy shrimps," such as *Eubranchipus*, occur seasonally in small, usually temporary ponds and differ from *Artemia* only in minor anatomical features.

Living specimens are best for study. They should be watched in the culture jar and can be mounted whole under a coverglass or between two thin slides if both sides are to be studied on a single specimen. Chloroform or ether (a few drops to 10 ml. of water) may be used, if desired, to narcotize or kill the specimens. Dissection of appendages or other parts should be done with mounted needles, under a binocular.

Body Plan. The body appears divided into three main regions, but because of difficulty in defining thorax and abdomen in a manner permitting comparisons with other Branchiopoda, it seems best to speak only of a head and trunk. The trunk has eleven limb-bearing somites, followed by two genital somites and six other limbless somites. The telson usually is fused to the last somite, but sometimes is distinct. Attached to the telson is a caudal furca, which is used for steering and for making sudden spurts through the water. It is, however, greatly reduced in polyploid specimens reared in strong brines.

Exoskeleton. The whole body is covered by an extremely thin exoskeleton, which nevertheless is remarkably impermeable to salts, water, and many other substances. This exoskeleton is molted about every 4–6 days throughout the adult life of several months.

Paired Appendages. The head bears the five pairs of appendages typical for Crustacea. The antennules are slender and tipped with a few special setae, or aesthetascs, believed to be chemosensory. The antennae are much thicker, do not bear setae, and are modified in adult males as large claspers. The large mandibles are without palps and act like a pair of mill wheels, rolling food into the mouth. Each of the much smaller maxillules bears a sheaf of setae, all curving medioanteriorly, and lying one above the other in a single, almost vertical plane. The maxillae are greatly reduced, being discernible with difficulty as small bumps, each bearing only three setae, located between the maxillules and the bases of the first trunk limbs. To see them and the maxillules, it is necessary to lift forward the large upper lip or labrum.

The trunk bears eleven pairs of leaf-like appendages, or phyllopodia. A typical one of these, from near the middle of the series, should be dissected off as close to the body as possible, then carefully spread out in a drop of water and mounted under a coverglass. The different parts arise from a central corm. Along the median border there is a series of five endites, of which especially the first two bear a long, regular fringe of filter setae, serving an important role in the process of filter-feeding. Distally the endites are succeeded by a large lobe, termed by Eriksson the telopodite. This bears strong, scraping setae, used in feeding on bottom detritus. Lateral to the telopodite is the highly movable exopodite, of special importance in swimming; it is fringed with long, feathered, paddle setae. Proximal to it, on the lateral border, is the fleshy metepipodite, serving probably as a gill and perhaps also in regulating the internal salt content. More proximal still is a very thin flap, the protepipodite.

Having determined the various parts on the flattened limb, their arrangement on the intact animal should be studied. It will be seen that the endites, telopodite, and epipodites all are directed backwards, forming valvular flaps closing off the space between their own limb and the limb next behind. Alternate expansion and contraction of the interlimb spaces, as the limbs beat forward and back, are largely responsible for the current of water which flows toward the animal, bringing food particles and oxygen. The food particles enter a wide median space, between the two walls of filter setae, from which they pass to the food groove, on the ventral midline of the trunk. In this groove the particles are swept forwards to the mouth.

396 ARTHROPODA

Digestive System. A short, vertical foregut, or esophagus, leads up from the mouth to open at right angles into the long midgut or stomach-intestine. In front of the point of entrance of the esophagus, the midgut expands into a pair of pouches. Secretion of digestive enzymes and absorption of food products probably occur both here and along the length of the midgut, which also plays an important part in osmoregulation. Posteriorly the midgut joins the short hindgut or rectum, provided with dilator muscles in addition to the circular muscles present on all parts of the gut. The rectum opens at the anus, between the furcal rami. Foregut and hindgut both are provided with a thin cuticular lining, continuous with the exoskeleton.

Circulatory System. The heart is a long tube (Fig. 190), dorsal to the gut, extending the length of the trunk. Blood enters it pos-

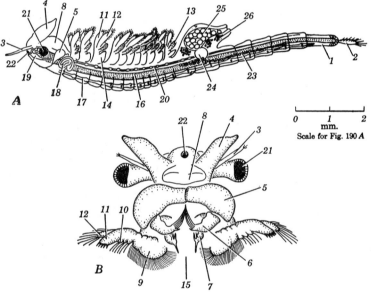

Fig. 190. *A.* Diagram of a female *Artemia*, shown in lateral view. (Redrawn from Lochhead.) *B.* Ventral view of the head and first trunk somite of a female *Artemia*; the labrum turned forwards and the first pair of trunk appendages spread laterally. *1*, Last trunk somite and the telson, fused together. *2*, Caudal furca. *3*, Antennule. *4*, Antenna. *5*, Mandible. *6*, Maxillule. *7*, Maxilla. *8*, Labrum. *9*, First endite of a phyllopodium. *10*, Fifth endite. *11*, Telopodite. *12*, Exopodite. *13*, Metepipodite. *14*, Protepipodite. *15*, Food groove. *16*, Midgut. *17*, Heart. *18*, Maxillary gland. *19*, Supraesophageal ganglion. *20*, Ventral nerve cord. *21*, Compound eye. *22*, Median eye. *23*, Ovary. *24*, Lateral pouch. *25*, Ovisac. *26*, Posterior shell gland.

teriorly and through valved lateral slits, or ostia, of which there is one pair in each of the posterior 14–15 somites. At the anterior end blood leaves the heart, entering the hemocoelic spaces of the head, and then circulating through the hemocoel of the trunk and appendages. Though there are no definite blood vessels, the blood nevertheless follows a more or less constant route. Movements of the trunk appendages undoubtedly aid in the circulation. Frequently the blood is colored red by hemoglobin. A single type of blood cell is present; these cells behave as amoebocytes when adhering to tissue surfaces. These blood cells originate from blood-cell-forming organs located near the base of each trunk limb (best seen in frontal sections).

Excretory System. Excretion probably is performed mainly by the pair of maxillary glands, which form rounded prominences on the sides of the body just behind the mandibles. Each consists of a blind, coelomic end-sac, followed by a coiled tubule which opens on the maxilla. The corresponding antennary glands of the nauplius larva are claimed by some investigators to be entirely absent in the adult, but a rudiment of the end-sac does seem to persist, its cells selectively taking up certain pigments. Uptake of pigment granules and other foreign particles also is performed by large phagocytic storage cells, located in many parts of the body, and perhaps most easily seen in the antennae. In well-fed individuals these cells also store fat and glycogen.

Nervous System. The supraesophageal ganglion, circumesophageal connectives, and two parallel chains of ventral ganglia, connected in ladder-like fashion by commissures, can best be studied in frontal sections. The supraesophageal ganglion receives nerves from the eyes and antennules. Antennal nerves join the circumesophageal connectives. A pair of ventral ganglia occurs in each somite, from the mandibular to the second genital, posterior to which there is only a slender pair of ventral nerves.

Sense Organs. The stalked compound eyes, each composed of numerous ommatidia, are conspicuous. The smaller, median or "nauplius" eye is also easily found. It is formed of three hemispherical cups and is located just in front of the supraesophageal ganglion. The aesthetascs on the antennules have already been noted. If sections are available, the paired frontal organs can be found, each consisting of a small group of what appear to be receptor cells, located under the hypodermis just ventrolateral to the median eye. Other presumed sense organs, including organs of taste on the mouth parts, have been described but are difficult to find.

Striking responses to light are easy to demonstrate on living specimens. Most obvious is a dorsoventral orientation, in which the ventral surface is kept constantly directed towards the source of light. Less frequent is an anteroposterior orientation, expressed as either negative or positive phototaxis. One method of demonstrating this is to keep specimens 2–3 hours in the dark in water made alkaline (to about pH 9) with sodium hydroxide, then to shine an intense and directional beam of light through the aquarium. For several minutes the animals will then display marked negative phototaxis.

Reproduction. In the male the testes are long and tubular, located usually in about the anterior four somites of the limbless part of the trunk. Sperm are produced in the more posterior parts, while the anterior part secretes a mucilaginous fluid. From the anterior end of each testis a coiled glandular duct leads ventroposteriorly to about the posterior border of the second genital somite. Here the duct turns and runs forwards to near the anterior border of the first genital somite, where it turns again and runs back to the tip of one of the two penes which project posteriorly from the second genital somite. In copulation the penes are everted or extended, evidently by blood pressure, to about four times their resting length, and the duct is then no longer kinked. Eversion of one or both penes often can be observed in a dying specimen confined under a coverglass. A retractor penis muscle runs to the tip of each penis from the midventral line.

The antennae in the male are modified as enormous claspers, ornamented and shaped so as to fit tightly around the female just in front of the ovisac. Clasping persists a number of days, the animals swimming around while thus united as a pair. Copulation occurs after each molt of the female. In some stocks of *Artemia* males are absent, reproduction then being parthenogenetic.

In the female the position of the ovaries is similar to that of the testes in the male. In the second genital somite each ovary is connected by a short duct to a lateral pouch, situated in the dorsal part of the prominent ovisac. Here the ova undergo maturation. When filled with ova, the lateral pouches are round and easily seen, but at other times they may be collapsed and more difficult to distinguish. Each opens into the median central part of the ovisac. Here the eggs are fertilized, acquire a shell, and undergo some embryonic development. Groups of shell glands, appearing like clusters of grapes, open into the median ovisac, and muscles are present by which the eggs can be kept moving. Shortly before a molt thick-shelled eggs are discharged from the ovisac; or, if the shell-gland secretion was very scanty, free-swimming nauplius larvae emerge, having hatched within the ovisac from very thin-shelled eggs.

Daphnia magna

By J. H. LOCHHEAD

Various species of *Daphnia* are known, in fresh-water habitats ranging from puddles to open lakes. *Daphnia magna* occurs in the northern hemisphere, in small ponds or ditches having a good growth of vegetation. It tends to be rather local in distribution, but when found may be very abundant. It has been chosen for study here because of its large size and because it is the species for which we have the best anatomical descriptions. Other species resemble it in most of the main essentials.

The anatomy is best studied on living specimens mounted under a raised coverglass, or on specimens mounted after being killed or narcotized with ether or chloroform (a few drops to 10 ml. of water). For a ventral view of the food groove and median parts of the trunk appendages, specimens submerged in a watch glass, with the valves of the carapace open, should be studied under a binocular microscope. For a closer study of the appendages these can be dissected off with needles under the binocular, after first removing one valve of the carapace.

The body is divisible into head and trunk, terminating in a large clawed structure termed the abreptor. The first five trunk somites bear appendages, and perhaps three more somites are present. A bivalved carapace entirely encloses the trunk, fused dorsally with its first two somites.

Most features of the adult female can be made out from Fig. 191. *1*, Exopodite of antenna; it is by means of the antennae that the animal swims in its characteristic flea-like jumps. *2*, Abductor muscle of antenna; a levator muscle is shown just posterior to this, and a second abductor some distance farther forward. *3*, Ventral border of left valve of carapace. *4*, Mandible; from the portion indicated by the label line the part into which muscles insert curves laterally and then dorsally; its extent is indicated on the diagram by broken lines; the medially directed chewing surface is seen just dorsal to the label line; just behind this, the maxillule, with three forwardly directed setae, is seen; a maxilla is absent, or perhaps is represented by a small bump covered with setules. *5*, First trunk appendage. *6*, Endobase of second trunk appendage. *7*, Exopodite of second trunk appendage. *8*, Endite of third trunk appendage, bearing a long row of filter setae which almost reach the roof of the food groove. *9*, Reduced distal endites and telopodite of third trunk appendage (compare with *Artemia*). *10*, Endite of fourth trunk appendage; most of its filter setae are drawn only at their tips where they project beyond the overlapping wall of filter setae of the third appendage. *11*, Exopodite of the fourth trunk appendage, forming a valvular flap at the distal exit from the fourth interlimb space; the third trunk appendage has a similar exopodite, the setae of which are seen projecting posteriorly, lateral to those of the fourth exopodite. *12*, Medial lobe of fifth trunk appendage; the exopodite of this appendage is concealed in the diagram by the abreptor. Not shown on any of the appendages are the laterally placed metepipodites, of which each trunk appendage has one. These are spongy bodies about the size of the compound eye or a little larger. The last four are heart-shaped, with the concavity directed forwards. The row slants dorsoposteriorly, so that the first one is farthest (ventral) from the food groove, the fifth one closest to the food groove. The metepipodites are usually regarded as gills. This view may be correct, but they also appear to be organs of salt absorption. They stain very distinctly with a variety of vital dyes, for example neutral red and methylene blue. *13*, Position occupied by "neck organ" during the first instar; this is a salt-absorbing organ, like the metepipodites, and it stains with the same vital dyes; it is absent in the adult. *14*, Roof of the food groove. *15*, Labrum, with some of its dorsoventral muscles. *16*, One of the "labral gland" cells. *17*, Left levator labrum muscle. *18*, Abreptor, which bears a paired row of spines and terminates in the spine-like caudal furca.

The trunk appendages beat constantly at a rapid rate, each one a certain amount out of step with the others. As the limbs move forward, the interlimb spaces are enlarged, particularly behind the third and fourth limbs, and especially towards their bases. This draws water into the median space between the two rows of limbs. Food particles pass up to the food groove, or first are filtered off on the filter setae. Filtration is so effective that even colloidal particles are secured. In the food groove the particles are swept forwards to the atrium oris under the labrum. Here the maxillules push the food between the mandibles, which roll it into the mouth. Unwanted ma-

DAPHNIA MAGNA 401

terial is removed from the atrium oris by a rotary action of the maxillules, the labrum first being swung ventrally out of the way. The abreptor also can be brought forwards to clean out unwanted material from the more anterior parts of the median space and food groove. Two long setae arising posteriorly on the endobase of the second appendage, and a third shorter seta farther forward, project somewhat

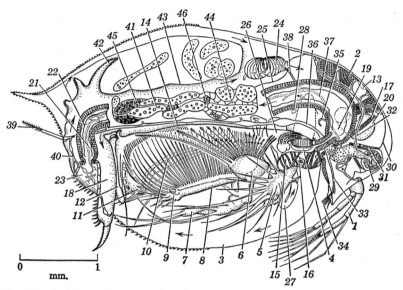

FIG. 191. Left half of an adult female *Daphnia magna,* viewed from the median sagittal plane. Items omitted include the fat cells, most of the muscles, and the setules on most of the setae. The numbers are explained in the text. (Compounded from various sources of which the more important are cited in the list of references.)

medially into the median space and help to prevent overlarge particles from reaching the food groove. Interesting modifications are present for combing the filter setae. Those of the fourth limb are combed by little tufts of laterally inclined setules at the tips of the overlapping filter setae of the third limb. The more posterior filter setae of the third limb are combed from the outside by rows of setules along the length of the endite of the fourth limb (a corresponding, probably non-functional, row of smaller setules is perhaps more clearly shown in the diagram on the endite of the third limb). The more anterior filter setae of the third limb are combed by laterally inclined setules on the tips of those setae that are arranged in a regular row on the endobase of the second appendage. The three most anterior setae on this endobase are not part of this series but are specialized for pushing food forwards to the maxillules.

During the back stroke of the appendages the third and fourth exopodites swing ventrally, permitting water to escape distally from the interlimb spaces. But these exopodites do not here create a swimming current, as do the trunk-limb exopodites of *Artemia*.

19, Beginning of midgut; the esophagus or foregut, which has a cuticular lining, is seen leading into this from the mouth; a spiny lower lip is seen between the end of the esophagus and the mandible. *20*, Left midgut caecum; its round opening into the midgut is evident on the diagram. *21*, Posterior part of midgut; the central part of the midgut has been cut away to permit the left ovary to be more clearly shown; the cells of the midgut (and also of its caeca) have a striated border, the depth of which is indicated on the diagram. *22*, Peritrophic membrane. *23*, Hindgut; this has a cuticular lining and is provided with well-developed circular muscles.

Because of its transparency *Daphnia* is a good subject for observations and experiments on the digestive system. Digestion occurs within the lumen enclosed by the peritrophic membrane, particularly in the region just anterior to the middle of the midgut. Absorption occurs all along the length of the midgut, but more especially in the anterior half, including the two caeca. Waves of reverse peristalsis carry food forwards from the region of maximum digestion. Water is rhythmically taken into and expelled from the hindgut, perhaps for respiratory purposes. Storage of reserve fat and glycogen occurs especially in the "fat cells," sometimes very conspicuous in the lateral parts of the trunk, in the trunk appendages, and even in the carapace. The degree of development of the fat cells varies greatly according to the sex, state of nutrition, and stage of the female reproductive cycle.

24, The heart, shown as though not cut by the section; its muscle fibers are faintly striated; elastic fibers form the lips of the single, lateral pair of ostia, through which blood enters the heart. *25*, One of the membranes delimiting a blood space; several other such membranes occur in other parts of the body. Some of the routes of blood circulation through the hemocoelic spaces are indicated on the diagram by arrows. The blood often is colored red by hemoglobin, dissolved in the plasma. A single type of amoeboid blood cell is present, displaying phagocytic activity towards foreign particles or microorganisms. The number of cells seen circulating in the blood varies greatly, since sometimes many of them adhere among the tissues.

26, End-sac of maxillary gland; this opens ventrally into a duct which loops up and down in the carapace, with altogether five sharp bends or kinks, finally opening on the food groove wall just behind the maxillule, medial to the distal third of the end-sac; the different sections of the duct react differently to vital dyes. 27, The fifth kink of the duct; the second kink is shown posterior to the end-sac; the small arrows indicate the normal direction of flow within the duct. It is presumed that the maxillary gland is excretory and that it perhaps aids in osmoregulation. Occasionally (in up to 2 per cent of the animals) the lumen of the duct on one side may be rendered bright red by a high concentration of oxyhemoglobin. 28, Reduced end-sac of antennary gland; this perhaps accumulates certain excretory products; it can be demonstrated by vital staining with brilliant cresyl blue, methylene blue, or neutral red; there is no duct. Scattered "nephrocyte" cells have been stated to occur in the body at times, but they may be the same as the fat cells.

29, Compound eye, formed by the median fusion of originally paired compound eyes; it contains twenty-two ommatidia. 30, Dorsal ligament suspending the compound eye; a ventral ligament also is shown. 31, One of the pair of dorsal muscle fibers for rotating the eye; a ventral pair is also shown, and a lateral pair lies concealed behind the brain and optic nerve; the eye is rotated in accordance with the direction of incident light; on a specimen confined under a coverglass it is rotated to face a light shining from the side, but presents its dorsoanterior surface towards a light shining from in front; in a free-swimming specimen such rotations are followed by reflex adjustment of the body position, by means of the antennae, bringing the body and eye back into the normal spatial relations with each other; this has the effect of keeping the top of the head (approximately the region between the antennary muscles) directed towards the source of light.

On the diagram a wide optic nerve is shown, connecting the eye with the supraesophageal ganglion, or "brain," which extends almost to the esophagus. From the midventral border of the brain a few small nerves run ventrally to a small group of ganglion or receptor cells, surrounding the tiny pigment spot of the simple median eye. Despite its small size and proximity to the brain, successful operational removal of the eye has been accomplished by Schulz in some interesting experiments. Running ventroanteriorly from the simple eye, a small nerve leads to the median frontal organ. 32, The left nerve supplying five groups of lateral frontal organs, all shown on the diagram as though mediodorsal in position, but actually scattered over the top and left side of the head. The frontal organs are thought to be sensory, but it is not known what sense is involved. On some specimens of *Daphnia magna* the lateral frontal organs enclose refractile bodies resembling the crystalline cones of the compound eye. *33*, Nerve to antennule; the special region of columnar epidermis shown just posterior to this nerve is of unknown significance. *34*, Presumed chemosensory setae or aesthetascs; four are shown, but there are actually nine on each antennule; the slightly longer, pointed seta, just in front of the aesthetascs, is probably a touch receptor. *35*, Nerve running into the base of the antenna. *36*, Ventral nerve cord, swelling at intervals into at least three ganglia. *37*, Lateral midgut nerve; the anterior connection of this is uncertain, but there are three connections from the nerve cord farther back (two shown on the diagram as cut stumps). *38*, Cardiac nerve; seemingly isolated nerve fibers are also shown on the heart wall, and what looks like a second cardiac nerve approaching the heart posteriorly; however, the existence of any nerve supply to the heart is disputed. *39*, Caudal seta, which judging from its innervation must be sensory. *40*, Abreptor nerve, with a branch to a plexus on the hindgut.

Very few of the nerves described above can be seen without special staining. Vital staining of the finer nerves and sense organs can be obtained with alizarine (a saturated solution, made up hot, with added urea to increase the solubility). Several different solutions adjusted to various pH's, chiefly alkaline, should be tried. Bubbling oxygen through the solution may also help.

41, One of six mature ova shown in the left ovary; this particular ovum is in the process of being squeezed through the very narrow oviduct leading to the brood chamber. In parthenogenetic reproduction, which is depicted here, egg-laying starts about ten minutes after a molt and all the mature ova pass into the brood sac at about 15-second intervals. At this time the ova are starting maturation, which is completed in the brood sac with the formation of only a single polar body and no reduction division (i.e., this is diploid parthenogenesis). Once in the brood sac the ova rather quickly round up. A total of two to fifteen (rarely as many as thirty) ova are laid at one time. In the brood sac they rapidly develop and are released, shortly before the next molt, as first-instar young, already having the adult form. Early instars last about a day each. The first batch of eggs is laid in the fifth, sixth, or seventh instar. Thereafter the instars each last 2 days or more. The average life span covers about seventeen instars.

42, Median dorsal process, which closes the brood sac behind, and which, judging from its shape, may serve to push fresh water into the brood sac whenever the abreptor returns to its normal position after being flexed forwards; in a quiescent individual the oxygen tension in the brood sac has been found to be rather low; the belief, often stated, that a nutritive fluid is secreted into the brood sac is incorrect, though it does appear to be true for certain other Cladocera. *43*, Depression in the floor of the brood sac marking the fourth trunk somite. *44*, Connective tissue of the ovary. *45*, Primordial germ cells of the ovary. *46*, A group of four oöcytes, of which one will grow at the expense of the other three to become a mature ovum.

Shown in the diagram in a characteristic position, near the postero-ventral border of the ovary, is the abortive oöcyte of a "winter egg." Dense, finely divided yolk makes such oöcytes appear black by transmitted light. At some stages of the ovarial cycle they sometimes are much larger than the one shown. But during most of the summer, if conditions are favorable, these oöcytes fail to develop further, and reproduction proceeds by parthenogenesis, the progeny all being females. Under certain "unfavorable" conditions (not well understood for *Daphnia magna*), some of the parthenogenetically reproducing individuals produce an occasional brood of males. More rarely there may be a mixed brood of males and females. Some of the females now produce haploid winter eggs in place of the diploid "summer eggs." The single special oöcyte in each ovary enlarges at the expense of other cells (besides the three others of its own group), until it occupies the whole posterior half of the ovary. Passage of these large ova into the brood sac occurs immediately after copulation. A portion of the brood sac surrounding them is specially modified into a thick-walled protective case termed an ephippium. Within this the two fertilized ova at first develop only as far as the stage in which mesoderm first appears. At the succeeding molt of the parent the ephippium splits apart from the rest of the cast skin, except that the spiny margins of the carapace remain attached to it. Ephippia of deep-water species of *Daphnia* float at the surface of the water, but those of *D. magna* sink to the bottom, where they often become entangled in masses by means of the curved carapace margins. Within such cast-off ephippia the winter eggs may remain in a resting state for months or even years, surviving such extreme environmental conditions as drying and freezing, and frequently becoming distributed to other ponds on the feet or feathers of birds. The young which eventually hatch are invariably females.

The male of *Daphnia magna* is only two-fifths the length of the female. It differs structurally from the female in a number of details. The antennules are longer than the labrum, and the tactile (?) seta on each is long and feathered. The anterior ventral margins of the carapace are fringed with spines and diverge widely to leave a large opening into the median space. On the tip of each first trunk appendage there are a clasping hook and an especially long seta. The two vasa deferentia open close together on a posterior papilla distal to the anus.

During mating the male clings by means of its claspers to the posteroventral margin of the female carapace, the sagittal planes of the two animals being at right angles to one another. Frequently a second male is present, clinging to the other side. Clasping may persist a day or more. In the actual act of copulation, the male abreptor is inserted into the female brood sac, and it is even possible that the terminal papilla is pressed into the opening of one oviduct.

Cyclops

By J. H. Lochhead

The taxonomy of the Genus *Cyclops* is complex, in that a considerable number of subgenera are distinguished, while some of the species are split into varieties or occur in several "forms," depending on the environment. The extent of the genus is in dispute, since certain subgenera of some authors are elevated to distinct genera by other authors. From the morphological point of view no one species has been adequately described. The present account is based on publications dealing with several species, more especially *Cyclops* (*Megacyclops*) *viridis*, *C.* (*Acanthocyclops*) *vernalis*, and *C.* (*Cyclops*) *strenuus*. Except in some small details the account should be found applicable to any species of the genus.

Among the Copepoda, the calanoids rather than the cyclopoids probably are the most primitive. But *Cyclops* is described here, chiefly because it is so easily available. Almost any small pond or lake will provide specimens, and even in midwinter some species continue reproduction. Many species readily become established in fresh-water aquaria. For experimental purposes individual specimens can be reared through all stages of the life history, in two or three drops of culture medium on a cavity slide, kept in a moist chamber. A medium based on manure, such as is commonly used for *Daphnia*, has been found suitable. The supply to the animals should be renewed daily or more often.

Living specimens can be mounted under a raised coverglass, or in a drop of water suspended in a wire loop as described by Williams. Dorsal and ventral views are easily obtained, but for a lateral view it may be necessary to cut off the antennules or to lay them back alongside the body. To kill or narcotize the specimen ether or chloroform can be used (a few drops to 10 ml. of water). If desired, the appendages can be dissected off with mounted needles, under a binocular microscope, preferably using a simple microdissector such as is referred to by Cannon. Care should be taken not to confuse copepodid larval stages with the adults. The copepodids are of adult form, but have fewer thoracic somites and are not sexually mature.

The anatomy of an adult female *Cyclops* is diagrammed in Fig. 192. *III-VI*, Third to sixth somites of the thorax; the first and second thoracic somites are fused with the head; a jointed waist occurs between somites five and six, rather than, as in the calanoids, between the thorax and abdomen. *1-4*, The four abdominal somites; the first two of these are distinct in the male but are fused in the female. *5*, Telson. *6*, Left ramus of caudal furca. *7*, Pleural fold, which extends downwards from the terga of the head and thorax. *8*, Rostrum. *9*, Antennule. *10*, Antenna, consisting of a protopodite of two segments and an endopodite of two segments; the exopodite, present in calanoids, is here absent. *11*, Mandible; this appendage has a spiny biting surface and a small palp which bears a few setae; this is in contrast to the condition in calanoids, where both the endopodite and the exopodite are well developed. *12*, Maxillule. *13*, Maxilla. *14*, Maxilliped; this is the first thoracic appendage; the position of its attachment, medial to the maxilla, is a peculiar feature. The homologies of the different lobes and segments on the maxillules, maxillae, and maxillipeds are not entirely clear. All three pairs of appendages are armed with spine-like setae and are used in feeding. Their appearance suggests a predacious feeding habit, and indeed *Cyclops* has been reported to feed on smaller Crustacea, rotifers, and sometimes even larger animals such as mosquito larvae and young fish. However, there is some evidence that the food more usually consists of diatoms, ciliates, flagellates, and possibly even fine detritus and bacteria. No filter apparatus is present, yet in some way not understood the animals do appear to be able to abstract surprisingly small food particles from the swimming current.

15, Second thoracic appendage, the first of four biramous swimming feet; each foot consists of a two-segmented protopodite, a three-segmented exopodite, and a three-segmented endopodite (not shown in the diagram); a median transverse plate unites the basal segments of each pair of swimming feet, so that the pair beats as a unit; the four pairs ordinarily are held pointing forwards, and in swimming are suddenly thrust back together. The sudden jumps which *Cyclops* makes are so rapid that it is difficult to observe the mechanism by which they are produced. It seems, however, that at least on occasion they are the result of simultaneous action by the antennules, antennae, and swimming feet, perhaps together with a flick of the abdomen and caudal furca.

The sixth thoracic appendages are much reduced, each consisting of only two small segments. A pair of first abdominal appendages perhaps is represented by small valvular plates which partly cover the openings of the genital ducts. In the female each valve is a very small plate or papilla, with one or two tiny setae. The diagram does not show this structure. In the male slightly larger plates are present, each with about three short setae.

Structures bordering the mouth are not shown in the diagram. Anteriorly there is a quite large labrum, packed with labral gland cells which appear to open dorsoposteriorly to the exterior by a common, median duct. Behind the mouth are a postoral bar and a pair of small, spiny paragnaths.

16, Anterior part of the stomach; the short, vertical esophagus and this part of the stomach together form the foregut, having a cuticular lining continuous with the exoskeleton. *17*, Posterior part of the stomach; the large cells of this region probably secrete the digestive enzymes and apparently also have an excretory function; they contain yellowish, refractile granules and at intervals are pinched off into the lumen of the gut, other cells growing to take their place; in a starved specimen the cast-off cells have been observed to clump together and then to pass posteriorly through the intestine, rectum, and anus. *18*, Intestine, belonging with the posterior part of the stomach to the midgut. *19*, Rectum or hindgut, provided with a cuticular lining and with special dilator muscles. Water is rhythmically taken into and expelled from the rectum through the anus, perhaps for respiratory purposes. Muscles attached to the front of the stomach alternately pull the stomach and intestine forwards and permit them to retract backwards, these movements being in time with those of the rectum. When retracted, the anterior part of the intestine may be coiled into a vertical sigmoid loop. These movements of the gut aid the circulation of the blood, apparently replacing the action of a heart, which is absent. Each time that the gut moves forwards, blood in the hemocoel is squeezed forwards dorsally. Each time that the gut moves backwards, blood is forced back ventrally.

The blood contains a single type of amoeboid blood cell. Often, in a ventral view of the animal, some of these cells can be seen creeping over the surface of the nerve cord and of adjacent nerves in the last two thoracic somites.

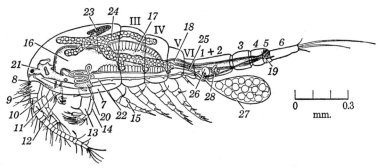

FIG. 192. Diagram of a female *Cyclops*, viewed from the left side; the internal organs appear through the transparent exoskeleton; the gut is shown as though cut in section. For explanation of the numbers see text. (Compounded from publications of several authors.)

Except perhaps for the rectum, no distinct morphological structure is present for respiration. Perhaps a respiratory exchange of gases occurs through most parts of the body surface. However, it may be that the dorsal part of the third thoracic somite is especially permeable. The exoskeleton and hypodermis in this region can be stained selectively in the living animal by a variety of oxidizable salts and leuco dyes. This may indicate respiratory activity, or it may indicate a region of active salt uptake.

20, The coelomosac of the maxillary gland. A coiled tubule leads from the coelomosac to the external aperture on the medial surface of the basal segment of the maxilla; not shown is a bladder-like enlargement of the tubule, near the external aperture. On living specimens the tubule is easily seen, but the coelomosac usually cannot be seen unless specially stained (as for example by keeping the animal for a week in a 0.5 per cent solution of trypan red). It is presumed that the maxillary gland functions in excretion, osmoregulation, or both.

Also shown on the diagram are two groups of cells which perhaps represent rudimentary coelomosacs of maxillulary and antennal glands; they are located respectively behind and in front of the esophagus, lateral to the nervous system. They stain with the same vital dyes as do the cells of the coelomosac of the maxillary gland, and perhaps normally they accumulate certain excretory wastes; the cells in the more anterior group have been observed to contain yellowish granules and to carry out slow amoeboid changes in shape.

21, The supraesophageal ganglion. Four different sense organs are shown innervated from this ganglion: dorsally, the left frontal organ, the two or more sensory cells of which are located laterally, close under the hypodermis; anteriorly, the median eye, made up of three pigment cups, two lateral and one ventral; ventroanteriorly, the left "corneal lens," which actually is not a lens for the eye, but rather a circular, thin area of the exoskeleton, under which are one to three unipolar sensory cells; ventrally, the left group of three large cells, probably sensory, located in the pleural fold. Of these four sense organs, only the eye can be seen easily without special vital staining. Keeping the animals for several days in a dilute solution of Congo red may stain the "corneal lenses." The following method has been used to stain the frontal organs: a specimen of *Cyclops* is washed in distilled water, then placed for 3–5 minutes in a 1 per cent methylene blue solution decolorized with sodium hydrosulfite, then placed in a drop of this solution on a slide until the dye has oxidized to an intense blue, and finally destained for 10–30 minutes in a lithium carbonate solution.

22, The ventral nerve cord, united with the supraesophageal ganglion by broad circumesophageal connectives. Ganglion cells occur in the connectives and in the cord as far back as the third thoracic somite, but not farther posteriorly, except perhaps for a few in the swelling in the fifth thoracic somite. Nerves for the antennae (not shown in the diagram) arise from the posterior portions of the connectives. Nerves for the fifth and sixth thoracic appendages arise in the fourth and fifth thoracic somites. Near the posterior limits of the sixth thoracic somite the nerve cord forks into left and right branches; the diagram shows the cut stump of the right branch and the full course of the left branch; this continues back to the telson, where a branch is given off to the left ramus of the caudal furca.

On each antennule one or more sensory setae of the type known as aesthetascs usually are present. They are hyaline, cylindrical setae, rounded rather than tapering at the tip. In females there often is only one such seta on each antennule, located in many species on the twelfth segment. In males the aesthetascs are typically very long (up to one-third the length of the antennule), and usually there are several.

23, Ovary, median dorsal in position. *24*, Left uterus; often the branches of the uteri are even more extensive than is shown in the diagram; in the uteri the ova acquire yolk and increase in size. *25*, Left oviduct. *26*, Seminal receptacle; the opening into the receptacle is medioventral in position. On each side a narrow, kinked sperm duct leads from the receptacle to the terminal part of the oviduct on that side. The kink probably serves to keep the duct closed, except when it is partly straightened out by a special muscle. Both the receptacle and the sperm ducts are lined with a cuticle, continuous with the exoskeleton, and they are present only after the final molt. Gland cells on the inner sides of the ducts probably secrete a lubricant. On each side a chitinous strut runs beside the duct and is easily mistaken for the latter. Another strut runs transversely between the two oviduct openings. Probably these three struts are remnants of the original joint between the first two abdominal segments. *27*, Left egg sac, attached to the vulva or opening of the left oviduct. *28*, One of a group of large gland cells, opening by fine pores in the exoskeleton and believed to secrete the material for the sac which surrounds the eggs.

The male *Cyclops* is somewhat smaller than the female and differs in several respects. Differences in the aesthetascs, in the first two abdominal somites, and in the first abdominal appendages have already been mentioned. A further external difference is in the form of the antennules, which in the male are slightly modified as claspers. Internally, the single, median testis is located in the position occupied by the ovary in the female. Spermatogenesis within the testis progresses from behind forwards, mature sperm being released anteriorly into the two vasa deferentia. The vas deferens on each side curves ventrally, then runs posteriorly as a thin-walled tube to about the level of the boundary between the third and fourth thoracic somites. Here it curves ventrally again and runs forwards in contact with the preceding part of the duct; the wall of this region is thick and the lumen narrow and sinuous. Below the anterior part of the testis the duct once more curves ventrally and now runs ventroposteriorly to one of the pair of seminal vesicles in the first abdominal somite. In the portion of the duct passing through the last four thoracic somites, the spermatozoa become enclosed within a thin membrane, together with a hyaline ground substance and some round, nucleated bodies.

The cylindrical spermatophore thus formed is then pressed into the seminal vesicle, where its walls conform to the available space, giving it its final oval or kidney-shaped form. A secretion present between the vesicle wall and the wall of the spermatophore probably serves later for attaching the spermatophore to the female.

Mating is favored by a purely chance contact between mature males and females. It can readily be observed by confining mature, unmated specimens together in a few drops of water, as for example when they are drawn into a pipette. The male at first clasps some part of the female's body by means of his antennules. Later the grip is shifted to the female's fourth pair of swimming feet (or the fourth and one or more other pairs), if the grip was not there in the first place. The two spermatophores are now removed from the seminal vesicles of the male and accurately attached close to the opening of the female seminal receptacle. Usually, or perhaps always, this transfer of the spermatophores is performed by two or more pairs of the male's swimming feet. Preliminary clasping may last only a few minutes or may persist up to a day (especially in species which commonly start clasping before the female has undergone the final molt to maturity). The actual transfer of spermatophores takes only a few seconds. A few minutes afterward the male releases his hold on the female.

Shortly after attachment of the spermatophores the spermatozoa which they contain are forced into the seminal receptacle, probably by swelling of the round, nucleated bodies. A few hours or days later the empty spermatophores drop off.

Usually about 1–5 days after mating the first pair of egg sacs are formed. On each side each of the ova emerges from the oviduct surrounded by a secretion believed to form one of the egg membranes. Spermatozoa are added as the ova pass the opening of the sperm duct, perhaps as a result of suction created by the passage of the ova. At the opening of the oviduct, under the rudimentary abdominal appendage, the ova become surrounded by the secretion of the large unicellular glands. Probably it is this secretion which is stretched and carried outwards by the emerging mass of ova, to form the membranous sac surrounding the eggs. The process of formation of the two egg sacs takes from 2 to 15 minutes.

Maturation and polar-body formation occur within the egg sac or, in some habitats, during the passage of the ova down the oviducts. After fertilization cleavage is rapid, and 1–5 days later free-swimming nauplius larvae are released.

Up to twelve or more pairs of egg sacs are produced by a single female, at intervals of 1½–6 or more days. Sometimes all these eggs are fertilized by sperm received at a single copulation.

There are six naupliar instars, followed by five copepodid instars and finally the adult instar. The adults apparently do not molt. Under favorable conditions, at room temperature, development from egg to adult takes from 8 to 50 days, depending on the species. The life span of single individuals in their natural habitat is probably from 4 to 9 months, those hatching in the early spring having the longest life. Old individuals may have broken setae, and they become covered with epizoons, such for example as vorticellids and green algae.

The adult, and more especially the copepodid instars, are able to survive in dried mud for several years, after drying up of the water in which they were living. At such times unicellular glands, especially noticeable in the swimming feet, probably secrete a protective membrane around the animal.

Lepas anatifera

By J. H. Lochhead

Lepas, one of the "goose-neck barnacles," has been chosen to represent the cirripedes here, because of its large size and relatively simple construction. However, no adequate account of its anatomy appears to have been published, and the present description and diagram have been pieced together from scattered references, not always on the Genus *Lepas,* supplemented by some personal observations.

Lepas is an inhabitant of the open ocean, where it occurs attached to floating wood or to ships. There are several species, difficult to distinguish, of very wide distribution. Occasionally pieces of wood, with attached *Lepas,* drift inshore and provide a source of live specimens. In a well-aerated aquarium such animals will live for a number of days.

In comparison with more typical Crustacea, the body of *Lepas* is very curiously modified. The long peduncle, which may be as much as 45 cm. in length, represents the anterior part of the head. This is attached by a narrow, S-shaped neck to the body proper, consisting of the rest of the head plus the thorax. An abdomen is represented only by a short caudal furca. A mantle is attached in the region of the neck and completely surrounds the body proper. This mantle develops from the carapace of the larva, although in the adult its attachment only to the preoral part of the head is in contrast with other Crustacea. Externally the mantle bears a number of calcareous plates. These are, anteriorly, a pair of large scuta; posteriorly, a median carina; and ventrally (at the free extremity of the animal), a pair of terga. The positions these plates occupy in relation to the body of the adult are the result of a complicated flexure in the neck region occurring at the time of metamorphosis. In the larva the scuta are ventral, the carina dorsal, and the terga posterior.

Inside the peduncle, near the middle of its length, is a pair of large cement glands, which, however, are difficult to see. Each consists of grape-like clusters of large cells (each 30–40 μ in diameter), from which a long duct leads to the base of the peduncle to open on a vestige of one of the larval antennules. These latter are embedded in the horny cement by which the animal is attached. They can be seen in young specimens, carefully freed from the substratum and boiled in 5 per cent potassium hydroxide. In *Lepas fascicularis* the attaching cement is inflated by gas to form a vesicular float.

A slit-like opening into the mantle chamber is present between the distal margins of the scutal and tergal plates. This can be tightly closed by contraction of the adductor scutorum muscle (*6* on Fig. 193). It also can be opened quite widely to permit extrusion of the thoracic appendages. The opening mechanism may be an erection of the mantle by blood, forced into it by contraction of muscles in the wall of the peduncle.

To see the body proper the calcareous plates should be removed with a thin razor blade, leaving as much as possible of the mantle still attached to the animal.

The six pairs of biramous thoracic appendages are all essentially alike, although the first pair is shorter than the others and set a little apart from them. Each appendage has a stout basal portion, to which are attached two long, many-jointed cirri. The cirri curve anteriorly and can be coiled up inside the mantle chamber. They are feathered with long setae, so that when extended the cirri form a net for catching plankton or particles of detritus. Attached near the base of the first thoracic appendage two "filamentary appendages" will be noted on each side, shown in broken lines on the diagram.

Undisturbed living animals carry out rhythmical "casting" movements, in which the body and thoracic appendages are alternately thrust out and then retracted. First the cirri are extended, apparently by blood pressure caused by contraction of the dorsal longitudinal body muscles. Then contraction of muscles between the oral cone and the adductor scutorum muscle causes the body to swing outwards, ventro-anteriorly, while at the same time the thorax is straightened and elongated, probably by a further increase in blood pressure. Retraction of the cirri and body follows, brought about by contraction of

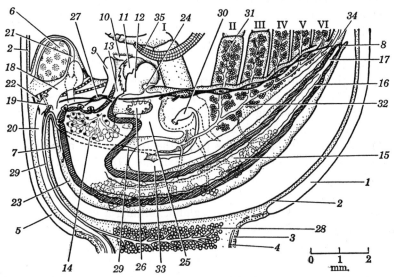

FIG. 193. Diagrammatic view of part of the right half of *Lepas anatifera*, seen from the median sagittal plane. The numbers are explained in the text. (In part original, in part compounded from publications of several authors.)

flexor muscles within the cirri and of ventral longitudinal muscles in the thorax. Food particles caught on the cirri by this casting action can be passed by them directly to the mouth parts. Sometimes quite large animals are caught, and in the laboratory *Lepas* will feed on animals even larger than itself.

The exoskeleton of the appendages, body proper, and inside of the mantle is shed at each molt, but there is no molting of the exoskeleton on the outside of the mantle (including the calcareous plates) or of that on the peduncle.

Further features of the anatomy are shown diagrammatically in Fig. 193. To conform with virtually all published figures of barnacles, the peduncle is represented as being below and the cirri above, although in nature *Lepas* normally hangs down the other way.

For a study of the internal anatomy, the easiest method is to make a median sagittal cut with a thin, sharp razor blade, although care is required to do even this approximately correctly. Under a binocular microscope many of the internal structures can then be found in one half or the other. The specimen should be submerged under water and can be manipulated with mounted needles. Some details may be better seen in a dissection from the left side, done under the binocular microscope with a thin sharp scalpel and mounted needles.

Structures shown on the diagram are as follows. I–VI, Bases of the six thoracic appendages. *1*, Carina. *2*, Mantle. *3*, Circular-muscle fibers in wall of peduncle. *4*, Longitudinal-muscle fibers in wall of peduncle; there are also some diagonal fibers. *5*, Median edge of right scutum. *6*, Adductor scutorum muscle. *7*, Region of body wall containing a network of elastic fibers, permitting stretching when the body is swung outwards during "casting." *8*, Right ramus of caudal furca.

9, Labrum. *10*, Mandibular palp; the teeth of the mandible can be seen vertically below it. *11*, Maxillule. *12*, Maxilla; the two maxillae are fused together in the midline to form a hind lip. *13*, Esophagus, or foregut, with a cuticular lining. *14*, Stomach; lateral to this is seen the right digestive gland, forming a grape-like cluster, orange in color; the dark patches are pits in the wall of the stomach, lined with very darkly pigmented cells; some of the pits are shown cut in section. *15*, Intestine; lateral to it are shown some lobes of the right testis. *16*, Peritrophic membrane. *17*, Hindgut, with a cuticular lining, terminating at the anus.

18, Median blood space of Cannon's "blood pump." *19*, Occlusar muscle; just dorsal to this the "rostral valve" can be seen at the beginning of *20*, the peduncular blood vessel. *21*, Right "scutal sinus," leading through a "scutal valve" in the blood pump (lateral to the median blood space) into the right "lateral vessel." *22*, One of the muscles the contraction of which is believed by Cannon to operate the blood pump. *Lepas* has no heart, but according to the work of Cannon, done chiefly on another genus but probably applicable to *Lepas*, there is a very definite blood circulation, aided by the contraction of muscles in various parts of the body, but especially due to alternate compression and release from compression of the blood pump. The route of circulation, as described by Cannon, is as follows: median space of the blood pump → peduncular blood vessel, almost to the attached end of the peduncle → back up the peduncle in parenchymatous spaces → sinuses in the mantle → paired scutal sinuses → lateral vessels in the blood pump → an intricate route through the body proper → back to the median space of the blood pump via a median dorsal sinus (*23* on the diagram). When prolonged high pressure is required in the body proper, as for example during erection of the penis, the flow of blood into the peduncle can be restrained by contraction of the occlusar muscle. Oxygenation of the blood probably occurs in the mantle and in the cirri; in each of the latter there are both an afferent and an efferent vessel. Perhaps the filamentary appendages are accessory respiratory structures.

24, Epidermal duct of maxillary gland. *25*, Bladder of maxillary gland, with cells of a secretory appearance. *26*, End-sac of maxillary gland, with a different type of secretory cell; the opening from the end-sac into the bladder is shown under the posteroventral corner of the stomach. The end-sac, and also some parts of the bladder, lie close to the surface, so that in a lateral view the area which they occupy appears somewhat translucent. Dissection is not likely to do more than reveal fluid-filled spaces. The shapes of the parts shown in the diagram are based on rather incomplete published descriptions. Presumably the gland is excretory, as in other Crustacea. Scattered "nephrocyte" cells have also been reported in various parts of the body.

27, Supraesophageal ganglion; leaving this anteriorly are three pairs of nerves, which run in a sheet of tough connective tissue; only the right nerves are shown in the diagram; largest and most lateral (ventral in the diagram) is the peduncular nerve which runs into the peduncle; much smaller is a nerve from the median eye; in the diagram this nerve is seen leaving the ganglion dorsoanteriorly, then disappearing lateral to the stomach, reappearing to supply the eye, located some distance dorsally on the anterior face of the stomach; a tiny ganglion is present on the nerve just before it reaches the eye; still smaller, and most medial of the three, is a nerve which ends on the anterior surface of the stomach just ventral to the eye. Despite its position deep within the body, the eye is perhaps more than a functionless remnant of a larval structure, since it has been observed that *Lepas* will react to shadows if they cross the rays of light arriving from certain directions. Leaving the supraesophageal ganglion posteriorly is a pair of circumesophageal connectives; the diagram shows the ganglion here as though it had not been sectioned medianly, so that the cut stump of the left connective is shown; the ganglia of the ventral nerve cord are shown in this same way; it will be noted that there is a ganglion in each of the first five thoracic somites; the first of these ganglia sends nerves (mostly not shown) to the mouth parts and the region of the adductor scutorum muscle, as well as to the first thoracic appendage; the last ganglion supplies the last two pairs of thoracic appendages, the penis, and the caudal furca.

28, Ovary. *29*, Oviduct; in the neck the two oviducts run close to the median line, whereas in the thorax each is close to the body surface; they are, however, very difficult to find. *30*, Glandular termination of oviduct, of bright yellow color and easily found by cutting open laterally the base of the first thoracic appendage; the slit-like female genital opening also is quite easily seen. After being laid the eggs are retained in the mantle cavity, those from each side being bound together into an "ovigerous lamella," probably by the secretion of the terminal part of the oviduct. Each lamella is attached to a long, low fold of the mantle, called the ovigerous frena, which will be found at the side of the cephalic region, near the mantle's attachment to the body. Development of the eggs proceeds in the ovigerous lamellae until free-swimming nauplius larvae are released. Quite frequently this occurs in specimens kept in aquaria.

31, Part of right testis; lobes of this occur at the sides of the intestine, in the ventral part of the thorax, in the basal parts of all the thoracic appendages, and in the posterior filamentary appendage; the lobes in the latter structure and in the base of the first thoracic appendage are omitted from the diagram. *32*, Vas deferens. *33*, Seminal vesicle; the cut stump of the left seminal vesicle is shown where the two vesicles come together, near *34*, the base of the penis. *35*, Tip of the penis. Sperm is deposited in sticky masses close to the openings of the oviducts in the mantle cavity of a neighboring individual, if there is one within reach; during erection the penis nearly doubles in length. Isolated individuals perhaps practise self-fertilization.

Heteromysis formosa

By J. H. LOCHHEAD

This species of mysid occurs in shallow water on the coasts of southern New England, Britain, and Norway. Near Woods Hole, Massachusetts, it most frequently is found in small swarms inside the shells of dead bivalves, often *Mactra*, dredged from 5 to 10 fathoms in August and September. In the females much of the body is an attractive rose red. The males are almost colorless. For a study of the general body plan other species of Mysidae may be used, since they agree with *Heteromysis* in most essentials.

Some of the more important structures are identified by numbers and letters in Fig. 194. *A* and *B*, Maxillule and maxilla, both of typical malacostracan form. *C*, First thoracic appendage. *iv*, Fourth somite of head, bearing the mandibles; just behind the mandible one of the paired paragnaths can be seen. In front of the mouth a large upper lip, or labrum, is present, not shown on the diagram. *v*, Fifth somite of head, bearing maxillules, one just visible here, squeezed between the paragnath and the exopodite of the maxilla. *vi*, Sixth somite of head, bearing maxillae, the endopodite and the exopodite of one of which are visible on the diagram. *1-8*, The eight thoracic somites; these all converge dorsally rather more than is shown on the diagram; the figure *1* is placed on the epipodite of the first thoracic appendage. I–VI, The six abdominal somites. All thoracic and abdominal somites are clearly distinct from one another. *9*, Outer flagellum of antennule, here cut short, as also are *10*, the inner flagellum of the antennule, and *11*, the flagellum of the antenna. *12*, Antennal exopodite, modified as a flattened "scale." *13*, Stalked compound eye. *14*, Rostrum. *15*, Mandibular palp. *16*, Carapace, not fused to the thorax dorsally, except to the first and part of the second somites. *17*, Exopodite of eighth thoracic appendage; the thoracic exopodites all beat upwards and backwards, creating swimming and feeding currents. *18*, Enlarged basal segment of exopodite; on the main diagram the first seven thoracic exopodites have been cut off, the cut stumps being represented by oval shaded areas. *19*, Propus of thoracic endopodite; these endopodites are used for walking and for holding large food masses. *20*, Epipodite of first thoracic appendage. *21*, Pleopod. *22*, Exopodite of uropod. *23*, Endopodite of uropod. *24*, Telson; the uropods and telson together form a broad tail fan.

Except in certain details, all the characters mentioned so far are typical for the so-called "caridoid facies," a hypothetical ancestral type from which all Malacostraca can be derived. *Heteromysis* was selected for study because it is an easily available species, lying rather close to this ancestral type. Its shrimp-like form is a further typical feature. In a number of details, however, *Heteromysis* has

420   ARTHROPODA

departed from the caridoid facies. The rostrum is reduced. Pleural folds are not developed at the sides of the abdominal somites. The pleopods are much reduced (those of the female crayfish being closer to the ancestral type). The carapace is fused to the first two thoracic somites and does not fully cover the last three thoracic somites. The endopodites of the first three thoracic appendages on each side differ

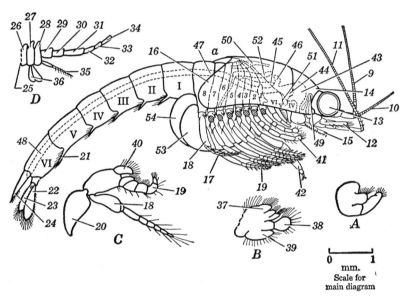

FIG. 194. Diagrammatic side view of an adult female *Heteromysis formosa*. (Compounded from publications of various authors.) *A*. Maxillule. *B*. Maxilla. *C*. First thoracic appendage. *D*. Diagram of a thoracic appendage in the "caridoid facies." The numbers are explained in the text. (*A*, *B*, and *C* redrawn from Sars.)

from succeeding endopodites, being modified for feeding; the first two thoracic appendages may be termed maxillipeds. The propus (or, according to Hansen, the carpo-propus) on the last five thoracic appendages is subdivided into several segments. The thoracic exopodites have unusually large basal segments. Epipodites are lacking except on the first thoracic appendage.

The complete thoracic appendage postulated for the caridoid facies is shown in Fig. 194*D*. *25*, The attachment to the body. *26, 27,* and *28,* Endites on the precoxa, coxa, and basis, which three segments together form the protopodite. *29,* Metabasis or preischium (uncertain whether this belongs to the protopodite or to the endopodite).

30, 31, 32, 33, 34, Ischium, merus, carpus, propus, and dactylus of the endopodite. *35,* Flagellar portion of exopodite. *36,* Divided metepipodite. Of these features the precoxa, preischium, and division of the epipodite into two parts are relatively rare in existing Malacostraca; but in *Heteromysis* Hansen identifies rudiments of precoxae and well-developed preischia.

A number of further features of interest may be noted on *Heteromysis*. *37,* The proximal endite of the maxilla, fringed with filter setae on which fine food particles collect; these particles are brought towards the animal in the feeding currents, enter between the bases of the thoracic appendages, and pass forwards in the medioventral food groove (which in females becomes divided by the brood pouch into left and right portions). *38,* Endopodite of maxilla, which helps to prevent particles from falling away from the mouth region. *39,* Exopodite of maxilla, the beating of which aids the outflow of the forward current from the food groove and of the respiratory current which entered dorsally at *a* (the outgoing current passes ventrally behind the paragnath). *40,* Proximal endite of first thoracic appendage, fringed with comb setae which remove the particles collected on the maxillary filter. *41,* Endopodite of second thoracic appendage, used, together with the endopodite of the first thoracic appendage, to help prevent food from falling away from the mouth region and sometimes to aid in pushing food towards the mandibles (normally this is done by the endites on the first thoracic limbs, maxillae, and maxillules, and by the paragnaths). *42,* Propus (or carpo-propus) of third thoracic appendage; the modification of this appendage for holding prey and larger food masses is a peculiarity of the Genus *Heteromysis;* other Mysidae, however, also are able to feed on relatively large animals, both living and dead, as well as to filter out fine suspended particles. *43,* Cardiac stomach, containing a gastric mill. *44,* Pyloric stomach. *45,* One of the longitudinal lobes of the hepatopancreas, which in life has a greenish or yellowish color; the other lobes are not shown. *46,* Dorsal midgut caecum. *47,* Intestine or midgut. *48,* Hindgut. *49,* Excretory antennal gland. *50,* Heart, of primitive elongated shape, but much shortened from the ancestral condition; two pairs of lateral ostia are present, at about the level of the third and fourth thoracic somites; arteries leave the heart approximately as in the crayfish. *51,* Line of fusion of the carapace with the cephalothorax; behind this line the branchial chamber is present; the beating of the epipodite of the first trunk limb, aided by the exopodite of the maxilla, keeps up a respiratory current through this chamber, as already described; gills are absent, respiratory exchange taking place through the inner wall of the carapace; blood lacunae in the carapace discharge into the pericardial sinus by a single opening on each side just above the ostia. *52*, Ovary; the oviduct leading down to the opening in the sixth thoracic somite is also shown. *53* and *54*, Oöstegites, of seventh and eighth thoracic appendages; these are outgrowths from the coxae; they form a brood pouch for the embryos, which develop there to a stage possessing all the appendages of the adult. An interesting feature of the embryos is their possession of caudal furcae, best developed in the penultimate stage. In the male a penis is present at the base of each eighth thoracic appendage and is unusually elongated in the Genus *Heteromysis*. The male also is distinguished by a conspicuous tuft of peculiar setae on each antennule, evidently sensory.

On the endopodite of each of the uropods a statocyst can be seen. This contains a statolith, the outer concentric layers of which are said to be composed of calcium fluoride! The eyes as well as the statocysts aid in dorsoventral orientation during swimming.

Crayfishes (and *Homarus*)

By J. H. Lochhead

Many species of crayfishes exist, belonging to two families, several subfamilies, and over thirteen genera. The preserved specimens sold for dissection in North America belong to several genera. The present account is sufficiently general to apply to any species and can also be used for the lobster (*Homarus*), which belongs to a third family placed in the same suborder as the two crayfish families. The habitats favored by crayfishes vary with the species. Most live under water, in streams, lakes, or swamps, but some live on moist land, where they construct burrows surmounted by the well-known "crayfish chimneys." Both plant and animal food is eaten by most species.

Preserved specimens, with the blood system color-injected, are most often used for laboratory study, but there are advantages to be gained from the use of living specimens. Not only do they permit observation of vital activities, but also many of the internal organs show up better in the fresh state than after preservation. The specimens can be narcotized with a saturated aqueous solution of chloroform, or they can be bled before dissection by cutting off the limbs with scissors. On specimens not bled in this way, the blood system can be color-injected, as will be described later.

Body Plan. Head and thorax are fused together to form a cephalothorax, the probable line of fusion being marked by a cervical groove. A sharp rostrum projects forward between the eyes. Covering the thorax is a hard carapace, fused dorsally to all the thoracic somites, and extending down on each side to form a gill cover, or branchiostegite. It should be noted that this is a hollow outgrowth of the body, having outer and inner epidermal walls, with a blood space between. Division of the thorax into somites is revealed by the paired appendages and by a series of midventral, triangular sternal plates, concealed in many species by hair-like setae on the limb bases. The abdomen is made up of six somites and is followed by a terminal telson. In the exoskeleton of each abdominal somite a dorsal tergum and ventral sternum are distinguished, joined together on each side by a ventrolateral projection termed the pleuron.

Exoskeleton and Hypodermis. The whole body is covered by an exoskeleton, secreted by a very thin, darkly pigmented epidermis or hypodermis. If a small portion of the hypodermis is examined microscopically, it will be seen to contain numerous chromatophores. They appear stellate in shape, with fine, branched processes when the contained pigment is dispersed, and look smaller and more rounded when the pigment is concentrated. The movements of the pigment are regulated by hormones, although in adult animals the color changes thus produced are barely visible through the thick exoskeleton. The exoskeleton is hardened by calcium salts, except at the flexible joints and in the period just after a molt. Molting of adult crayfishes occurs, in at least some species, only twice a year. Many species are restricted to waters with a calcium content above a certain minimum.

Paired Appendages. The appendages vary much in appearance, but are thought nearly all to be derived from a single basic type. Reference to the figure given under *Heteromysis* (Fig. 194) will show the parts of this basic, biramous appendage. In the simplest case there is a protopodite, divided into two segments, the coxa and basis, from the latter of which arise a medial endopodite and a more lateral exopodite. In addition, on the appendages of the head and thorax, the coxa may bear a lateral epipodite, and both the coxa and basis may be produced into medial lobes or endites. On the thoracic appendages the endopodite usually is divided into five segments, the ischium, merus, carpus, propus, and dactylus ("Can Bad Indigestion Make Crayfish Penitent Drinkers?"). Two additional segments, the precoxa and the preischium, are distinguished on some appendages by Hansen.

Because of the way some of the appendages overlap, it is best to study them in succession from behind forwards. They are numbered, however, from before backwards. To see the basal parts of most of those on the head and thorax, the branchiostegite and the ventroposterior covering of the head should be removed from one side. This can be done by cutting with scissors from the dorsoposterior margin of the branchiostegite, along a line high on the side of the thorax, then curving ventrally a short distance in front of the cervical groove. The small portion of exoskeleton of the head thus included has to be carefully freed from underlying tissues with the aid of a sharp scalpel. So far as possible the appendages should be studied *in situ*. If it is necessary to remove an appendage for closer study, this is best done by cutting with a scalpel as close to the body as possible, while pulling the appendage with forceps held in the other hand. Afterwards the corresponding limb of the other side can be studied *in situ*. Observations of limb movements on living specimens ought also to be made.

The most posterior appendages are termed the uropods. On these the endopodites and exopodites are modified as flat, rounded blades. When expanded, the two uropods, together with the telson, form a wide tail fan, by means of which the animal can dart backwards. Farther forward are the pleopods or swimmerets of the first five abdominal somites. The third, fourth, and fifth pleopods are simple, biramous, and best developed in the female, in which sex they carry the eggs; the attaching cement for this purpose is said to be secreted by so-called tegumental glands in the exopodites and endopodites, and sometimes adjacent regions. In both sexes the pleopods beat so as to cause a current of water, but the functional significance of this is uncertain. The second pleopods resemble the succeeding three pairs in the female, but in the male crayfish are modified as stiff copulatory appendages. In the male *Homarus* the only modification of the second pleopods is a stiff extra process on each, medial to the endopodite. The first pleopods are much reduced in the female (in certain species sometimes absent), while in the male they again are modified as stiff copulatory appendages. (In crayfishes of the southern hemisphere the first pleopods are absent in both sexes.)

On the thorax the last five pairs of appendages are called pereiopods. They all are uniramous, exopodites being absent. Each of the first three pairs terminates in what is termed a chela, so constructed that the dactylus can bite against a prolongation of the propus. The term cheliped is appropriate for these three pairs of appendages but usually is restricted to the large first pair. This pair comes into use chiefly as weapons of defense or in predatory feeding, while the two smaller pairs are used for cleaning the body and for picking up small particles of food. The large first pereiopod, as compared with succeeding pairs, has undergone some torsion, bringing the dactylus into a medial rather than a lateral position. A clear concept of this torsion can be gained by determining the axis of flexure at each joint of the first two pereiopods and indicating these axes by means of pins appropriately inserted. Incidentally it will be found that the line of flexure differs at every joint, a feature which gives the limbs a wide range of mobility. The large chelipeds further differ from the other pereiopods in that the basis and ischium are firmly fused together (though occasionally this condition is observed on the other pereiopods too). Perhaps associated with this fusion is the fact that it is especially the large chelipeds which can be cast off, or autotomized, when the limb is wounded or firmly held. The break on such occasions always occurs at a specific fracture plane in the basi-ischium, possibly representing the plane of fusion between the two segments, though more probably lying just distal to this. On the remaining pereiopods, fracture planes also occur, just distal to the joint between basis and ischium, or, in Hansen's scheme, at the line of fusion between the ischium and preischium. Autotomy of these limbs, however, is relatively rare, and lobsters seem unable to cast them off without an assisting pull applied distally. After autotomy of any limb, loss of blood is minimized by a diaphragm and valve which may be seen closing the end of the stump. Other features of interest to observe on the pereiopods include special setae and spines on the dactylus and propus of the fifth or fourth and fifth legs, used by the female to clean the abdomen and pleopods before egg-laying; also, in males of certain species of crayfish, large spiny projections from the ischia of the third and/or sometimes an adjacent pair of pereiopods, used to lock under the legs of the female during mating. The male and female genital openings also should be noted, located on the medioposterior borders of the coxae of the fifth or of the third pair of pereiopods respectively.

426   ARTHROPODA

The three anterior thoracic appendages, anterior to the pereiopods, are called maxillipeds. Of these, the third pair forms a protective cover for the other mouth parts; it also partially closes off the channel in which the respiratory current flows. Its endopodites assist in tearing and manipulating the food, often passed to them by the chelate second and third pereiopods. Its exopodites, together with those of the other two pairs of maxillipeds, often beat in such a way as to give further impetus to the emerging respiratory current. In the lobster the coxa of each third maxilliped is produced into two small endites. The second maxilliped on each side resembles the third but has a much smaller endopodite; there are no endites. The first maxilliped has a much-reduced endopodite, two large endites, on the coxa and basis respectively, and a leaf-like epipodite folded longitudinally.

Each thoracic appendage, except the first and the last, bears a gill, attached to the coxa. Further details concerning these will be given under the respiratory system.

The head bears the five pairs of appendages typical for Crustacea. Proceeding forwards, these are as follows: the maxillae (or second maxillae), the maxillules (or first maxillae), the mandibles, the antennae (or second antennae), and the antennules (or first antennae). On the maxilla of each side, the coxa and the basis each bears a prominent two-lobed endite; there is a small, unsegmented endopodite, and the exopodite is much expanded dorsoposteriorly or is fused with an epipodite, so as to form a leaf-like flap termed the scaphognathite or gill bailer. This structure, which is important for respiration, partially fits into a groove formed by the longitudinally folded epipodite of the first maxilliped. The maxillule is small and curves closely around the mandible; it lacks an exopodite and epipodite, has a small endopodite, and has relatively large endites attached to the coxa (or precoxa) and to the basis. The mandible has a small, three-segmented palp, of which probably the two distal segments represent the endopodite, while the proximal segment represents the basis or the fused basis and coxa. The much larger biting portion then represents either the coxa or the precoxa. Not much actual chewing is done by the mandibles, which serve rather to roll food into the mouth and to hold it firmly while fragments are torn off by the third maxillipeds. On the antenna there are a coxa, a basis, a scale-like exopodite, and an endopodite consisting of three large segments plus a long, terminal flagellum; the most proximal segment of the endopodite appears to be divided longitudinally, permitting greater flexibility of movement; a rounded projection on the ventral surface of the coxa contains the opening from the green gland; the left and right exopodites can be spread out to act as horizontal rudders, especially during a downward dive through the water. The antennule consists of three basal segments, followed by two small flagella; these are not thought of as being exopodite and endopodite, since the primitive crustacean antennule was uniramous.

Respiratory System. On each side of the thorax a gill chamber is present, bounded externally by the branchiostegite, internally by the body wall, and to a large extent closed ventrally by the appendages. Small chinks between the appendages open into the chamber ventrally, but the only considerable openings are one at the ventroposterior angle of the branchiostegite and a channel which leads out near the anterior end of the scaphognathite. In a living specimen the rapid beat of the scaphognathite can be observed after cutting out a narrow strip of the branchiostegite just posterior and parallel to the ventral end of the cervical groove. A strong current of water is driven forward in the channel leading out from the gill chamber, by the alternate beating of the posterior and anterior portions of the scaphognathite. Water enters the gill chamber mainly through the posterior opening, as can be demonstrated by introducing from a pipette a few drops of carmine suspension near this opening. Once every few minutes the scaphognathite reverses its beat for a few strokes so as to wash loose silt or other debris, which may have settled on the gills. Hair-like setae on the inner wall of the branchiostegite perhaps aid in cleaning the gills that brush against them.

The gills are of the type known as trichobranchs, each consisting of numerous filaments arranged around a central axis, somewhat like a bottle brush. Depending on their position of attachment, three series of gills are distinguished, podobranchs, arthrobranchs, and pleurobranchs. On a single thoracic somite the following gills may be present: one podobranch, arising from the coxa of the appendage; two arthrobranchs, arising one just ventroanterior to the other, both from the membranous joint which attaches the coxa to the body; and one pleurobranch, arising higher on the side of the body. In addition there may be a leaf-like lamella, or mastigobranch, accompanying the podobranch, and a tuft of coxal setae (sometimes termed a setobranch). Not all these structures are present on all thoracic somites. The actual distribution is shown in the table on p. 428.

ARTHROPODA

	Podo-branch	Mastigo-branch	Anterior Arthro-branch	Posterior Arthro-branch	Pleuro-branch	Coxal Setae
2nd maxilliped	1	1[1]	1[3]			
3rd maxilliped	1	1[1]	1	1[4]		1
1st pereiopod	1	1[1]	1	1[4]		1
2nd pereiopod	1	1[1]	1	1[4]	1[2,5]	1
3rd pereiopod	1	1[1]	1	1[4]	1[2,5]	1
4th pereiopod	1	1[1,2]	1	1[4]	1[2,5]	1
5th pereiopod					1[2]	

[1] Not in crayfishes of the southern hemisphere.
[2] Not in Cambarinae.
[3] Not in *Homarus;* rudimentary in some crayfishes of the southern hemisphere.
[4] Rudimentary or absent in some crayfishes of the southern hemisphere.
[5] Rudimentary or absent in most crayfishes.

The axial stems of the podobranchs perhaps represent modified epipodites, especially since in most crayfishes of the southern hemisphere a few gill filaments occur on the epipodite of the first maxilliped. It has even been suggested that the arthrobranchs and pleurobranchs are derived from epipodites of precoxae which have fused with the body. However, at least in their appearance, it is the mastigobranchs rather than the gills which most closely resemble epipodites. In crayfishes the lamella of each mastigobranch is longitudinally pleated and is fused along most of its length with the axis of the accompanying podobranch. In lobsters the mastigobranchs and podobranchs are separate. No function for the mastigobranchs is known. The coxal setae probably help to close off the gaps between appendages which lead up into the gill chamber.

Circulatory System. Parts of this system are among the first items to be encountered in an internal dissection. Thus it seems desirable to describe the circulatory system now, although some parts will not be seen until later. The internal dissection is best done with the animal under water, particularly when dissecting very soft structures, such as the hepatopancreas and gonads.

If a living specimen is being used, the circulatory system may be color-injected. This is best done by boring a hole mid-dorsally in the carapace, halfway between the posterior border and the cervical groove, and using a syringe needle just large enough to fill the hole. This method usually fills the pericardial sinus and branchiocardiac sinuses, as well as the heart and arteries. If it is desired to avoid having the heart surrounded by the color mass, it is possible to inject the heart directly, using a very fine needle, after first having cut a small window through the carapace and dorsal wall of the pericardial sinus.

To see the heart and neighboring arteries, it is necessary to remove the carapace and the dorsal exoskeleton of the head. For this, suitable lateral scissors cuts should be made and the exoskeleton then carefully loosened from underlying tissues with a sharp scalpel, starting at the posterior margin of the carapace. Considerable care is necessary not to damage certain arteries and muscles located close under the hypodermis. The dorsal exoskeleton of the abdomen should be removed in the same way, starting at the posterior end.

The heart lies in a space called the pericardial sinus. It is suspended by three elastic ligaments on each side and is stretched laterally by these ligaments during diastole. Blood enters the heart from the pericardial sinus through valved slits, termed ostia, of which there are a dorsal, a lateral, and a ventral pair. Thin-walled arteries, whose connections with the heart are very easily broken, leave the heart anteriorly and posteriorly. An internal valve is located at the start of each. From the anterior dorsal margin of the heart a slender ophthalmic artery runs forward in the midline, close under the hypodermis. On the anterodorsal surface of the stomach, not far behind the eyes, this artery expands slightly to form a structure termed the cor frontale, then forks into branches which supply the brain, eyestalks, and antennules. Two tiny muscles, the posterior basal ocular muscles, arise by tendons from the dorsal integument, close on either side of the midline, a short distance posterior to the cor frontale. These two muscles run ventroanteriorly, converging towards each other, pass through the wall of the cor frontale, and emerge to join a median tendon which runs ventrally to its insertion on the integument between the antennal coxae. The belly of each muscle lies within the cor frontale, and the swelling of these two muscles when they contract must exert pressure on the surrounding blood. It is thought that the whole arrangement acts as an accessory heart, increasing the flow of blood to the brain, eyes, and antennules. A further remarkable feature of the cor frontale is that it also encloses a part of the stomatogastric nervous system (to be described later). A tiny pair of anterior basal ocular muscles arises between the eyestalks and inserts on the same median tendon as does the posterior pair, but without traversing the cor frontale.

A large antennary artery leaves the heart on each side close to the origin of the ophthalmic artery and runs lateroanteriorly to supply the gonad, branchiostegite, body wall, thoracoabdominal muscles, hepatopancreas, stomach and its muscles, adductor muscle of the mandible, antennal gland, antenna, antennule, eyestalk, rostrum, and brain. From the ventroanterior margin of the heart a pair of hepatic arteries originates, each running a course parallel to the more dorsal antennary artery, and supplying the hepatopancreas. Ventroposteriorly a median dorsal abdominal artery leaves the heart. The beginning of this artery is swollen as a thin-walled bulb, from which a pair of arteries arises supplying the gonads and genital ducts. The root of one of these arteries is much enlarged, and from this root a large branch, the sternal artery, which will be considered presently, passes vertically downwards. From the thin-walled bulb the dorsal abdominal artery runs posteriorly, dorsal to the intestine, giving off a pair of lateral segmental arteries in each abdominal somite to supply the intestine, posterior parts of the hepatopancreas, abdominal muscles, pleopods, uropods, and telson. The sternal artery passes either to left or right of the intestine and pierces the ventral nerve cord, below which it divides into a small, ventral abdominal artery and a larger, ventral thoracic artery. The ventral abdominal artery supplies the nerve cord, the last two pairs of pereiopods, and a small portion of the ventral part of the abdomen. The ventral thoracic artery supplies the nerve cord, first three pairs of pereiopods, the mouth parts, and portions of the antennal glands.

All the arteries branch repeatedly, the fine, ultimate branches sometimes being termed capillaries, because they often have a diameter as small or even smaller than that of a mammalian capillary. However, such a use of the term seems inadvisable, since we have no proof of physiological exchanges through the walls of these vessels, nor do they connect with venules as do true capillaries in a closed blood system. Instead, they open into tissue spaces, in which the blood can bathe the tissue cells directly. Some of these spaces are no more than the chinks between cells. These, however, are continuous with wider spaces, termed sinuses, as well as with the main cavity of the body. All these spaces develop from the primary body cavity, and since they contain blood they are collectively spoken of as the hemocoel. In the hemocoel the blood follows a more or less definite course of circulation, eventually reaching, in the thorax, the sternal sinus, which surrounds the ventral nerve cord and ventral thoracic artery. From here the blood enters the gills, passing up the outer side of each gill in an afferent branchial sinus, then descending on the inner side of the gill in an efferent branchial sinus. A small branch from the afferent sinus leads out to a superficial plexus in each gill filament, from where a channel starting in the center of the filament returns the blood to the efferent sinus. From the gills the blood on each side of the body passes to the pericardial sinus through six branchiocardiac sinuses (five in the lobster). These run dorsally within the body wall, which in this region is very thin and membranous, being protected by the overlying gills and branchiostegite. From the pericardial sinus the blood enters the heart, as already described.

The nervous supply to the heart is described under the nervous system, but if only a single specimen is available and it is desired to see these nerves, they should be looked for before the dissection has progressed too far.

The blood may have a slightly bluish tint in life, imparted to it by a respiratory pigment, hemocyanin, dissolved in the plasma. Three types of blood cells are present: (1) explosive corpuscles, which initiate blood-clotting by an explosive cytolysis; (2) thigmocytes, which phagocytize foreign particles and which further contribute to blood-clotting by throwing out long, fine processes; (3) amoebocytes, which are much less phagocytic than the thigmocytes, but which display considerable amoeboid activity, as can be observed in a gill filament of a living animal. Blood allowed to flow out under water usually coagulates rapidly, the clot having the consistency of a soft gel. The blood cells are derived, at least in part, from a pair of blood-cell-forming organs on the anterodorsal surface of the cardiac stomach, as well as from small nodules in the connective tissue in various parts of the body. The organs on the stomach are not easily seen, other than histologically. They are thin, semitransparent, and formed of discontinuous lobules in the crayfish, but are more solid and continuous in the lobster.

Muscular System. Muscles will be considered now, because many of them have to be cut during the dissection of other structures. Only the most important ones will be mentioned. Some will not be seen until later in the dissection, but are nevertheless included here.

Most noticeable are the muscles of the abdomen. Of these, the relatively slender extensors originate dorsolaterally in the thorax, a short distance behind the cervical groove, and run back through the pericardial sinus, to insert on the six terga of the abdominal somites. The much more powerful flexor muscles originate near the floor of the thorax, along most of the length of the upper surface of an exoskeletal invagination, the endophragmal shelf; in the abdomen the fibers become arranged in groups and follow a complexly twisted course, inserting on the six sterna and in the telson. These muscles are able to act as flexors because their points of insertion lie ventral to the exoskeletal hinges between the successive abdominal somites. The hinges are placed on the sides of the abdomen, at the level where the terga and pleura meet. If the exoskeleton is carefully chipped away around one of these hinges, a small ball-and-socket type of joint will be disclosed. The great length of the abdominal flexors probably favors the strength of their contraction, while their subdivision into separately twisted bundles may serve to reduce the friction between fibers.

On each side of the stomach is the large adductor muscle of the mandible. This originates dorsally by a large head attached to the exoskeleton, and quickly tapers to a strong tendon running anteroventrally, to insert about midway along the posterior border of the mandible. Contraction of the two adductors rotates the mandibles in such a way that they bite together.

Connecting the stomach with the exoskeleton are a number of paired muscles. Largest of these are three pairs chiefly responsible for the action of the teeth in the gastric mill (q.v.), namely the anterior, posterior, and posterolateral gastric muscles. These are all dorsal in position. A number of pairs of much smaller, dilator muscles also run between the stomach and various parts of the exoskeleton. In the wall of the stomach itself numbers of small, intrinsic muscles run between various ossicles. One author maintains that these serve chiefly to bring about movements whereby the teeth and other internal parts of the gastric mill are kept clean.

Under the endophragmal shelf will be found the proximal muscles of the pereiopods. Most of them have their origin on parts of the endophragmal skeleton (q.v.). Two small muscles insert on the proximal rim of each coxa, four insert on the proximal rim of the basis of each large cheliped, and two insert on the proximal rim of the basis of each of the other pereiopods. It is of interest to note that the muscle used in autotomy is the levator anterior basis in the large cheliped, and simply the levator basis in each of the other pereiopods.

This muscle can best be found by cutting a window in the dorso-anterior part of the coxa, enlarging the window if necessary after the muscle has been seen. The autotomizer muscle in the cheliped inserts by two separate tendons.

The more distal muscles of the pereiopods both originate and insert within the appendages. Some of them have provided material for important studies of neuromuscular physiology, referred to briefly under the nervous system.

Reproductive System. To display the gonad it is necessary to remove the heart and the tough, membranous floor of the pericardial sinus. In the female, care is necessary not to damage the thin-walled oviducts which lie immediately below the pericardial floor and which may be confused with the transverse bands of fibers in that floor.

In the crayfish there is a single, Y-shaped gonad, very small in immature specimens, becoming much larger at maturity, particularly in the female. The paired lobes of the gonad extend forwards on either side of the stomach, while the unpaired lobe may extend back a short distance into the abdomen. In a male the testis often resembles the hepatopancreas (q.v.) in its whitish color and soft consistency, and thus rather easily becomes indistinguishable if at all squashed during dissection. In some species of crayfishes the three distal extremities of the testis have the form of little, bulbous swellings, of which the posterior one sometimes is found twisted below the intestine.

In the lobster the ovary is H-shaped, the short cross bar being placed just in front of the pericardial sinus. The paired longitudinal parts may extend two-thirds the length of the animal.*

The testis of the lobster also is H-shaped, or the cross bar may be lacking so that there are separate paired testes.

* In fresh lobsters the color of the ovary provides interesting evidence regarding the reproductive history of the individual. For 4 or 5 years the ovary slowly matures, going through color phases of bright yellow, flesh color, salmon, light olive green, and a rich dark green. Copulation then occurs (immediately after a molt), and 1 or 2 months later the eggs are laid. The ovary now is grayish white with green flecks (the remains of eggs that failed to be discharged). After 5 or 6 weeks the ovary turns light green and the flecks become orange. The eggs which were laid are retained on the pleopods for about 11 months, at the end of which time the larvae hatch. Shortly after that there is another molt and copulation again occurs. Meanwhile the shade of green of the ovary steadily darkens, until a second batch of eggs is laid, 2 years after the first laying. The ovary is again grayish white for a few weeks, but this time has both orange flecks and green flecks (remains of ova which failed to be discharged at the first and second layings respectively).

The oviducts are paired and start below the pericardial sinus, each passing outwards to the body wall, inside of which it then runs downwards to open on the inner aspect of the coxa of the third pereiopod. Each duct is wide and ribbon-like as it leaves the ovary, but quickly tapers to a small cylindrical tube. Often the ducts are overlooked because they are thin-walled and transparent. But at the time of egg-laying the walls actually become quite glandular, contributing an albuminous or mucilaginous secretion.

The vasa deferentia are paired and leave the gonad at the same level as do the oviducts. In the crayfish each vas deferens is a small, coiled, relatively thick-walled tube, often pinkish in color, which opens at the tip of a flexible papilla on the inner surface of the coxa of the fifth pereiopod. Usually the coils are especially numerous in the upper part of the duct, and care is needed not to be deceived as to the point where the duct leaves the testis. The walls of the duct secrete a gelatinous material which surrounds the sperm as it passes out. In the lobster the duct first runs outwards from the testis to the body wall, then curves ventrally and thickens along its dorsoposterior margin. This thickened part of the duct is glandular and coils in an S-shaped fashion. It is followed by a small swelling, representing a sphincter, and this in turn by a thinner ductus ejaculatorius. Microscopic examination of some of the contents of the vas deferens may reveal some of the peculiar spermatozoa, small rounded cells, from the periphery of each of which radiate a number of long, fine, curved processes.

A seminal receptacle is present between the bases of the fourth and fifth pairs of pereiopods in the female of the lobster and of certain crayfish genera. This is not an internal structure but is simply a small pocket in the exoskeleton. Frequently it is termed the annulus. It is thought to have evolved independently in the lobster and in crayfishes respectively.

During copulation the male holds the female on her back by means of his chelipeds and other pereiopods, aided in some crayfishes by the spiny projections on certain of the ischia. The copulatory pleopods, which normally lie forwards tightly pressed against the ventral surface of the thorax, are lowered to about a 45° angle, and then are held in that position by one of the fifth pereiopods carefully crossed under the body. The flexible papillae, on which the vasa deferentia open, fit into the grooves of the first pleopods, and issuing sperm, surrounded by a gelatinous secretion, flows along these grooves. If there is a seminal receptacle, the tips of the pleopods are inserted into this and the sperm passes in, being securely retained there by the

CRAYFISHES (AND HOMARUS) 435

gelatinous secretion which hardens into a plug. Where there is no seminal receptacle, the sperm is deposited in packets, termed spermatophores, on various parts of the female's body, but particularly between the bases of the last two pairs of pereiopods. In crayfishes copulation occurs when both male and female have completely hardened exoskeletons, but in the lobster copulation always occurs shortly after a female molt, when her exoskeleton is soft or at most of a leathery consistency. In many species of crayfishes the first pleopods of the mature males alternate between two forms at successive molts, only one of these forms being adapted for copulation.

Egg-laying occurs quite some time after copulation: 10–45 days later in some crayfishes without any seminal receptacle, a few days to several months later in crayfishes with a seminal receptacle, and up to a year or more later in the lobster. The sperm remain viable all this time, and indeed have been found to survive at least 2 years in the receptacle of the lobster. During egg-laying the female lies on her back, with the abdomen curled forwards so tightly that the space in which the pleopods occur is closed except for an opening in front. The eggs emerge from the oviducts in a mucilaginous secretion and are carried back into the temporary brood chamber by a current of water caused by the beating of the second and perhaps other pairs of pleopods. The ova are fertilized by sperm released by an unknown mechanism from the receptacle or from spermatophores, and are attached to the pleopods by a secretion which hardens into a cement. After some weeks or months, during which the developing embryos are continually supplied with fresh water by movements of the pleopods, the young hatch, as schizopod larvae in the case of the lobster, or as miniatures of the adult in the case of a crayfish. Newly hatched crayfish remain attached to the maternal pleopods for a further short period of time. Molting of ovigerous females is specially delayed until after the release of the young.

Digestive System. The digestive tract is divided into a foregut, comprising the esophagus, cardiac stomach, and pyloric stomach; a midgut, comprising the hepatopancreas and part of the intestine; and a hindgut, comprising the rest of the intestine and the rectum. The internal epithelium of the foregut and of the hindgut is ectodermal and secretes a cuticular lining which is cast at each molt.

All parts of the digestive tract should first be studied *in situ* in the animal. To facilitate this study the body wall of the cephalothorax should be cut away on one side, if this has not already been done. Care has to be taken to avoid breaking the delicate connections of the hepatopancreas and intestine behind the stomach. After a general study has been completed, the different portions of the digestive tract may be removed from the body for study of parts not otherwise visible. However, unless a second specimen is to be available, this should not be done until certain parts of the nervous system (q.v.) have been studied. The stomach and the hindgut should be slit open midventrally, and their internal structure observed under water. A small part of the hepatopancreas should be mounted for study under the microscope.

From the mouth the short, vertical esophagus ascends to open into the floor of the cardiac stomach. The opening is guarded by a median anterior and a pair of lateral valves, all beset with setae pointing into the stomach. On the anterior wall of the cardiac stomach a large oval area is present on each side; before each molt this area secretes a calcareous disk known as a gastrolith. Immediately after the molt the two gastroliths, being no longer protected by the overlying cuticular lining, are dissolved in the stomach. Probably the calcium salts thus released are absorbed and used for prompt hardening of certain essential structures such as the mandibles, tips of the chelae, and teeth of a structure termed the gastric mill.

It is as the site of this gastric mill that the cardiac stomach is chiefly important. The grinding elements of this mill are three teeth, one dorsal and two lateral, much more used for chewing the food than are the mandibles or other mouth parts. The mechanism for moving the teeth and other parts of the cardiac stomach is very complex, involving some thirteen pairs of muscles and a framework inside the stomach wall composed of over two dozen distinct plates, ossicles, and bars. Small setae on many parts of the stomach lining point towards the region where the three teeth bite together.

In the cardiac stomach the food not only is ground up but also is digested by enzymes from the hepatopancreas, which arrive probably via route 3 described in the next paragraph. Only after the food has been ground into fine particles or is in solution can it pass into the next division of the stomach. Larger particles have to be rejected through the esophagus.

A constriction marks the boundary between the cardiac stomach and the much smaller pyloric stomach. Internal folds and ridges of the stomach lining, beset with numerous setae, form a complex system of filters and channels, through which small particles are admitted from the cardiac stomach into the pyloric stomach and thence either into the intestine or into the left or right hepatopancreas. Several main routes are present. (1) A route which starts just behind the dorsal tooth and passes back as a dorsal groove of the pyloric stomach, termed the midgut filter, which presently divides into left and right channels opening dorsally into the midgut. (2) A pair of routes each of which starts below one of the lateral teeth, curves laterally around a medioventral projection termed the cardiopyloric valve, then joins with its fellow behind this to pass back medianly between a pair of plates which forms a so-called muscular press, about midway between the upper and lower limits of the pyloric stomach; from here this route opens into the midgut. (3) A route which starts behind the esophageal opening, as a ventral groove of the cardiac stomach; this groove soon forks and each division passes ventrolaterally around the cardiopyloric valve, and enters one of the two so-called gland filters, which lie within a pair of prominent ventral swellings of the pyloric stomach; at this point materials in solution, or particles small enough to pass between the tiny setae in the gland filters (spaced about 10 μ apart), pass on into channels leading to the openings of the left and right divisions of the hepatopancreas; particles too large to pass through the gland filters can pass dorsally by converging channels which join route 2 in front of its entrance into the midgut.

The walls of the pyloric stomach are stiffened internally by some nine plates and bars, complexly articulated with one another. About half a dozen muscles serve to regulate the size of the various channels and to squeeze particles through some of the filters.

The region which appears to be the posterior part of the pyloric stomach is the beginning of the midgut. This lacks a cuticular lining and possesses a cuboidal rather than a columnar type of epithelium. Its junction with the pyloric stomach slants ventroanteriorly. Into it a very delicate and easily broken hepatopancreatic duct opens on each side, just dorsoposterior to the swelling of the pyloric stomach which houses the gland filter. Four valvular flaps, dorsal, lateral, and ventral, project into the midgut from the pyloric stomach. A short caecum projects dorsoanteriorly from the midgut, pressed so closely against the pyloric stomach that it frequently is overlooked. In crayfishes, there usually can be seen, just posterior to this point, a slight, transverse ridge which marks the end of the midgut and the beginning of the hindgut. The latter runs back as the intestine almost to the telson, where it enlarges slightly as the rectum. The anus opens on the ventral surface of the telson. The internal surface of the hindgut is thrown into six longitudinal folds and is covered throughout with a cuticular lining. In the lobster the condition is rather different.

Here the midgut includes the whole intestine, which runs back to the last abdominal somite, where it meets the hindgut modified as a relatively large rectum; a posterior midgut caecum is given off dorsally just before the junction with the rectum and runs back most of the length of the sixth abdominal somite. In larval lobsters the midgut is said to be short and the hindgut long, as in crayfishes.

The very large hepatopancreas is divided into left and right divisions, although these are united under the pyloric stomach. The two divisions represent outgrowths from the midgut and are invested by a very thin membrane. When this membrane is broken, the tiny terminal tubules which make up the bulk of the organ will be seen to fray out if the specimen is under water. The total internal surface offered by these tubules is very large, since they are extremely numerous and the two divisions of the hepatopancreas extend nearly the whole length of the cephalothorax and into part of the abdomen. The color of the hepatopancreas varies in fresh specimens from whitish, through yellowish, to various shades of green, depending on the season and the stage of the molt cycle.

Histologically the tubules and ducts of the hepatopancreas show four types of cells. These are (1) numerous columnar cells with a striated border, important in absorption and in storage of reserve food substances; (2) less abundant, vacuolated, often pear-shaped cells, which secrete the digestive enzymes; (3) still less frequent, tiny cells near the basement membrane, which are thought to replace spent gland cells; (4) circular and longitudinal muscle fibers said to occur outside the basement membrane, but of such minute size as to be almost undetectable in ordinary sections.

The hepatopancreas has several important functions. It is the chief site of absorption, the absorptive surface of the rest of the midgut being very small. It secretes all the digestive enzymes. Its cells store reserve food substances in quantities varying with the seasons and with the stages of the molt cycle. Probably it acts to regulate the amount of sugar and possibly other constituents in the blood. It is the chief organ of excretion for dialysable dyes injected in excess into the blood. Its cells contain a complete set of enzymes for breaking down purines through uric acid and urea to ammonia.

Excretion and Osmoregulation. The principal organs for excretion are the antennary glands, frequently referred to as the green glands. They are located, one on each side, anterior to the ventral portion of the cardiac stomach. In the crayfish each consists of a ventral, glandular part, connected posterodorsally with a more dorsal bladder, which drains anteriorly into a duct leading ventrally to the opening already seen on the ventral surface of the antennal coxa. The glandular part is firm and shaped like a planoconvex lens, with the flattened surface directed upwards. The bladder is likely to be overlooked, since it usually is collapsed and its walls are very thin and membranous. The distended bladder may be considerably larger than the glandular part and pressed posteriorly against the stomach, hepatopancreas, and other organs.

The glandular part has a complex internal structure, difficult to make out completely even in sections. It begins blindly as a dorso-anterior coelomosac, having complexly infolded walls. This part opens anteriorly into the labyrinth, made up of a maze of channels occupying the peripheral parts of the organ. The labyrinth ultimately opens into a long and very much convoluted tubule, which occupies most of the central portion of the organ and leads finally to the opening into the bladder. If the entire gland is removed to a watch glass, the limits of its main regions usually can be recognized.

The cells of all three of these divisions of the glandular part of the nephron are secretory in appearance. Branches from the antennal and ventral thoracic arteries provide an unusually copious blood supply, their minute terminal branches discharging into tiny hemocoelic spaces which surround all the ramifications of the gland. The bladder is non-secretory and receives from the antennal artery a much more restricted arterial blood supply.

It is reported that no valves or sphincters occur between any of the internal parts of the gland. A valve is present only at the external orifice. Here a soft operculum covers most of the external opening. However, the actual excretory pore is only a thin, crescentic slit near the anterior border of this operculum, the concavity of the crescent being directed forwards. Any attempt to inject or inflate the bladder will fail, unless the needle or tube is aimed accurately at the crescentic slit rather than at the operculum as a whole.

In the antennal gland of the lobster certain important differences must be noted. The coelomosac occupies most of the dorsal aspect of the glandular part, with the labyrinth surrounding it peripherally and extending under it ventrally. A tubule is absent, probably correlated with the fact that the lobster secretes a scanty, isotonic urine, in contrast to the copious, hypotonic urine secreted by the fresh-water crayfish. From the anteriomedial portion of the labyrinth a white projection extends ventroposteriorly; this projection contains a number of parallel tubes which open into the duct from the bladder near its external orifice. There is no direct opening into the bladder, and thus the bladder can be filled only by the backing up of urine through its duct. This duct runs downward from the bladder in front of the glandular part of the organ, then turns and runs ventroposteriorly to the external opening, located in the center of the operculum.

The work of the antennal glands in excretion and osmoregulation is supplemented by the hepatopancreas and the gills. The roles of the hepatopancreas in excretion have already been mentioned. A passive exchange of ions and water molecules occurs between the blood in the gills and the surrounding water, and there is evidence also that the gills are able to take in certain ions against a concentration gradient. When dyes which are non-dialysable or of large molecular size are injected into the blood, they are accumulated by certain cells in the gills, as well as by the cells of the coelomosacs in the antennal glands.*

Endoskeleton. The exoskeleton not only is continued internally as linings for the fore- and hindgut, but also projects internally as so-called apodemes, secreted by inpocketings of the epidermis. Collectively, these apodemes form the endoskeleton. Some of them take the form of tendons for the attachment of certain muscles. Others are in the form of hard, skeletal parts. Of these the most conspicuous are those in the floor of the thorax, which form a structure called the endophragmal shelf. This shelf is a latticework made up of curiously shaped apodemes, of which two pairs project internally at each junction between somites. The median pair forms an arch over the sternal sinus, in which the nerve cord and the ventral thoracic artery lie. Each member of the lateral pair is Y-shaped when viewed from above, with the stem of the Y directed outwards and the arms of the Y meeting the tops of successive endosternal arches. Functions of the endophragmal shelf are to protect the ventral nerve cord, to contribute one component in the joint for the articulation of each of the pereiopods, and to provide attachment for a number of muscles, particularly those of the appendages below, and the large flexor abdominis above.

Before each molt certain parts of the endoskeleton are resorbed, making the molt mechanically possible. All other parts of the endo-

* The injection of dyes into the hemocoel can provide a striking demonstration of the separate identity of the parts of the antennal glands. Suitable for the purpose in a large crayfish would be 0.25 ml. each of 0.5 per cent aqueous solutions of cyanol and Congo red. Examination 14-24 hours after injection should show the lumen of the labyrinth to be a vivid blue, and the cells of the coelomosac to be a bright red.

skeleton are cast at the molt along with the exoskeleton to which they remain attached.

For those who wish to study in detail the endoskeleton and the ossicles of the stomach, three types of preparation can be employed. Boiling in 5 per cent potassium hydroxide to remove all non-skeletal tissues is the method of preparation most often used. This has the disadvantage, however, of removing the cuticulin from the skeleton, leaving only the underlying chitin. A perfect molted skin is superior in this respect, but lacks a few connecting parts of the endoskeleton. If a really complete specimen is desired, it can be obtained by prolonged tryptic digestion, followed by soaking in a fat solvent.

Nervous System. Time may not permit identification of some of the smaller nerves to be described here. They usually can be seen, however, in a fresh specimen submerged under water, observed with the aid of a binocular microscope. If preferred, the specimen can be submerged in alcohol to make the nerves white.

The cerebral or supraesophageal ganglion lies in the midline, just ventral to a point between the eyestalks. From this a number of nerves radiate out, of which four pairs are larger than the others, namely: (1) the nerves to the antennules; (2) the optic nerves from the eyes, each expanded within the eyestalk into a series of four optic ganglia; these latter will be seen when the dissection for the sinus gland is carried out, as described in the section on endocrine glands; (3) tegumentary nerves from epidermal sense organs on the dorsal surface of the head; (4) the nerves to the antennae, with small branches to the excretory bladders.

Larger than any of these nerves are two circumesophageal connectives, which run from the cerebral ganglion to the subesophageal ganglion, located behind the esophagus underneath the anterior end of the endophragmal shelf. A slight swelling on each connective as it passes to one side of the esophagus marks the position of the connective ganglion. Two minute nerves belonging to the so-called stomatogastric system leave this ganglion. One of them passes forwards and medially, to meet its fellow from the other side under the stomach. Here a minute median branch is received from the cerebral ganglion, and another branch passes over the anterior surface of the stomach towards its anterodorsal surface. Here this branch pierces the ophthalmic artery, expands as a tiny ganglion inside the cor frontale, and then leaves the artery to proceed posteriorly on the dorsal surface of the stomach. Other branches of the stomatogastric system, too minute to be easily seen, ramify over the surface of the stomach and esophagus, pass outwards to certain areas of the integument, and extend back dorsally as far as the pericardial sinus.

If the esophagus is gently pulled forwards, a slender tritocerebral commissure may be seen, running just behind the esophagus from one connective to the other.

To see the regulatory nerves of the heart, which issue from the subesophageal ganglion, a special method of preparation has to be used. First, the esophagus should be cut, just above the connectives. Then the stomach and the hepatopancreas should be carefully pulled forwards and removed. The adductor muscles of the mandibles also should be removed. This procedure should reveal the cardiac inhibitors, a pair of nerves which emerge from a squarish median hole in the endophragmal shelf, just behind the esophagus and just in front of the most anterior point of origin of the flexor abdominis muscles. Each of these nerves passes outwards and backwards, crossing the extensor abdominis muscle, then ascends dorsoposteriorly inside the body wall to the pericardial sinus, where it breaks up into an internal plexus. The pair of cardiac accelerators emerges through a second median hole in the endophragmal shelf, which is much larger and more rounded than the first one. This second hole, however, is covered by the flexor abdominis muscle, the anterior origin of which should be carefully cut, at least on one side, and the muscle cautiously bent back. Each accelerator nerve follows a course parallel to the more anterior inhibitor and eventually joins the same pericardial plexus.

Details of the pericardial plexus can be made out only with great difficulty. It is known, however, that it connects dorsolaterally with the heart. Impulses arriving via either accelerator or inhibitor fibers serve only to modify the beat of the heart. Normally this beat is maintained at a steady rate by a spontaneous, rhythmic discharge of impulses from a ganglionic group of scattered nerve cells on the inner dorsal surface of the heart. There are about sixteen such cells in the crayfish, nine in the lobster. They can be seen by slitting open a fresh heart midventrally, pinning it out flat, and staining for about 10 hours in a refrigerator with 0.005 per cent methylene blue in physiological saline, acidulated to pH 4.0–5.0 with hydrochloric acid.

Since the subesophageal ganglion and succeeding thoracic ganglia lie under the endophragmal shelf, the beginner may find it easiest to identify the ventral nerve cord first of all in the abdomen, and then work forwards. To do this, the left and right flexor muscles in the abdomen should be separated with the fingers, aided slightly with a scalpel. On looking down between these muscles in the midline, the ventral nerve cord will then be seen, lying just above the slender ventral abdominal artery. In each of the first five abdominal somites the cord is slightly expanded as a ganglion, each of which gives rise to three pairs of lateral nerves, the third pair appearing to come off slightly behind the ganglion. In the sixth abdominal somite a terminal ganglion is present which supplies the telson as well as the last somite.

On following the nerve cord forwards into the thorax it will be found to dip down into the sternal sinus. Here it can be traced farther forwards to the subesophageal ganglion by gradually cutting or chipping away the endophragmal shelf. The two most posterior thoracic ganglia are very close together, just behind the point where the sternal artery passes between the two separated halves of the nerve cord. In front of this separation, three distinct ganglia are present, one in each of the somites bearing the first three pairs of pereiopods. In the somite bearing the third maxillipeds a ganglion is present indistinctly divided from the subesophageal ganglion. This latter represents the fused ganglia of the somites bearing the mandibles, maxillules, maxillae, and first two pairs of maxillipeds.

The ventral nerve cord may now be lifted out from the animal and placed in water in a petri dish for study under the binocular microscope. Roots of the lateral nerves will show up clearly, and it will be seen that the cord and its ganglia really are paired, though the left and right divisions are for the most part rather completely fused. In the connectives between ganglia only nerve fibers are present, whereas in the ganglia nerve cells also are present. Among the nerve fibers it should be possible to distinguish four especially large ones, located near the midline on the dorsal surface. These are the so-called giant fibers, which conduct impulses more rapidly than do the other fibers. Their stimulation brings about the "escape reflex," in which the antennae, pereiopods, and pleopods are thrust forwards, while the flexor muscles of the abdomen contract strongly to produce a powerful flip of the tail. If not seen in a whole mount, the giant fibers can be located in a histological cross section.

The innervation of the appendages will not be described here, but certain unique features about the nerves in the pereiopods deserve special mention. Like most nerves, each of these nerves contains many (upwards of 10,000) sensory fibers, but unlike most nerves each contains extremely few motor fibers and is further remarkable in containing a few special inhibitory fibers. Determinations of the number of efferent fibers in the nerve as it passes through the ischium have indicated about twelve motor and three inhibitory fibers. These have to supply the great number of muscle fibers in seven different muscles, which they do by repeated branching. Moreover, each of the fibers in a single muscle may be supplied by as many as four motor nerve fibers and sometimes by two inhibitory nerve fibers. The efferent nerve fibers in the main nerve are all much larger than are the sensory fibers, a fact which makes possible their isolation for physiological experimentation.

Sense Organs. A pair of stalked compound eyes is present, each composed of upwards of 10,000 photoreceptor units called ommatidia. Under a binocular microscope the exoskeleton of the eye can be seen to be modified into many corneal lenses or facets, mostly square in shape, each being the outermost portion of a single ommatidium. The inner parts of these ommatidia and their connections with the optic nerve can be made out in a general way in a histological preparation, or in an eye bisected longitudinally with a razor blade (if a fresh specimen is being used the eye should first be immersed briefly in boiling water). To ascertain the whole structure, however, would require a more prolonged microscopical study of histological preparations.

Each ommatidium is surrounded along its whole length by a sleeve of cells containing black pigment and near its inner end by some cells which contain a reflecting pigment. In a light-adapted eye the black pigment is distributed along most of the length of the cylindrical sleeve, so that light entering the ommatidium is not able to pass into neighboring ommatidia. In a dark-adapted eye the black pigment is concentrated at the two ends of the sleeve, leaving the reflecting pigment exposed, and permitting light to pass from one ommatidium to another. These differences in the distribution of black pigment are easily seen in eyes slit open longitudinally after the animals have been killed by immersion in water at 80°C. for 10–15 seconds. The two different states also can be detected in intact animals by briefly illuminating the eye, in the dark, with a bright beam of light. If the eye is light-adapted, it will appear black, especially towards the periphery. If the eye is dark-adapted, it will still appear black around the periphery, but in a large circular area in the center it will be a brilliant orange-red. This color results from the fact that the exposed reflecting pigment near the inner end of each ommatidium is reflecting the light back through the central structure termed the rhabdome, containing the pigment, visual red.

Distal to the rhabdome, between it and the corneal facet, lies the crystalline cone, which plays the principal role in the optics of the ommatidium.

Even after loss of the eyes, crayfishes are able to avoid light. The sensitive tissue concerned is in the sixth abdominal ganglion, although the actual receptor cells have not been identified. Possession of such a caudal photoreceptor is perhaps of value to the crayfish when backing into a supposedly dark crevice; if a second opening into the crevice should be present, light entering through this opening might act as a stimulus and warn the crayfish of the unsafe nature of the retreat.

Identified as probable chemoreceptors are certain small, blunt-tipped setae, with thin walls, and without lateral setules, known usually as aesthetascs. These receptors occur on some of the mouth parts, and more particularly on the ventral surface of the more distal segments of the outer flagella of the antennules. In this latter position the setae are in small tufts of three or four, two tufts to each segment, and each seta has a characteristic node about halfway along its length. Chemosensitivity has been demonstrated on various other parts of the body, but the nature of the receptors in such locations remains uncertain.

Setae which respond to touch, and perhaps sometimes to water currents, are abundant on various parts of the body. These setae are small, pointed, usually feathered with setules, and at the base each has a thin-walled, bulb-like swelling of the exoskeleton, permitting free movement of the seta in any direction.

A specialized use of such setae is met with in the statocysts, one of which is located in the proximal segment of each antennule. Each of these statocysts consists of a sac, opening to the outside, within which are numerous sand grains and the sensory setae. The external opening is a minute slit, concealed by a tuft of setae, located on the dorsal surface of the proximal segment, about one-third of the distance from the anterior end. In large specimens a fine pin can be pushed through the opening into the membranous sac, which occupies most of the proximal two-thirds of the segment. To see this sac it is necessary to cut away the exoskeleton on the two sides of the segment, using a razor blade. The sac should then be cut open dorsolaterally, to disclose the sand grains and the rows of minute sensory setae, arranged in a horseshoe-shaped fashion on the floor of the sac, and, in the lobster, in a small tuft projecting down from the dorsal surface. A binocular microscope is required to see the details. Most or all of the sand grains are cemented to the setae, though many of the delicate strands of cement are likely to become broken during the dissection. Like all other sensory setae, the ones here are supplied by nerve fibers, which in this case come from a small branch of the antennulary nerve, entering the sac posteriorly. If a crayfish commences to molt in captivity, it may be possible to repeat Kreidl's classical experiment, by placing the animal in filtered water with a supply of minute iron filings on the bottom. The lining of the membranous sac, with its contained sand grains, is cast at the molt, and after the molt the animal rubs its antennules on the bottom to refill the sac. In this case iron filings, instead of sand grains, will enter the sac. If the animal is now blinded by black lacquer painted over the eyes, and is suspended by threads back upwards in the water, it can be induced by a magnet to carry out righting movements with its legs. Instead of responding to the direct pull of gravity, it now responds to the resultant of this force and the force exerted by the magnet.

Proprioceptors presumably must be present to make possible the maintenance of posture and of muscle tonus, particularly in the pereiopods. The only sense receptors of this type for which we have any clear evidence are some so-called movement receptors, located in the membranous joints between successive segments of the legs. These receptors have been demonstrated physiologically in a number of decapods, but are little known morphologically, and have not yet been looked for in the crayfish or lobster. They respond to movement, but not to tension or pressure.

Endocrine Glands. Internal regulation in the crayfish or lobster is partly nervous but depends also on a variety of hormones. Activities affected by these hormones include dispersion and concentration of the pigment in chromatophores, migration of retinal pigment in the eyes, frequency of molting, deposition of the gastroliths, growth of the ovaries, amount of blood sugar, and rate of heart beat. Hormones influencing all these activities are secreted by the sinus glands, one of which is located in each eyestalk. In addition at least two chromatophorotropic hormones are secreted by unidentified cells in certain parts of the central nervous system.

The sinus glands are extremely difficult to locate in preserved specimens. In a fresh specimen they are more easily found, because of a characteristic bluish color. If, in such a specimen, the exoskeleton, the hypodermis, and the dorsal retractor muscle are carefully removed from the dorsal surface of the eyestalk, as far out as the base of the eye, the sinus gland will then be seen, lying on the posterodorsal surface of the second and third optic ganglia. The gland is roughly stellate in shape, with blunt lobes. It can be seen more clearly after removal of the thin sheath which covers it and the optic ganglia. Another gland may then be revealed, named the x-organ, which has been suspected to be endocrine. In the crayfish this organ is very small and is located some distance proximal to the sinus gland. In the lobster the x-organ is larger, shaped like a bunch of grapes, and located on the anterior surface of the optic ganglia, just opposite the sinus gland.

Callinectes sapidus
By J. H. Lochhead

Among the decapod Crustacea, the crabs have departed most from the primitive "caridoid facies." They are thus of special interest as highly modified forms which have proved very successful. *Callinectes sapidus* is one of the largest of the crabs on the Atlantic and Gulf coasts of the United States, where it is usually known as the "blue crab." It also ranges as far south as Uruguay. The males commonly reach a width of about 17 cm. (measured from the tips of the lateral spines), while the females are somewhat smaller. Other species of the genus occur in various regions, such as the west coast of Africa and the Atlantic and Pacific coasts of South America (northwards to Cape Hatteras and southern California).

Callinectes sapidus sometimes penetrates from the ocean into fresh water, and indeed seems to be most abundant in brackish water, as for example in Chesapeake Bay. The females, however, always return to waters of relatively high salinity before the hatching of larvae from their eggs. In Chesapeake Bay, and perhaps in some other localities, this involves a migration of several hundred miles, by the young crabs travelling to the waters of low salinity where mating occurs, and by the mature females returning to the salt waters suitable for larval development. The adults are able to feed on a variety of living plants and animals, but their principal food seems to be dead animals. In common with most other members of the Family Portunidae, the last pair of pereiopods are modified as paddles by means of which the animals can swim very rapidly. Indeed, it has been claimed that one species of portunid is capable of catching mackerel in the open ocean.

With the large chelae the animals can strike with extreme rapidity, in contrast to the slower crayfish or lobster. The bite, however, seldom does more than draw blood. Painful bites can be avoided if the animal is picked up by the base of one of the flippers, held between the thumb and finger.

The account to be given here will deal mainly with those features in which the crab differs notably from the crayfish or lobster. It is assumed that one of the latter will already have been studied by the student. In most features the account will be applicable to almost any other brachyuran.

Body Plan. The cephalothorax is greatly depressed and is much expanded laterally. The pereiopods are thus borne wide apart, and the last five thoracic sterna on the ventral surface are very broad. The first three thoracic sterna, and the last two of the head, become progressively narrower as one passes forward; the three most anterior ones (that is, the last two cephalic and the first thoracic) are fused into a minute triangle. The thoracic sterna posterior to these three also are fused, but the lines of fusion remain distinct. Smaller episterna are present between the outer ends of the last five thoracic sterna and the coxae of the pereiopods, and a minute pair of episterna also can be distinguished on the somite bearing the third maxillipeds.

A little more than halfway back on the dorsal surface of the cephalothorax, a shallow, transverse groove extends a few millimeters to either side of the midline, then abruptly turns lateroanteriorly. This is termed the cervical groove and usually is regarded as representing the posterior boundary of the head. The short, pointed rostrum is situated slightly below the dorsal anterior border of the cephalothorax.

The abdomen is much reduced and is bent under the thorax. If it is straightened out, however, one can distinguish the six somites and the telson, although the first somite is overlapped dorsally by the thorax, and in the male the third through fifth somites are fused. The somites have calcified terga and pleura and membranous sterna. Three shapes of abdomen are met with, narrow in the male, triangular in the immature female, and broad and rounded in the mature female. In the first two of these the abdomen fits into a depression on the ventral surface of the thorax and is held there by a pair of tubercles on the fifth thoracic sternum, which fit into grooves near the tip of the sixth abdominal somite.

Paired Appendages. Uropods are absent, as are the third through fifth pleopods in the male, and the first pleopods in the female. The remaining pleopods in the female are relatively small and lack large setae until the mature stage following the final molt. In this stage the eggs become attached to specially long setae on the endopodites. The first and second pleopods of the male are copulatory, as in the crayfish, and those of the first pair are strikingly long and slender.

In the pereiopods the segments are named as in the crayfish. In all pairs the basis and ischium are fused. All pairs autotomize readily, the fracture plane probably being just distal to the proximal end of the ischium, or, in Hansen's scheme, at the line of fusion of the ischium and preischium. The male genital openings are on flexible papillae on the last coxae, as in the crayfish, but the female openings are in the sixth thoracic sternum, near the midline. The chelipeds are jointed in an interesting manner, permitting them to be folded close against the body, from which position they can be thrust forward very suddenly, as already noted.

The maxillipeds, maxillae, maxillules, and mandibles are all constructed on essentially the same plan as in the crayfish. The various parts can easily be identified by a comparison with the latter animal, although there are some notable differences in shape and relative proportions. The greatest difference lies probably in the epipodites of the three pairs of maxillipeds, which in the crab are very long, fringed with feathered setae, and project back into the gill chambers. They will be seen later when the respiratory system is studied. Use of the appendages in feeding is much as in the crayfish. The terminal three segments of the third maxillipeds form a flexible palp, used to clean the eyes, antennules, antennae, and portions of the other mouth parts.

The antennae are attached just medial to the eyes. Each has a large basal segment fused to the head, and an endopodite composed of two small segments followed by a moderately long flagellum. A flattened immovable process, attached laterally to the basal segment, may represent the exopodite. A movable, calcified operculum on the basal segment, just above the prominent ridge in front of the mouth parts, covers the opening of the antennary gland. The antennules are placed medial to the antennae. Each consists of a three-segmented basal portion, tipped by two tiny flagella. The most proximal segment of the base is very broad and is recessed into a pit on one side of the rostrum. The two succeeding segments can be folded back on the proximal one.

During much of the time the antennules flick jerkily in the direction of any approaching current of water. The currents are perceived by touch receptors on both flagella of the antennules, on the antennae, and perhaps on other neighboring structures. Correlated with this activity of the antennules the exopodites of the maxillipeds beat rapidly so as to draw a current of water across the front of the head, first from one direction, then from another. It has been suggested that the animal thus is able to test out its surroundings and to move in the appropriate direction when an attractive chemical stimulus is discovered. The necessary chemoreceptors usually are presumed to be located on the antennules.

Exoskeleton and Hypodermis. The exoskeleton is of the same type as in the crayfish. It is molted periodically as long as growth continues, but in the female, and perhaps in the male, there normally is no further molting after the animal has become mature. Apparently the females rarely live more than 3 years.

For a dissection of the internal parts the student should begin by removing the exoskeleton from most of the dorsal surface, trying at the same time not to remove the richly pigmented hypodermis. With scissors a cut can be made all the way around the cephalothorax, about a millimeter or so from the periphery. The dorsal exoskeleton then can be pried up, starting at the anterior margin, using a scalpel to free it progressively from the underlying hypodermis. Just in front of the cervical groove certain muscles attach to two calcified internal projections of the exoskeleton about 3 mm. apart, one on either side of the midline. These have to be cut. Behind them there are wisps of muscle attaching to the exoskeleton along two diverging lines, one on either side, which run lateroposteriorly from the positions of the ligaments. These and a few other small muscles attaching to the exoskeleton also have to be cut.

In the hypodermis, red, black, and white chromatophores can readily be distinguished under a binocular microscope. The melanophores are reported to be remarkably rich in riboflavin.

Muscular System. Certain muscles will be among the first items seen, so this system will be described first. However, no dissection should yet be done which might damage other structures. Each of the two calcified internal projections already mentioned provides a common origin for the medial posterior gastric and the anterior dorsal pyloric muscles. After the hypodermis has been cautiously removed, these two paired muscles can be followed forwards to their insertions on the stomach. The anterior dorsal pyloric muscle divides into two components, which run anteroventrally to their insertions on the morphologically dorsal surface of the pyloric stomach. This surface, however, is posterior in the crab, because of the way the pyloric stomach is directed ventrally. The many other muscles inserting on the stomach will not be described here. In essentials they are as in the crayfish. Just lateral to the two calcified internal projections of the dorsal exoskeleton, the posterior adductors of the mandibles arise, each by a small head which quickly tapers to a long tendon running ventroanteriorly to the posterior border of the mandible. The wisps of muscle that were noted along a line running lateroposteriorly will be found to connect the carapace with the top of an inward projection of the white, calcified body wall. Below this projection, the large proximal muscles of the fifth pereiopod will be found, especially important because of the modification of that appendage for swimming. Farther forward, where the tips of the gills converge to an apex, there is a second inward projection of the body wall, on which and in front of which more strands of muscle insert from the carapace. A short distance in front of this the head of the small external (or minor) abductor of the mandible will be seen, surrounded by lobes of the hepatopancreas. This muscle quickly tapers to a ligament, which runs anteroventrally to insert near the tip of a conspicuous white apodeme, projecting dorsolateroposteriorly from the mandible. Also attaching to the tip of this apodeme is the large lateral adductor of the mandible, which arises by two heads, slightly lateral to the position of the eyes, underneath the overlying hepatopancreas. The sheet-like internal (or major) abductor of the mandible inserts along the posterior border of the mandibular apodeme and arises on a medioventral part of the endoskeleton.

Lateral to the inner, dorsal limits of the body wall, the carapace extends far out to the side as the branchiostegite. As in the crayfish this is a hollow outgrowth from the body, but here the internal cavity is much more extensive, especially towards the front. In it there are found extensions of the gonads, hepatopancreas, and bladder, as well as certain muscles. After removal of the dorsal exoskeleton, the outer, dorsal surface of the branchiostegite is represented only by the hypodermis. This hypodermis is attached to the thin inner surface of the branchiostegite, which forms the roof of the gill chamber, by numerous, regularly spaced, little columns of connective tissue. In addition, there are a number of wider columns of muscle tissue, arranged in about three rows, running approximately anteroposteriorly. Contraction of these dorsoventral muscles would raise the roof of the gill chamber, but the functional significance of this is unknown.

In the pereiopods four muscles insert on the basis, not only in the chelipeds as in the lobster, but in the more posterior pairs as well. The levator anterior basis, inserting by a strong tendon on the anterodorsal edge of the proximal rim of the basis, is the muscle used in autotomy.

The abdomen is provided with flexor and extensor muscles, running from somite to somite as in the crayfish. As might be expected, however, they are much reduced, and they have no fibers arising farther forward than the last thoracic somite.

Circulatory System. The circulatory system is surprisingly similar to that of the crayfish, the main difference being in the way venous blood is conducted to the gills. This difference probably is correlated with the fact that the thoracic appendages of the crab are attached far out from the midline.

The heart is broad and fills most of the pericardial sinus. Its three pairs of ostia are anterodorsal, lateral, and dorsoposterior in position. It is suspended by five pairs of elastic ligaments, plus a single, median posterior ligament. The arteries which leave the heart, each with an internal valve at the base, are the same as in the crayfish and have an essentially similar distribution to the various organs. Perhaps the major difference worthy of note is the presence of a pair of branches from the superior abdominal artery, which supply the dorsal proximal muscles of the more posterior pairs of pereiopods, the muscles running from the carapace to the top of the body wall, and the hindgut caecum (to be described later). The antennal artery on each side pierces the head of the posterior adductor mandibularis muscle, just after leaving the heart.

Details of the system of sinuses are not easily made out, but a description will be given here to complete the functional picture. The sternal sinus, into which venous blood comes from most parts of the body, is incompletely divided into left and right portions, and it extends up to the floor of the pericardial sinus, surrounding not only the nerve cord and ventral thoracic artery, but also the thoracic portion of the hindgut. From its dorsolateral margins, just below the ventrolateral limits of the pericardial sinus, five channels on each side, termed branchial sinuses, run ventrolaterally to open into the infrabranchial sinus. This runs longitudinally, just above the dorsolateral margins of the thoracic coxae, from the level of the second maxilliped as far back as the fifth pereiopod. Blood returning from these last seven thoracic appendages does not go to the sternal sinus, but instead enters the infrabranchial sinus directly. From here nine afferent branchials enter the gills, each passing up the outer side of its gill. An efferent branchial comes down the inside of each gill, and five

CALLINECTES SAPIDUS 453

branchiocardiac sinuses on each side run up in the body wall to enter the pericardial sinus by three openings.

From the posterolateral corner of the pericardial sinus, a tough, membranous pouch on each side projects into the posterior part of the branchial chamber. The function of these pouches is unknown.

The blood contains hemocyanin; it has the same three types of blood cells as in the crayfish; and it clots in a similar fashion. A blood-cell-forming organ is located on the dorsal part of the stomach, but cannot be distinguished in an ordinary dissection.

Reproductive System. The ovary has two lateral divisions, connected by a cross bar just behind the pyloric stomach; these divisions run both forwards and backwards, above the hepatopancreas. The anterior portion on each side curves laterally, skirting the cardiac stomach, and in the fully developed ovary continues laterally behind the anterior margin of the cephalothorax to a point just posterior to the lateral spine. The posterior portion runs rather directly backwards, and when fully grown may extend back into the first abdominal somite.

A short distance behind the cross bar, under the pericardial sinus, there is a pair of short oviducts, entirely concealed within the ovary. The duct on each side enters an oval-shaped seminal receptacle, near its medial anterodorsal border. The receptacle extends downwards almost to the sternal floor of the thorax. A short vagina leads from its posteroventral border to the opening on the sixth thoracic sternum. Both the vagina and the ventral part of the receptacle have a cuticular lining, continuous with the exoskeleton.

The appearance and size of the ovary and of the seminal receptacle vary greatly with the stage of development. In immature specimens the ovary is thin and white, and its arms may be very short. The seminal receptacles are white and not larger than the pyloric stomach. Immediately after copulation, which occurs just after the final molt, the ovary is still small and white, but the seminal receptacles are enormously distended, sometimes being as large as the heart, and the gelatinous "sperm plug" which each contains makes it pink in color. A short time later, the receptacles are much smaller and are again white, the sperm plugs which accompanied the sperm apparently having been absorbed. The ovaries now have started to grow and are becoming orange in color. Their growth continues for several months, their arms first reaching full length and then broadening, until, shortly before egg-laying, they almost completely cover the hepatopancreas. The ova which they contain at this stage are crowded with orange yolk and are large enough to give a granular appearance to the ovary.

Fertilization of the ova evidently occurs at or shortly before egg-laying, either in the oviducts or in the seminal receptacles. Examination of the inside of a receptacle reveals a mass of white sperm stored on a medioventral ridge.

After laying, the eggs are attached to setae on the endopodites of the pleopods and are carried for a few weeks, until the zoeae hatch. The mass of attached eggs, or "sponge," is at first a light orange but steadily darkens as embryonic development progresses, until it is almost black. The ovary at this time may be somewhat smaller than before, but remains bright orange because of numerous mature ova which it still contains.

After the zoeae have hatched, the fact that hatching has occurred can be detected from the remnants of egg shells or attaching stalks, which a careful search will reveal on the endopoditic setae of the more anterior pleopods. More indirect evidence that eggs have hatched is provided by the presence on the gills of mature nemertean parasites of the Genus *Carcinonemertes*, which always first develop in the egg sponge.

Somewhat later, probably usually in the same summer, a second batch of eggs is laid, and now the ovary appears collapsed and is gray or brownish in color, with few or no orange areas. Enough sperm is present in the receptacle, from the single copulation, to fertilize even further batches of eggs, but it appears that two batches are the normal limit, since from this point on the ovary degenerates.

In the mature male, the testis has two narrow, rather convoluted arms, running a course closely similar to that of the lateroanterior arms of the ovary. These are connected by a cross bridge posterior to the pyloric stomach, shortly behind which the testis connects on each side with the much-coiled vas deferens. The connection is made through a very small vas efferens, less than 3 mm. long, entirely concealed within the testis and the coils of the vas deferens. The extent to which the vas deferens is coiled and differentiated into distinct regions varies greatly with the stage of development. When fully developed it comprises four main divisions. First there is a region where the duct forms a coiled, white mass, close to the mid-dorsal line, between the stomach and the pericardial sinus. Within this region the duct slowly increases in size up to a diameter of nearly 1 mm. There follows a region where the duct is still larger and its coils are embedded in a solid mass, pebbled pink in appearance, and up to about 1 cm. in diameter. This mass passes forwards ventral to the testis and the preceding portion of the duct, close against the lateral wall of the stomach, then doubles back ventral to itself. The third region of the duct is a considerably coiled, translucent greenish tube, up to about 2 mm. in diameter. It starts a short distance behind the stomach and passes back lateral to the intestine and then posterior to the pericardial sinus. The fourth region of the duct is a slender, translucent tube, which starts in about the seventh thoracic somite, lateral to the intestine. It passes anteroventrally, then turns and runs a relatively straight course lateroposteriorly to the opening at the tip of a papilla on the last thoracic coxa. This papilla lies permanently inserted into the groove of the first pleopod. A part of the second pleopod also fits into the groove and is said to act as a piston, pushing sperm along towards the vagina.

If some of the contents from different parts of the testis, or from the first, narrow portions of the vas deferens, are mounted in sea water and examined under a high magnification, it may be possible to see the spermatozoa. The mature spermatozoan has a central part 1.5–2 μ in diameter, plus several radiating, finely pointed, cytoplasmic processes. Earlier stages in spermatogenesis lack the cytoplasmic processes and are larger.

The first region of the vas deferens secretes the thin capsules for the spermatophores and probably also at least a part of the material for the sperm plug. The spermatophores are stored in large numbers in the later, thicker portions of this region of the vas deferens. Each spermatophore usually is shaped somewhat like a bird's egg and is from 200 to 500 μ long. During copulation the spermatophores break up, probably by mechanical action while passing along the groove in the first pleopod.

The second region of the vas deferens secretes further material for the sperm plug. Yet another constituent of the sperm plug may be added by tegumental glands in the first pleopod, which probably open into the groove of this appendage.

For about 3 days before a female's final molt, she usually is carried around by a male, which clasps her, back uppermost, against his ventral surface by means of his three pairs of walking legs. Evidently the males can distinguish when the female is approaching this particular molt, or else only such females submit to being carried.

After the female has molted she turns over on her back, extends her abdomen, and copulation occurs. This persists for perhaps as much as a day or two, the male moving around while holding the female in place with some of his walking legs.

Digestive System. The esophagus, cardiac stomach, pyloric stomach, and intestine are essentially as in the crayfish. The esophagus, however, runs dorsoposteriorly, and in conformity with this the pyloric stomach is directed vertically downwards and joins the intestine at a 90° angle. The cardiac stomach does not secrete gastroliths. The midgut is less than a centimeter long. The terminal part of the hindgut is only slightly dilated, hardly deserving to be distinguished as a rectum. The hepatopancreas is divided into left and right divisions, each of which extends far out around the anterior margin of the cephalothorax, with many small lobes. Each division also extends deep into the floor of the cephalothorax and runs posteriorly below the pericardial sinus, sometimes as far back as the fourth abdominal somite.

In place of the single, very small midgut caecum met with in the crayfish, there is a pair of quite long, though very narrow, caeca, arising just behind the pyloric stomach. The two come up vertically together under the ophthalmic artery, just posterior to and together with the anterior dorsal pyloric muscle. Between this muscle and the medial posterior gastric muscle each caecum curves lateroanteriorly and runs out under the gonad (if present), close to the wall of the stomach. Presently it turns more laterally, passes ventral to the gonad and antennary artery, and turns again to run anterolaterally outside and ventral to the artery. Finally it turns laterally once more to end in coils just dorsal to about the first large lobe of the hepatopancreas that is visible in dorsal view. In color the midgut caeca are a translucent white, often with contained air bubbles in specimens that are being dissected. Their epithelial cells, like those of the rest of the midgut, are absorptive in function and have a striated border.

Behind the heart, another much-coiled caecum will be seen, mostly dorsal to the intestine. From these coils a straight part of the caecum runs back, usually on the right side of the intestine, and finally opens into the intestine where this shows an annular thickening in the second abdominal somite. Unlike the median posterior caecum of the lobster, this is not a branch of the midgut, but is a hindgut caecum, with a cuticular lining like that of the rest of the hindgut. It has an especially rich blood supply, but its function appears to be unknown.

Excretory System. The paired antennary glands are on the ventral surface of the head, back of a position between the antennae and the eyestalks, and partly concealed by the overhanging stomach. Each is squarish in outline, pale greenish or yellowish in color, and often appears as though composed of tiny, coiled tubes. Its microscopic structure is essentially as in the lobster, the dorsal central part being occupied by the much-lobed coelomosac, which opens ventroanteriorly into the still more lobed labyrinth occupying the periphery and ventral parts of the organ. A tubule, such as occurs in the crayfish, is absent. A horseshoe-shaped opening at the inner posterior corner of the gland, not often seen, opens upwards from the labyrinth into the bladder. The bladder is a remarkable organ, of very large size, yet likely to be overlooked because its walls are thin and transparent and if seen at all are likely to be mistaken for connective tissue. In a fresh specimen it can be rendered more opaque and thus more easily seen, if, as the dissection progresses, it is flooded locally with 70 per cent alcohol acidified with nitric acid in the proportion of three parts per hundred. From the main vesicle of the bladder, located above the gland, a number of lobes extend out in various directions. Most prominent are an epigastric lobe, rising at the front and side of the stomach towards the dorsal surface, and a hepatic lobe which passes posteriorly under the insertion of the lateral adductor of the mandible and then turns laterally between the muscle and the hepatopancreas. Smaller lobes push their way between the hepatopancreas and the stomach and around the esophagus. The cells of all these lobes of the bladder, when viewed in sections, are secretory in appearance.

The main vesicle of the bladder opens to the exterior by a duct, which starts above the middle of the dorsal anterior margin of the gland, and passes forwards below the ventral one of two transverse endophragmal struts seen just behind the antenna. The duct opens on a thin membrane, corresponding to the operculum of the crayfish, but here covered over by a second and different kind of operculum. This outer operculum is calcified, can be raised and lowered by special muscles, and is thought to be the modified antennal coxa. The actual opening of the duct is under the ventral, medial corner of the outer operculum. An injection into the duct, to reveal the details of the bladder, would be most desirable, but unfortunately the opening is so minute that any such attempt almost certainly would puncture the thin, surrounding membrane.

The urine of the crab is isotonic with the blood, in conformity with the absence of a tubule in the antennal gland. When in brackish or fresh water, the crab excretes a copious isotonic urine, and makes good the salt loss by an active and selective uptake of ions by the gills.

Certain injected substances are actively removed from the blood by cells of the hepatopancreas, and by special cells in the gills, as in the crayfish. Excretion in the antennal glands is not wholly passive in crabs, for it is known that some ions, such as sulfate and magnesium, are concentrated into the urine.

Respiratory System. Study of the respiratory system has been delayed to this point, because of the manner in which branches of the reproductive, digestive, and excretory systems extend into the cavity of the branchiostegite.

Along its lateral margin, the outer wall of the branchiostegite turns very sharply downwards and inwards to meet the body wall just above the insertions of the thoracic appendages. The inner wall of the branchiostegite, forming the outer wall of the gill chamber, curves much more gradually. Above the appendages, the branchiostegite fits so tightly against the body wall that at first sight the two appear to be fused together. However, if the branchiostegite is strongly pressed upwards, while the specimen is held up to the light, a thin slit all around the lower margin of the branchiostegite will be disclosed. This slit normally is tightly shut, and a fringe of fine setae on the rim of the branchiostegite further occludes the opening. Only above the base of the cheliped is an opening left, leading posteriorly into the hypobranchial part of the gill chamber. Such an anterior inhalant opening is in contrast to the condition in the crayfish. The opening is very large and nearly circular when the chelipeds are raised and extended forwards. When the chelae are held in the normal position, folded against the body, the opening is reduced to a wide slit, which becomes further narrowed by a flange on the base of the epipodite of the third maxilliped whenever the two maxillipeds are opposed in the midline. A fringe of setae on the basi-ischium of the cheliped filters part of the water which enters the slit.

Water which enters through the above opening finds itself in the hypobranchial division of the gill chamber. It then passes upwards, through the gills, into the hyperbranchial division, whence it flows forwards into the exhalant passage. This opens above the specially shaped distal end of the endopodite of the first maxilliped. Even when the third maxillipeds are tightly closed over the other mouth parts, a pair of small anterior openings remains, through which the exhalant currents can escape. The lateral edge of each exhalant passage is sealed by the exopodite and endopodite of the first maxilliped, by the exopodite of the second maxilliped, and by the exopodite of the third maxilliped with its lateral fringe of setae. The flattened base of the epipodite of the first maxilliped helps to separate the hypobranchial space from the exhalant passage. The effective beat of the scaphognathite is normally from below upwards, and this is thought to be a further factor ensuring the correct circulation of the water. As in the crayfish, the emerging current of water may be given further impetus, or deflected to either side, by the beating of the exopodites on the maxillipeds.

At periodic intervals the beat of the scaphognathite is reversed, to aid in cleaning the gills. Such reversal is also made use of in water of low oxygen content, for taking in surface-oxygenated water or even air through the anteriorly placed exhalant openings. When a partially submerged animal thus takes in air, it emerges as bubbles through the large inhalant openings.

If the thin roof of the gill chamber has not already been removed, this should now be done. The shape of the gill chamber should be observed, and its relations to the body wall, branchiostegite, inhalant openings, and exhalant passage should be fully understood. It probably will prove necessary to cut away the anterior exoskeleton bit by bit as far medially as the antennae.

Overlying the gills is the long epipodite of the first maxilliped, fringed with feathered setae. The shorter but similar epipodites of the second and third maxillipeds lie below the gills. By sweeping up and down the length of the gills these epipodites help to keep the gills clean. The first is moved largely by its own independent muscles. The other two for the most part share the movements of the protopodites of their respective appendages.

The gills are not trichobranchs, like those of the crayfish, but are phyllobranchs. Each consists of a central axis bearing an anterior and a posterior row of closely set, leaf-like lamellae (except on the arthrobranch above the second maxilliped, which bears a posterior row only).

There are eight gills on each side, distributed on the somites of the different appendages as follows: second maxilliped, one podo- and one arthrobranch; third maxilliped, two arthrobranchs; first pereiopod (or cheliped), two arthrobranchs; second and third pereiopods, one pleurobranch each.

Endophragmal Skeleton. Parts of the endophragmal skeleton will be met with in an ordinary dissection, and a more complete study

460   ARTHROPODA

can be made from preparations such as were suggested for the crayfish. The skeleton is, however, extremely complex, and a description of it here does not seem worth while, particularly as there is no endophragmal shelf to conceal the ventral nerves and arteries.

Nervous System. In the fresh condition the nerves are rather transparent, but they become more opaque if the specimen is submerged for 1 minute or longer in 70 per cent alcohol. The finer nerves sometimes stain well when methylene blue or toluidin blue is applied locally to the fresh tissues.

The general plan of the nervous system is the same as in the crayfish, the one great difference being the fusion of all the ventral ganglia in the cephalothorax into a single mass. The supraesophageal ganglion lies immediately behind the rostrum. It can be better seen after removal of a movable, transverse apodeme, which runs just above it between the two eyestalks. Nerves that can be seen connecting on each side with the supraesophageal ganglion are the antennulary, optic, oculomotor, tegumentary, and antennary. Posteriorly there are two fine nerves, not easily found, which connect with the stomatogastric system.

The circumesophageal connectives are long, over two-thirds of this length being behind the esophagus. The tritocerebral commissure, connective ganglion, and connections with the stomatogastric system are essentially as in the crayfish.

The fused ventral ganglion lies near the floor of the cephalothorax, in the fourth and fifth thoracic somites. Its dissection from the dorsal surface is difficult, because of the way it is covered dorsally and at the sides by a complex arrangement of small muscles, tough membranes, and lobes of the hepatopancreas. For most purposes a dissection from the ventral surface will prove adequate. This can be done by cutting out a broad central strip from the sternal plates, and removing some of the proximal muscles of the pereiopods, above which the ganglion lies. A conspicuous hole near the center of the ganglion is for the descending sternal artery. From the periphery of the ganglion, nerves radiate out on each side to supply the series of appendages, from the mandible to the fifth pereiopod inclusive. The nerves for the first five of these appendages are very small. Those for the mandible and maxillule are fused together. The nerve for the third maxilliped is somewhat larger, and that for the cheliped is largest of all. Posteriorly a small double nerve passes back in the midline into the abdomen. From the dorsal surface of the ganglion, at the same level as the nerves to the third maxillipeds, arises a pair of fairly large cutaneous nerves. Each of these runs lateroanteriorly, just behind the nerve to the second maxilliped, then curves upwards inside the body wall, to supply the lining of the carapace and the roof of the gill chamber.

The innervation of the heart is not easily made out, but will be described for those who are especially interested. Accompanying each cutaneous nerve outwards from its origin, in contact with its posterior border, is a small cardiac inhibitor nerve. This presently leaves the cutaneous nerve and turns abruptly posteriorly, to run back inside the body wall to the pericardial plexus. Two other pairs of small nerves, the cardiac accelerators, arise a short distance behind the cardiac inhibitors, at about the level of the nerves to the chelipeds. On each side, the anterior of these two runs outwards dorsal to and between the nerves for the third maxilliped and the cheliped. After passing upwards, it then curves posteriorly and runs back ventral and parallel to the cardiac inhibitor nerve, to join the pericardial plexus. The more posterior accelerator nerve runs out dorsal to the nerve for the cheliped, curves upwards through a lobe of the hepatopancreas, then passes back at a considerably deeper level than the other two, to join the pericardial plexus from below.

Parts of the pericardial plexus lie over the openings from the branchiocardiac sinuses into the pericardial sinus. Connection of the pericardial plexus with the heart is made dorsolaterally. The intrinsic ganglion inside the dorsal wall of the heart contains about nine nerve cells.

Sense Organs. The compound eyes are constructed on fundamentally the same plan as in the crayfish. For protection from mechanical injury they can be laid back into sockets in the front of the head.

The statocysts of the megalops larva function like those of the crayfish, but in the adult crab they do not contain sand grains or other otoliths. The dorsal slit leading into the statocyst is entirely closed, except immediately after a molt. The line marking the position of this slit is easily seen, however, running transversely across the outer two-thirds of the dorsal surface of the relatively very broad first antennulary segment. If the dorsal exoskeleton of this segment is carefully cut away, the sac of the statocyst will be disclosed. Three swellings project into this sac, each bearing specialized setae. Those on the small posteromedial swelling are the ones which bore sand grains in the megalops larva. The setae on the large lateral swelling are of uncertain function, but a vertical row of setae on the anteromedial swelling is of special interest because these setae perhaps react to movements of the liquid in the statocyst. At least this would seem to be possible from their structure. They are extremely thin, relatively long, fringed with setules only near their tips, and are attached at their bases in such a way as to permit great mobility. Observation shows that very slight movements of the liquid in the sac are sufficient to cause these setae to sway.

Tactile setae like those of the crayfish occur on many parts of the body. The receptors of *Callinectes* have not been closely studied, but in related crabs various other structural types of sensory setae have been found and have been tentatively identified as receptors for touch and for thigmotaxis. Chemoreceptors are found on the tips of the pereiopods and also somewhere in the region bathed by the respiratory current. The chemoreceptors on the pereiopods seem to be cylindrical protoplasmic processes from single nerve fibers, lodged in pores in the exoskeleton. There is some dispute as to whether the aesthetascs on the outer flagellum of the antennules are chemosensory. Setae on both flagella of the antennules and on the antennae appear to be current receptors. Proprioceptive movement receptors in the joints of the limbs have been demonstrated physiologically in *Callinectes*, as described under the crayfish.

Endocrine Glands. As in the crayfish, the eyestalks contain a pair of sinus glands, which secrete several hormones. The gland in either stalk can be seen in a fresh specimen after the dorsal exoskeleton and hypodermis have been carefully cut away. It is a bluish white, pear-shaped organ, lying within the stalk in line with a notch in the proximal border of the eye.

Two chromatophorotropic hormones can be extracted from various parts of the central nervous system. The cells secreting these hormones have not been identified.

Spirobolus marginatus

By J. B. Buck and M. L. Keister

Spirobolus is a large, widely distributed millipede found under logs and debris in damp woods. In life, its body segments are dark bluish gray in color, handsomely margined posteriorly with orange-brown, and its legs are bright pink. Adults have about fifty segments and may attain a length of 10 cm. They feed on decaying wood and leaves.

Living specimens of *Spirobolus* form excellent material in which to study the metachronal rhythm of segmental appendages, seen also in the polychaete worms. If an active individual is observed while walking on a flat surface, regular, successive waves of convergence and divergence of its legs will be seen to pass forward. Careful observation of a chilled specimen will permit one to work out the number of legs cooperating in each wave, the number of segments between waves, the cycle of activity of the individual leg and its relation to the action of preceding and succeeding legs on the same and opposite sides, and other aspects of metachronal activity. The rhythm may be reversed by pressing on the animal's head.*

After handling a live specimen, the pungent, brown secretion of segmental scent glands may be found on the fingers.

External Anatomy. The body of *Spirobolus* consists of a head, and a long trunk composed of a series of nearly uniform, articulated, ring-like segments. A typical trunk segment is almost perfectly circular in cross section. The exoskeletal covering is formed by the fusion of a tergum, extending dorsolaterally and comprising more than nine-tenths of the circumference of the ring, with two ventral, rectangular sterna, of which the anterior is much the larger. Each ring or annulus is considered to represent two consecutive single segments, which have become fused in the course of evolution. Evidence of this double nature is seen externally in a faint medial tergal groove, and in the fact that almost all segments have two sterna, each with a pair of legs and a pair of spiracles. Internally, the duality is indicated, for example, by the fact that the heart has two pairs of ostia per segment. In this account, "segment" will be used to refer to the double segment or ring, as well as to the first few segments, which appear to be single.

Each tergum snugly overlaps the anterior margin of the contiguous tergum behind, and is in turn partially overlapped by the posterior

---

* The complexity of this locomotory activity calls to mind the familiar ode to an opisthogoneate relative of *Spirobolus:*

> "A centipede was happy, quite,
> Until a frog, in fun,
> Said, 'Pray, which leg goes after which?'
> This raised his mind to such a pitch
> He lay distracted in a ditch
> Considering how to run."

margin of the tergum in front. Because of these articulations, the body of *Spirobolus* has considerable flexibility, in spite of the extreme hardness and rigidity conferred upon the exoskeleton by the carbonates with which it is impregnated. By coiling ventrally into a tight spiral, the animal shields the delicate membranous areas between the sterna of adjacent segments.

The legs consist of seven segments, of which the proximal six are short and rather uniform. The distal segment is modified into a single, curved claw. The leg segments are sometimes given the same names applied to the insect leg, and sometimes a special set of names. The basal segments (coxae) articulate with the posterior margins of the sterna. The more distal segments of the leg extend out horizontally from the lateral, rather than ventral, surface of the coxa, curving gradually downward, so that only the tip of the terminal claw is in contact with the substratum. The segment preceding the claw bears a lateral and two medial large bristles near its distal end and a few additional bristles along its medial edge, but otherwise the external surfaces of the legs and trunk are nearly glabrous.

The head of *Spirobolus* is covered anteriorly and dorsally by a hard, roughly hemispherical cap, the epicranium (Fig. 195). Its anteroventral edge, which is bilobed and bears tooth-like processes and a transverse row of bristles, is thought to correspond to the labrum of insects, though here it is not separate or movable. The head capsule is deeply incised at about the middle of its lateral margins. Antennae with seven short subsegments articulate at the anterior ends of these indentations and can be withdrawn into them for protection. In life, the tips of the antennae tap the substratum continuously and rapidly as the animal walks. Dorsal to the bases of the antennae are "aggregate" eyes, consisting of triangular patches of about forty dark, separate, simple eyes or ocelli.

The ventral surface of the head is almost covered by the broad, square, labium-like gnathochilarium. Between the labrum and the gnathochilarium is a pair of remarkable bisegmented, grinding mandibles, working transversely, as in most arthropods. Each mandible is divided into a massive basal segment, which is immovable, though supplied with powerful muscles, and a distal movable gnathal segment. The parts of the latter, from tip to base, are a large, conical, movable tooth, a brood scoop, an oval plate studded with bristles, and finally a heavily ridged section very reminiscent of the surface of certain types of mammalian molars. When the gnathal portions

SPIROBOLUS MARGINATUS

of the mandibles are removed, additional sifting plates and strainers can be seen on the inner surface of the labrum.

Overlapping the head posteriorly is a large convex tergum, the collum, which will here be considered to belong to the first trunk segment (Fig. 195). It is about twice as long as succeeding terga, but extends laterally only about halfway down the sides of the body and is not fused ventrally with a sternum. The second tergum likewise

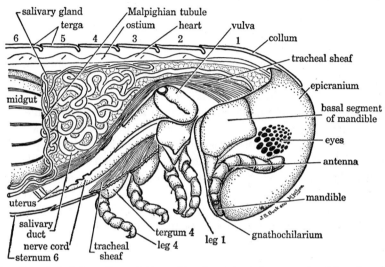

FIG. 195. *Spirobolus*. Lateral view of head and first six segments of female, with exoskeleton of right side of trunk removed. Salivary ducts simplified. ×7.

lacks the usual sternum, but its posteroventral borders are produced medially into two large spines which almost meet in the midventral line. Anterior to these spines, underlying the posterior part of the collum, are two pairs of walking legs. They will be referred to as the first and second pairs of legs, without attempting to associate them with a particular segment or segments. Each pair articulates with a thin, yoke-like sternum, whose arms are embedded in the membranous area ventral to the collum. The legs themselves have unusually long coxae.

Between the sternum bearing the second pair of legs and the spines of the second tergum are the genital openings. The two sperm ducts open at the tip of a muscular, triangular penis, covered by a wrinkled membrane. The penis is often found wholly retracted into a transverse, medioventral slit. In the female, the genital openings are associated with a pair of large, almond-shaped, hardened, yellowish vulvae, situated far laterally between the base of the second sternum and the anterior edge of the second tergum (Fig. 195). Both sets of organs will be described later in connection with the reproductive systems.

The third, fourth, and fifth segments have more or less typical sterna (although that of the third is tilted almost vertically). Each bears a single pair of legs and, anterior and lateral to the leg bases, a pair of small, slit-like spiracles. In the male, the coxae of the third, fourth, and fifth pairs of legs are produced ventrally into club-shaped knobs (decreasing in size from anterior to posterior), each with a small posterodistal concavity.

The sixth segment has two sterna, each with a pair of spiracles and a pair of legs which articulate on its posterior edge. The sixth segment is the first to have scent glands, the openings of which are tiny circular pores in the middle of the tergum in the midlateral line. All succeeding segments back to the preanal one (except the seventh segment of the male) have two pairs of typical legs, two pairs of spiracles, and a pair of scent glands, and are externally indistinguishable.

In the male, the seventh segment has no walking legs, and the sterna are represented by a small, anteromedian plate. The narrow ventral extensions of the posterior edge of the tergum join in the midline, and anterior to them is a large oval opening revealing a pair of remarkable processes somewhat resembling folded frog's legs (Fig. 196). The "thigh" regions of these leg-like structures are the paired gonopods, or copulatory appendages, each ensheathed in various sclerotized plates and processes. The gonopods are regarded by some authorities as modified legs, although the issue is confused by the fact that in some species typical walking legs occur in addition. If the internal organs are to be dissected, the vulva and gonopod are best studied in connection with the reproductive system. If not, each can be carefully pulled out for examination and separated from the powerful muscles which attach them internally. The gonopods will be found to consist of a two-segmented structure, the distal end of which forms a hollow scoop, partly closed by a movable blade, and is reminiscent of the crustacean chela. The vulva is a flattened, ovoid capsule, bearing on its lateral surface a sinuous crevice and, at its dorsal (internal) end, a membranous trapdoor and a large opening by which it communicates with the uterus.

The body of *Spirobolus* terminates in a pair of large, clamshell-shaped plates, which meet to form a vertical, slit-like anus. The rectal chamber enclosed by these plates molds the ovoid fecal pellets. The anal segment has no appendages or typical sternum. It is overlapped anteroventrally by a curved plate which extends backward from under the ventral part of the preanal segment. The preanal segment also lacks legs, and it appears that its tergum is continuous all the way around, forming a ring.

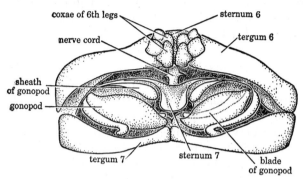

FIG. 196. *Spirobolus*. Ventral view of segments 6 and 7 of male. ×9.

Internal Anatomy. The exoskeleton of *Spirobolus* is so hard, and preserved specimens are so brittle, even when fixed by injection (as they should be, for proper internal preservation), that a conventional type of dissection should not be attempted. However, a very good idea of the internal anatomy can be gained by a regional, rather than systemic, attack, particularly if one practices first on one of the less critical regions. The type of preparation most useful in dissecting *Spirobolus* is obtained by first dividing the body, in the fashion suggested below, into sections short enough to be straight and to be easily manipulatable, then cutting through the body wall on both sides, with a fine, sharp pair of scissors, just above the midlateral line, taking care not to cut into the gut. By this process there are obtained a number of dorsal and ventral "half-shell" preparations which can be dissected from the inside out.

Carefully divide the body into four parts by cutting with a razor between terga 12 and 13, 20 and 21, and 27 and 28. Keep carefully in mind the original arrangement of these pieces, so that the continuity of the various organ systems can be maintained. The region of segments 13–27 includes a stretch of midgut and other internal organs which is relatively uniform.

Cut the piece composed of segments 13–20 into dorsal and ventral "half-shells" as described above. In some instances, the dorsal half of the preparation will lift off cleanly, leaving the gut and sometimes the heart in place in the ventral half. In other instances, the gut may adhere to the dorsal body wall and have to be gradually and carefully teased loose by breaking the white, thread-like arteries and tracheal bundles which bind the parts together. Cut the piece comprising segments 21–27 into two mirror halves by slicing it in the median sagittal plane with a sharp razor. Use these left and right halves mainly for comparison purposes while becoming familiar with the general anatomical arrangements. Use also the anterior and posterior cross-sectional views which can be obtained by looking into the cut surfaces of segments 12 and 28. Dissect slowly and cautiously, with the specimen immersed in water, by teasing and shifting with needle points rather than by cutting. Dissect each section completely before passing on to another, rather than trying to trace a given system in its entirety at one time. In some body regions, gray spongy masses of "fat" will have to be cleared away in order to expose the desired structures, and in others, masses of gritty white crystals (not carbonates) may obscure the view. It should be noted that there is some individual variation in the location of some of the viscera. The segment numbers given are therefore sometimes approximate.

The Middle Region. By dissecting between and within successive terga, the method of articulation of the body segments and the arrangement of the body-wall muscles can be seen (Fig. 197). The hardened parts of the exoskeleton are joined together as a continuous covering by thin, flexible intersegmental cuticle, which folds in between the contiguous segments. The anterior edge of each tergum turns inward at right angles to form a flange, which is overlapped by the posterior edge of the tergum ahead. To this flange are attached three sets of segmental muscles. The innermost layer (closest to the gut) consists of longitudinal fibers which run directly from the flange of the segment in which they lie to the flange of the segment behind. External to the longitudinal muscles, there are two layers of diagonal muscles, the inner of which is limited to the dorsal region. Patches of fat are often interspersed between the layers.

Among the muscles in the midlateral line in each segment are the paired scent glands. Each is a simple, dark-speckled sac about 1 mm. in diameter, attached to a short, wide duct bearing rings of circular muscle in its walls.

The heart consists of a slender tube running the entire length of the body. It is found sometimes in the pericardial space between the muscles near the mid-dorsal line, and at other times adhering to the mid-dorsal surface of the gut, partly buried in fatty or glandular tissue. In some regions, the heart is empty and transparent, in others full of cream-colored coagulated blood, and occasionally with the gritty white crystalline material.

The ventral surface of the heart is smooth, but the dorsal one is marked in the middle of each segment by a dilation (Fig. 197). There are two pairs of dorsal ostia in each segment, the smaller pair being close to the anterior limit of the segment, the second and larger pair

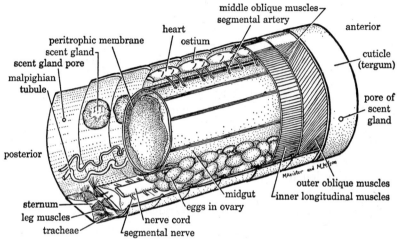

FIG. 197. *Spirobolus*. Dorsoposterolateral stereogram of midbody segments of female. ×5.

in the dorsal dilation, near the center of the segment. Two pairs of large lateral arteries are given off in each segment, close to the ostia. The heart wall is composed mainly of a very thin layer of circular-muscle fibers, on the outer surface of which run extremely minute tracheae and a syncytium of what appear to be multipolar nerve cells (visible under an immersion lens).

The midgut, which extends approximately through segments 7–32, is a thin-walled, large-bore tube. It is striped longitudinally on its external surface with narrow, thin muscle bands, but most of its bulk is circular muscle. Internal to its smooth endothelium is a delicate cuticular lining, the peritrophic membrane. The analogous structure in insects is believed to protect the endothelial cells from abrasion. The gut is coated externally with varying amounts of fatty tissue, in which run arteries from the overlying heart and tracheae and nerves from below.

Laterally, the body wall is ordinarily padded with masses of fatty tissue. Between the surface of this material and the gut, on each side, run two slender, much contorted tubules (Fig. 197). It will be found later that these two apparently separate tubules are actually the ascending and descending limbs of one of the two very long Malpighian tubules. They are ordinarily white and thin-walled, but occasionally brownish and rather stiff. (If the specimen is a mature female, some of the eggs will have to be removed in order to trace the tubules.)

Ventral to the gut lie the gonads. In the midbody region, the testes consist of a pair of contorted brown tubules running closely applied to the dorsal surface of the nerve cord. The ovary varies in appearance, depending on development, but ordinarily contains a large mass of eggs in various stages, with the large, mature, brownish orange, shelled ones becoming increasingly frequent toward the anterior end. They must be removed to expose underlying structures.

The ventral nerve cord lies under the gonads. It is somewhat flattened and nearly uniform in width and gives off two pairs of nerves to the body wall and viscera in each segment. It shows no outward sign of being longitudinally double.

Beneath the nerve cord (which should be removed), in the anterior part of each segment, can be seen the transverse, hard, anterior edges of two pairs of hollow apodemes, or exoskeletal invaginations. These apodemes lead inward and anteriorly from the spiracles and are roughly T-shaped. The leg muscles are attached to the cross bar of the T, and profuse bundles of tracheae arise from its anterior and lateral surfaces. When examined under the compound microscope, these tracheae are seen to have spiral thickenings, but otherwise to differ strikingly from insect tracheae by being unbranched, apparently non-tapering, very long, and arranged in enormous parallel bundles free from other tissue. They range in diameter from 3 to 10 μ and tend to occur in bundles containing only one size.

Posterior Region. The heart can be traced in the posterior part of the body to the anterior margin of the preanal segment, where it ends blindly. The midgut continues posteriorly to about segment 32, where it suddenly constricts (the constriction is covered with a sleeve of fatty or glandular-appearing tissue) and joins the hindgut (Fig. 199). There is a sphincter muscle at this point. The two Malpighian tubules continue posteriorly, but a few segments behind the junction of the midgut with the hindgut, one limb of each tubule turns forward and enters the gut at the constriction and in the midlateral line. The other continues back nearly to the anal segment.

Posterior to the constriction, the outer surface of the hindgut is covered with a few broad strips of longitudinal muscle, instead of numerous narrow strips as in the midgut. The wall is more or less thrown into annular folds, depending on the degree of contraction during fixation. The inner lining is deeply folded longitudinally (Fig. 200), rather than smooth as in the midgut. In the neighborhood

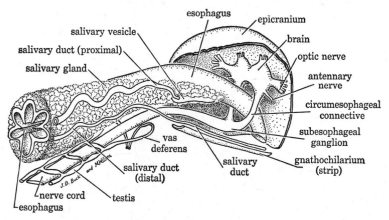

FIG. 198. *Spirobolus*. Posterolateral stereogram of head and first four trunk segments of male. ×6.

of segments 36 and 37, there is a pronounced sphincter, following which the gut expands into a spherical chamber about two segments long. This is followed by another, but less pronounced, constriction. Between this point and the rectal chamber, the hindgut is large bore and usually full of fragments of rotten wood. Its wall is thin and shows conspicuous annular folds. The hindgut is often packed with nematode worms, which may entirely occlude the sphincters.

The gonads continue under the gut to within a half-dozen segments of the posterior end, the ovary containing eggs progressively less mature. For a few segments, the testes consist of rather slender, beaded tubules with two pairs of diverticula per segment. Posterior to this, they are simple tubules.

Anterior Region. Anterior to segment 13, the dissection should proceed with special care, as many important structures are contained in the forepart of the body (Figs. 195, 198).

The heart runs forward and terminates under the collum in a slender artery which passes into the head. In segments 2–5, there is only one pair of ostia per segment. The midgut narrows abruptly, usually in about segment 6, at which point there is a sphincter muscle. This point is marked externally by a snarl of fine, white Malpighian tubules. The gut continues forward into the head as a very slender

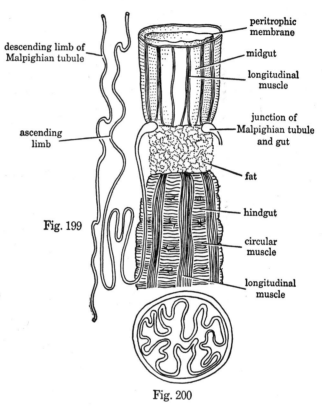

Fig. 199. *Spirobolus*. Dorsal view of intestine in the region of segments 29-36. ×5.

Fig. 200. *Spirobolus*. Cross section of hindgut. ×5.

esophagus with a hexagonally pleated inner epithelium (Fig. 198) and a glistening, thin, chitinous, stomodeal lining. However, the esophagus appears thick, because it is hidden by a pair of large white or orangish salivary glands, which in turn are partially covered with a solid mass of contorted salivary ducts, resembling the surface of the primate cerebral cortex (Fig. 195). Both glands and ducts continue forward on the lateral surfaces of the esophagus into the head. Lateral to the salivary glands near the head, are two tremendous, glistening sheaves of tracheae, the pooled contributions from the spiracles of several segments posterior to the fourth. These tracheae run mainly to the mandibular muscles. As will be visible later, the tracheae from segments 3 and 4 form two large sheaves running beneath the nerve cord into the head.

The tracing of the Malpighian tubules and particularly of the salivary ducts requires extraordinarily careful dissection. By unravelling, it can be shown that all the Malpighian tubing on one side of the esophagus—a length of about 70 mm.—is part of a single continuous loop. The two limbs of this loop pass posteriorly together around the sides of the esophagus and back along the lateral body walls as already seen. It is therefore clear that there are only two Malpighian tubules. Each of these, after originating at the boundary of the midgut and hindgut, runs posteriorly for a short distance, then anteriorly to the salivary gland region, then back almost to the hind end of the body.

The salivary gland ducts likewise consist of a pair of continuous loops. The two limbs of each loop, which run laterally along the esophagus into the head, can be seen by removing the two large dorsal tracheal sheaves. The tracing of the ducts into the head should be deferred until the epicranium is chipped off to expose the brain. The smaller, more ventral limbs of the loops are rather easily followed onto the internal surface of the gnathochilarium, where they open close to large bosses near the middle of its distal edge. The two larger limbs continue lateral to the esophagus and connect with the lateral walls of a small transparent bladder close under the esophagus, at the anterior ends of the two salivary glands (Fig. 198). It appears, then, that salivary secretion elaborated in the glands on the esophagus is conducted forward to a reservoir in the head, then passes into the paired salivary ducts, which turn back for a long excursion on the esophagus before finally returning to the head and emptying on the gnathochilarium.

The two sexes differ markedly in the anterior region, although, in both, all the anterior organs are bound together by a profusion of large nerves and tracheae. In the female, the ovary begins to taper in about the twelfth segment, and by the eighth has been reduced to a bore sufficient to accommodate only one egg at a time. Here it joins the uterus, a stiff, muscular tube, which runs forward to the third segment, where it bifurcates. Its two arms turn dorsolaterally and end in expanded, oval, fibrous capsules, enclosing the vulvae, which lie lateral to the esophagus below the large dorsal bundles of tracheae. Each vulva has a membranous flap covering an opening in its proximal (dorsal) end. The vulvae have muscles attached to their bases and are apparently extrusible. Little seems to be known about sexual processes in *Spirobolus*, but it appears that the eggs squeeze out between the capsules and the vulvae just posterior to the second pair of legs, receiving, at the same time, sperm from the vulvae, which have been previously charged by the male. The vulvae, therefore, act as spermathecae. During egg-laying, the female first regurgitates material and molds it into a cup by movements of her head and legs. She then lays an egg in this cup, closes it over with her head, and polishes it into a spherule with her legs. As the young millipedes develop, they feed upon the capsule and emerge as immature forms, with only seven pairs of legs.

As already seen, the testes begin at the posterior end as straight, slender tubes, transform to beaded tubes with lateral diverticula in the region of segment 35, and then to thicker, convoluted ducts. Farther forward they become increasingly linked by bridges, and anterior to segment 12 they become a single ladder. In segment 7 of the male, a great chitinous process, sheathed in powerful muscles, rears on either side of the esophagus, just anterior to the place where the midgut flares out. These processes are the bases of the gonopods. The Malpighian tubules pass medial to them before proceeding posteriorly. When the esophagus and dorsolateral tracheal sheaves between the gonopods are removed, the testes can be seen (Fig. 198). They run forward to segment 3, then narrow somewhat to form a pair of sperm ducts, each of which curves out far laterally around a posteriorly directed, internal spine of tergum 2, then runs medially below the nerve cord to open at the tip of the penis, just anterior to the main spines of tergum 2. The tracing of the sperm ducts to and through the penis is best accomplished in ventral dissection, particularly if the penis is retracted. The ducts run among powerful muscles of legs and penis. By analogy with other millipedes it appears that the male charges its gonopods with sperm from the sperm ducts by flexing the anterior end of the body so as to approximate the ventral surfaces of segments 2 and 7.

Copulation in *Spirobolus* has been observed to take place at night with the male and female rearing up and coming in contact by their ventral surfaces, with the anterior end of the male curved over the head of the female. Presumably, this brings segment 7 of the male close to segment 2 of the female, so that sperm may be transferred from the gonopods to the vulvae. This interpretation is supported by the fact that the cavities of the gonopods will accommodate the vulvae of the opposite sides.

The head is best dissected by carefully chipping away the top and sides of the epicranium. Mandibular muscles fill most of the posterior and lateral space in the head and should be picked away to expose the brain, salivary ducts, and esophagus. The brain (Fig. 198) consists of two anteroposteriorly flattened supraesophageal ganglia, from which run short fans of nerves to the eyes, and a pair of nerves to the antennae. Two connectives embrace the esophagus and join the brain to an apparently single subesophageal ganglion, from which the nerve cord runs posteriorly out of the head. The esophagus leaves the preoral cavity, or space enclosed by labrum, gnathochilarium, and mandibles, as a simple tube which runs dorsally, then passes below the brain and out of the head. Its anterior part does not appear to be differentiated into a specialized pharynx.

Periplaneta americana

By J. B. Buck and M. L. Keister

Cockroaches are among the most ancient and successful of insects. Because of their efficient sense organs, diversity of feeding habits, and well-developed legs and wings, they are able to thrive in a great variety of habitats. Many species are omnivorous, and some are equipped with symbiotic intestinal protozoans which permit digestion of cellulose. Within historic times, a number of species have become thoroughly established as pests in human habitations.

Periplaneta americana is the largest domiciliary roach distributed over most of the United States. The adults are a rich mahogany brown in color and up to 40 mm. long. Unlike many species of roaches, both sexes are fully winged in the adult stage. If living specimens are available, observe the movements of the long, sensitive antennae, and test the visual and olfactory responses and the powers of climbing and locomotion on various surfaces. Roaches have been used very extensively for experimental work, particularly in the physiology of insect heart, muscle, and nerve. They have also been used in the study of insect behavior and have been shown to be capable of learning to run a maze.

The dissection directions given below for *P. americana* agree well with published descriptions of the anatomy of the slightly smaller *Blatta orientalis*. They can also be used for the study of the small *Blatella germanica*, and the giant Florida roach, *Blaberus craniifer*, although these species show a number of structural deviations from *P. americana*. Some of the major anatomical differences between *B. craniifer* and *P. americana* have been listed at the end of these directions, because the large size of the former species (50–60 mm.) makes it favorable dissection material. The physiological information reported in the following pages has been culled from the literature, and includes some work on common species other than *P. americana*.

External Anatomy. *Periplaneta* is conspicuously flattened dorsoventrally, a fact which contributes materially to its elusiveness. As in most insects, the body is divided clearly into head, thorax, and abdomen. The head is pear-shaped, as seen from the front, and flattened in the anteroposterior axis. The head exoskeleton consists dorsally and posteriorly of a hemispherical cap, the epicranium, which is traversed by an inconspicuous median longitudinal suture. Anteriorly, there are two broad, flattened plates: the clypeus, fused dorsally with the frons, or anterior face of the epicranium, and the labrum, or upper lip, articulated with the ventral edge of the clypeus. The bases of the long, many-jointed antennae arise from the dorsolateral edges of the frons and are partially enclosed posteriorly by the large, kidney-shaped compound eyes, which occupy the most lateral surfaces of the head. Dorsolateral to the antennal bases are a pair of white spots, which are thought to be the remnants of ocelli, or simple eyes, such as are still functional in some insects.

By reflecting the labrum, which bears interesting feeding bristles on its inner surface, the pair of robust, toothed mandibles, or chewing jaws, is exposed. These mandibles are moved laterally by powerful muscles which arise from most of the internal surface of the epicranium. Lateral and posterior to the mandibles are the paired maxillae, and posterior to these the lower lip, or labium (fused second maxillae), consisting, like the maxillae, of numerous parts jointed together. The space bounded anteriorly by the labrum, laterally by the mandibles and maxillae, and posteriorly by the labium, is the preoral food cavity. Projecting ventrally into this cavity is an elongated, fleshy, median process, the hypopharynx or lingua. After the arrangement of the mouth parts has been seen (which is best done from the ventral aspect, with the head tipped forward), the parts should be removed if more detailed study is planned. The maxilla and labium show evidence of being derived from the generalized arthropod appendage by the development of various specialized plates, filter bristles, sensory processes, and accessory manipulators, which aid in tasting and feeding. Names for these various parts can be found in any standard textbook of entomology.

The head is joined to the thorax by a short, slender neck, covered by thin, flexible cuticle, and protected by four pairs of cervical plates. The first thoracic segment (prothorax) is covered dorsally by an enlarged, flattened, shield-like skeletal plate (tergum), the pronotum. The second and third thoracic segments (meso- and metathorax) are each about half as long as the prothorax. Both the latter segments bear dorsolaterally a pair of wings. The two anterior wings (tegmina) are stiffened and thickened and serve as protective covers for the underlying membranous, pleated, functional pair. If a living roach is held under the microscope with a strip of reflecting material inserted under a wing or a tegmen, the circulation of the blood in these structures can be observed by strong reflected light. Ventrally the thorax is covered by numerous irregularly shaped sclerites which serve as protective plates and articular surfaces for the three pairs of legs. The medioventral thoracic sclerites, or sterna, are reduced to small shield-shaped plates.

The three pairs of legs are similar in structure, and each consists of five major segments: the most proximal (basal) flattened, ham-shaped coxa; the very short, triangular trochanter; the large, laterally directed femur, which has a slot along its medial (posterior) edge which accommodates the next most distal segment (tibia) when the leg is folded; the slender, elongated tibia, with a profuse array of spines; and the five-partite tarsus, of which the terminal division bears a pair of claws and an adhesive pad (arolium). The legs are well supplied with hairs, some of which have been shown to be touch receptors, and some to be sensitive to chemical stimulation.

Two pairs of small, slit-like spiracles, or valvular openings into the tracheal system, are found in the thorax. Each anterior spiracle is elliptical in shape, has a flap-like cover hinged along its anterior edge, and is situated far dorsally in the lateral membranous area between pro- and mesothorax, just beneath the pronotum. The posterior spiracles are crescentic and lie dorsolaterally in the membrane between the meso- and metathorax.

Basically, the abdomen of insects consists of eleven segments, each covered dorsally by a single broad tergum, and ventrally by a similar sternum. However, the posterior part of the abdomen of the roach has undergone extensive modification, so that only ten terga and seven sterna (nine in the male) are externally recognizable. The lateral, or pleural, regions connecting the dorsal and ventral sclerites, are membranous, in contrast to the separate plates (pleura) found in some insects. The characteristic waxy coating of the cuticle is the secretion of unicellular glands copiously distributed beneath the cuticle over the entire body surface, particularly in the anterior halves of the terga and sterna.

The actual locations of various internal structures according to segment are subject to some uncertainty, because of the overlapping of the skeletal plates, and because of varying degrees of telescoping of the abdomen. Accordingly, in the following descriptions, structures are arbitrarily assigned to the segment the tergum of which they usually underlie.

The first six abdominal segments are similar in both sexes and are easily recognized, except that the first sternum is much reduced. There is a pair of oval spiracles in each of the first eight abdominal segments. The most anterior pair is near the lateral edges of the first tergum, while the others are at the extremities of small, posteriorly directed, membranous sacs situated in the anterodorsal corners of the pleural regions. The anus opens in an unpigmented membranous area at the posterior end of the body, dorsal to the external genitalia.

The segmentation is much obscured in the posterior region, because of modification of various terga and sterna into accessories of the reproductive systems. Any stretching of the abdomen to display the segments should be done with care, as otherwise the heart and other internal organs may be ruptured. In the male, the ninth tergum is almost completely overlapped by the narrow eighth, except at its lateral edges, which extend ventrally in the pleural region. The tenth tergum is flatter, broader, and more transparent than any of the anterior ones, and is deeply notched in the middle of the hind margin. Originating under the anterolateral corners of the tenth tergum is a pair of long, many-jointed, sensory cerci, which project posterolaterally. The cerci have been shown to be auditory organs, probably receptive to vibrations up to a frequency of about 800 cycles per second, and also to be concerned in the "evasive" reflex of the roach to air currents. No eleventh tergum, as such, is present. Sternum 9 is broad and bears a pair of unjointed, bristled styles near the corners of its posterior margin. It is the most posterior typical sternum and underlies terga 8, 9, and part of 10. The more posterior sterna are highly modified into genitalia, which will be considered later. The abdomen ends rather bluntly. Its fleshy posterior surface is partly covered by a pair of triangular plates, the paraprocts, extending transversely on each side between the cercal region and the anus. Each bears a median vane, lying in the horizontal plane.

In the female, terga 8 and 9 are very narrow and are almost hidden under 7. Tergum 10 is broad and flat and is bilobed, though not as markedly as in the male. The posterior part of the seventh sternum is greatly enlarged and is divided longitudinally into a pair of convex, vertical, gynovalvular plates, shaped somewhat like clamshells. These enclose a chamber in which the egg capsule, or oötheca, is formed. The oötheca is expelled through a vertical, posteriorly directed slit at the end of the chamber. Styles are lacking, but cerci and paraprocts are present. However, the median horizontal vanes of the paraprocts are larger and darker than in the male. The more posterior sterna have been highly modified in form and position in connection with the reproductive system. Their interrelationships will be considered later.

Internal Anatomy. To display the viscera, the wings should be removed with fine scissors and the thoracic and abdominal terga cut through close to their lateral margins, but leaving enough to identify the segments. Each tergum should be removed individually, beginning posteriorly, and extreme care should be taken to tease away all adherent membranes and tissues close to the exoskeleton, leaving the underlying organs uninjured.

Circulatory System. The heart is a mediodorsal, thin-walled, transparent tube (Fig. 201). It extends most of the length of the body, close beneath the exoskeleton, which is translucent and permits observation of heart beat in the uninjured animal. The heart is invested by a diaphanous pericardial membrane and lies in an ill-defined pericardial space among the broad, thin bands of longitudinal muscle which cover the trunk dorsally. Near the posterior margin of the metathorax is the most conspicuous of several segmental enlargements, a dorsal dilation of the main heart tube. This dilation is drawn out laterally into flaps from which run delicate alary muscles, serving to anchor the heart in position laterally, and probably also aiding its dilation. Ventral to these flaps, in the lateral walls of the heart, is a pair of diagonal, slit-like ostia, or intake valves. Similar segmental cardiac enlargements, with ostia and alary filaments, occur in the abdomen (Fig. 201). The heart terminates in the anterior part of the tenth segment, into which it sends a pair of arteries. Among the alary filaments are numerous whitish chains of pericardial cells, or nephrocytes, having the ability to absorb colloidal particles from the

480 ARTHROPODA

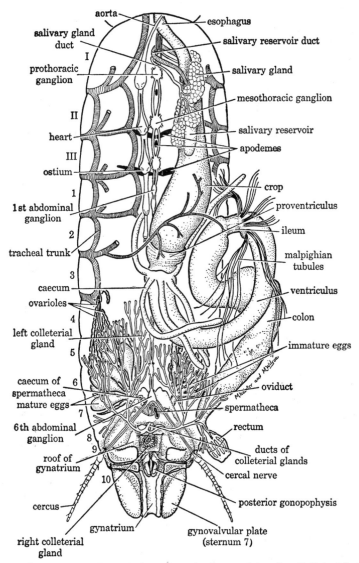

Fig. 201. *Periplaneta*. Dorsal view of trunk viscera of female. Colleterial glands and Malpighian tubules greatly simplified. Heart displaced to the left, and shown in only two abdominal segments. Terminal ganglion displaced slightly anteriorly, and posterior part of gut exteriorized. Both mature and immature ovaries shown. Gynatrium spread laterally to expose gonopophyses. Roman numerals indicate thoracic segments; arabic numerals indicate abdominal segments. ×4.5.

blood. They can be well seen in living roaches injected intra-abdominally on the previous day with dilute trypan blue or Congo red. The contractions of the heart drive the blood in waves, usually anteriorly. The blood reaches the tissues through a pair of flat, relatively short, inconspicuous lateral arteries in each of several segments (not readily visible in preserved material), and through an aorta which runs on the dorsal surface of the esophagus through the neck into the head and terminates without branching just anterior to the circumpharyngeal nerve connectives. The blood is returned to the pericardial space from all parts of the body through indefinite tissue spaces and re-enters the heart through the ostia.

Physiological experiments with roach-heart preparations have indicated that insect cardiac muscle differs from vertebrate cardiac muscle in having a much shorter refractory period and in giving graded, rather than all-or-none, responses. Such experiments, and also observations on heart beat, action of ostia, and so on, in the living roach, are best carried on in a ventral dissection.

Tracheal System. By carefully dissecting away the thin bands of longitudinal muscle lateral to the heart in the thorax and abdomen, large tracheal trunks can be found looping in from the spiracles in each segment and running close along the ventrolateral wall of the heart. In the prothorax, two pairs of particularly large trunks run forward into the head, one beside the aorta, the other ventrolateral to the esophagus. Numerous branches, linking together spiracles in contiguous segments, and supplying various internal organs, can easily be seen, and further parts of the tracheal system will be encountered in all subsequent dissections. Microscopic examination of almost any organ of the roach will show the profuse supply of tiny tracheae which ramify through the tissues. Deposits of fat are particularly prominent in the spiracular regions.

Digestive System. The alimentary canal begins in the preoral food cavity, continues as a narrow, tubular pharynx which runs between the circumpharyngeal nerve connectives, and leaves the head as a slender esophagus. Study of the cephalic portions of the digestive system should be postponed until the head is dissected in studying the brain.

By carefully clearing away the heart, muscles, and tracheae in the mid-dorsal region of the thorax, the underlying esophagus and paired salivary glands can be exposed. The esophagus is a narrow, muscular, ridged tube, which dilates, at about the meso-metathoracic boundary, into a large, bulbous, thin-walled crop, which may be full of food.

The white, hair-like filament on the dorsal midline of the esophagus is the recurrent nerve. It runs into a triangular ventricular ganglion near the junction of the esophagus and crop. Two nerves from the posterior corners of this ganglion run along the ventrolateral surfaces of the crop for some distance.

The paired salivary glands are opaque, whitish, lobulated masses extending from the posterior quarter of the prothorax to about the posterior border of the mesothorax. When the glands are carefully freed from other tissue and from the esophagus, it will be seen that each is composed of a number of more or less distinct pebbly white lobes, interconnected by white, thread-like tubes, which unite into a single duct. Between the lobose portions of each gland and the esophagus, posteriorly, is an elongated, flattened bladder, or reservoir, with a fibrous wall. This reservoir also empties anteriorly into a duct but is not directly connected with the gland. By pinning the gut to one side, the ducts from the gland and reservoir of the other side can be traced as they run steeply anteroventrally near the median line among large muscles and tracheae. The duct of the reservoir is the more anterior and larger of the two. Both ducts are peculiar in having spirally thickened cuticular linings like tracheae, although in the duct from the gland this structure is concealed, over most of its length, by a thick, white epithelium. Both ducts pass under the ventral nerve cord and turn forward, where they join the corresponding ducts from the other salivary gland and reservoir. Then the two common ducts unite with each other and run into the head through the occipital foramen, the opening in the posterior surface of the epicranium. The details of their course are best left for completion with the dissection of the nervous system. The respective functions of the salivary gland and bladder are not entirely understood, but it is known that the salivary secretion is almost neutral and contains a powerful amylase.

If live roaches are available, the anterior parts of the alimentary canal can be prepared for study simply by immersing a specimen in water and pulling off its head, which brings with it the esophagus, crop, and salivary glands.

The thorax is mainly occupied by the powerful leg muscles. Their arrangement, however, is too intricate to permit a satisfactory dissection to be made without using several specimens.

The study of the abdominal viscera requires the removal of the thin covering bands of longitudinal muscle and the careful teasing away of patches of chalky-white, friable "fat" wherever it invests or obscures the structure under observation. However, since parts of the reproductive system lie in the fat in the dorsolateral regions of the abdomen, no dissection beyond that absolutely necessary to display the gut should be attempted at this time.

The crop extends from the meso-metathoracic boundary to about the posterior limits of the third abdominal segment. Its wall, as can be seen in a mounted fragment under the compound microscope, consists of an outer layer of circular-muscle fibers, a layer of sparse bundles of longitudinal muscle and fine tracheae, and an innermost layer of squamous endothelial cells. Often, also, its chitinous stomodeal lining can be seen within as a wrinkled, transparent membrane.

Posteriorly, the crop narrows and fits into a compact, muscular, turnip-shaped gizzard or proventriculus, which, like all the digestive system, is well supplied with large tracheae. The narrow posterior end of the proventriculus is produced to form a cylindrical stomodeal valve which protrudes into the lumen of the succeeding part of the alimentary canal, the ventriculus, which is the beginning of the midgut.

The ventriculus has about the same diameter as the widest part of the proventriculus but is relatively thin-walled. From its anterior end, which forms a ring or shoulder around the valve of the gizzard, arise eight floppy, transparent, pyloric caeca, each about 8 mm. long. The caeca are arranged in a rosette around the gut, some of them extending anteriorly and some posteriorly. The midgut itself is lined internally with unusually tall columnar cells, and its lining is thrown into folds or ridges. It also encloses a delicate, transparent, inner tube, the peritrophic membrane, probably a secretion of the intestinal epithelium. The function of the peritrophic membrane is supposedly to protect the cells of the endothelium from abrasion by hard particles in the food. It has been shown to be permeable to digestive enzymes.

Posterior to the gizzard, the ventriculus makes a wide counterclockwise coil, extending posteriorly into the sixth abdominal segment, then curving forward into the fourth. At the most anterior point of this coil, there occurs a narrow annular region from which arises a tangle of extremely fine, whitish threads (yellowish in fresh preparations), the Malpighian tubules. These are insinuated over and between most of the abdominal viscera. By combing the tubules out, they can be seen to arise in six groups, of about a dozen each. The Malpighian tubules are commonly assigned the role of extracting soluble wastes from the blood with which they are bathed, concentrating these wastes (e.g., into "urate" crystals, which can be seen in their lumina), and passing them into the gut for subsequent defecation.

They have also been reported to be extremely rich in ascorbic acid and vitamins of the B group, and to contain a dipeptidase. In a fresh specimen dissected in insect Ringer's solution, the contractions of the Malpighian tubules and of most of the parts of the digestive system are easily observed.

Immediately posterior to the level of entry of the Malpighian tubules is a short, narrow, muscular region of the gut, the ileum. Following this is a large, thin-walled, brownish, flaccid tube, the colon. At the point of junction is an internal constriction, the ileocolic valve. The colon makes a couple of loops, then runs back to the ninth segment, where it constricts into a muscular rectum, which in turn empties to the outside through a slit-like anus under the tenth tergum. In the female, the rectum and the posterior part of the colon are concealed in a snarl of slender, glistening, yellow-brown, dichotomously branching tubules which are not actually connected to the gut. These tubules make up the left colleterial gland (the right one being much reduced). With care the tubules can be parted medially into two groups, permitting the removal of the alimentary canal. In any case, the dissection of the posterior gut in both sexes should be conducted with care, so as not to destroy parts of the underlying and surrounding reproductive system.

By cutting across the posterior part of the crop, and dissecting the rectum free from the fleshy lobes surrounding the anus, then making sure that the gut has been freed along its entire length from the numerous large tracheae which supply it, the whole tube can be lifted out into a dish of water. Here the anatomy and arrangement of the parts discussed may be observed with greater clarity.

Upon dissecting the outer cup-shaped muscular wall of the gizzard (proventriculus) away from its inner sclerotized lining, six shiny, brown, irregularly shaped cavities, flanked by thin, dark, longitudinal supporting rods, will be exposed. The cavities extend into six hard teeth, which project into the lumen of the gizzard. Normally, the cavities are occupied by fleshy pads which project inward from the inner surface of the muscular sheath of the gizzard. If the gizzard is split longitudinally by cutting between two of the teeth, the interior of the grinding mill is exposed. Posterior to each tooth is a "cushion" covered with fine setae, and between the teeth are longitudinal ridges, also with setae, in definite number and arrangement.

By cutting through the gut at the ileocolic valve, groups of gold-colored setae can be seen projecting into the lumen.

Careful dissection of the rectum discloses that its wall is folded in on itself anteriorly and has six longitudinal outpocketings, the rectal papillae or glands.

Digestion in the roach is believed to occur mainly in the crop, where food which has been broken up by the mandibles and mixed with salivary secretion is temporarily stored. Juices containing lipase and proteinase, formed in the pyloric caeca, are passed forward into the crop (the stomodaeal valve being closed to prevent backflow). The wall of the crop is impermeable to water, and although some absorption of fat may occur there, most of the products of digestion are absorbed after the food has passed back into the midgut. The teeth and filters in the proventriculus aid in breaking down the food particles during this passage. It is uncertain whether the midgut epithelium secretes (aside from forming the peritrophic membrane) as well as absorbs, although cytological changes interpreted as secretory have been observed. The hindgut serves primarily for temporary storage of solid wastes and as the principal site of absorption of water. In the latter function, the rectal glands are thought also to be important.

Reproductive System of the Male. After the rectum and the fleshy lobes around the anus have been removed, the posterior end of the abdomen may be spread laterally with care, to display better the copulatory apparatus. This will be seen to consist of an amazing and asymmetrical array of hooks, spines, plates, teeth, knobs, pads, and bristles, together with a number of powerful muscles extending into the abdomen (Fig. 204).* The long (3 mm.), slender, sinuous, hard structure which lies farthest to the left in the spread preparation is the titillator. In the intact animal, its distal end, which bears a small, corkscrew-shaped barb, lies in the medioventral line. The titillator is one of a complex of processes called the left phallomere. The lobe closest to the titillator, which ends bluntly in two black knobs and bears a dorsolateral vane and a ventrolateral hairbrush-like pad along its length, is called the pseudopenis. In the intact animal, its distal end lies slightly to the left of the midventral line, overlapping the titillator. Medial to the pseudopenis are several minor lobes of the left phallomere of which the most ventral, which bears a long spur, is called the asperate lobe. In the mid-dorsal position is a thin, blade-like process terminating in a saw-toothed edge and two large teeth pointing to the left, and bearing a sickle-shaped spur

---

* The functions of these parts and those of the female concerned in copulation have only recently been made clear (Gupta, 1947). We have adopted Gupta's terminology.

about midway along its posterior edge. This structure, which originates not far ventral to the base of the right cercus, is the serrate lobe of the right phallomere. Underlying the right phallomere, in the right ventral position among the other external genitalia, is the large, fleshy, ventral phallomere, bearing on its ventral surface a hardened brown plate shaped like the sole of a slipper.

The testes are small, pale, lobulated masses, about 2½ by 1 mm., lying far laterally and dorsally in the fourth and fifth abdominal segments. They are embedded in fat just internal to the spiracles and to the robust dorsoventral segmental muscles which aid tracheal ventilation by compressing the abdomen. The vasa deferentia run posteriorly from the testes as delicate white tubes. In *P. americana* the sperm are said to be formed and stored while the animal is still in the last nymphal (immature, wingless) stage, after which the testes partially degenerate. This may account for the fact that the testes are difficult to find in the adult male, and that the vasa deferentia can rarely be followed throughout their courses.

Most of the remainder of the male reproductive system is concentrated in the region of a large, median, dandelion-shaped mass of tubules in segments 7, 8, and 9 (Fig. 204). This mass, which will be called the utricular gland, consists of two types of finger-like utricles or tubules, whitish or yellow in color, and rather uniform in diameter. The longer tubules tend to be peripheral in location, and the shorter ones central, as seen from the dorsal surface.

To expose the internal genitalia further, the large muscle running diagonally from the left margin of tergum 8 to the base of the right phallomere must be cut, and the fat, muscles, Malpighian tubules, and tracheae overlying the utricular gland carefully cleared away. Hidden in the interior of the mass of utricles is the slightly enlarged anterior end of the thick-walled, fibrous, ejaculatory duct. It can be exposed by picking off the individual utricles. In doing this, extreme care should be taken to preserve the paired vasa deferentia, which enter the anterodorsal surface of the base of the ejaculatory duct. When the gland is sufficiently denuded, it will be seen to have overlain the large terminal ganglion of the nervous system in the seventh segment. This ganglion is thought to represent the fusion of ganglia of several of the posterior segments. From it a number of pairs of nerves supply various viscera, and in particular one small and one very large nerve run diagonally backwards to each cercus. The cercal nerves are much favored as objects for neurophysiological investigation. The two vasa deferentia, running backward from the testes, dip ventrally under the cercal nerves, turn medially almost to the midline, and then, as already seen, pass forward into the mass of utricles.

The utricular gland should now be dissected out sufficiently so that it can be turned over to the right, off the large ganglion. To do this, it is necessary to cut a large branch of the left cercal nerve which runs to the ventral surface of the gland. This surface of the utricular gland presents much the same appearance as the dorsal, except for the presence of the ejaculatory duct. Grouped posteriorly among the short utricles in the central part of the gland are a number of mottled, white tubules, slightly larger and more bulbous than the short utricles (Fig. 204). These are the seminal vesicles, which store the sperm. If a vesicle is picked off and teased open in water on a slide, the large spermatozoa can be recognized microscopically, even in preserved material. Before copulation the sperm are packaged into a single spermatophore.

The relations of the muscular ejaculatory duct to the utricular gland can be further clarified by picking off some of the utricles on the ventral surface of the gland. The duct itself can be traced back along the midline by cutting the right cercal nerves, reflecting the terminal ganglion to the left, and gently clearing away fat, nerves, and tracheae ventral, lateral, and posterior to the nerve cord. At the same time, other structures will be exposed. The most prominent of these is the slender, flattened, white, unpaired phallic or conglobate gland. This is about 9 mm. long overall, and lies slightly to the right of the nerve cord. Its anterior end, in the anterior part of segment 6, is an inflated region about 1.5 by 0.75 mm., which may partly overlie the penultimate ganglion, or droop to the right. The gland tapers posteriorly, passing under the terminal ganglion.

To either side and ventral to the nerve cord run two large longitudinal muscle strips. The one on the right curves posteriorly through the base of the main mass of the left phallomere and is inserted in the base of the titillator. The muscle band on the left of the nerve cord also runs into the base of the left phallomere. Between these two muscle bands run the ejaculatory duct and the duct from the phallic gland. The former, which lies close to the muscle bundle on the right, runs posteriorly into the fleshy base of the ventral phallomere (i.e., the process with the slipper-shaped plate). Here it opens to the outside. As is consistent with its origin by invagination of the ectoderm, the ejaculatory duct is lined with cuticle. Near the distal end of the duct, this cuticle is thickened into a slender, brown, canoe-shaped sclerite. The duct from the phallic gland, which lies closer to

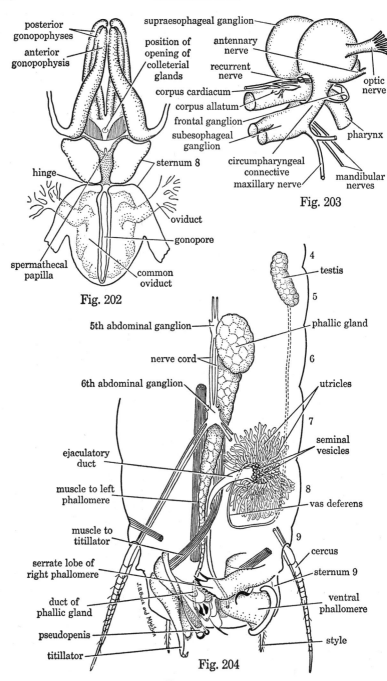

Figs. 202–204

the muscle bundle on the left, passes directly back to an opening in the axilla between the pseudopenis and the asperate lobe of the left phallomere.

Reproductive System of the Female. After the alimentary canal has been removed, the paraprocts and the fleshy tissue underlying the position of the anus should be dissected away, taking care not to encroach on the nearby viscera. This operation exposes the dorsal region of the two clamshell-like gynovalvular plates of sternum 7. By spreading apart the plates laterally, the gynatrium, or copulatory and oötheca-forming chamber, can be seen (Fig. 201). The plates are lined with a thick, fleshy layer which is thrown into lip-like folds on either side of the dorsal slit where the plates meet. Likewise, the ventral lining is invaginated to form a longitudinal fold which permits distention of the chamber during the formation of the oötheca.

As previously mentioned, the sterna posterior to 7 have been highly modified in shape and position. This modification consists of the invagination of sterna 8 and 9 to form the anterior part of the gynatrium. Attached to the inside surface of the anterior part of the dorsal wall of the gynatrium, is a robust, median complex of posteriorly directed processes somewhat resembling the jaws of a pair of pliers. This structure actually consists of three pairs of forceps-like structures, the gonopophyses, the blades of which are individually movable by means of muscles attached to their bases. The large, fleshy, dorsal pair, and the thin, hardened pair which they enclose, are called the posterior gonopophyses. They are regarded as derivatives of the appendages of the ninth segment. The third, dark, narrow, j-shaped pair, which are overlain dorsally by the posterior gonopophyses, are the anterior gonopophyses. They are derivatives of the appendages of segment 8. They will be seen more clearly in a later stage of dissection (Fig. 202). The gonopophyses, collectively, correspond to structures used as ovipositors in many insects. In the roach, however, they serve to shape the oötheca. They are capable of considerable anteroposterior movement and hence may vary somewhat in position in different individuals.

FIG. 202. *Periplaneta.* Gonopophyses and parts of anterior and dorsal walls of gynatrium. Internal view, with structures laid out flat. ×11.

FIG. 203. *Periplaneta.* Posterolateral view of brain and associated structures. ×23.

FIG. 204. *Periplaneta.* Dorsal view of reproductive system of male. Posterior part of abdomen spread apart laterally. Numbers indicate abdominal segments. ×8.

After the alimentary canal is removed, it can be seen that the slender tubules of the left colleterial gland fill much of the space on both sides of and ventral to the gut, as well as dorsal to it (Fig. 201). The tubules are mainly yellow, but vary somewhat in color in different regions, and may be quite dark. They are remarkably uniform in diameter throughout their courses. Because of the brittleness of the tubules, it is not practicable to trace every branch of the gland, but by separating a mass from the main gland and unravelling it in a dish of water, the profuse dichotomous branching can be seen. The connection of the colleterial glands with the other genitalia will be studied shortly.

The ovaries lie laterally in segments 4, 5, and 6, embedded in fat. If the specimen is mature, clusters of ovoid eggs, about 2 mm. in length, will be seen laterally in segment 6. From the distal ends of some of these eggs, segmented, tapering rods of pale tissue extend forward, sometimes as far as segment 3. These are the ovarioles, containing strings of immature eggs. A number of them are bound together into a rather flat mass on each side of the abdomen, and fit between the gut and the lateral body wall. They are fairly brittle and hence must be dissected with care. However, the colleterial glands should be traced first.

Gradually pick away the tubules of the (left) colleterial gland, beginning at its anterior limits, and leaving undisturbed, for the present, tubules which run near the ovary or far ventrally. As the bulk of gland tissue and fat in segments 5, 6, and 7 is reduced, the nerve cord becomes visible ventrally, together with the dark edges of various plates making up the roof of the gynatrium. The gynatrium itself is roughly triangular in dorsal aspect. Its apex is usually about at the middle of the anterior margin of segment 6, and its sides extend almost to the bases of the cerci. However, its position and that of structures on its dorsal surface vary somewhat with the reproductive stage of the specimen.

Clearing off the tip of the gynatrium, in the anterior part of segment 6, exposes the terminal ganglion of the nerve cord. The region immediately behind it should be dissected with special care. Embedded just under the posterior edge of the ganglion is the spermatheca, a small, sausage-shaped body about the length of the ganglion (Fig. 201). It is about twice the diameter of the colleterial tubules and has a thick, white wall with a cuticular lining, which is often dark brown. By cutting the left cercal nerve (when it is finally exposed), and turning back the terminal ganglion, it can be seen that the spermatheca has a short, contorted duct, which is joined by a similar short, coiled caecum before it turns ventrally into the gynatrium.

As the yellowish tubules of the left colleterial gland are gradually removed, there is exposed a white mass of soft, friable tubules of slightly smaller diameter but showing a similar dichotomous branching. These tubules belong to the right colleterial gland (Fig. 201). By progressively following and removing both sets of tubules, each can finally be traced to a single short duct. These ducts unite to form a common duct of somewhat greater diameter. This runs directly ventral through a fleshy mass filling a shallow, semilunar depression in the dorsal wall of the gynatrium, at a point about a millimeter posterior to the opening of the spermatheca, and opens into the gynatrium dorsal to the bases of the posterior gonopophyses.

The ovaries and oviducts, lying lateral to the gynatrium, should now be dissected. Each of the mature eggs, of which there may be up to eight per ovary, occupies the base of an ovariole and is invested basally by a delicate white filigree of cells in chains. Usually four eggs on each side are loosely fitted into the flared end of a flat, white tube, forming a structure somewhat resembling a four-fingered glove. The ovarioles containing the other mature eggs usually arise separately. All the ovarioles of one ovary fuse posteriorly to form a single, fibrous oviduct, which turns toward the midline and passes under the anterior end of the gynatrium (Fig. 202). Here the oviducts from both ovaries unite to form a soft flabby sac, which has many tracheae in its walls. This, in turn, opens immediately into the gynatrium through a large, fleshy-lipped, medial gonopore. The gonopore is a vertical slit in a shield-shaped plate derived from the eighth sternum. This plate (Fig. 202) is inclined posteroventrally and forms the anterior wall of the gynatrium. Dorsoanteriorly, it is hinged to two other, smaller, paired plates, also derivatives of the eighth sternum. These smaller plates form the anterior part of the dorsal wall of the gynatrium and are bound posteriorly to the bases of the posterior gonopophyses. For adequate study of these structures, the gynatrium and the entire seventh sternum should be removed as a unit. The posterior gonopophyses can then be withdrawn anteriorly from the oöthecal part of the chamber and folded back (dorsally and anteriorly) together with the dorsal wall of the gynatrium. The anterior, genital part of the gynatrial chamber can then be separated from the posterior, or oöthecal, part (sternum 7). When the genital portion is then opened out flat (Fig. 202), it exposes the following structures: the main plate of sternum 8, with its central gonopore;

the ventral surface of the posterior gonopophyses, with which are associated the paired, j-shaped anterior gonopophyses; and a papilla between the small, dorsal pair of plates derived from sternum 8. On this papilla opens the spermathecal duct. In the part of the dorsal wall of the gynatrium which overlies the bases of the posterior gonopophyses may be found the opening of the colleterial glands. If Fig. 202 is traced on paper and the paper folded at the hinge line, the normal interrelations of the various structures may become clearer.

Mating behavior in roaches is apparently quite variable. According to the most recent description of copulation in *Periplaneta*, the male and female back up to each other, and by the use of the titillator, the male pries apart the gynovalvular portions of sternum 7 of the female sufficiently to permit the insertion of his genitalia into the gynatrium. At the same time, the anterior and dorsal walls of the gynatrium shift posteriorly, enlarging the cavity and making the gonopore and spermathecal papilla more accessible. The pseudopenis is inserted into the gonopore of the female and rotated 90°, so that it is held in position. The basal parts of the two anterior gonopophyses are held by the lateral teeth and by the sickle-shaped hook of the serrate lobe of the right phallomere, respectively, while their distal parts are clamped by basal plates of the right phallomere. The ventral phallomere shifts to the right, allowing the lips of the ejaculatory duct to dilate and come into a position to expel the spermatophore directly onto the spermathecal papilla. The sequence of the preceding events is not completely clear, but it is known that the spermatophore is not attached during the first hour of the hour and a quarter which copulation usually lasts. After the spermatophore is attached, the secretion of the phallic gland is poured out and hardens in a couple of hours, forming a case around the spermatophore and cementing it firmly to the spermathecal papilla. The sperm move into the spermatheca in the course of the next 20 hours or so, after which the spermatophore is discarded.

In egg-laying, the eggs enter the gynatrium alternately from the two oviducts, receive sperm from the spermatheca, and are coated with the secretions of the colleterial glands as they pass into the oöthecal chamber. In life, the tubules of the left colleterial gland contain a white, opalescent fluid filled with calcium oxalate crystals. In contrast, the translucent tubules of the right gland contain a clear, watery solution. The secretion of the left gland has been shown to contain a protein, that of the right gland a dihydroxyphenol. The contents of the two glands are secreted together around the eggs as they issue through the gonopore. In the presence of oxygen and of an enzyme (a polyphenoloxidase), the phenolic compound is oxidized to the corresponding quinone, which then combines with the protein from the left gland to form the dark, non-chitinous, extremely resistant, tanned protein of the oötheca. In the gynatrium, the colleterial secretion is molded by the walls of the oöthecal chamber and by the gonopophyses into a purse-shaped capsule with a zipper-like suture along its dorsal midline. Each hard, dark oötheca contains sixteen eggs, standing nearly vertically in two staggered rows, the eggs on the right side having come from the left ovary and vice versa. Each egg is precisely oriented in all three axes, both with respect to the other eggs in the oötheca and to its former position in the ovary. The capsule requires a day or more for completion, during which time it protrudes farther and farther out of the oöthecal chamber.

The Nervous System. The two large, bilobed, white suprapharyngeal ganglia, or brain, lying almost directly between the bases of the antennae, can be exposed by carefully chipping away the epicranium and clypeus and dissecting away the underlying muscles (Fig. 203). The brain rests upon an X-shaped endoskeletal plate, the tentorium, which cannot yet be seen, but which should be kept in mind as the dissection proceeds. The anterior arms of this plate are attached to the anterolateral regions of the epicranium, close to the bases of the mandibles, and the posterior arms to points lateral to the occipital foramen.

From the dorsolateral regions of the brain arise a pair of large optic nerves which can be traced through their short courses to the compound eyes. Dissection easily shows the association of their fibers with the ommatidia, and also much of the gross structure of the compound eye (see discussion of the compound eye in the description of the crayfish). A pair of short antennary nerves are given off laterally from the small anteroventral lobes of the brain, and two stout connectives run posteroventrally from the large posterodorsal lobes of the brain. The latter nerves encircle the pharynx and pass through a central aperture in the tentorium to the subesophageal ganglion, which lies just inside the ventral rim of the occipital foramen. The pharynx, accompanied along part of its course by the lingual muscles, runs dorsally between the anterior arms of the tentorium, then passes posteriorly between the nerve connectives. The short region of the pharynx posterior to the connectives has folds in its lining and is sometimes called the posterior pharynx. It joins the esophagus, which passes out of the head between the posterior arms of the tentorium.

Large tracheae are also found in the region of the brain. Several pairs of nerves run forward from the subesophageal ganglion to the mandibles, maxillae, and labium. In tracing the labial nerves, look also for the median common salivary duct, which opens on the posterior surface of the base of the lingua. Examination of the cephalic nerves is facilitated by removal of the labrum and of the labium with its internal cuticular sheet. First, however, trace the labial nerves.

From each circumpharyngeal connective, a small nerve passes forward and, turning medially, enters a small, triangular frontal ganglion lying on the pharynx close under the frontal region of the epicranium (Fig. 203). From this ganglion, the recurrent nerve passes back along the dorsal midline of the pharynx beneath the brain. Just posterior to the connectives, it is associated with a pair of minute, lateral ganglia. Slightly farther posteriorly, it sends a nerve on each side to two associated endocrine organs lying lateral to the esophagus. The more elongated, anterior of these organs is the corpus cardiacum, of unknown function in the roach, but having the power, even in high dilution, of causing a contraction of red chromatophores of Crustacea. The rounded, more posterior, corpus allatum is known in other insects to produce hormones, which, among other functions, inhibit metamorphosis during larval molts and control egg production in the adult. The glands are fairly difficult to find in preserved roaches but show up clearly in fresh tissues, where the corpus cardiacum is pale blue and the corpus allatum pale yellow. Posterior to these endocrine glands, the recurrent nerve runs along the dorsal surface of the esophagus into the thorax, as already seen.

One other head structure of interest can be observed at this time, although it is not directly concerned with the nervous system. This is an accessory circulatory organ, a small vesicle which is found just posteromedial to the base of each antenna and is connected with its counterpart by a strand of muscle passing under the brain. These are the ampulliform organs, which aspirate blood from a sinus surrounding the brain and pump it into the antennae. The white, corkscrew-shaped vessel, which conducts the blood outward, can be found under the transparent membrane surrounding the base of the antenna.

The double nerve cord, leaving the subesophageal ganglion, passes posteriorly through the occipital foramen and runs along the midventral line. To expose the cord, a complex system of diagonal muscles criss-crossing the ventral thoracic wall must be removed. These muscles, which are concerned with movements of the legs and thoracic segments, are supplied by numerous nerves from the three large thoracic ganglia. Two types of leg innervation have been distinguished physiologically, one controlling the rapid muscle twitches involved in locomotion, the other concerned with the slower postural movements. Many of the muscles are anchored at one end to supporting knobs invaginated from the exoskeletal plates. These supports, or apodemes, correspond functionally and in derivation to the endophragmal skeleton of the crayfish. In both meso- and metathorax, there is a conspicuous, median, tubular apodeme protruding between the ventral nerve trunks. Extending laterally from beneath the nerve cord, in the posterior portion of the meso- and metathorax, are two pairs of flattened, wing-like apodemes (Fig. 201).

The ventral nerve cord is flanked by large tracheae through most of its course. In both sexes, there are six abdominal ganglia, each with a number of pairs of lateral nerves. They appear to vary somewhat in position, probably due to different degrees of telescoping of the abdomen or to shifts caused by movements of the genitalia. It has, in fact, been stated that the latter are facilitated by slack in the cord between ganglia. Both sexes have one ganglion almost at the junction of the metathorax and first abdominal segment, and the male has a second near the posterior limits of the first abdominal segment. In the male, ganglia appear to be lacking in segments 3 and 5, and in the female in either 6 or 7, depending on the position assigned to the terminal ganglion.

In dissecting roaches, confusion is sometimes occasioned by the presence of parasitic worms, of which a number of species are harbored (see Miall and Denny). Thread-like filarias, which seem to favor the fat bodies lining the ventral body wall near the nerve cord, are easily mistaken for convoluted gland ducts.

SUPPLEMENTARY NOTES ON *Blaberus craniifer*

Aside from numerous minor differences from *Periplaneta americana* in anatomical proportions and shapes, *Blaberus* shows the following differences in structure. There are large tracheal loops surrounding posteriorly the segmental enlargements of the heart. The salivary glands are diffuse and transparent, the reservoirs are very large and delicate, and the gland duct runs in contact with the reservoir duct. The crop and proventriculus are rotated 90° to the right and appear asymmetrical. The proventriculus is much reduced, and there is no distinguishable ileum. There are ten thick, tapered caeca of assorted lengths. Spiracle 1 opens on the tergum, 2–7 on pleura, and 8 on a tergal extension.

The testes are large and are situated in segments 6 and 7. The right paraproct of the male has a hook directed anterolaterally. The titillator is replaced by a short, flat, setose plate. The ejaculatory duct opens in the base of the pseudopenis of the left phallomere, which is thick, knob-like, and dorsally grooved, and terminates in a horseshoe-shaped pad of bristles. The right phallomere appears to be lacking. The ventral phallomere is a process shaped distally like a calf's foot and enclosed in a voluminous membranous sleeve. It receives the duct of the phallic gland. Posterior to the utricular gland, there is an accessory gland, apparently not present in *P. americana*, which has a duct opening with the ejaculatory duct.

In the female, there are four spermathecae. The uterus is a large sac, opening, together with the oviducts, into the anterodorsal end of the gynatrium. In gravid females, it accommodates a huge (4 cm.), j-shaped oötheca, containing forty-eight eggs, which displaces most of the abdominal viscera. There is no gynovalvular portion of sternum 7 or true oöthecal chamber.

Drosophila melanogaster (Larva)

By J. B. Buck and M. L. Keister

Because of their transparency, many dipterous larvae are very suitable for study of insect anatomy, cytology, physiology, and development as seen in the living specimen. *Drosophila* has been chosen for description because it is widely available and can be raised under controlled conditions, but similar observations can be made on young blowfly larvae, and larvae of mosquitoes, small species of *Chironomus*, and particularly *Sciara*. *Drosophila* larvae go through four developmental periods (stages). The last two molts take place after the larva has stiffened and become motionless, so that the last two larval skins are not shed but remain covering the pupa.

The general technique of preparation involves pressing the larva between slide and coverglass, so that it is held firmly for microscopic study. Early third-stage larvae (at 25°, about 4 days after hatching) are at the best age, since they are large enough (3–4 mm.) to handle easily, but have not yet developed so extensive a fat body as to obscure seriously the other viscera, as often happens in older larvae.

Slender, actively feeding larvae are preferable to plump, sluggish larvae crawling on the wall of the culture bottle. However, a good deal can be seen even in older larvae, particularly if they are first fasted for a day or two by being left on wet filter paper in a tightly sealed dish.

In preparing the mount, the larva is moistened with water, oriented transversely to the longitudinal axis of the slide, and covered with a no. 1 coverglass. With the preparation on the stage of a dissecting microscope, the coverglass is held by the edges with thumb and index finger of one hand and pressed down until the larva is sufficiently flattened and is oriented as desired. Orientation is easily accomplished by rolling the larva by moving the cover laterally, and the longitudinal extension of the body is aided by sliding the cover gently back and forth in the direction of the longitudinal axis of the larva. The desired degree of flattening is achieved when the internal hydrostatic pressure has risen sufficiently to evert the finger-like spiracles at the anterior ends of the two main tracheal trunks (Fig. 205). A drop of melted paraffine (or, better, beeswax) is then put along the edge of the coverglass at three or four points to cement it in position. Then, after running a small drop of water under the edge of the coverglass to maintain humidity, the mount is ringed with Vaseline to prevent drying. (If an aquatic larva is used, most of the space under the coverglass should be filled with water.) The larvae in such preparations often remain in good condition for a day or longer. Several mounts should be made, both to allow for possible escapes and to give views from a number of orientations. Observations can be made with all powers of the compound microscope.

General Topography and Tracheal System. The body of the larva consists of eleven segments, exclusive of the pointed anterior region containing the black mouth hooks (which is not considered a true segment). The first three body segments are destined to form the adult thorax but at this stage also contain head structures which have been invaginated into the trunk segments. The boundaries of the segments are marked externally by slight constrictions and by annular bands of minute, backward-pointing denticles studding the transparent cuticle. Small patches of similar denticles surround cutaneous sense organs sparsely distributed over the body surface.

The prominent tracheal system consists essentially of two large, longitudinal trunks, lateral to the mid-dorsal line. (These trunks are likely to collapse and become fluid-filled under the pressure of the coverglass.) Each trunk gives off segmental branches both laterally

498   ARTHROPODA

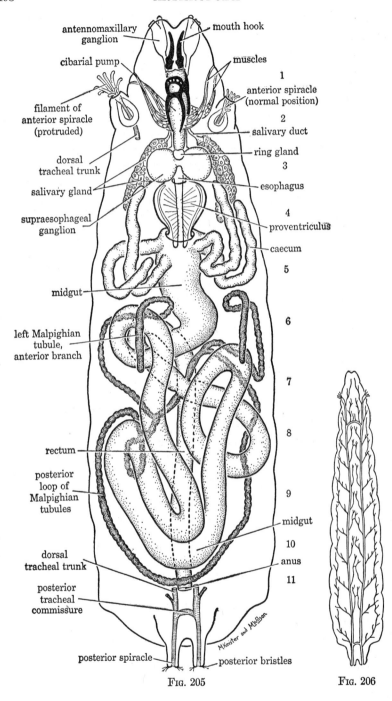

Fig. 205   Fig. 206

and medially. The lateral branches are interconnected to form a small, longitudinal trunk slightly ventral to each midlateral line. The paired medial branches of segments 3–10 are often joined transversely to form bridges between the trunks. In addition, there are two large direct connections between the trunks in segments 2 and 11 (Fig. 206). A profuse arborescence of fine tracheae from all the main branches covers most of the viscera. The minute, terminal tracheal capillaries are well seen in the body wall and on the surface of the gut.

Each dorsal tracheal trunk opens to the exterior, anteriorly and posteriorly, through projecting spiracles. Each posterior spiracle consists of a horny, brown spiracular plate, pierced by three slit-like openings. These openings admit air to a relatively simple atrium filled with a felt-like layer of cuticular hairs. A retractor muscle is attached to the inner surface of the spiracular plate. Externally, the rim of the plate bears four tufts of minute bristles covered with a hydrophobic material, which protects the spiracle from entrance of water. When the larva burrows in its semiliquid food, the bristles of both spiracles cooperate in retaining a single air bubble. By analogy with the breathing adaptations of diving beetles, this bubble may serve as a physical lung providing gaseous oxygen for the submerged larva. A very interesting feature of the internal anatomy of the region surrounding the posterior spiracles is the presence of large cells with clear, sinuous canals within their cytoplasm. These unicellular, perispiracular glands are believed to secrete the hydrophobic material which coats the bristles and spiracular plates.

In first- and second-stage larvae, the anterior spiracles are simple. In third-stage larvae, each consists of seven to nine finger-like processes, which are open at the tip (Fig. 205). These processes are ordinarily kept retracted within a pouch. Neither they nor the spiracular chambers show the spiral thickenings in the cuticular lining which are seen in all the other parts of the tracheal system. Though the anterior spiracles are open to the air, the facts that they are ordinarily retracted, and that the anterior end of the larva is buried in the semiliquid food almost continuously, indicate that the posterior spiracles are the principal functional ones. During pupation, the anterior spiracles are everted, and they are then generally considered to become functional. However, little physiological work has been done on the tracheal system of *Drosophila*.

FIG. 205. *Drosophila* larva. Dorsal view, with most of tracheae, and all of fat bodies and heart omitted. ×25.

FIG. 206. *Drosophila* larva. Dorsal view of tracheal system. ×13. (After Rühle.)

The most conspicuous structures in the body are probably the fat bodies, which are long, irregular ribbons, darkly granular by transmitted light, white by reflected light. The fat bodies extend most of the length of the body in two masses, one on each side of the mid-dorsal line. They vary in extent, depending on the age and nutritional history of the larva. Small patches of fat are found laterally in each segment. Near these patches, in segments 3–10, are minute rudimentary spiracles marking the positions where functional lateral spiracles occur in many insect larvae. The fat body is regarded as a storage site for reserves to be used during the pupal transformations. This view is consistent with its reduced state in starved larvae.

Various regions of the body wall are interconnected by delicate strands of striated muscle, many of which can be seen twitching or slowly contracting at intervals.

Heart and Circulation. The heart is a transparent tube extending the length of the body in the mid-dorsal line. Because of its transparency and its exceedingly thin wall, its presence is recognized in most regions only by the commotion among the other viscera caused by its rapid, irregular pulsations. Where not overlain or underlain by more opaque viscera, however, the wall of the heart can be seen during systole and appears to be fibrous. In the last abdominal segment, the heart flares out and ends in three lobes. In each of the notches between the lateral lobes and the large median lobe is a slit-like ostium. In some specimens, the rapid alternate opening and closing of these valves can be seen beautifully. No other openings have been found in the wall of the heart, except at the anterior end, in the region of the brain, nor any arteries. Paired internal swellings are sometimes seen on the inner surfaces of the wall. They clap together, after the pulse wave, in a manner suggesting that they act as internal valves in preventing backflow.

The normal direction of blood flow is from posterior to anterior in the heart, and from anterior to posterior in the rest of the body. The direction can be ascertained by watching the movement of small granules occasionally seen in the blood. Ordinarily, no blood cells circulate through the heart. However, clear, irregularly shaped blood cells are found in various parts of the body, clinging to the body wall or viscera, and changing shape and position slowly. A particularly large aggregation is found in the clear posterior region of the body behind the viscera.

Flanking the heart are segmental pairs of large, oval or spindle-shaped, binucleate, coarsely granular cells with large vacuoles. They are attached to the delicate alary fibers supporting the heart. These pericardial cells, or nephrocytes, are the first cells to take up dyestuffs introduced into the blood. In fact, injection of very dilute neutral red solution into the larva several hours before examination is one of the best ways to demonstrate the nephrocytes, since they then show up clearly as paired, dark-red spots in rows paralleling the heart. Microscopic examination shows that all the color is concentrated in the vacuoles. Because the nephrocytes seem to pick up and concentrate foreign soluble and colloidal material in the blood, it has been generally believed that their normal function is the removal of injurious substances from the blood.

Digestive System. The mouth is a simple opening at the extreme anterior end. It is overlain by two large, black, rod-like hooks with serrated, ventrally curved tips. These hooks are built into a complicated supporting system of hard skeletal rods and platelets, which is controlled by numerous muscles. The posterolateral plates partly enclose an invaginated preoral cavity, the cibarium, whose suctorial nature is indicated by the large dorsal mass of muscle fibers running perpendicular to the lumen. The cibarium joins the tubular pharynx, at the anterior end of which the slender salivary duct enters ventrally. This duct, which has spirally ridged walls like a trachea, is formed by the fusion of two short, similar ducts from the paired salivary glands. Each salivary gland is an elongated, oval sac lying laterally, usually in segments 2 and 3. Its lumen is of variable size, depending on the amount of secretion therein. This secretion seems to accumulate as pupation approaches and has been shown to be extraordinarily rich in glutamic acid, but its function is still unknown. Associated with each salivary gland is a narrow strip of fat.

The walls of the salivary glands are made up of very large polygonal cells whose large nuclei contain the famous "giant salivary gland chromosomes" so important in modern genetical advances. Short regions of these chromosomes can often be recognized in the living gland as delicate, contorted, cross-hatched threads. For careful examination of the chromosomes, a larva should be "decapitated" in a drop of Ringer's solution by drawing crossed teasing needles across the body just behind the jaws. The viscera then usually gush out, and the glands can be cut loose from their ducts, put into a drop of acetocarmine stain, and squashed under a coverglass. Chromosomes like those of the salivary glands are also found in the gut cells and in several other larval tissues. They are believed to be produced by a process of internal multiplication of the normal somatic chromosomes. Such "internally polyploid" nuclei seem to be characteristic of larval tissues which are destined not to persist in the adult fly (imago), whereas the imaginal tissues which start development in the larva (for example, the nervous system and gonads) have normal diploid nuclei.

Posterior to the pharynx, a slender, short esophagus leads to a bulbous, muscular proventriculus in segment 4. The lumen of the proventriculus is ordinarily shut tightly, although occasional peristaltic swallowing movements are seen. Just behind the proventriculus, the midgut begins. It immediately receives four tubular caeca, which project anteriorly. The walls of the caeca are composed of large, flat cells with large discoid nuclei and wide transparent peripheral zones. Finely divided material can sometimes be seen sloshing back and forth in the lumina in response to contraction of the walls. Posterior to the entrance of the caeca, the wall of the gut is cytologically similar to that of the caeca, except that the prominent, clear borders surrounding the tile-like epithelial cells grow less noticeable. Food material can be seen to be held away from the actual wall of the gut by the transparent, tubular peritrophic membrane, which extends back from the posterior rim of the proventriculus. The midgut is very long and makes numerous coils in the posterior half of the body. Since much of the course of the gut is often obscured by the fat bodies, its study in the living larva may profitably be supplemented by preparations made by cutting off one or both ends of a larva in a drop of Ringer's solution, then pulling out the alimentary canal. It is also very helpful to sprinkle a vital dye (e.g., neutral red, methylene blue, litmus) on the food an hour or so before examining the larva. This not only outlines the course of the gut clearly, but also may give a rough measure of the hydrogen-ion concentration in different regions.

At the point where the midgut joins the hindgut, there is a slight constriction, from each side of which arises a Malpighian tubule. The hindgut makes a short anterior turn, then runs straight back to the anus in the midventral line. Each Malpighian tubule divides shortly after its origin into two narrow, beaded-appearing, pale yellow-green branches, one passing anteriorly to the region of the caeca, the other passing posteriorly. The posterior arms of the two tubules run among the coils of the mid- and hindgut, then join to form a continuous loop.*

* This loop may not occur in all strains, since it was apparently lacking in the larvae studied by M. Strasburger.

The distal ends of the anterior arms tend to bend back abruptly. Their cells are often filled with an opaque, white material and present a strikingly different appearance from those in other regions of the tubules. Through most of the length of the Malpighian tubules, the cells contain an assortment of refractile particles. These are birefringent, although they do not look like crystals. Refractile particles are also sometimes seen in the lumina of the tubules. Little is known about the excretory activities of the Malpighian tubules in *Drosophila*, but by analogy with other insects, the particles in the lumen may be crystals of calcium carbonate and uric acid and are probably different from the intracellular refractile material (see also remarks on the Malpighian tubules of the roach).

The proximal regions of the Malpighian tubules are capable of peristaltic movement.

The cells of the Malpighian tubules have a so-called brush border, consisting of closely packed, fine, passively movable filaments arranged perpendicularly to the lumen. Such borders are common in excretory and absorptive epithelia.

The gut anterior to the proventriculus is empty of food, and the proventriculus is never observed to permit regurgitation. It seems likely, therefore, that the midgut is the principal site of both digestion and absorption, and that digestive juices are produced mainly in the caeca, as in the roach (*q.v.*). The hindgut is usually empty, though in recently fed specimens, peristaltic waves carrying masses of feces to the anus can be observed.

Nervous System. The larval nervous system of *Drosophila* includes the following: a pair of large supraesophageal ganglia lying dorsal to the esophagus and medial to the salivary glands, and connected by a commissure; a pair of connectives encircling the esophagus; a large, triangular ganglionic mass ventral to the proventriculus, apparently representing the fused ganglia of all the thoracic and abdominal segments; a profuse brush of delicate nerves which leave the ventral ganglionic mass and supply the muscles and viscera; and a pair of ganglia lying in the extreme anterior tip of the larva and supplying various sense organs. No typical longitudinal ventral chain of ganglia is present in the larva.

Lying dorsal to, and between, the two dorsal ganglia is the small, median ring gland. This gland provides a hormone which brings about the larval molting and is necessary for the development of the imaginal tissues.

Reproductive System. The gonads are poorly developed in the larva and consist only of a pair of small oval (male) or tiny round (female) masses of glassy-clear, small cells, many of which are in active division. The gonads are set in windows in the fat body, about a third of the distance from the posterior end of the body. No ducts or accessory glands are present.

Imaginal Disks. Space does not here permit a consideration of the changes occurring during pupation and metamorphosis in *Drosophila*, but some mention should be made of the imaginal disks which become clearly visible late in the third larval stage. These disks are flattened masses of clear tissue, some of which attain approximately the size of a cerebral ganglion. They represent the *anlagen* of many structures which will continue development during the pupal stage and form parts of the adult fly. There is a disk for each eye, each antenna, each of the six legs, each of the two wings, each of the two balancers, the external genitalia, and various other structures. As might be expected, there are a number of disks in segments 1, 2, and 3, which give rise to the appendages of the head and thorax of the adult insect. Some of these, lying lateral, ventral, and dorsal to the general region of the salivary glands, appear to be coiled and remind one vaguely of the shape of the human external ear.

The imaginal disks possess the capacity of independent differentiation. Numerous experiments have shown that disks transplanted to another region of a given larva or into another individual may continue to develop into their specific adult form. One of the easiest of such experiments to perform involves dissecting an imaginal eye disk from the dorsolateral surface of the brain of one *Drosophila* larva and injecting it into the posterior region of another larva, through a capillary glass needle. If the transplant "takes," the adult fly emerges from the puparium with a complete extra eye lying loose among its abdominal viscera. The transplant has fully formed ommatidia, often parts of a brain, or even an antenna. If a disk from a strain of one eye color is transplanted into a host of another eye color, the transplanted eye often develops the color which is characteristic of the donor type.

11.

Chaetognatha and Echinodermata

CHAETOGNATHA

Sagitta elegans

By T. H. WATERMAN

Sagitta elegans (Figs. 207–210) is a typical member of the Phylum Chaetognatha in being a solitary, marine, free-swimming, carnivorous animal. Its glassy, bilaterally symmetrical body is spindle-shaped and externally distinguished by two pairs of lateral fins and a caudal fin, all of which are horizontal in orientation, like the flukes of a whale. In addition a number of paired active raptorial jaws or hooks are a prominent feature of the head region. The length of this species of "arrowworm," as chaetognaths are commonly called, varies from 30 to 45 mm. in mature specimens. Frequently *Sagitta* occurs in enormous swarms in the plankton and, being a voracious feeder, forms a critical link in the food chains of other important animals. This species is the only neritic chaetognath in the western North Atlantic, where it is abundant over the continental shelf from the Grand Banks to Cape Cod. It is also common along the coast of northern Europe; fairly similar species are cosmopolitan in distribution.

Many of the features of *Sagitta's* morphology and physiology may best be observed in fresh, living material. However, specimens carefully preserved in weak (4 per cent) formalin in sea water are quite adequate for general study. Cleared specimens, stained with borax carmine and methyl blue and prepared for microscopic examination as whole mounts, are also useful. Fine details of structure can be adequately studied only in appropriately prepared serial sections.

External Structure. The body of *Sagitta* is divided into three regions: head, trunk, and tail. The first of these is delimited from the trunk by a neck-like decrease in diameter, while the tail begins

506 CHAETOGNATHA AND ECHINODERMATA

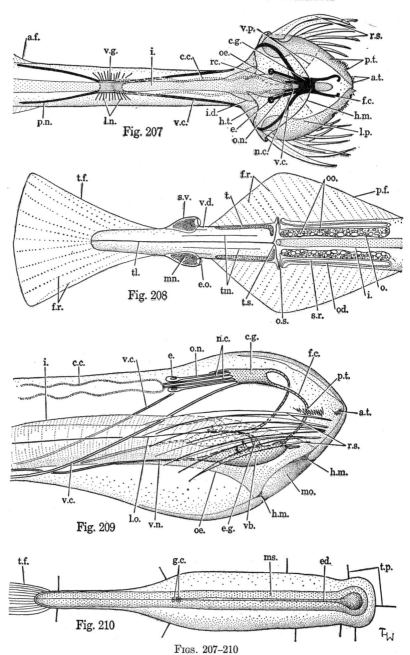

Figs. 207–210

SAGITTA ELEGANS 507

FIG. 207. Dorsal view of the anterior third of *Sagitta* with its raptorial spines partly spread. ×20. (Compounded after Burfield's and Kühl's monographs.)
FIG. 208. Dorsal view of the posterior third of *Sagitta*. ×20. (Compounded after Burfield's and Kühl's monographs.)
FIG. 209. Dorsolateral view of the head of *Sagitta* to indicate the central nervous system and the sheathed position of the spines when the hood is drawn maximally over the head. ×30. (Compounded after Burfield's and Kühl's monographs.)
FIG. 210. Dorsal view of a *Sagitta* just hatched. ×200. (After Doncaster.)

ABBREVIATIONS FOR LABELS ON FIGS. 207–210

a.f. : anterior fin
a.t. : anterior teeth
c.c. : corona ciliata
c.g. : cerebral ganglion
e. : eye
ed. : endoderm
e.g. : esophageal ganglion
e.o. : external opening of seminal vesicle
f.c. : frontal connective
f.r. : fin rays
g.c. : germ cells
h.m. : hood margin
h.t. : head-trunk septum
i. : intestine
i.d. : intestinal diverticulum
l.n. : lateral nerves
l.o. : lateral esophageal nerve
l.p. : lateral plate
mn. : mass of spermatozoa
mo. : mouth
ms. : mesoderm
n.c. : corona ciliata nerve(s)
o. : ovary
od. : oviduct

oe. : esophagus
o.n. : optic nerve
oo. : ova
o.s. : opening of seminal receptacle
p.f. : posterior fin
p.n. : posterior nerve
p.t. : posterior teeth
rc. : opening of retrocerebral organ
r.s. : raptorial spines
s.r. : seminal receptacle
s.v. : seminal vesicle
t. : testis
t.f. : tail fin
tl. : tail septum
tm. : secondary tail septa
t.p. : tactile processes
t.s. : trunk-tail septum
vb. : vestibular ganglion
v.c. : ventral ganglion connective
v.d. : vas deferens
v.g. : ventral ganglion
v.n. : ventral esophageal nerve
v.p. : margin of ventral plate

just caudad of the point where the anus opens on the midventral surface, three-fourths or more of the body length from the anterior end. The head is quite flat dorsoventrally; the trunk, being less flattened, is ovoid in cross section, while the tail is circular in section.

The head bears the mouth anteroventrally within a depression, called the vestibulum, whose depth and shape can change markedly while prey is being caught and ingested.

When the animal is swimming, but not feeding, the head is almost entirely covered by a prepuce-like hood formed by a double layer of epidermis. Only a circular opening controlled by a sphincter muscle is present in the hood under these conditions. This opening is located just over the mouth. When *Sagitta* is seizing prey and feeding, the hood is drawn back from the head to the neck region where it originates.

Anterior and dorsal to the mouth are two sets of cuticular teeth. These are arranged in two curved rows on either side, five to eight anterior ones and eleven to fifteen posterior. Behind these small teeth and originating laterally just posterior to the mouth are the raptorial jaws or seizing hooks. These large, tapering, curved, sharp rods which may be nearly as long as the head, number from eight to twelve on each side. At rest these jaws, or piercing spines, of *Sagitta* lie against the sides of the head in a groove just beneath the edge of the lateral plates (described below). Like the teeth, the jaws consist of a transparent brownish, chitinous material; they look individually like the jaw element of certain polychaetous annelids. In a live specimen the gape and closure of these jaws are striking features of the animal's activity.

Dorsally a pair of complex eyes is borne on the head. Between these eyes and extending posteriorly onto the trunk is a ridge of ciliated epithelium called the corona ciliata. This structure, of unknown function, is in the shape of an elongated ellipse, whose major axis is ten to twelve times the length of the minor.

The more anterior of the two pairs of horizontal fins is borne on the trunk. The posterior pair is partly on the caudal end of the trunk, partly on the anterior end of the tail. The median caudal fin is terminal on the tail. Each of these five fins is stiffened by two series of rod-like fin rays. None of the fins are independently movable, but they function for balancing and locomotion when the whole body moves by means of contractions of the longitudinal muscles.

Laterally on the body surface between the posterior, horizontal, paired fins and the tail is borne a pair of seminal vesicles. These vase-shaped structures taper posteriorly and have their external openings anterolaterally. They and the paired, lateral openings of the oviducts and seminal receptacles which occur at the level of the broadest part of the posterior, paired fins will be mentioned further in the section on reproduction.

General Body Organization. Internally the body of *Sagitta* is also divided into three regions: head, trunk, and tail. In each are parts of the coelom separated from one another by two transverse septa. These septa are not alike in their embryonic origin, and it is not yet clear whether the three body regions of *Sagitta* consist of segments comparable with those found (in greater number) in annelids, arthropods, or other definitely metameric animals. The body cavity of *Sagitta* is an enterocoele since it develops embryologically from outpocketings of the gut endoderm. The adult coelom is peculiar because in some body regions, particularly parietal, it has no clearly defined peritoneum. This may result from the differentiation of the longitudinal-muscle bundles of the body within the cells of the mesodermal pouches, since the nuclei and undifferentiated protoplasm of these cells apparently line the coelom while the distal structurally specialized regions of the same cells contain the longitudinal-muscle fibers.

The body cavity in the head extends into the fold comprising the cephalic hood. In the trunk region the coelom is large and divided bilaterally into two compartments by the digestive tract and the dorsal and ventral mesenteries which support it. The latter are, however, freely perforated so that the two compartments, which are filled with a clear coelomic fluid, are not isolated from each other. A definite pattern of movement set up by cilia on the cells lining the coelom circulates this body fluid. The diffuse arrangement of these cilia in *Sagitta* little resembles the typical organization of these organelles in mollusks and annelids but is more like that found in echinoderm epithelia.

The fluid content of the tail coelom is also circulated by ciliary activity. As will be mentioned later, spermatozoa in various stages of maturation are present here in mature animals. This third division of the coelom is completely separated into bilateral halves by a longitudinal sagittal partition. In addition each of these halves is further subdivided anteriorly by a longitudinal parasagittal partition.

Aside from the components of the coelomic circulation mentioned above *Sagitta*, like the other Chaetognatha, has no blood vessels, heart, or other parts of a vascular circulatory system.

No specialized respiratory surfaces for the exchange of oxygen and carbon dioxide, no definite convection mechanisms for circulating water over the body surface, or any respiratory pigment are known to occur in *Sagitta*. The relatively small size (largest arrowworm is about 90 mm. long) of these animals may be an important factor here.

Although several structures of unknown function have been described as excretory organs in chaetognaths, *Sagitta* is not known to have specific organs for this purpose.

The skeletal elements of *Sagitta* are mainly in the head region. Like the cuticle, which in most parts of the body forms a thin, histologically structureless covering, these elements are produced by the epidermis, on the surface of which they lie. The head skeleton consists of two pairs of plates, one lateral and the other ventral, and the various teeth and jaws mentioned above. All these are covered by the thin cuticle, although the tips of the teeth and jaws do not have this covering when fully developed.

The elongate lateral plates lie on the lateral margins of the dorsal surface of the head; they run its full length, and their position in various phases of activity largely determines the shape of this region of the animal's body. They are narrow and thicker at their rostral end; posteriorly they become less rod-like, thinner, and broader. These lateral plates have two functions: (1) support of the two sets of teeth and the raptorial spines, (2) site of origin for many of the muscles in the head. The ventral plates, which also function as a point of attachment for certain head muscles, are shorter, triangular, thin structures located ventrolaterally on the head with their anterior borders just behind the level of the mouth.

These skeletal elements may be conveniently studied in preparations which have been stained with Congo red, although they can be made out, with more difficulty, in unstained live or well-preserved specimens. Careful observation will show that the jaws or spines each have a central canal loosely filled with "pulp." This canal ends blindly and subterminally, however, and hence could not serve for the injection of poison into prey as has been suggested.

The only skeletal elements in the trunk and tail are the supporting rays of the fins. As mentioned above, these are present in a double row of rods, of epidermal derivation, that serve to stiffen each fin. The fin itself consists mostly of a layer of "ground substance" separating the two series of fin rays. Neither histological structure nor embryonic origin has been demonstrated for this layer.

The muscular system of *Sagitta* is very complicated in the head region, where elements active in movements of the neck, hood, jaws, teeth, vestibulum, and mouth are well developed. Thirty-three (fifteen pairs plus three unpaired) muscles are present. They function mainly in the capture and swallowing of prey which is often relatively large and active (see discussion of feeding below). In the trunk and tail the main musculature consists of paired longitudinal dorsal and ventral elements which provide the power for the animal's locomotion. All this somatic musculature is cross striated, although in some body regions, perhaps active in tonic contraction, the arrangement of muscle bands is quite peculiar and irregular.

Locomotion of this chaetognath, as in most other members of the phylum, is brought about by a dorsoventral longitudinal flexion of the body. The contractile waves responsible for this movement are effected by the two dorsal and two ventral longitudinal muscles of the trunk and tail contracting alternately. The "wavelength" of these contractions is relatively long and their amplitude small. By their means *Sagitta* is driven headfirst through the water in a short spurt, a centimeter or less at a time. A powerful contraction, like those which occur at longer intervals in a series of weaker ones, will move the animal 5–10 cm. ahead at one time. Note that, although locomotion of this sort may be quite adequate to maintain the chaetognath's position in the water, to capture food, or even to carry out vertical migrations, it is ineffectual for any sustained horizontal movement. Hence *Sagitta* is a truly planktonic animal, i.e., one whose swimming is too weak or random to make it independent of water currents.

The only apparent function of the cephalic hood of *Sagitta* is concerned with locomotion and the "streamlining" of the animal's body. The head, with its anteroventral vestibulum and mouth, its size, and its spreading armament of spines, is a notable deviation in the otherwise tapered, streamlined body. When *Sagitta* is swimming, however, the hood is drawn almost completely over the head as described above. In this condition the otherwise projecting parts are restrained so that the anterior end of the animal is smooth and rounded like the head of a torpedo. When prey is encountered, the hood is suddenly drawn back to the neck region, and the elaborate skeletal and muscular elements involved in seizing food spring into position with remarkable speed.

In feeding habits *Sagitta* is carnivorous and voracious. It is able to engulf fish larvae, small crustaceans, and other living prey of a size comparable to its own. In so doing the posterior ends of the lateral skeletal plates spread far apart. This markedly stretches the head laterally and erects the raptorial jaws on either side into a gaping array of formidable spines. These clamp together, piercing the prey as soon as contact is made. By this embrace the food is brought into contact with the anteroventral aspect of the head, where the two sets of small teeth help to hold and maneuver it into the mouth. Both vestibulum and mouth cooperate in this action by opening widely to encompass the prey.

The digestive tract of this form is a simple straight tube, consisting of an esophagus in the head and an intestine in the trunk. The only noteworthy structural features of this system are the diverticula in the neck region. Two lateral intestinal diverticula project forward towards the head, while two esophageal pockets of the same sort, but lying dorsally and ventrally, project backwards. Apparently this curious coupling arrangement has an accordion-like action that absorbs the stresses caused by the marked distortion of the head when the jaw apparatus is used.

The intestinal wall consists of three layers: (1) an epithelium made up partly of secretory cells and partly of another cell type occasionally ciliated, (2) a basal membrance, a histologically structureless, supporting lamella continuous with a similar layer in the dorsal and ventral mesenteries of the gut and in turn continuous with the basal membrane underlying (and produced by) the epithelium of the body surface, (3) a thin layer of circularly arranged muscle fibers also continuous with similar layers covering the mesenteries and forming the parietal boundaries of the coelom. No compound glands of endodermal origin (such as "liver" or "pancreas") are associated with the digestive tract of *Sagitta*. The physiology of digestion and the digestive enzymes of these animals have not yet been studied.

The nervous system of *Sagitta* consists of two unpaired and two sets of paired ganglia, their paired connectives, and radiating nerves. Near the dorsal surface of the head a median cerebral ganglion lies at about the same level as the mouth. Paired frontal connectives lead ventrally to the vestibular ganglia, which are located on either side of the anterior end of the digestive tract. Just dorsal to these and connected to them are the small esophageal ganglia. The only ganglion not in the head is the unpaired, ventral ganglion lying in the midline of the trunk one-third of the way from head to tail. This is connected to the cerebral ganglion by a pair of lateral commissures.

The histology of the ganglia of *Sagitta* is peculiar, since the peripherally located ganglion cells have no axons or dendrites which have been demonstrated, and the center of the ganglion appears to be very finely granular regardless of the plane in which it is cut. This latter appearance is thought to result from the presence of an extremely fine meshwork of interlinked fibers, but this has not been actually demonstrated. The peculiarities observed may rather be the result of inadequate neurological techniques.

There are only two types of sense organs whose functions seem fairly clear in this chaetognath. Tactile organs, which consist of circular or ovoid aggregates of special epidermal cells, occur as small projections scattered over the surface of the body. From the center of each of these patches of cells arises a bundle of very fine bristles at right angles to the surface. Contact with or other movement of these structures is believed to effect stimulation of these organs, but physiological evidence is lacking.

The paired eyes lie just posterior to the cerebral ganglion in the dorsal part of the head. Each is connected to this ganglion by an optic nerve. A single eye consists of a dorsoventrally flattened structure circular in surface view. Within it a set of pigmented partitions, lying in the major planes of symmetry of the body, divides the organ into five sections. Laterally there is a single section comprising half the eye; medially there are four comprising the other half. In each sector of the eye are a number of light-sensitive cells which fill up the space between the pigment cup and the outer capsule of the whole eye. These sensory cells are arranged in a simple retina, which interestingly enough is inverted like those of Turbellaria, Nematoda, some Annelida, and certain other animals. As the five separate parts of this eye have no image-forming mechanism within themselves, the whole organ is comparable to an aggregate of five ocelli. Such a structure could be sensitive to light intensities and in a rough way to the direction of illumination or shadows.

Other organs in which sensory functions have been assumed but not demonstrated occur in *Sagitta*. The corona ciliata mentioned above is connected to the cerebral ganglion by a pair of nerves and has been thought to be a chemoreceptor. In another chaetognath, however, the homologous structure apparently has an excretory function. Another organ, called the retrocerebral organ, a bilocular glandular structure located in the posterior part of the cerebral ganglion and opening by a pore on the midline of the dorsal surface of the head just anterior to the corona ciliata, has also been thought by some to be sensory in function. Again physiological evidence is much needed.

Sagitta is hermaphroditic, but cross fertilization effected through a mutual exchange of spermatozoa is apparently the rule. The testes are paired and lie in the anterolateral region of the tail coelom compartments. Prospective male germ cells are continually budded off from the testes of a mature animal, and the maturation of spermatozoa takes place while the cells are circulating in the coelomic fluid. Stained smears of this material or histological sections may be used to study the details of this process. Paired, unconvoluted vasa deferentia run from the coelom in the neighborhood of the testis posteriorly to the seminal vesicles, whence the sperm are shed into the sea water in the process of their transfer to another individual.

The paired ovaries lie in the posterolateral parts of the trunk coelom. They are filled in a mature individual with developing oögonia and oöcytes. These have originated from primordial germ cells which are readily identifiable at least as early as the gastrula stage of development (as are also the primordial male germ cells).

The lateral surface of each ovary is partly enveloped in the crescentic oviduct which opens by a lateral pore just dorsal to the posterior paired fin at the level of the anus. There is no patent opening through which eggs may pass from the ovary to the oviduct. Within the oviduct is a second tubular structure which ends blindly at the upper end of the gonoduct. This is the sperm receptacle; it opens to the exterior by a pore within that of the oviduct. After sperm exchange by two individuals has occurred, apparently with the aid of spermatophores, the seminal receptacle is filled with the elongate ripe spermatozoa.

At this point fertilization is accomplished by a remarkable process in which a fertilization tube is established between the seminal receptacle and a ripe ovum lying in the lateral part of the ovary. This tube is intracellular within two so-called accessory cells provided by the oviducal wall (compare pollen tube formation in angiosperms!). Through this tube a sperm swims to the egg, which it enters through a micropyle. The zygote then creeps through the oviducal wall into the cavity of the oviduct and thence is shed into the outside sea water.

The eggs, which are pelagic, are 0.2 mm. in diameter. Cleavage is holoblastic and radial. Gastrulation occurs by invagination, and mesoderm origin, as mentioned above, is by archenteric pouches. The mouth does not form from the embryonic blastopore but develops independently. Hence, *Sagitta* is a deuterostome like the chordates and echinoderms. Hatching occurs in about 2 days. The resulting larva is basically like the adult animal, into which it gradually matures by a straightforward differentiation. Thus *Sagitta's* larval development provides no useful clues as to its phylogenetic affiliations. The life cycle of *Sagitta* is such that five or six generations occur within a year. Various stages may be observed in preserved plankton or in fresh material when that is available.

ECHINODERMATA

Asterias forbesi

By W. M. REID

Asterias forbesi, which is the common sea star or starfish on most of the eastern seashore, ranges from Maine to the Gulf of Mexico and may be found from low-tide level to a depth of about 50 meters. As is typical of the more specialized groups of Asteroidea, this species intergrades with closely related species. This description may also be used for the more northern *Asterias vulgaris* and for the frequently described *Asterias rubens* from English shores. Specimens are found either singly or in large aggregations on rocky or shelly bottoms. *Asterias forbesi* has been studied extensively because of its economic importance as a predaceous oyster-eating organism.

It is very desirable to have living specimens to demonstrate external structures and activities of the animal. Dissection may also be carried out on a living form after anesthetizing it with about 15 ml. of 10 per cent magnesium chloride. Unless anesthetized or killed in fresh water, autotomy, the spontaneous severance of arms from the central disk, will occur while dissecting. If preserved specimens must be used, alcohol is superior to formalin as a preservative since the latter softens the skeleton. Although not essential, it is possible to obtain sea stars with the water vascular system injected in a contrast color. Dry specimens, prepared by killing in alcohol and treating with boiling water, are used to demonstrate the skeletal system. Individual plates may be prepared by macerating a specimen in fresh sodium hypochlorite. Commercial bleaching fluids, if fresh, provide a convenient source of this solution. Decalcified stained sections of a ray are essential for study of the nervous system.

External Characteristics. This species is known for great variability in color, with cream, pink, orange, and purple being common. In preserved specimens colors fade to a yellow or brown. The five rays or arms are poorly separated from the central disk. Four- to seven- or eight-rayed specimens are known, but they are probably the result of regeneration after injury.

The mouth is located in the center of the surface, which is usually applied to the substrate. It is used as a reference point in distinguishing the oral or actinal surface from the aboral or abactinal surface on the opposite side. The oral opening is guarded by five groups of spines or mouth papillae. Radiating out from the mouth in each arm are ambulacral grooves which are provided with sucker-tipped tube feet. The tube feet in each arm are located in four rows except at the tip of the arm, where a single modified terminal tentacle is found. Through the action of internal muscles the grooves may be deepened and narrowed for protection of the tube feet. A double row of spines along the edges of the groove also serves the same purpose. The terminal suction disks at the end of the tube feet collectively attach the animal strongly to surfaces. This firm adhesion may be illustrated by suddenly lifting an animal attached to a glass plate and noting the strength required to remove it. Some of the tube feet usually break, leaving the sucker portion still attached. On ventral surfaces the tube feet are thrust forth to anchor and then shorten. However, study of animals on a sandy bottom shows that on horizontal surfaces movement is accomplished by a coordinated lever-like movement of many tube feet. Individual action resembles the movement in mammalian legs.

The aboral or abactinal surface is marked by the presence of the asymmetrical subpentagonal madreporic plate. This plate is bright red in life but fades in preserved specimens. With magnification numerous closely set grooves may be seen radiating out from the center. The two rays adjacent to the madreporite are known as the bivium, while the other three constitute the trivium. If the madreporic plate is placed away from the observer, the arm to the right is numbered I, the one to the left II, while III, IV, and V follow in counterclockwise fashion. The anus is located in the interradius between IV and V but is seldom found externally.

The aboral surface is studded with irregularly arranged blunt spines. Examine the surface with magnification. Attached to these spines and between them are large numbers of pincher-like pedicellariae. Draw a thread across some of them and note the movements of the jaws. Remove a few, mount on a slide, and examine under the compound microscope. The crossed pedicellariae have jaws arranged like the blades of a pair of scissors (Fig. 211). They are smaller than the

ASTERIAS FORBESI

uncrossed or major pedicellariae. Minute sets of muscles which are attached to the single basal and the two valvular calcareous plates operate to open and close the jaws.

In living specimens membranous sacs, dermal papulae (also known as gills or dermal branchiae), will be seen to extend and withdraw. They are thought to be chiefly respiratory in function. Inject about 5 ml. of carmine suspension in sea water into the animal. Because of ciliary action the carmine particles may be seen circulating in the

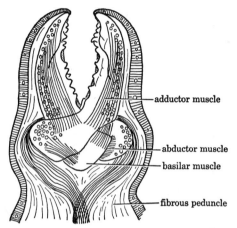

FIG. 211. Crossed pedicellaria of *Asterias glacialis*. (Redrawn after Cuénot.)

gills. Set this animal aside to be examined 8 hours later. Some of the several types of cells which have been described from the coelomic fluid are phagocytic and take up the carmine particles. These cells will be found migrating to the outside through the thin-walled gills. This migration may be watched by picking off some of the gills with forceps and examining them under the compound microscope. It has been concluded from such observations that these cells are excretory in function. Examine a few drops of fluid from the coelomic cavity under the microscope for a further study of these cells. As a part of the clotting process some of the cells tend to agglutinate on the slide.

The entire external surface of the body, including spines, pedicellariae, tube feet, and gills, is ciliated. The direction of ciliary movements has been plotted by using carmine particles. It has been suggested that these ciliary currents are useful in removing debris from the animal or in transporting food toward the mouth.

Situated at the tip of each ray near the base of the terminal tentacle is the so-called eye, which contains specialized sense cells evaginated from the ectoderm. Many of these cells develop a red pigment which makes the spot visible in living specimens.

Skeleton. The skeleton is composed of a large number of tube-shaped calcareous ossicles lodged in a leathery perisoma. The plate arrangement in the rays may be seen in a dry specimen. Transect one ray and remove the aboral surface from another on such a specimen. The ambulacral groove is bounded on the two sides by a series of closely fitting aboral plates which unite at a peak in the center. From the inside they form the ambulacral ridge. Each of these flattened plates is channeled by a half pore which, together with the corresponding half pore of the next plate, makes a channel through which the tube foot extends. These pores alternate in position, the first one being close and the second further from the ambulacral ridge. Below these aboral ossicles is a single row of spine-bearing adambulacral plates. The aboral surface consists of a framework of smaller ossicles.

Dissection. Open up the aboral surface of the three arms of the trivium by making a cut along both lateral edges of each arm, beginning at the tip and continuing to the central disk. Loosen the underlying organs from the suspending mesenteries and lift up the free flaps. Cut away half the aboral surface of the central disk next to the trivium, taking care not to damage the madreporic plate or the attached organs near the anus between rays IV and V. Cut the tough skeleton and mesentery in the two interradial areas between rays I and V, and rays II and III.

Coelom. The spacious body cavity is known as the perivisceral coelom. It is lined with ciliated epithelium which maintains a very active circulation of the albuminous body fluid. The water vascular system, described later, constitutes another part of the coelom. Other derivatives of the embryonic coelom persist in the adult as a system of tenuous tubes, most of which are difficult to demonstrate except in sectioned material. These tubes include the axial sinus, the inner and outer perihemal canals near the nerve ring, five radial canals aboral to the radial nerves, a pentagonal aboral perihemal sinus on the aboral disk, and five pairs of genital sinuses. The function of this so-called perihemal system has been much discussed. It has been considered both a blood vascular system and an excretory system. Because of the uncertain function of these canals they should be regarded as coelomic derivatives, and the term perihemal applied with caution.

Digestive System. The mouth leads into a narrow esophagus (Fig. 212) which expands into the sac-like cardiac stomach. The cardiac stomach may be everted through the mouth in the more specialized

ASTERIAS FORBESI

Asteroidea and may often be found in this condition when the animal has been feeding. It is retracted by paired retractor muscles found on the ambulacral ridge of each arm. Aborally the cardiac stomach is separated by a constriction from the pyloric portion, the latter being the part seen on top in the dissection. The large double brownish glands located in the rays are the hepatic caeca or gastric glands, which

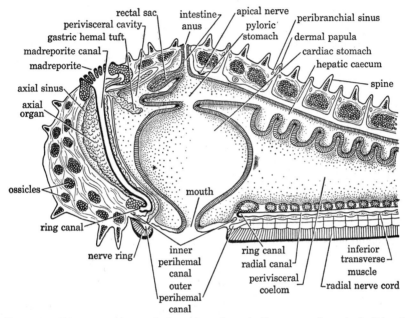

FIG. 212. Diagrammatic vertical section through the ray and central disk of *Asterias*. (Redrawn after Chadwick.)

secrete powerful proteolytic enzymes. Moreover, being hollow and ciliated, they also assist in circulation of digested food. Ducts on the aboral surface of each portion unite to empty into the pyloric stomach.

The pyloric stomach narrows into the cone-shaped intestine which is connected to the anus. A short distance before the anus the intestine is joined by the expanded rectal sac which varies in the number of lobules composing it.

Reproductive System. A pair of gonads is present in each arm. They are located aborally in the proximal portion of the ray. They lie free in the perivisceral coelom except for attachment near the genital pores, a pore for each gonad being located close to the interradial line. Both the gray testes and the orange ovaries are more prominent before spawning. Fertilization and development, explained more fully below, are external. Large streams of eggs and sperm are released early in the summer.

Water Vascular System. The stone canal or madreporic canal is a roughened tube, composed of calcareous rings, leading from the madreporic plate to the ring canal surrounding the mouth. The stone canal is almost completely enveloped by a sac-like covering which forms the so-called axial sinus. Hidden by the stone canal and the undissected portion of the aboral disk, but within the axial organ, is the glandular axial organ. The tissue of the axial organ extends into the various perihemal canals as a part of the so-called hemal system. The function of the hemal system, like that of the perihemal system, is not well understood, and the term hemal is misleading. Some investigators ascribe rhythmic movements to parts of the axial organ, so that it is considered a vascular system. Others have postulated that it is excretory, that gonadal tissue is formed here, or that lymphocytes are produced within it.

Locate the union of the stone canal with the ring canal. Nine small vesicles, Tiedemann's bodies, project inward from the ring canal. A tenth Tiedemann's body is missing, and the corresponding position occupied by the stone canal. Tiedemann's bodies probably produce the amoeboid cells found in the water vascular system. Five radial canals concealed under the ambulacral ridge go from the ring canal to the tip of each of the arms. Each tube foot is supplied with liquid from the radial canal by a lateral canal, which may be seen by spreading the two double rows of tube feet in an injected specimen. By laying aside the hepatic caeca a double row of bulb-like enlargements will be found on either side of the adambulacral ridge. These are the ampullae each of which is connected with a tube foot. The ampullae contain reserve fluid which, by contraction of muscle bands within their walls, forces fluid into the tube feet. The foot is forced to elongate because of hydrostatic pressure within it. Successive rings of connective tissue strands prevent lateral expansion of the tube foot. As the longitudinal muscles of the tube foot contract, the fluid is again returned to the ampulla. A valve within the lateral canal prevents fluid from being forced back into the radial canal.

Nervous System. Certain major divisions of the nervous system may be demonstrated in gross dissection and in sectioned materials. However, cellular details in many areas have never been demonstrated, since the nervous tissue of echinoderms does not respond to the usual histological methods. By spreading apart the two double rows of tube feet the radial nerve cord may be seen between them. The radial nerve cords, and in the mouth area the nerve ring to which all radial cords connect, make up the central nervous system. This system is continuous with a sensory-association nerve plexus underlying the entire epithelium. Although this layer is considerably thick-

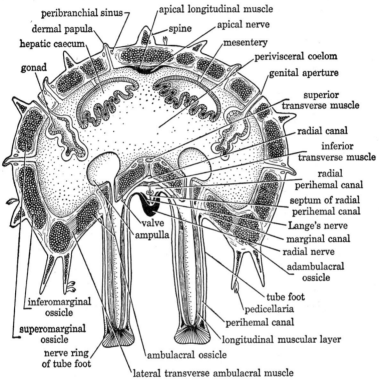

Fig. 213. Diagrammatic cross section of the ray of *Asterias*. (Redrawn after Chadwick.)

ened to form a nerve ring in the tip of each tube foot, it thins out to a thickness of 10–15 μ in most areas. Numerous bipolar sensory cells have been demonstrated in this layer.

Examination of a cross section of a ray will reveal the V-shaped radial nerve cord (Fig. 213). The oral portion of this cord contains nerve cells of sensory and association fibers. Major nerve trunks may be seen in this area by locating the cross sections of the numerous fibers which appear as minute dots. Further from the oral surface and next to the perihemal canal will be found the so-called deep nerve or Lang's nerve. Motor neurons which innervate the tube feet and ampullae have been located within this layer. The coordinated activity of the tube feet is under the control of the central nervous system. This fact may be readily demonstrated by experimental destruction of selected portions of the central nervous system. However, the dermal papulae, spines, and pedicellariae are locally coordinated by the nerve net contained within the epithelial plexus.

Development. Because of the transparency and ease with which fertilization can be controlled, echinoderm eggs have been used extensively for embryological and physiological studies. The sea star, sea urchin, and sand dollar are most commonly employed for this purpose. A brief description of the methods used in fertilization of all three forms is included here. The larval forms are described in embryology textbooks.

Ripe gonads are found in *Asterias* and *Echinarachnius* in the spring and early summer months. In *Arbacia* the breeding season is later, and gravid gonads may be found until early fall. Since sexes cannot be distinguished externally and since the gonads may be shrunken and empty if the animal has already spawned, it is usually necessary to open several animals.* Dissecting instruments should be washed in tap water after dissecting each animal to prevent undesirable contamination with sperm.

Sea stars are opened by cutting along the aboral margin of a single arm, and the gonads removed. When a ripe specimen has been obtained, sever the dissected arm and save the animal in a sperm-free aquarium for future work. As previously mentioned, ovaries are orange while testes are gray. Sea urchins may be opened by a meridional cut or by cutting the peristomal membrane away. The gonads, which hang by mesenteries in the aboral half, are red in gravid females and gray in males. With sand dollars, cut across the animal with heavy scissors, thus exposing the gonads. The ovaries are purple while the testes are gray.

* A simple method of obtaining ripe eggs and spermatozoa from unopened *Arbacia* and *Echinarachnius* has recently been developed. A specimen is placed with the aboral surface down in a Syracuse watch glass and 0.5–2 ml. of 0.5 M potassium chloride is injected into the perivisceral coelom through the peristomial membrane with a hypodermic syringe. The ripe sex cells are extruded into the watch glass through the genital pores. This material may then be diluted as desired. The method has the following advantages: (1) only mature reproductive cells, uncontaminated by body fluids or immature cells, are produced; (2) eggs or spermatozoa may be obtained from the same specimen several times if the animal is properly stored in running sea water between injections; (3) specimens are not mutilated and may be used later for dissection studies.

Remove the ovaries from a female and suspend them in a double layer of cheesecloth over a finger bowl containing sea water. After sufficient numbers of eggs have shed, discard the cheesecloth. By using a clean pipette draw off approximately 200 eggs and add them to about 50 ml. of sea water. Wash once by decanting the surface water after the eggs have settled to the bottom. Examine to see whether a large proportion of the eggs have intact germinal vesicles (sea urchins only). If so, set aside until mature.

To obtain normal development the sperm suspension must be very dilute, or polyspermy will result and the embryos will be abnormal. Add 1 drop of "dry" sperm (sperm from an isolated testis which has had no water added) to 25 ml. of sea water. Mix the suspension and add a few drops of this fluid to the eggs. Agitate the mixture. After 5 minutes pour off the water and wash several times with fresh sea water. The rate of development depends upon available oxygen and the temperature, so that development may vary considerably from the accompanying table. Keep the eggs cool for normal development. Development through metamorphosis into a radially symmetrical animal depends upon a satisfactory food supply such as certain species of diatoms.

APPROXIMATE DEVELOPMENTAL RATE OF ECHINODERM LARVAE MAINTAINED AT ROOM TEMPERATURE

	Asterias forbesi (hours)	*Arbacia punctulata* (hours)	*Echinarachnius parma* (hours)
Polar-body formation	¾–1	..	1½
Two cell	2–3	¾	2¾
Four cell	3	1¼	3
Eight cell	4	1¾	3¼
Sixteen cell	5	3	6¼
Early blastula	7	6	10
Late blastula	10	10	21
Late gastrula	18–21	Prism 20	Prism 38
Bipennaria	24–30	Pluteus 48	Pluteus 48

Ophioderma brevispinum
By W. M. REID

The brittle star, *Ophioderma brevispinum*, is olive green to almost black. It is found in shallow water and down to a depth of 280 meters. It is widely distributed, having been reported from Cape Cod to Brazil.

The more southern forms are sometimes regarded as different subspecies named *O. brevispinum olivaceum*. Specimens are often common among sea weeds, where they hide during the daytime and feed at night.

Brittle stars may best be studied by comparison with the sea star. Since they are not suitable for gross dissection, some parts of the anatomy have been worked out only through a study of serial sections. The following description applies to structures visible with the naked eye and with the aid of a dissecting microscope. A single living or alcohol-preserved specimen will suffice for study.

External Form. It will be noted that pedicellariae, ambulacral grooves, dermal papulae, and suckers upon the tube feet are missing.

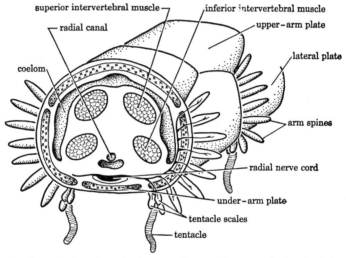

FIG. 214. Cross section through the ray of an ophiuran at the level of the intervertebral muscles. (Modified after Delage and Hérouard.)

The tube feet, called tentacles in this group, are largely sensory in function, but they are capable of passing a particle of food along the rays to the mouth. The flexible arms are sharply marked off from the pentagonal disk. The arms resemble the vertebral column in the tail of a vertebrate in structure and function. Internally they consist of a column of calcareous vertebrae, each one being connected to the next by two pairs of muscles (Fig. 214).

By well-coordinated activity of four arms the animal is able to move about swiftly. The fifth, unused arm may either lead or follow behind. The temporary anterior and posterior pairs of arms cooperate in alternately bearing the weight of the animal and shoving the body forward, then throwing themselves forward. If an attachment point is available, an arm may become coiled around it and pull the body forward.

Dermal Skeleton. The more than 1600 species of described ophiuroideans are distinguished largely by the constant arrangement of

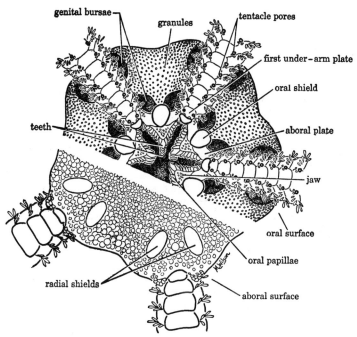

FIG. 215. Oral and aboral surface of the disk of *Ophioderma brevispinum*.

dermal plates which form characteristic patterns. The more typical of these plates are present in *Ophioderma brevispinum*, although this species is atypical in that all of the aboral and part of the oral plates are hidden by uniform granules (Fig. 215) which range from 100 to 180 per square millimeter.

The arms, which in this species are almost circular in cross section except for flattening on the oral surface, have four sets of plates, two lateral, one upper-, and one under-arm plate. The upper-arm plates, which are broadly in contact with each other, are oblong and nearly twice as wide as long. Proceeding toward the tip they become more rounded on the sides, with the outer edge being more curved and the inner edge narrow. Near the lateral edge of the third upper-arm plate are about ten small scales.

The under-arm plates are nearly square with rounded corners. They are usually slightly longer than wide and are broadly in contact with each other. The first under-arm plate is much wider than long and has rounded sides. Although the second plate is much longer than the first, it is not quite as long as the third. Near the lateral margins of each plate are pores through which tentacles extend. Just proximal to each pore are two short spines known as the tentacle scales. These should not be confused with the longer arm spines, of which there are seven or eight. The inner tentacle scale is nearly half the length of the under-arm plate, while the outer one is about half as long and covers the base of the lowest arm spine. The arm spines are constant enough in size, number, and position to be of diagnostic importance. In this species they are all relatively short and appressed to the arm. The under spine is slightly shorter than the others. The species name, *brevispinum*, is derived from this characteristic.

On the aboral surface the granules extend out on the sides of the arm so that the first three upper-arm plates form a narrow ridge running toward the center of the disk. If the granules are scraped from the disk surface, two oval plates will be found lateral to the attachment of each arm. These are the radial shields which are often of taxonomic importance.

The largest plates on the oral surface, the oral shields, are located in the interradial areas. These are oval with the long axis directed radially. The madreporic plate is a modified oral shield which is slightly larger and more irregular than the others. The minute pores communicating with the madreporic canal are difficult to locate. Lying on both sides of each oral shield are two triangular aboral plates with rounded corners. Except for these fifteen plates the oral surface is granulated in this species. Pointing forward from the oral shields are the five jaws. On the inner tip of each jaw inside the mouth cavity is a row of five blunt teeth. Unless the mouth is well opened, they will not all be seen since they are arranged one under the other. Although the teeth may be turned outward through the oral opening to enlarge the mouth, they are not used in mastication. Tooth papillae, typically present, are missing in this species. The edge of the jaw is fringed by a row of oral papillae. They are located on the surface of the jaw in contrast to the teeth, which are inside the mouth. There

OPHIODERMA BREVISPINUM

are about seven on each side of the jaw, the one next to the arm plate being the smallest while the second is the widest and longest of all.

Genital Bursae. Distal to the oral shields is a pair of prominent grooves known as the genital bursae. A second pair is found on the edge of the disk next to the arms. The ten pairs of bursae distinguish this genus from other genera in which the outer five pairs are missing. The genital products are discharged by the gonads into these bursae. In some viviparous species such as *Amphipholis squamata* these bursae act as brood pouches. In *Ophioderma brevispinum*, as with most species, development is external with the reproductive products being discharged into the water. The bursae open and close rhythmically in a movement which has been interpreted as a respiratory activity.

Dissection. The only practical dissection may be made by removing the aboral disk. This view shows the gonads and the sac-like stomach.

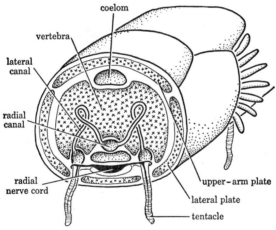

FIG. 216. Cross section through the ray of an ophurian at the level of the tentacles. (Modified after Delage and Hérouard.)

Hepatic caeca, intestine, and anus are missing. Neither the gonads nor the digestive system extends into the arms. All the teeth and the vertebrae of the proximal portion of the arm together with the muscular attachments may be seen.

The water vascular system differs from that of the sea star only in minor respects. Ampullae (Fig. 216) and Tiedemann's bodies are missing, while the madreporic plate and madreporic canal are oral. The central nervous system is highly developed with ganglionic enlargements in the arms along the radial cords. These supply the intervertebral muscles with nerve connections.

Pull off an arm and examine the cross section. Locate four muscles arranged as an upper pair and a lower pair (Fig. 215). Simultaneous contraction of the right lateral muscles or the left lateral pair brings about side motion. Contraction of the upper muscles raises the arm, while the lower muscles draw the arm orally. Vertical movement is more limited than lateral movement in the Order Ophiurae, of which this is a representative, because of mechanical limitations imposed by the structure of the vertebrae. The vertebral articulations consist of a complicated system of knobs and grooves which may be demonstrated by placing an arm in sodium hypochlorite and later examining the plates under a microscope. In the Order Euryalae the vertebrae are shaped like an hourglass, permitting more vertical movement than is found in the Ophiurae.

Arbacia punctulata

By W. M. REID

The purple urchin, *Arbacia punctulata*, ranges from Cape Cod to Florida and has been reported off the coast of Cuba and Yucatan from the littoral zone to 230 meters. In some localities it may be seen in its natural habitat in tide pools, while in other areas it must be obtained by dredging. Although it is commonly known as the purple urchin, it varies somewhat in color from brick red, dull red, purplish brown to almost black. Urchins live together in aggregations. They move about slowly by means of spines and tube feet on shelly or rocky bottoms.

The larger green urchin, *Strongylocentrotus dröbachiensis*, is found abundantly north of Cape Cod. Although it is readily distinguished from *Arbacia* by its color, smaller spines, and various specialized features of the test and lantern, the following description may be used with minor modifications for *Strongylocentrotus*.

Three kinds of preparations may be used to advantage. External features are best studied on living animals, but specimens preserved in alcohol may be substituted. A dry test with organic matter removed will be needed for study of the plates. This may be prepared by placing the animal in a fresh commercial bleaching reagent (sodium hypochlorite). Further cleaning may be completed with a tooth

ARBACIA PUNCTULATA 529

brush. A preserved specimen whose skeleton has been softened in 2 per cent nitric acid for 24–48 hours will show many delicate internal structures. If only the aboral surface is submerged in the acid, the rigid oral skeleton will hold the animal together for study.

External Structures. The internal organs are enclosed in a globular test or skeleton made of many interlocking calcareous plates. The

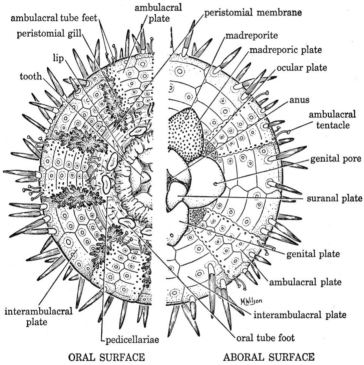

FIG. 217. Oral and aboral surfaces of the test of *Arbacia punctulata,* with spines partly removed.

test is somewhat compressed so that two poles are distinctly recognizable. It is flattened at the oral end, which is readily identified by the presence of five protruding teeth. Four valve-like plates at the opposite pole surround the anus, making up the area known as the periproct (Fig. 217).

Examine the oral region under magnification, noting the collarshaped lip surrounding the teeth. A circular ring of muscles in the lip draws the mouth opening together like a purse string. Surrounding the lip is the ciliated peristome, a membranous covering which extends to the edge of the calcareous test. Perforating the peristome outside the lip are five pairs of large oral tube feet which have also been called oral or ambulacral suckers, papillae, or tentacles. Two other modifications of tube feet are present. Just outside the peristome are five double rows of ambulacral tube feet, their prominent suckers being held open by embedded calcareous plates. These form meridional rows extending in the direction of the aboral surface. Near the equator or ambitus, tube feet lacking suckers, known as tentacles, are present. Both ambulacral tube feet and tentacles may be extended beyond the spines in living specimens. On the outer edge of the peristome, in the interambulacral area, will be found five pairs of finger-like, branching, peristomal gills which appear to be respiratory in function. If the ambulacral tube feet are carefully removed at the edge of the peristome, a small bean-shaped spheridium will be found in the center between each double row of tube feet at the base of the first ambulacral spines. This sense organ, a modified spine, is believed to act in geotactic responses. Experimental removal slows the righting reaction.

Pedicellariae. Three types of pedicellariae have been described for *Arbacia* (Figs. 218, 219, 220). (1) Numerous ophiocephalous pedicellariae are clustered on the peristome. The three broad blades give the whole head a serpent-like appearance in a side view. The supporting calcareous plate in the stalk comes up to the head. (2) The smaller, more active triphyllous pedicellariae are found in among the spines as well as on the periproct and the peristome. The stalk, although highly variable in length, is generally shorter and the valves more slender than in the ophiocephalous type. (3) Tridentate pedicellariae are largely confined to the oral surface and are relatively rare. Large tridentate forms are more abundant in specimens collected from the southern part of the range than in those from the north. Structurally they resemble closely the triphyllous type, but their jaws are narrower toward the tip. These jaws range from the size of a tridentate jaw to ten times this long. A complete series of intergrades may be demonstrated between tridentate and the triphyllous types, so that the term triphyllous is generally reserved for the abundant small forms.

Spines. Primary spines are the only type found on *Arbacia*. Secondary and tertiary or miliary spines which are reduced in size are found on *Strongylocentrotus*. In *Arbacia* the spines, which all bear minute longitudinal flutings, are shortest in the aboral areas and longest at the equator. Those close to the peristome are often spatulate with a wider tip than base. Occasional blunt spines are found in this area. Each spine has on its proximal end a hollow socket which fits onto a corresponding tubercle on the test. Remove a spine and

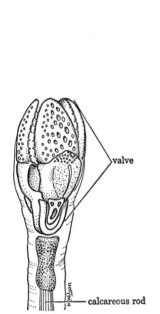

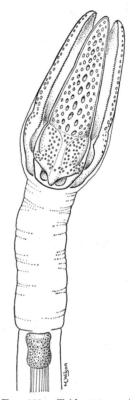

FIG. 218. Ophiocephalous pedicellaria of *Arbacia punctulata*.

FIG. 219. Triphyllous pedicellaria of *Arbacia punctulata*.

FIG. 220. Tridentate pedicellaria of *Arbacia punctulata*.

examine it and the tubercle under magnification. The circular attachment of the spine to the tubercle consists of two rings of muscle as well as an outer epidermal covering. The innermost muscular ring is used in rigid fixation of the spine. It becomes set when pressure is applied to the spine. The outer ring moves the spine in locomotion. The entire test, spines, pedicellariae, and spheridia as well as the coelomic cavities within are covered with ciliated epithelium.

Test. Examination of a clean test will show the interlocking calcareous plates which are used extensively in echinoid taxonomy. The description applies only for *Arbacia.* The four suranal plates of the periproct are usually missing in such preparations. Five meridionally arranged ambulacral rays point from the peristomal region upwards to the periproct. Each possesses two double rows of perforations through which the tube feet connect with the water vascular system inside the test. Between these ambulacral areas are five double rows of interambulacral plates. The ambulacral areas correspond to the ambulacral grooves, while the interambulacral areas correspond to the sides of the rays of the sea star. The sea urchin thus may be compared with a sea star in which the five arms have been pinned upward to cover all the aboral surface except the area around the periproct. This region is therefore homologous to the entire aboral surface of the sea star, while the remainder of the sea-urchin surface is homologous to the oral surface of the sea star. Growth of the test occurs through the production of new plates in the ambulacral area near the periproct and because of an increase in the size of plates already formed.

Around the hole left by the four suranal plates is found, with occasional variants, a ring of ten plates. The larger triangular plates at the end of the interambulacral areas are called genital, gonadal, or basal plates. Each of them contains a pore through which the reproductive products are expelled. One of them, the madreporic plate, is larger than the others and contains numerous minute perforations. Between the genital plates at the tip of the ambulacral area are the smaller ocular or radial plates. Each is perforated by a small pore through which the terminal tentacle projects.

The body of the test consists of ten rows of interambulacral and ten rows of ambulacral plates. The double row of interambulacral plates is separated by a prominent zigzag suture. Plates at the equator are the longest and may have as many as four convex, highly polished tubercles for spine articulation. The width becomes progressively less in the periproctal region, where only one spine per plate is found. A similar progressive narrowing occurs in the peristomal region. The double row of ambulacral plates which are narrowest near the periproct becomes increasingly larger toward the peristome, so that they are about equal in width to the adambulacral plates in this region. Each ambulacral plate bears a single tubercle nearest the center of the ambulacral region and a set of double pores toward the outside. Examination of the inner surface shows that each of these ambulacral plates is compound, resulting from the fusion of

ARBACIA PUNCTULATA

several little plates, each pair of pores representing an individual plate. The first pair of ambulacral plates nearest the peristome is detached from the test. They may be demonstrated on the inner surface of the peristome where they support the oral tube feet. Five pairs of prominent projections, the auricles, are seen inside the peristomal edge.

Internal Structure. With a fine saw or a pair of scissors make an equatorial cut after the spines have been removed. Before separation

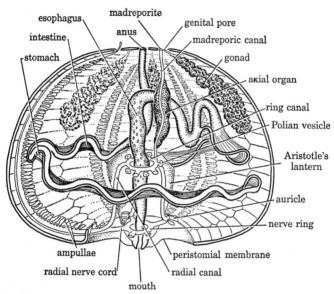

FIG. 221. Diagrammatic vertical section of *Arbacia*. (Modified from Petrunkevitch.)

of the two halves carefully sever the mesenteries (Fig. 221), holding loops of the stomach and the intestine to the walls and locate the non-calcareous madreporic canal. The canal connects the madreporic plate region with the ring canal which encircles the esophagus just as it leaves the lantern. The madreporic canal may be freed from the mesentery which stretches between the esophagus and the aboral region. After demonstrating these structures the mesentery and the madreporic canal may be cut, and the two portions of the test separated.

If a second specimen is available, soak the aboral surface in 2 per cent nitric acid. Cut a circle around the outside of the periproct with a pair of fine scissors and then make five meridional cuts through the interambulacral areas. Dissections should be carried out with the specimen completely submerged in water.

Digestive System. A complex structure of calcareous plates and muscles known as Aristotle's lantern (Fig. 222) surrounds the mouth and oral end of the esophagus. It serves to crush and grind the food

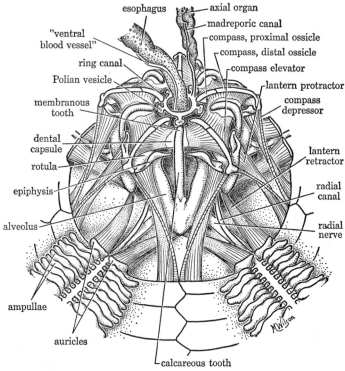

Fig. 222. Aristotle's lantern, from *Arbacia punctulata*.

of the omnivorous urchin, whose diet consists of small crustacea, sponges, and detritus. A detailed description of this intricate mechanism is given below.

Trace the tubular digestive tract from the esophagus as it leaves the lantern through (1) an almost complete clockwise turn, the stomach (cardiac stomach of some writers), (2) a counterclockwise loop, the intestine (pyloric stomach), (3) a sharp aboral bend, the rectum to (4) the anus in the middle of the four suranal plates. As the esophagus leaves the lantern area, it gives off a smooth, hollow, ciliated tube, the siphon. The siphon parallels the stomach on the inner side and again rejoins the digestive tract. Since currents of water pass from the esophagus to the intestine through the siphon, it is thought to act as a supplementary respiratory and flushing organ. The stomach is festooned on the outer edge and attached by delicate strands of mesentery to the outer test wall in the interambulacral areas. Mesenteries are absent in the ambulacral areas and on the siphon side. Fecal pellets may often be seen within the intestine. They are ejected from the anus, which is opened as the hinged suranal plates swing outwards. It is said that these fecal pellets are passed along by pedicellariae until they are dropped over the edge of the test.

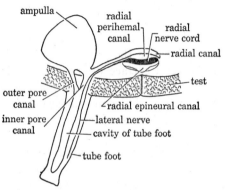

FIG. 223. Diagrammatic cross section through the tube foot of an echinoidean. (Redrawn from Chadwick.)

The Ambulacral System. The perforated madreporic plate, the madreporic canal, and the ring canal which rests upon the lantern membrane just outside the esophagus have already been described. With a strong light and some magnification five grape-shaped bodies, the Polian vesicles, may be located connected to the ring canal in the position corresponding to the Tiedemann's bodies in *Asterias.* Between them the radial canals go under the rotulae (see lantern) where they may be traced as the lantern is dissected. Without dissection they may be found again passing between the auricles to the ambulacral areas. The tube foot is connected with the rest of the water vascular system by two canals, one of which passes through the inner and the other through the outer pore canal of the ambulacral plate (Fig. 223). The outer canal connects directly with the ampulla. The inner canal joins the lateral canal at its point of merger with the ampulla. This double connection with the ciliated ampulla makes possible better circulation between it and the tube foot than is found in the sea star. The radial canal continues aborally, giving off lateral canals until the terminal tentacle is reached.

The Gonads. The five interambulacral aboral areas of the test are filled with reproductive organs in gravid specimens. The gonads are reddish brown in the female and gray in the male. In specimens that have shed, the genital pores and a ring-shaped coelomic sinus, the genital rachis, may be seen. Sexes do not differ externally. Development is treated briefly under the sea star.

Aristotle's Lantern. The lantern is an intricate mechanism used in chewing food materials and sometimes for locomotion (Fig. 222). Examine first the parts in a dry specimen. Five V-shaped jaws or pyramids hold the teeth in place (Fig. 224). Each limb of the V-shaped pyramid is a separate plate known as an alveolus or half pyramid (Fig. 225). In an isolated pyramid break apart the two alveoli along the median suture and notice the grooved dental slide in which the tooth is held. The lateral surface of the alveolus is greatly roughened by horizontal corrugated ridges which make an effective sawing surface. The outer edge of this same surface serves for attachment of a short but powerful interpyramidal (comminator) muscle closing the jaws and used in tearing off food. The tooth is pointed at the oral tip and tapers off aborally into a long membranous extension which is usually missing in dry specimens. The aboral end of each tooth is enveloped in a large, delicate, balloon-shaped dental capsule seen only in recently collected living specimens. Perched on the aboral end of each alveolus is a small epiphysis (Fig. 226) one end of which curves inward towards the tooth. It may be readily broken loose at the line of suture, revealing a pitted surface on the end of the alveolus. In *Strongylocentrotus* and in the often-figured *Echinus* the two epiphyses join and act as additional support for the tooth. The cavity between the upper ends of the alveoli, the foramen magnum, is comparatively large in all three species. Radiating outward from the esophagus are five fine, conspicuous curved rods, the compasses or radii. Each compass is made up of an inner proximal ossicle and an outer distal ossicle which has a rounded tip in *Arbacia* but a bifurcated one in *Strongylocentrotus*. Beneath each compass and joined to it at the inner end is a heavier rod-shaped plate, the rotula or brace (Fig. 227). The outer end of the rotula joins the alveoli in the interpyramidal area. There are, therefore, 40 pieces in the skeletal apparatus of the lantern: 5 teeth, 10 alveoli, 10 epiphyses, 5 rotulae, and 10 compass ossicles.

Muscles may best be located on preserved specimens. Two delicate muscle strands are inserted on the expanded tips of each compass.

ARBACIA PUNCTULATA 537

These compass depressors or radial compass muscles take origin from the interambulacral area of the edge of the test. A group of circular muscles (compass elevators) connects each compass with the next

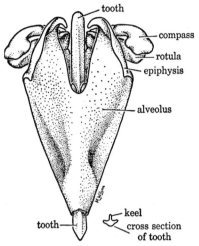

Fig. 224. Lantern ossicles of *Arbacia punctulata*.

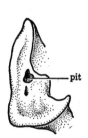 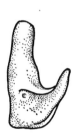 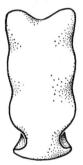

Fig. 225. Top of half-pyramid, with epiphysis removed.

Fig. 226. Epiphysis, from above.

Fig. 227. Rotula, from above.

one. The entire lantern is covered with a membrane known as the lantern membrane. This membrane separates the perivisceral coelom from the lantern coelom, the latter being connected with the dermal branchiae found on the peristome. Elevation of the compasses by the combined action of the circular compass muscles causes this lantern membrane to be lifted and the lantern coelom to be enlarged with consequent retraction of the gills. The combined action of the compass depressors decreases the volume of the cavity and expands the gills. The compass depressors are probably also used in pulling the lantern to an upright position.

Protractor muscles originate on the aboral end of each alveolus and are inserted in the interambulacral area. They serve to extend the mouth parts. Retractor muscles which open the jaws take origin from the lower outer face of the alveolus and are inserted on the auricles of the test. By pulling upon the rotula with a pair of forceps it can be demonstrated that the rotula is held firmly in articulation with the epiphysis by an external and an internal pair of rotula muscles. They bind the lantern together and probably assist in rocking movements. There are thus 60 muscles associated with the lantern itself: 10 compass depressors, 5 compass elevators, 10 protractors, 10 retractors, 5 interpyramidal, 10 internal rotula, and 10 external rotula muscles.

Nervous System. With a good light and some magnification the major nerve cords may be demonstrated. The radial cord is readily located, as it passes through the base of the auricles under the radial canal. It appears as a broad, whitish cord. Carefully cut away the alveoli on one side with scissors and trace the radial nerve inward to reveal its union with the pentagon-shaped peripharyngeal nerve ring. Histological preparations are required to demonstrate the nerves supplying tube feet and lantern muscles.

Echinarachnius parma

By W. M. REID

Echinarachnius parma, the common sand dollar, ranges from Labrador to New Jersey on the Atlantic coast and in the Pacific Ocean from Vancouver to Japan. It is common in isolated areas on sandy bottoms from low-tide level to 1600 meters. Living specimens vary from flesh red to deep brownish red. When the animal is injured or preserved, the color changes to greenish brown. Although they are otherwise of little economic importance, young specimens frequently are eaten by haddock.

This species should be studied after the sea urchin and compared with it. Fundamental differences in structure are correlated with adaptation to a sandy rather than a rocky bottom. A dry test and a living or alcohol-preserved specimen should be available.

External Structure. The test is extremely flattened, the mouth opening central on the oral surface, while the anus is near the margin of the test. The anus is usually aboral but near the margin in young specimens, marginal in older specimens, and oral in oldest specimens. The madreporic plate is centrally located on the aboral surface and is surrounded by four genital pores. A fifth genital pore and gonad are missing, as the space usually occupied by them is utilized by a portion of the intestine.

Both surfaces are covered with two kinds of minute spines. Short spines with an enlarged tip are covered with cilia, while the longer ones lack cilia at the tips. The longest spines are found in the interradial areas near the peristome and at the margin of the disk. Ambulacral areas are marked by indistinct grooves on the oral surface. These give off two side branches near the margin. On the aboral surface bilobed respiratory tube feet are located in five double rows. These double rows flare apart in the middle, giving a petal-like appearance to the ambulacral zone which is best seen on the dried test. Very numerous minute locomotor tube feet are found both in the ambulacral and interambulacral areas of both surfaces. These feet with pink-tipped suckers are used effectively in feeding on sand which is encrusted with microorganisms, and, in combination with the spines, are used in locomotion through the sand. Among the spines will be found a single type of two-jawed pedicellaria.

Dissection and Internal Anatomy. Dissection may best be carried out by breaking into the aboral surface and picking or cutting away the plates with heavy forceps or a small saw. The test may be softened by soaking overnight in 2 per cent nitric acid. In the margin of the disk numerous supporting pillars connect the oral and aboral surfaces, giving additional strength and partially dividing the interior.

Immediately under the aboral wall are the extensive gonads which extend into all available spaces in sexually mature individuals. Reproduction is treated briefly under the sea star. Carefully remove the gonads to reveal the siphon and the ribbon-shaped digestive system which, except for the marginal anus, coils in a manner similar to that in the sea urchin.

The madreporic canal is often torn away from the madreporic plate in dissection. Attached to it is the prominent axial sinus containing the dark axial organ. Radiating outwards from the lantern are the darkened radial canals. They branch into extremely fine lateral canals which supply the numerous tube feet.

The Lantern. Although the lantern is well developed, it is very different from the lantern of *Arbacia* (Fig. 228). It has a greater horizontal diameter and has a star-shaped cross section, whereas the lantern of *Arbacia* is shaped like an inverted cone and is circular in cross section. In the clypeastroids, of which this is a representative, the lantern is said to be procumbent with the five unequal jaws hold-

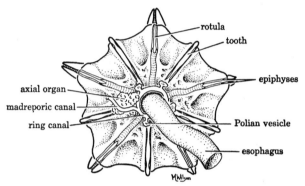

FIG. 228. Aboral view of Aristotle's lantern, from *Echinarachnius parma*.

ing the teeth in a plane nearly parallel to the oral test wall. Each jaw consists of two alveoli with fluted wings which hold the blade-shaped tooth in the dental slide between them. The wings of the alveoli project sharply outward, giving the entire jaw the shape of an arrowhead. The alveoli from adjoining jaws come together to form the points of the star. Resting between these alveolar tips near the point are three small calcareous bars. The center bar is the rotula, while the other two are the small epiphyses. Compasses are absent. Well-developed, paired retractor muscles are inserted on the lower surface of the jaw suture near the tooth, while the origin is on an auricle. A weakly developed protractor muscle is attached higher up on the jaw with its origin also on the auricle between retractors. Very strong interpyramidal muscles are visible on the oral surface. They appear to work as antagonists to the retractors, making the lantern a well-developed grinding apparatus. The teeth, however, are unable to extend as far outside the mouth opening as those in the sea urchin because of their procumbent position and the lack of a flexible peristome.

Thyone briareus

By W. M. Reid

The sea cucumber, *Thyone briareus*, is found in limited areas from the intertidal zone down to 20 meters. It ranges from Vineyard Sound south to the Gulf of Mexico. It lives on either a sandy or a muddy bottom, usually being completely buried except for the posterior end; the anterior end is occasionally protruded in feeding. Since it is seldom possible to view this species in its natural habitat, feeding and respiratory movements may best be studied in an aquarium with a sandy bottom. By contractions of circular and longitudinal muscles animals so placed gradually work themselves down until only the posterior end is left protruding.

The musculature must be relaxed for successful dissection. Injection of 5–10 ml. of 10 per cent magnesium chloride into the coelomic cavity will accomplish relaxation after an hour. Tentacles then may be gradually worked out at the anterior end. Specimens for preservation must be killed with the tentacles extended. Colored fluids may be used to inject the respiratory trees through the cloaca or the water vascular system through the Polian vesicle.

Autotomy and Regeneration. The radical autotomy (complete evisceration) for which the holothurians are noted rarely occurs in nature with *Thyone* even under conditions of extreme fouling. However, it may be readily induced by various chemical stimuli. Most specimens autotomize if placed in a solution of 0.1 per cent ammonium hydroxide made up in sea water. After the animal has remained in the solution for about a minute, violent circular muscular contractions usually occur. If these movements do not begin spontaneously, hold the animal up by the posterior end for a few seconds. The entire anterior end of the animal with the tentacles, calcareous ring, and the digestive system is violently expelled. These structures are regenerated in miniature in 2 or 3 weeks and gradually regain their original positions. Various tissues are resorbed in the process so that the regenerated animal is much smaller than the original form.

External Morphology. *Thyone briareus* varies in color from dirty olive to almost black. The tentacles have a grayish appearance. The tube feet or podia are rather evenly distributed over the entire surface of the body, though on some specimens it is possible to group them into five ambulacral areas. They often have a pinkish tinge because of the presence of hemoglobin in the water vascular system. Suckers are well developed on the ventral surface but are often lacking on the podia near the dorsal radii.

The secondarily acquired radial symmetry is somewhat obscured by a superimposed bilateral symmetry associated with the elongated structure and the burrowing habits of this group. The mouth may be considered anterior and the cloaca posterior. The concave surface which is uppermost when the animal is held by both ends is dorsal,

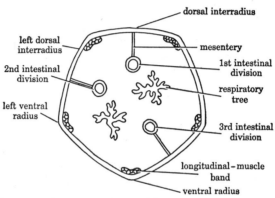

FIG. 229. Diagrammatic cross section of *Thyone*, showing arrangement of radii and mesenteries.

and the convex lower surface is ventral. In relaxed specimens radial orientation may be established by ridges formed by longitudinal muscle bands located in the radii (Fig. 229). On either side of the mid-dorsal line are two of these bands, the right dorsal radius and the left dorsal radius. Another, the ventral radius, is found in the mid-ventral position. The other two radii spaced between the dorsals and the ventral radius are the left ventral radius and the right ventral radius. The five interradii are dorsal, right dorsal, left dorsal, right ventral, and left ventral.

The cloacal aperture is built around a calcareous framework described below. Five groups of sensory papillae surround the opening. Anteriorly a crown of ten radially arranged tentacles surrounds the mouth. These tentacles are borne on the crown of the introvert, which is a band of tissue lacking the tube feet and musculature present in the rest of the body wall. The two ventral tentacles are considerably smaller than the rest. Opposite them, between the two dorsal tentacles, the genital papilla is found. It is generally larger in males than

THYONE BRIAREUS 543

in females, but this characteristic is too unreliable to be used in external determination of sex.

Dissection. Make a longitudinal cut into the coelomic cavity in the right ventral interradius. Cut just below the right ventral radius, being careful to protect the intestinal mesentery in the middle of the interradius. The specimen may then be pinned down under water in a dissecting pan.

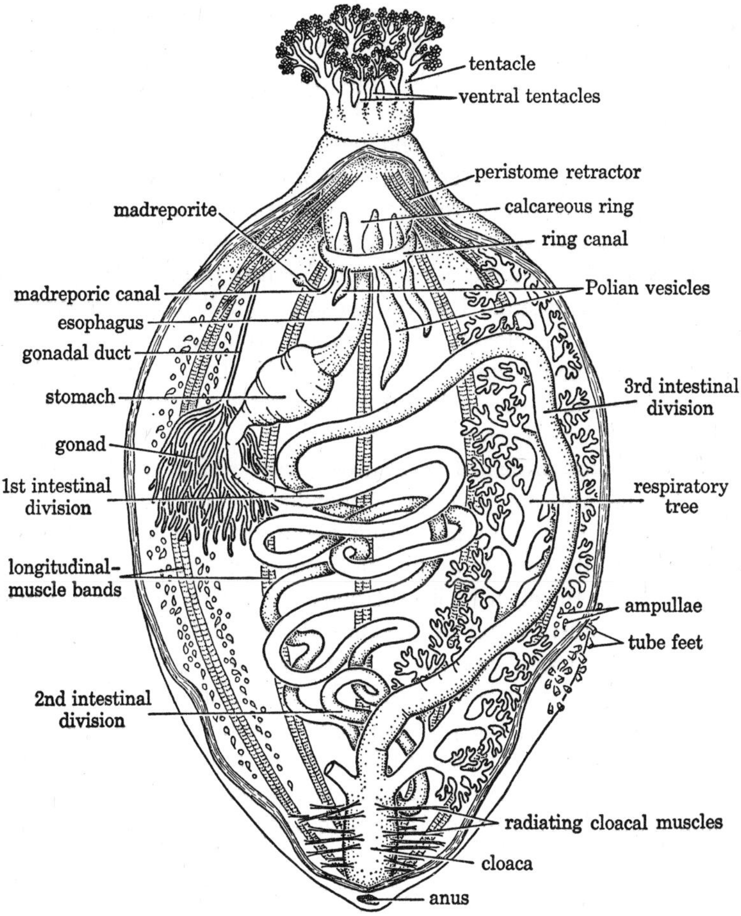

FIG. 230. Dissection of *Thyone briareus* from the right ventral side. (Modified after Coe.)

Digestive System. The mouth opening is surrounded by a lip leading into a pharynx which penetrates a ring of calcareous plates (Fig. 230). After a slight constriction the pharynx joins a short esophagus. This in turn leads to the muscular stomach which joins the coiled intestine. The intestine is many times the length of the animal and has three major divisions. The first portion attached to the stomach progresses toward the posterior end and is held in position by a mesentery attached to the dorsal interradius. The second division progresses forward with attachment to the left dorsal interradius. The third division is attached to the right ventral interradius and is connected to the muscular cloaca by a short rectum.

During the evening hours the anterior end is sometimes extended, and the tentacles are expanded into long delicate strands as they take part in feeding movements. The tentacles are waved about to pick up plankton and other food particles and then are alternately drawn into the mouth. Here the food particles are scraped off, and the sand and other debris rejected. Analysis of the digestive system contents shows the food to be protozoa, small worms, crustacea, diatoms, and algae fragments along with detritus from muddy bottoms.

Respiratory System. Respiration is accomplished by means of a pumping device, the cloaca, which brings about an exchange of fluid in highly branched respiratory trees within the coelom. Red corpuscles containing hemoglobin within the water vascular canals probably also assist in oxygen transport.

The two finely branched respiratory trees which extend from the cloaca to the anterior end are suspended by delicate mesenteries. Muscle cells within the walls bring about periodic local pulsations. The cloacal wall contains a ring of circular muscles, while radiating muscles connect the cloaca with the body wall. Contraction of the circular muscles causes water to be expelled through the cloacal aperture if it is open or forced into the respiratory tree if closed. Rhythmic closing and opening of the aperture may be readily observed on living animals, the speed of the process being dependent upon the oxygen content of the water. The contraction of the radiating cloacal muscles causes the coacal chamber to enlarge and water to be sucked in. Occasionally the musculature of the entire body may contract, causing a forceful spouting.

Reproductive System. Sexes are separate, and fertilization and development are external. The male and female reproductive organs can be distinguished only by microscopic preparations. Each type of gonad consists of a series of fine filaments growing along the anterioposterior axis on either side of and attached to the dorsal mesentery. The smaller anterior filaments are immature, while the more posterior ones which are mature are pigmented and greatly distended. Larger specimens usually possess larger numbers of tubules, over 1000 having been reported from a large male. A basal portion which is able to produce new tubules has been demonstrated in sectioned material. These filaments empty their products into a common chamber connected to the genital duct which conducts the eggs or sperm cells forward to the genital papilla.

Water Vascular System. In living specimens the water vascular system stands out with a deep reddish color because it contains cells

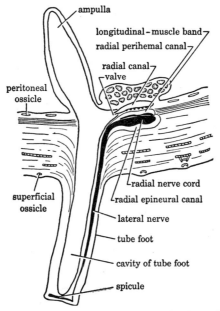

FIG. 231. Diagrammatic cross section through the tube foot of a holothuroidean. (Redrawn after Delage and Hérouard.)

bearing hemoglobin. To see these cells and the ciliary activity within the system, scrape off a few of the filamentous red ampullae and examine them under the microscope. In favorable specimens the ring canal may be seen just below the calcareous ring. From it hang one to three Polian vesicles. The internal madreporic plate is a calcareous mass hanging by a slender madreporic canal which is in turn attached to the ring canal. From the area of the ring canal radial canals run forward, each giving off a pair of lateral canals which serve two of the oral tentacles. Ten ampullae near the calcareous ring contain reserve fluid for these tentacles. The radial canals then bend backwards and descend to the posterior end embedded between the longitudinal muscle bands. Radial canals give off numerous long lateral canals which connect with tube feet and ampullae (Fig. 231).

Muscular System. Five paired bands of longitudinal muscles extend the length of the animal in the radial zones. Joined to these near the anterior end are the powerful retractors attached to the calcareous ring. These retractors serve to pull in the entire introvert. Equally strong but more diffuse circular muscles line the body wall under the longitudinal bands. These muscles not only make possible the worm-like movements used in burrowing, but also come into play in spouting and in autotomy.

Nervous System. The circular and radial nerves correspond in relative positions with those in the sea star. They are difficult to distinguish in gross dissections but may be seen in histological preparations. Likewise perihemal and hemal systems have been described from sectioned materials.

Calcareous Plates. In all holothurians the calcareous skeleton is much more dispersed than in the other classes of echinoderms. It consists of isolated calcareous plates distributed in various parts of the body. They are considered of major significance in the taxonomy of the group. Plates may best be isolated in sodium hypochlorite bleaching reagent and studied in water mounts. In *Thyone,* which represents one of the more advanced forms, calcareous deposits are found in the tentacles, the tube feet, the madreporite, the calcareous ring, and the anal ring, while skin "tables" are confined to the two ends. These thumbtack-shaped tables consist of plates perforated by eight holes. At right angles to these plates two to four spires taper to sharp teeth at the end opposite the plate attachment. The spires are connected with one or two cross rods. The calcareous ring consists of five radial and five interradial plates. The radial plates are notched anteriorly for passage of the radial nerves, while posteriorly they have two long posterior projections. The cloacal aperture is supported by a ring of five plates, each with a posterior projection, the anal tooth. Terminal plates with many perforations are found in the sucker portion of the tube feet, while the side walls are supported by semicircular bracelet-shaped ossicles.

12.

Enteropneusta and Chordata

ENTEROPNEUSTA

Saccoglossus kowalevskii

By L. H. KLEINHOLZ

Saccoglossus kowalevskii (Fig. 232) (also formerly known as *Dolichoglossus kowalevskii*) is a soft, worm-like, burrowing animal living in sand flats near the low-tide mark of the Atlantic coast from Massachusetts Bay to North Carolina. The animals are reported to reach a length of 15 cm. or more. Their burrows are frequently marked by characteristic coils of sand castings between 1–2 cm. high. Specimens which have just been dug up from a sand flat may be partly covered by a fragile tube consisting of sand particles held together by a mucous secretion. Living individuals possess a distinctive, mildly unpleasant, iodoform-like odor. The body, whose external surface is ciliated, is coiled posteriorly in a number of loose turns. The internal anatomy, especially in small individuals, cannot be studied very satisfactorily by gross dissection of the living or preserved specimen and is best examined in histologically prepared sections. References to such anatomical details may be found in the standard textbooks and monographs mentioned in the bibliography.

The body of the animal is divided into three readily recognizable regions, the anterior proboscis, the collar, and the posterior trunk region. The proboscis is a yellowish white to yellowish pink conical structure, about five to ten times longer than it is wide. The anterior tip is bluntly rounded; the wide base of the proboscis narrows abruptly at the posterior end to form a short, thin proboscis stalk or peduncle. The mouth opening, which reportedly cannot be closed, is ventrally located at the base of the proboscis but is usually covered by the free anterior margin of the collar which also overlaps the proboscis stalk.

548 ENTEROPNEUSTA AND CHORDATA

The collar, immediately posterior to the proboscis, is cylindrical in shape and reddish orange in color (especially in males) with a white

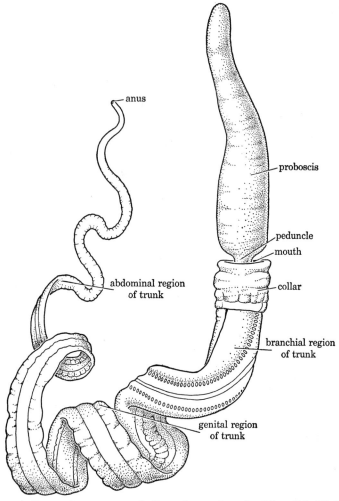

Fig. 232. External anatomy of *Saccoglossus kowalevskii*. (Modified after Bateson.)

ring near the posterior border. The free posterior margin of the collar overlaps the anterior border of the trunk.

The trunk, which is the longest portion of the body, is itself divisible into three regions. The anterior branchial region is characterized by the presence of paired, transverse gill slits arranged in longitudinal series to each side of the mid-dorsal line. The gill slits lead from the external surface into the pharyngeal cavity. The second or genital region of the trunk is less sharply recognizable. Gonads are not restricted to the genital region but may also extend somewhat anteriorly into the branchial region; the last pair of gill slits is considered, more or less arbitrarily, to mark the boundary between these two trunk regions. The sexes of *Saccoglossus* are separate; the reproductive organs are gray in the female and yellow in the male. The gonads open to the outside by genital pores that are usually too small to detect on the external surface. In many genera of the Balanoglossida the body wall is extended laterally into prominent genital ridges or wings; these ridges are also present in *Saccoglossus*, but it is claimed that they arise in this species simply because of the distention produced by the maturing gonads.

The third, greenish yellow abdominal region of the trunk tapers rapidly toward the posterior end on which lies the terminal anus.

CHORDATA

Amaroucium constellatum

By L. H. KLEINHOLZ

The Genus *Amaroucium* is a compound ascidian, numerous species of which are widely distributed but whose taxonomic distinctions are not equally well defined in all regions of their occurrence. The description given here is of *A. constellatum;* the zooids of other common Atlantic coast species differ slightly in their general morphology. *Amaroucium constellatum* is found along the Atlantic coast, ranging from New Hampshire to the Gulf coast of Florida. It may be found on wharf-pilings as well as on the sea bottom, to depths ranging from 2 to 60 meters.

The colonies occur as lobulated, hemispherical, or dome-shaped masses, up to 8 cm. in diameter and 3–5 cm. high, resembling a head of cauliflower. The surface of the colony is smooth and gelatinous in texture. The test, which may range from yellowish to orange-red in color, is somewhat opaque, but the zooids are conspicuous because of their deep color. Living material is preferred, but preserved colonies are also satisfactory for demonstrating the structure of the zooids. Living material can be studied by cutting a 2–3-mm.-thick section vertical to the surface of the colony; this section, immersed in a petri dish containing sea water, can then be examined with a compound or dissecting microscope. If a mass of the colony is squeezed over a finger bowl containing a small amount of sea water, one can obtain zooids as well as embryonic stages and tadpole larvae. By allowing the expressed material to settle, complete zooids can then be selected and removed with a pipette for detailed study with the compound microscope. With preserved material zooids should be teased out of a small portion of the colony and mounted in a drop of alcohol or glycerine. Individuals are frequently broken by either method of removal from the test, but such portions of zooids are useful for study, since, as will be described below, most of the organs are contained in the thoracic and abdominal regions of the zooid.

The zooids of *A. constellatum* (Fig. 233) are longer and stouter than in some of the related species. A complete zooid may measure 14–17 mm. in length and may be about 1 mm. thick. In general the organ systems of the zooid are similar to those described later for the solitary ascidian, *Molgula*. The body is divisible into three distinct regions: a thorax, an abdomen, and a postabdomen, the latter region being the one frequently broken off in removing zooids from the test. The thorax is occupied chiefly by the branchial or pharyngeal portion of the digestive tract; the abdomen encloses most of the rest of the digestive tract and is traversed along its length by the gonoducts; the postabdomen contains the gonads, the heart, and the epicardium.

The margin of the incurrent or branchial siphon is divided into six lobes, each of which may be partly cleft into two. The mouth region, which begins shortly posterior to the branchial siphon, opens into the pharynx or branchial chamber. Near the junction of the mouth and pharynx is a circlet of about twelve tentacles. The endostyle is a groove, located in the midventral wall of the pharynx, made up of longitudinally coursing tracts of ciliated and mucus-secreting cells. Anteriorly the endostyle is continuous with the peripharyngeal bands, also ciliated, which encircle the anterior end of the pharynx. The dorsal lamina, which is present in the mid-dorsal line of the pharynx of *Molgula*, is replaced in *Amaroucium* by a longitudinal series of tongue-like processes, the dorsal languets, each of which arises from a transverse pharyngeal vessel a little to the left of the mid-dorsal line. The pharyngeal perforations, the stigmata, are arranged in ten to thirteen rows along the length of the pharynx. In each row, composed of sixteen to eighteen stigmata on each side, the three or four perforations nearest the endostyle are smaller than the others of the same row. The role of the pharynx in filter-feeding and in respiration is described in the section on *Molgula*.

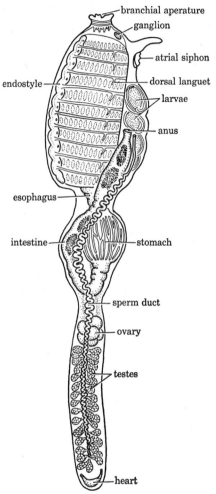

FIG. 233. Diagram of a zooid of *Amaroucium constellatum*. (Modified after Van Name.)

The pharynx opens posteriorly into a prominent esophagus, considerably narrower than the branchial chamber. In the abdominal region of the zooid the posterior end of the esophagus leads into a widened stomach, the external wall of which is thrown into numerous (twenty to twenty-five or more) narrow, closely spaced, longitudinal, orange-red folds that are characteristic for this species. The stomach passes posteriorly into a narrower intestine which makes a single loop at the posterior end of the abdomen and then continues anteriorly into the thoracic region. The terminal portion of the intestine, the rectum, ends in a slightly lobed anal orifice which opens at the level of the posterior third of the pharynx into an atrial cavity, the cloacal chamber. The atrial cavity in turn communicates with the outside by an atrial or excurrent siphon, from the anterior border of which projects a simple finger-like atrial languet.

The central nervous system is represented by a small ganglion, located dorsally between the incurrent and excurrent siphons. It is little more than a remnant of the typically chordate central nervous system that is present in the larval stage, but which becomes considerably reduced as a result of metamorphosis of the tadpole larva to the adult zooid.

The reproductive system, consisting of a single pair of male and female gonads and their ducts, is located for the most part in the postabdominal region of the zooid. The ovary, containing large eggs, occupies the anterior portion of the postabdomen. The testis, located posterior to the ovary, is composed of a number of small ovoid lobules which discharge their products into a sperm duct. The sperm duct passes as a comparatively thick, sinuously coiled tube in an anterior direction through the postabdomen and abdomen to terminate dorsally in the thorax at the level of the rectum and posterior to the excurrent siphon. The oviduct is a delicate tube and is more difficult to see. The atrial cavity, into which the gonoducts open, serves as a brood pouch in which embryos and larvae in various stages of development may be found. Larvae which hatch are discharged to the external environment through the excurrent siphon.

The circulatory system is difficult to study in detail. The heart may be recognized as a U-shaped tube near the posterior end of the red-tipped postabdomen. In the living zooid slow contraction waves can be seen passing along the heart wall. The vessels to the viscera cannot be readily seen, except in the pharynx, where the mid-dorsal and the transverse vessels are evident.

The postabdomen of zooids in the Family Synoicidae, to which *Amaroucium* belongs, may be regarded as a stolon, invaded by the gonads and heart, from which buds are formed asexually. The epicardium, a tubular outgrowth from the posterior wall of the branchial sac, extends posteriorly through the abdomen to the posterior region of the postabdomen, and contributes to the asexual formation of such buds; the epicardium may, however, be difficult to detect.

Molgula manhattensis

By L. H. KLEINHOLZ

Molgula manhattensis is widely distributed along the Atlantic coast, being especially common from Massachusetts to Chesapeake Bay. The animals are found in shallow water, on wharf-pilings, where they may occur in abundant masses, on anchored and submerged objects, and on eelgrass. Large individuals are 20–25 mm. in their largest dimension. When a living individual, which has been resting undisturbed in sea water, is picked up and gently squeezed between the fingers, a fine jet of sea water is discharged, sometimes to a distance of 2 or more feet; this explains the common name of "sea squirt" for such ascidians. Living animals are best for study, but preserved specimens are moderately satisfactory.

External Anatomy. The rather firm test or tunic is, in some living individuals, a translucent greenish yellow color. The viscera can be discerned through this test, especially if the animal is held over a bright light. This is readily accomplished by inverting the shade of a goose-neck desk lamp and setting upon it the finger bowl containing the animal. In many specimens, however, the outer surface of the test has become so covered by debris (sand, bits of shell, and other materials) entangled on hair-like processes of the tunic that not only does the translucency disappear, but even the external form of the animal may be thoroughly obscured. In such cases much of the foreign material can be scraped off the tunic in order to view the internal organs.

Study living individuals which have been allowed to remain undisturbed for several minutes in finger bowls containing just enough sea water to cover the extended siphons. The two extended siphons arise close together and are of unequal length. By adding finely powdered carmine particles to the surface of the water it is possible to determine which is the incurrent or branchial siphon and which the excurrent or atrial siphon. Observe that the margin of the incurrent siphon is divided into six lobes and that the aperture of the excurrent siphon is square.

The sensitivity of the test to mechanical stimuli can be investigated by touching selected regions of the test and of each siphon with the tip of a dissecting needle or bristle. The test is believed not to contain sensory nerves but to transmit the pressure of such mechanical stimuli to nerves lying in the mantle tissue directly underneath the test. Two types of reflex behavior have been observed in the responses of some ascidians to mechanical stimulation and may be tested in *Molgula*. Direct reflexes result from mechanical stimulation of the body surface; a gentle stimulus results in the contraction of the siphon nearer the point of stimulation; if the stimulus is stronger, not only the siphon rim near the stimulated point, but also the opposite siphon will close; a vigorous mechanical stimulus results in contraction of both siphons and the body. Crossed reflexes result from stimulation of the interior surfaces of the siphons; a delicate stimulus on the inside of one siphon results in a closure of the other siphon, the stimulated siphon remaining wide open; with a stronger stimulus the stimulated siphon remains open, the opposite one closes, and the body contracts, bringing about a discharge of water from the interior. These reflexes are protective for the incurrent siphon, the intake of large objects or noxious substances being avoided; for the excurrent siphon they serve as excretory reflexes for ejecting feces and reproductive cells. The effect upon these reflexes of removal of the dorsal ganglion, described below under the central nervous system, may be studied.

The incurrent siphon marks the anterior end of the animal; the atrial or excurrent siphon arises from the dorsal surface. In an animal with a translucent test the endostyle can be distinguished as a white, thread-like groove lying in the midventral line of the pharynx. With these reference points it is possible to orient the right and left, dorsal and ventral, and anterior and posterior aspects of the animal.

The test can be removed by cutting through it longitudinally with fine scissors (care being taken not to injure the body wall or mantle that lies directly underneath) and stripping the incised portions from the animal. By examining a mounted piece of the test microscopically, several different types of cells as well as "vessels" in which blood corpuscles normally circulate may be recognized. Actually the test is a non-living cuticular secretion, chemically similar to cellulose, of the epidermal cells of the mantle; it has been secondarily invaded by these vessels and cells.

The muscle fibers which contract the body and siphons are less well developed in *Molgula* than in more contractile species (*Ciona*). The fibers are especially conspicuous around the siphons, where they form sphincters, but over the rest of the mantle they are less closely spaced and may cross each other in transverse and oblique strands.

Internal Anatomy. With the test now removed, the general body organization can be more clearly seen through the mantle (Fig. 234). The largest structure is the branchial sac or pharynx at the anterior end of the digestive tract. The space between the digestive tract and

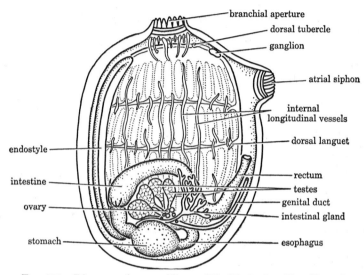

FIG. 234. Diagram of an ascidian. (Modified after Van Name.)

the inner wall of the mantle is the atrium or peribranchial chamber. Into this space, near the base of the excurrent siphon, open the reproductive ducts and the intestine; the numerous slits or stigmata of the pharyngeal wall also open into this peribranchial chamber; the various materials entering the atrium can be discharged to the outside through the excurrent siphon. The beating of the translucent heart, which is located on the right side of the animal near its posterior end, can be clearly seen with the aid of a hand lens. An ovary and a testis, intimately associated to form a gonadal mass, are present on each side; on the left side the gonadal mass is partly enclosed within the intestinal loop. Posterior to the gonadal mass of the right side is the conspicuous excretory organ, a brownish yellow sac.

The Digestive System. The alimentary tract begins as a short, tubelike mouth, the surface of which is lined by the inturned test of the incurrent siphon. The mouth opens into the pharynx, whose ventral wall is attached to the mantle. It may be advisable at this point to remove a large portion of the mantle from the left side by using fine scissors and forceps, at the same time being careful not to damage the pharynx. Near the junction of the mouth and pharynx is a circlet of branched tentacles of varied size. They act as a mechanical grid through which the incurrent stream of water must pass; this grid is well provided with sensory nerves, as can be determined by dropping a foreign particle, such as a sand grain, into the aperture of the incurrent siphon and observing its forceful reflex ejection. The sensitivity of this region to various salts and chemicals can be tested by introducing a sample of the solution to be tested into the mouth of the incurrent siphon. It is advisable to tie a layer of cheesecloth over the mouth of the medicine dropper by which the solution is administered in order to prevent the current of the introduced solution from acting as a mechanical stimulus.

The endostyle is the conspicuous groove lying in the midventral wall of the pharynx. This groove is composed of longitudinally arranged bands of ciliated and mucus-secreting cells which participate in the filter-feeding activity of the animal. Anteriorly the endostyle is continuous with the peripharyngeal bands, also ciliated, which encircle the oral end of the pharynx and then meet in the mid-dorsal line of the pharyngeal wall to form the dorsal lamina. The dorsal lamina passes posteriorly and ends at a small opening into the esophagus. Like many sedentary animals, the ascidians concentrate food particles that are suspended in the water. Mucus, secreted by cells located in the endostyle, is distributed dorsally to the dorsal lamina by the beating cilia which line the inner surface of the pharynx; the inner wall of the pharynx thus becomes coated with mucus in which become trapped the food particles drawn in through the incurrent siphon; the mixture of food and mucus is transported dorsally and then along the dorsal lamina posteriorly into the esophagus. This method of food concentration and food transportation can be studied by immersing a living animal in a sea-water suspension of carmine particles and observing, after an interval, the accumulation of the carmine particles along the dorsal lamina. The endostyle of protochordates has been considered to be homologous with the thyroid gland of vertebrates; this homology has been based largely on the fact that a portion of the endostyle in the ammocoete larva of cyclostomes is involved in forming thyroid tissue during metamorphosis into the adult cyclostome.

In addition to the mid-dorsal lamina and the midventral endostyle, the pharyngeal wall has on each side six longitudinal folds which are conspicuous because they bear internal longitudinal vessels.

It will already have been observed that the pharynx is perforated by a large number of minute clefts, the stigmata. If a piece of the pharyngeal wall is excised from a fresh specimen and is mounted on a slide in a drop of sea water, these stigmata can be examined in greater detail under high magnification with the compound microscope. The stigmata will be seen to be curved and actually to be spaces in a meshwork of small blood vessels. The stigmata are lined with actively beating cilia which are the effective agents in creating the inflowing current of sea water. As might have been suspected, the close association of blood vessels and a current of sea water indicates that the pharynx has a respiratory function as well as its filter-feeding function. The water entering the pharynx through the incurrent siphon passes through these stigmata into the atrial cavity.

The regions of the rest of the digestive tract are not well differentiated externally. The very short esophagus opens into the tubular stomach, the external surface of which is covered by a brown digestive gland; this gland may be considered to mark the anterior and posterior extent of the stomach region. The intestine passes posteriorly from the stomach and, on the left side, makes a somewhat more than semicircular loop anteroventrally and is then retroflexed so that the rest of the intestine lies almost in contact with the inner curve of this first loop. The anus opens into the atrial cavity, just posterior to the base of the excurrent siphon.

Reproductive System. The hermaphroditic gonadal structures are found on the right and on the left sides of the animal, adhering to the inner wall of the mantle. On the left side the ovary is located in the secondary loop of the intestine and is an elongate, flask-shaped structure whose longest axis is vertical. In the living animal the ovary, as seen with the low power of the dissecting microscope, is translucent and greenish in color; the surface has a pebbled appearance due to the presence of follicles of varied size, depending upon the degree of maturation of the egg. The left testis, opaque and white in color, borders and overlaps the ovary. The right pair of gonads is located near the middle of the right side; the long axis of the ovary runs transversely across the animal; the testis overlaps the posterior border of the ovary. The gonoducts, each consisting of two tubes, the oviduct and the vas deferens, open into the atrial cavity near the base of the excurrent siphon.

Excretory System. The excretory organ is a small, brownish white, conspicuous sac attached to the inner wall on the right side of the body, slightly posterior to the gonadal mass of that side. Excretory wastes in the form of a brownish fluid and some solid concretions are found within the sac. It possesses no duct or opening to the exterior, so that its function is apparently accumulation of the waste products.

Circulatory System. The heart is an elongate, translucent, tubular structure located on the right side, between the gonad and the excretory organ. It can be readily recognized in the living animal by its wave-like, peristaltic beat. Two blood vessels leave the heart, one at each end. The branches of these vessels cannot be readily discerned in their finer anatomical details. The dorsal vessel proceeds from the heart, giving off a branch to the test, and then divides into a number of sinuses supplying the viscera; these collect again to form a vessel that runs longitudinally on the dorsal wall of the pharynx. The dorsal longitudinal vessel sends lateral branches to the pharynx, these branches in turn becoming more finely divided to constitute the vascular network of the pharynx. The ultimate capillaries of the pharynx unite to form a longitudinal vessel on the ventral wall of the pharynx; this, joined by branches from the test, in turn opens into the ventral end of the heart.

The activity of the tunicate heart is unusual in the animal kingdom. This peculiarity, which consists of a periodic reversal of the beat, is readily observable in living *Molgula*. The beat starts at one end of the heart and passes as a peristaltic wave toward the other end. After a number of pulsations in one direction, the beat slows and a slight pause occurs, followed then by beats in the opposite direction. This is followed in turn by a pause after which the direction of the peristaltic wave reverses again. The beat may also be seen occasionally to start simultaneously at both ends of the heart. The physiological basis of this reversal of beat is not clearly understood beyond the indication that pacemakers are located at each end of the heart and alternate in initiating the peristaltic contraction wave.

The contraction during which blood leaves the dorsal end of the heart and proceeds toward the viscera is called the advisceral beat, while the abvisceral beat is one directed toward the ventral end of the heart. Since the blood vessels do not possess valves, the reversal in direction of heart beat results also in a reversal of blood flow in them, so that "arteries" become "veins" and "veins" become "arteries" with each of these periodic changes in direction.

Central Nervous System and the Neural Gland. In the adult tunicate the central nervous system is reduced to a small, oval, so-called cerebral ganglion and some nerve fibers. The cerebral ganglion is located dorsally in the mantle between the two siphons, to which it supplies nerves. The role of the cerebral ganglion in regulating the direct and crossed reflexes of the siphons can be studied in the living animal either by removing the ganglion or by severing its nerve fibers. Ganglion removal can be accomplished with the aid of watchmaker's forceps and fine scissors or knives fashioned from dissecting needles. The animal is fastened to a wax-bottomed tray or finger bowl with the siphons uppermost. (The animal can be anesthetized to prevent undue contraction of the siphons by keeping it in ice-cold sea water.) A slit is made through the test in the intersiphonal region, the mantle is incised, and the ganglion removed. Physiological isolation of the ganglion can be accomplished by similarly exposing the ganglion and then by means of a fine knife cutting around the margin of the ganglion and also making a cut to free it ventrally; in this fashion all nerves proceeding to and coming from the ganglion are severed, without actually removing the ganglion from the animal.

Closely associated with the ganglion is a small glandular mass, the neural gland. In most ascidians this gland lies ventral to the ganglion and has therefore been called the subneural gland. In several genera, however, including *Molgula*, the gland is dorsal to the ganglion. A small tube or duct leads from this neural gland to the pharynx, into which it opens on a rounded prominence, the dorsal tubercle. This tubercle is located in the mid-dorsal line between the circle of oral tentacles and the anterior end of the dorsal lamina. The function of the neural gland in tunicates is unknown; it has been speculatively reported as being an excretory structure, a mucus-secreting gland, and a lymph gland, with little supporting evidence for any of these varied views. It has been considered to be the homologue of the vertebrate hypophysis. Interestingly enough, endocrine effects similar to those obtained from different pituitary hormones have been observed when extracts of the neural gland and dorsal ganglion have been tested on vertebrates. Among the hormone-like effects reported have been oxytocic and vasopressor ones, comparable to those obtained with preparations of the posterior lobe of the pituitary; chromatophorotropic ones, comparable to that obtained with extracts of the intermediate lobe of the pituitary; and gonadal-follicle-stimulating effect, comparable to that obtained from the anterior lobe of the pituitary. Some of these observations may be questioned because the experiments were not performed critically. The significance of these observations in explaining the physiology of the neural gland is dubious, since there is no evidence to suggest that any of these same functions are maintained in the ascidians by this gland.

Developmental Stages of *Molgula*. In *Molgula manhattensis* the gametes are discharged into the external sea water, fertilization and development occurring there. In *M. citrina*, however, fertilization is internal, the embryonic and larval stages being retained in the atrial cavity of the parent individual. Early developmental stages of *M. manhattensis* can be obtained by artificial fertilization. The tests of two individuals are removed; the ovaries of one individual and the testes of the second individual are removed with the aid of a binocular dissecting microscope. The ovaries and the testes are transferred respectively to a Syracuse dish and a depression slide, both containing sea water. Eggs are removed from the ovaries with a fine dissecting needle and are carefully washed free of body fluid by several changes of sea water. The diameter of the egg, exclusive of follicle, is 0.11–0.15 mm. The testes are minced with very fine scissors, and the resultant suspension is diluted with several drops of sea water. A few drops of this sperm suspension is then added to the dish containing the eggs, and after a time the eggs are washed in changes of sea water. Development is rapid; the tadpole larva, formed in about 24 hours, swims about freely for 1–10 hours; at the end of this free-swimming period the tadpole larva settles down, becomes attached to the substratum, and gradually metamorphoses into the adult form.

The Ascidian Larva

By L. H. Kleinholz

The larval stage of an ascidian, commonly called the tadpole larva because of its resemblance to the amphibian tadpole, is usually studied because during this stage an ascidian possesses the particular structures that characterize it as a chordate. It is interesting to note that, although solitary and colonial ascidians had been studied and described since the seventeenth century, it was not until the latter half of the nineteenth century, as a result of Kowalevski's investigation of larval structure, that their affinity to the chordates was recognized; up to that time tunicates had been considered to be closely related to the mollusks.

THE ASCIDIAN LARVA

The description given below applies particularly to the larva of *Amaroucium constellatum* (Fig. 235). Both living and preserved larval material may be used for study, the former being preferable unless stained whole mounts of larvae are available. If a mass of the living colony is squeezed over a dish of sea water, the expressed zooids,

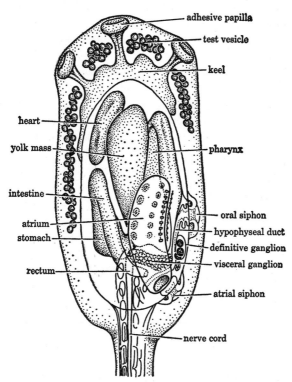

FIG. 235. *Amaroucium constellatum,* anatomy of tadpole. (Redrawn after Grave.)

tadpoles, embryos, and eggs settle to the bottom while most of the more slowly settling jelly-like debris can be decanted off with several changes of sea water. The larvae lie motionless for a few seconds after having been squeezed from the colony, but soon begin to swim about. They can be removed with a pipette to a slide for study. Individuals can be fixed in Bouin's solution for staining as whole mounts or for the preparation of sections for more detailed anatomical and histological study. A second method of obtaining living larvae in the laboratory is to place a mass of the colony in a jar of sea water at night; many larvae are discharged shortly after sunrise, although some will also be released during the day and can be collected with a pipette from the sides of the container. If only preserved material is available, tadpoles can be obtained by teasing apart with dissecting needles a small portion of a colony; preserved tadpoles thus obtained can be cleared for study by mounting them in a drop of glycerine.

For purposes of orientation, that portion of the larval body containing the adhesive papillae is called anterior; the tail is considered posterior; the sensory vesicle containing the conspicuously pigmented sense organs is dorsal in position. A striking rearrangement of parts occurs during metamorphosis, when the anterior portion of the tadpole becomes attached to the substratum and is transformed into the basal portion of the zooid.

The fully developed tadpole larva is slightly more than 2 mm. in length, the body portion alone being about 0.7 mm. long. The body and tail are enclosed in a translucent, non-cellular tunic or test which, in the tail region, is flattened dorsoventrally and expanded horizontally to constitute the wide tail fin. The mantle, which lies underneath the tunic, also completely envelops the tadpole, but can be seen well only in sectioned material. Three adhesive papillae, which consist of stalked outgrowths from the body wall, are enlarged at their anterior portions into cup-shaped disks. It is by means of these papillae that the tadpole larva attaches itself to the substratum at the end of its free-swimming period. A large number of test vesicles, lying between the mantle and the external surface of the tunic, are divided into four unequal groups by the stalks of the adhesive papillae. The test vesicles become distributed over the entire external surface of the tunic about 24 hours after the beginning of metamorphosis and are subsequently concerned with the formation of the common tunic substance of the colony.

Located dorsally in the posterior part of the body, between the anterior oral siphon and the posterior atrial siphon, is an oval sensory vesicle into the cavity of which project two pigmented sense organs. The larger, posterior, cup-shaped structure is a functional photoreceptor, the eye. The smaller anterior black sphere is the statolith contained in the static organ. A visceral ganglion, the cortex of which is composed of a single layer of large nerve cells, extends ventrally from just beneath the sensory vesicle to connect with the nerve cord. The hollow nerve cord, readily visible in the tail, lies slightly to the left of the notochord because of a permanent twisting of the tail 90° to.the left. Near its anterior end the nerve cord bends sharply dorsally and slightly to the right to join the ventral end of the visceral ganglion.

All the aforementioned parts of the nervous system degenerate during metamorphosis of the tadpole. A definitive ganglion, which persists as the nerve center of the zooid, is a small oval structure located slightly to the left of the sensory vesicle.

A subneural gland, arising as an outgrowth of the middle portion of the hypophyseal duct, is not readily seen in whole mounts of the larva. The hollow hypophyseal duct lies ventral to the definitive ganglion, its cavity connecting anteriorly with the cavity of the oral siphon, while posteriorly the duct ends blindly near the atrium. The anatomical relationships of these structures are best studied in sectioned material.

The pharynx occupies a conspicuous anterodorsal position in the body of the tadpole. The ventral wall of the pharynx is closely associated with a large, cone-shaped yolk mass; the dorsal wall of the pharynx contains the endostyle; the lateral walls of the posterior portion of the pharynx are in contact with the inner walls of right and left horns of the atrial cavity with which the pharyngeal cavity communicates by means of three rows of gill slits. This relationship can be readily seen in sections through the pharyngeal region. The middle portion of the atrium connects with the atrial siphon. The stomach and the intestine are located ventral to the posterior portion of the conical yolk mass. The intestine passes posterodorsally, entering the left horn of the atrium, where it terminates in an anus near the base of the atrial siphon.

The heart differentiates from the pericardial sac, which is a thin-walled oval structure lying ventral to the anterior portion of the yolk mass and just anterior to the loop formed by the intestine as it turns dorsally.

The notochord, which occupies the anterior two-thirds of the tail axis, enters the body and passes anteriorly to a point ventral to the sensory vesicle, where it terminates in contact with the pharynx. Two bands of muscle, one dorsal and one ventral, constitute an envelope within which are included the caudal portions of the nerve cord and notochord.

Bibliography

GENERAL

Brandt, K., and C. Apstein. 1901–41. *Nordisches Plankton. Zoologischer Teil.* Lipsius and Tischer, Kiel.

Bronn, H. G. 1859–1948. *Klassen und Ordnungen des Thier-Reichs.* Winter'sche Verlagshandlung, Leipzig.

Delage, Y., and E. Hérouard. 1896–1903. *Traité de zoologie concrète.* Schleicher Frères, Paris.

Grassé, P. 1948–49. *Traité de zoologie, anatomie, systématique, biologie.* Masson et Cie, Paris.

Grimpe, G., and E. Wagler. 1925–44. *Die Tierwelt der Nord- und Ostsee.* Akademische Verlag., Leipzig.

Harmer, S. F., and A. E. Shipley. 1895–1909. *Cambridge natural history.* Macmillan Co., London.

Hyman, L. H. 1940. *The invertebrates.* McGraw-Hill Book Co., N. Y.

Korscheldt, E., and K. Heider. 1893–1909. *Lehrbuch der vergleichenden Entwicklungsgeschichte der wirbellosen Thiere.* Fischer, Jena.

Kukenthal, W., and T. Krumbach. 1923–38. *Handbuch der Zoologie. Eine Naturgeschichte der Stämme des Tierreiches.* Walter der Gruyter and Co., Berlin.

Lang, A. 1891, 1896. *Textbook of comparative anatomy* (Translated by H. M. and M. Bernard). Macmillan Co., London.

Lankester, R. 1900–09. *Treatise on zoology.* A. and C. Black, London.

Parker, T. J., and W. A. Haswell. 1940. *A textbook of zoology.* Macmillan Co., London.

Schulze, P. 1922–40. *Biologie der Tiere Deutschlands.* Borntraeger, Berlin.

Sedgwick, A. 1898–1909. *A student's text-book of zoology.* Swann Sonnenschein and Co., Ltd., London.

PROTOZOA

General

Bishop, E. L. 1943. Studies on the cytology of the Hypotrichous Infusoria. I. The relation of structure to regeneration. *Jour. Morph.*, 72, 441–472.

Calkins, G. N., and Francis M. Summers. 1941. *Protozoa in biological research.* Columbia Univ. Press, N. Y. 1148 pp.*†

Claparède, E., and J. Lachmann. 1868. *Études sur les Infusoires et les Rhizopodes.* H. Georg, Genève. 330 pp.†

Galtsoff, P., et al. 1937. *Culture methods for invertebrate animals.* Comstock Publ. Co., Ithaca, N. Y. xxxii + 590 pp.

* indicates that the reference possesses excellent illustrations.
† denotes that the reference possesses an extensive bibliography.

Hegner, R. W., and W. H. Taliaferro. 1925. *Human protozoölogy.* The Macmillan Co., N. Y. 597 pp.
Hyman, L. H. 1940. *The invertebrates.* McGraw-Hill Book Co., N. Y. 726 pp.*†
Kahl, A. 1930–35. Urtiere oder Protozoa. In *Dahl's Die Tierreichwelt Deutschlands.* Teil 18–30. Gustave Fischer, Jena. 886 pp.*†
Kent, W. S. 1880–82. *A manual of the Infusoria.* Vol. III. David Bogue, London. LI plates.†
Kudo, R. R. 1946. *Protozoology.* 3rd ed. Charles C. Thomas, Springfield, Ill. xiii + 778 pp.*†
Lund, E. E. 1941. The feeding mechanisms of various ciliated Protozoa. *Jour. Morph.,* **69,** 563–571.†
Needham, J. G. 1937. *Culture methods for invertebrate animals.* Comstock Publ. Co., Ithaca, N. Y. 590 pp.
Penard, E. 1902. *Faune Rhizopodique du bassin du Lèman.* Henry Kündig, Genève. 714 pp.
Penard, E. 1922. *Études sur les Infusoires d'eau douce.* H. Georg, Genève. 331 pp.†
Smith, G. 1933. *The fresh-water algae of the United States.* McGraw-Hill Book Co., N. Y. xi + 716 pp.
Stein, F. M. 1867. *Der Organismus der Infusionsthiere.* Abt. II. Wilhelm Engelmann, Leipzig. 355 pp.†
Wenyon, C. 1926. *Protozoology.* Bailliere, Tindall and Cox, London. xvi + 1563 pp.

Euglena

Jahn, T. 1946. The euglenoid flagellates. *Quart. Rev. Biol.,* **21,** 246–74.
Johnson, L. 1944. Euglenae of Iowa. *Trans. Amer. Micr. Soc.,* **63,** 97–135.
Mast, S. 1941. Chapt. 5 in *Protozoa in biological research,* ed. by G. Calkins and F. Summers. Columbia Univ. Press, N. Y. xli + 1148 pp.

Peranema

Mast, S. 1941. Chapt. 5 in *Protozoa in biological research,* ed. by G. Calkins and F. Summers. Columbia Univ. Press, N. Y. xli + 1148 pp.
Pitelka, D. 1945. Morphology and taxonomy of flagellates of the genus *Peranema. Jour. Morph.,* **76,** 179–92.

Volvox

Mast, S. 1941. Chapt. 5 in *Protozoa in biological research,* ed. by G. Calkins and F. Summers. Columbia Univ. Press, N. Y. xli + 1148 pp.
Rice, N. 1947. The culture of *Volvox aureus* Ehrenberg. *Biol. Bull.,* **92,** 200–209.
Smith, G. 1944. A comparative study of the species of *Volvox. Trans. Amer. Micr. Soc.,* **63,** 265–310.

Trypanosoma

Hoare, C. 1938. Morphological and taxonomic studies on mammalian trypanosomes. V. The diagnostic value of the kinetoplast. *Trans. Roy. Soc. Trop. Med. Hyg.,* **32,** 333–42.
Lwoff, M. 1938. L'hématine et l'acide ascorbique, facteurs de croissance pour le flagellé *Schizotrypanum cruzi. Compt. rend. acad. sci. Paris,* **206,** 540–42.

Wenyon, C. 1926. *Protozoology.* Bailliere, Tindall and Cox, London. xvi + 1563 pp.

Trichomonas

Kirby, H. 1931. Trichomonad flagellates from termites. II. *Eutrichomastix,* and the subfamily Trichomonadinae. *Univ. Calif. Publ. Zool.,* **36,** 171–262.

Kirby, H. 1944. Some observations on cytology and morphogenesis in flagellate protozoa. *Jour. Morph.,* **75,** 361–421.

Trichonympha

Cleveland, L. 1947. Sex produced in the protozoa of *Cryptocercus* by molting. *Sci.,* **105,** 16–17.

Kirby, H. 1932. Flagellates of the genus *Trichonympha* in termites. *Univ. Calif. Publ. Zool.,* **37,** 349–76.

Amoeba proteus

Mast, S. O., and W. L. Doyle. 1935. Structure, origin, and function of cytoplasmic constituents in *Amoeba proteus.* I. Structure. *Arch. Protistk.,* **86,** 155–180.*†

Mast, S. O., and W. F. Hahnert. 1935. Feeding, digestion, and starvation in *Amoeba proteus* (Leidy). *Physiol. Zool.,* **8,** 255–272.

Mast, S. O., and N. Stahler. 1937. The relation between luminous intensity adaptation to light, and rate of locomotion in *Amoeba proteus* (Leidy). *Biol. Bull.,* **73,** 126–133.

Schaeffer, A. A. 1916. Characters of *A. proteus* Pallas (Leidy), *A. discoides* spec. nov., and *A. dubia* spec. nov. *Arch. Protistk.,* **37,** 204–228.†

Arcella vulgaris

Bles, E. J. 1929. *Arcella.* A study in cell physiology. *Quart. Jour. Micr. Sci.,* **72,** 527–648.†

Actinophrys sol

Leidy, J. 1879. *Freshwater rhizopods of North America.* Govt. Printing Office, Washington. 324 pp. 48 plates.†

Actinosphaerium eichhornii: see Actinophrys

Monocystis lumbrici

Hesse, E. 1909. Contributions à l'étude des monocystidées des Oligochaetes. *Arch. zool. exp.,* **43** (Ser. 5, **3**), 27–299.†

Gregarina blattarum

Ellis, M. 1913. A descriptive list of the cephaline gregarines of the New World. *Trans. Amer. Micr. Soc.,* **32,** 259–296.

Schneider, M. A. 1875. Contributions à l'histoire des grégarines. *Arch. zool. exp.,* **4,** 493–604.

Sprague, V. 1941. Studies on *Gregarina blattarum* with particular reference to the chromosome cycle. *Ill. Biol. Monogr.,* **18,** 1–57.*

Plasmodium vivax

Moulton, F. R. 1941. Human malaria. *Amer. Assoc. Advancement Sci., Pub.* 15. Washington, D. C. 398 pp.

Russell, P. F., L. S. West, and R. D. Manwell. 1946. *Practical malariology.* Saunders, Philadelphia. 684 pp.†

Wilcox, Aimee. 1943. Manual for the microscopical diagnosis of malaria in man. *Nat. Inst. Health Bull.* 180. 39 pp.*†

Opalina obtrigonoidea

Cosgrove, W. B. 1947. Fibrillar structures in *Opalina obtrigonoidea* Metcalf. *Jour. Parasit.,* **33**, 351-357.†

Metcalf, M. M. 1923. The opalinid ciliate infusorians. *Bull. U. S. Nat. Mus.* 120. Govt. Printing Office, Washington. 484 pp.*†

Tyler, A. R. 1926. The cultivation of *Opalina. Sci.,* **64**, 383-384.

Didinium nasutum

Mast, S. O. 1909. The reactions of *Didinium nasutum* (Stein) with special reference to the feeding habits and functions of trichocysts. *Biol. Bull.,* **16**, 91-118.†

Coleps hirtus

Noland, L. E. 1925. A review of the genus *Coleps* with descriptions of two new species. *Trans. Amer. Micr. Soc.,* **44**, (1), 3-13.†

Prorodon griseus

Tannreuther, G. W. 1926. Life history of *Prorodon griseus. Biol. Bull.,* **51**, 303-320.†

Paramecium caudatum

Jakus, M. A. 1945. The structure and properties of the trichocysts of *Paramecium. Jour. Exp. Zoöl.,* **100**, 457-485.†

Jennings, H. S. 1931. *Behavior of the lower organisms.* Columbia Univ. Press, N. Y. 366 pp.†

Mast, S. O. 1947. The food-vacuole in *Paramecium. Biol. Bull.,* **92**, 31-72.*†

Colpoda cucullus

Taylor, C. V., and A. G. R. Strickland. 1935. Some factors in the excystment of dried cysts of *Colpoda cucullus. Arch. Protistk.,* **86**, 181-190.

Frontonia leucas

Bullington, W. E. 1939. A study of spiraling in the ciliate *Frontonia* with a review of the genus and a description of two new species. *Arch. Protistk.,* **92**, 10-66.

Colpidium colpoda

Burbanck, W. D. 1942. Physiology of the ciliate *Colpidium colpoda.* I. The effect of various bacteria as food on the division rate of *Colpidium colpoda. Physiol. Zool.,* **15**, 342-362.

Spirostomum ambiguum

Bishop, Ann. 1923. Some observations upon *Spirostomum ambiguum* (Ehrenberg). *Quart. Jour. Micr. Sci.*, **67**, 391–434.†

Stentor coeruleus

Dierks, K. 1926. Untersuchungen über die Morphologie und Physiologie des *Stentor coeruleus*. *Arch. Protistk.*, **54**, 1–91.*†

Johnson, H. 1893. A contribution to the morphology and biology of the stentors. *Jour. Morph.*, **8**, 467–562.*†

Stylonychia pustulata

Summers, F. M. 1935. The division and reorganization of the macronucleus of *Aspidisca lynceus* Mull., *Diophrys appendiculata* Stein and *Stylonychia pustulata* Ehrbg. *Arch. Protistk.*, **85**, 173–208.†

Euplotes patella

Pierson, B. F. 1943. A comparative morphological study of several species of *Euplotes* closely related to *E. patella*. *Jour. Morph.*, **72**, 125–165.†

Turner, J. P. 1933. The external fibrillar system of *Euplotes* with notes on the neuromotor apparatus. *Biol. Bull.*, **64**, 53–66.

Vorticella campanula

Noland, L. E., and H. E. Finley. 1931. Studies on the taxonomy of the genus *Vorticella*. *Trans. Amer. Micr. Soc.*, **44**, 81–123.†

Zoöthamnium arbuscula

Furssenko, A. 1929. Lebencyclus und Morphology von *Zoöthamnium arbuscula* Ehrenberg. *Arch. Protistk.*, **66**, 376–500.*†

Podophrya collini

Root, F. M. 1914. Reproduction and reaction to food of the suctorian *Podophrya*. *Arch. Protistk.*, **35**, 164–194.†

PORIFERA

General

Bidder, G. G. 1923. The relation of the form of a sponge to its currents. *Quart. Jour. Micr. Sci.*, **67**, 293–323.

Bidder, G. G., and G. S. Vosmaer-Roell. 1928. *Bibliography of sponges, 1751–1913*. Cambridge Univ. Press, Cambridge. 234 pp.*

Hyman, L. H. 1940. *The invertebrates*. McGraw-Hill Book Co., N. Y. 726 pp.*†

Minchin, E. A. 1900. Phylum Porifera. Chapt. III in *A treatise on zoölogy*, ed. by E. Ray Lankester. Adam and Chas. Black, London. 178 pp.†

Morgan, A. H. 1930. *Field book of ponds and streams*. G. P. Putman's Sons, N. Y. 448 pp.†

Parker, G. H. 1914. On the strength of water currents produced by sponges. *Jour. Exp. Zoöl.*, **16**, 443–446.

Pratt, H. S. 1935. *A manual of the common invertebrate animals*. Blakiston, Philadelphia. 854 pp.†

Rentschel, E. 1923-25. Parazoa in *Kükenthal und Krumbach's Handbuch der Zoologie.* I. Winter'sche, Leipzig. Pp. 307-418.

Microciona prolifera

Galtsoff, P. S. 1925. Regeneration after dissociation (an experimental study on sponges). I. Behavior of dissociated cells of *Microciona prolifera* under normal and altered conditions. *Jour. Exp. Zoöl.,* **42,** 183-221.

Galtsoff, P. S. 1925. Regeneration after dissociation (an experimental study on sponges). II. Histogenesis of *Microciona prolifera* Verr. *Jour. Exp. Zoöl.,* **42,** 223-255.†

Wilson, H. V. P., and J. T. Penny. 1930. The regeneration of sponges (*Microciona*) from dissociated cells. *Jour. Exp. Zoöl.,* **56,** 73-147.

CNIDARIA

General

Carlgren, O. 1940. A contribution to the knowledge of the structure and distribution of the cnidae in the Anthozoa. *Acta Univ. Lund,* N.S., **36** (3), 1-62.

Fraser, C. M. 1937. *Hydroids of the Pacific Coast of Canada and the United States.* Univ. of Toronto Press, Toronto. 207 pp.*†

Fraser, C. M. 1944. *Hydroids of the Atlantic coast of North America.* Univ. of Toronto Press, Toronto. 451 pp.*†

Fraser, C. M. 1946. *Distribution and relationships in American hydroids.* Univ. of Toronto Press, Toronto. 464 pp.

Hargitt, C. W. 1905. The medusae of the Woods Hole region. *Bull. U. S. Bur. Fish.,* **24,** 21-79.*

Hyman, L. H. 1940. Observations and experiments on the physiology of medusae. *Biol. Bull.,* **79,** 282-296.

Hyman, L. H. 1940. *The invertebrates.* McGraw-Hill Book Co., N. Y. 726 pp.*†

Kramp, P. L. 1943. On development through alternating generations, especially in Coelenterata. *Vidensk. Medd. Naturh. Foren. Kjøb.,* **107,** 13-32.

von Ledebur, J. F. 1939. Ueber die Atmung der Schwämme und Coelenteraten. *Ergeb. Biol.,* **16,** 262-291.

Mayer, A. G. 1910. *Medusae of the world.* Vols. 1-3. Carnegie Institution, Washington. 735 pp.*†

Weill, R. 1934. Contribution a l'étude des cnidaires et de leurs nématocystes. *Trav. stat. zool. Wimeraux,* Tomes 10-11, 1-701.*†

Woollard, H. H., and J. A. Harpman. 1939. Discontinuity in the nervous system of coelenterates. *Jour. Anat.,* **73,** 559-562.

Yonge, C. M. 1944. Experimental analysis of the association between invertebrates and unicellular algae. *Biol. Rev.,* **19,** 68-80.†

Tubularia

Barth, L. G. 1944. The determination of the regenerating hydranth in *Tubularia. Physiol. Zool.,* **17,** 355-366.

Liu, C. K., and N. J. Berrill. 1948. Gonophore formation and germ cell origin in *Tubularia. Jour. Morph.,* **83,** 39-59.

Lowe, E. 1926. The embryology of *Tubularia larynx* (Allm.) *Quart. Jour. Micr. Sci.,* N.S., **70,** 599–627.

Hydractinia

Ballard, W. W. 1942. The mechanism for synchronous spawning in *Hydractinia* and *Pennaria. Biol. Bull.,* **82,** 329–339.

Collcutt, M. C. 1897. On the structure of *Hydractinia echinata. Quart. Jour. Micr. Sci.,* **40,** 77–99.

Schijfsma, K. 1935. Observations on *Hydractinia echinata* (Flem.) and *Eupagurus bernhardus* (L.). *Arch. néerl. zool.,* **1,** 261–314.

Schijfsma, K. 1939. Preliminary notes on early stages in the growth of colonies of *Hydractinia echinata* (Flem.). *Arch. néerl. zool.,* **4,** 93–102.

Obelia

Berrill, N. J. 1948. A new method of reproduction in *Obelia. Biol. Bull.,* **95,** 94–99.

Hammett, F. S., and D. W. Hammett. 1945. Seasonal changes in *Obelia* colony composition. *Growth,* **9,** 55–144.

Gonionemus

Joseph, H. 1925. Zur Morphologie und Entwicklungsgeschichte von *Haleremita* und *Gonionemus.* Ein Beitrag zur systematischen Beurteilung der Trachymedusen. *Z. wiss. Zool.,* **125,** 374–434.

Thomas, L. J. 1921. Morphology and orientation of the otocysts of Gonionemus. *Biol. Bull.,* **40,** 287–298.

Hydra

Ewer, R. F. 1947. The behavior of *Hydra* in response to gravity. *Proc. Zool. Soc. Lond.,* **117,** 207–218.

Hyman, L. H. 1930. Taxonomic studies on the hydras of North America. II. The characters of *Pelmatohydra oligactis* (Pallas). *Trans. Amer. Micr. Soc.,* **49,** 322–333.

Jones, C. S. 1941. The place of origin and the transportation of cnidoblasts in *Pelmatohydra oligactis* (Pallas). *Jour. Exp. Zoöl.,* **87,** 457–475.

Jones, C. S. 1947. The control and discharge of nematocysts in hydra. *Jour. Exp. Zoöl.,* **105,** 25–60.

McConnell, C. H. 1938. The hatching of *Pelmatohydra oligactis* eggs. *Zool. Anzeiger,* **123,** 161–174.

Reis, R. H. 1948. Spontaneous change of form of young *Pelmatohydra oligactis. Trans. Amer. Micr. Soc.,* **67,** 70–81.

Schultze, P. 1922. Der Bau und Entladung der Penetranten von *Hydra attenuata* Pallas. *Arch. Zellforsch.,* **16,** 383–438.

Aurellia

Bethe, A. 1942. Ueber das Zusammenwirken zweier Randkörper bei Medusen und einige allgemeine Fragen der Koordination. *Z. vergl. Physiol.,* **29,** 394–417.

Bullock, T. H. 1943. Neuromuscular facilitation in Scyphomedusae. *Jour. Cell. Comp. Physiol.,* **22,** 251–272.

Hein, W. 1900. Untersuchungen ueber die Entwicklung von *Aurelia aurita*. *Z. wiss. Zool.*, **67**, 401–438.
Kinosita, H. 1941. Initiation of entrapped circuit wave in a scyphomedusa, *Mastigias papua*. *Jap. Jour. Zool.*, **9**, 209–220.
Lowndes, A. G. 1942. Percentage of water in jelly-fish. *Nature*, **150**, 234–235.
Percival, E. 1923. On the strobilization of *Aurelia*. *Quart. Jour. Micr. Sci.*, N.S., **67**, 85–100.
Thiel, M. E. 1936–38. Scyphomedusae. *Bronn's Klassen und Ordnungen des Tier-reichs*. 2, Abt. 2, Bch. 2, Lief. 1–5. 848 pp. (incomplete).*
Thill, H. 1937. Beiträge zur Kenntnis der *Aurelia autita* (L.) *Z. wiss. Zool.*, **150**, 51–96.

Metridium

Pantin, C. F. A. 1942. The excitation of nematocysts. *Jour. Exp. Biol.*, **19**, 294–310.
Parker, G. H., and A. P. Marks. 1928. Ciliary reversal in the sea-anemone, *Metridium*. *Jour. Exp. Zoöl.*, **52**, 1–6.
Ross, D. M., and C. F. A. Pantin. 1940. Factors influencing facilitation in Actinozoa. The action of certain ions. *Jour. Exp. Biol.*, **17**, 61–73.
Stephenson, T. A. 1928, 1935. *The British sea anemones*. Ray Society Monographs, London. 2 vols., 148 pp., 426 pp.*†

Astrangia

Boschma, H. 1925. On the feeding reactions and digestion in the coral polyp *Astrangia danae*, with notes on its symbiosis with zoöxanthellae. *Biol. Bull.*, **49**, 407–439.
Smith, F. G. W. 1948. *Atlantic reef corals*. Univ. of Miami Press, Coral Gables. 112 pp.*
Vaughan, T. W., and J. W. Wells. 1943. Revision of the Suborders, Families, and Genera of the Scleractinia. *Geol. Soc. Amer., Spec. Papers*, 44. 363 pp.
Yonge, C. M. 1940. The biology of reef-building corals. *Gr. Barrier Reef Exp.*, **1**, 353–389.

Renilla

Eisen, G. 1874. Bidrag till kännedomen om Pennatulidslägtet Renilla Lamk. *Kungl. Svenska Vetenskapsakad. Handl.*, **30**, 1–15.
Parker, G. H. 1920a. Activities of colonial animals. I. Circulation of water in *Renilla*. *Jour. Exp. Zoöl.*, **31**, 343–367.
Parker, G. H. 1920b. Activities of colonial animals. II. Neuromuscular movements and phosphorescence in *Renilla*. *Jour. Exp. Zoöl.*, **31**, 475–515.
Wilson, E. B. 1883. The development of *Renilla*. *Trans. Roy. Soc. Lond.*, **174**, 723–815.

CTENOPHORA

Pleurobrachia

Gudger, E. W. 1943. Some ctenophore fish catchers. *Sci. Monthly*, **57**, 73–76.
Heider, K. 1927. Vom Nervensystem der Ctenophoren. *Z. wiss. Biol.*, Abt. A, Bd. 9, 638–678.
Hyman, L. H. 1940. *The invertebrates*. McGraw-Hill Book Co., N. Y. 726 pp.*†

Lowndes, A. G. 1942. Ciliary movement and the density of Pleurobrachia. *Nature*, **150**, 579–580.
Mayer, A. G. 1912. Ctenophores of the Atlantic Coast of North America. *Carnegie Inst. Publ.* 162. 58 pp.

PLATYHELMINTHES

Polychoerus

Costello, H. M., and D. P. Costello. 1938. A new species of *Polychoerus* from the Pacific coast. *Ann. Mag. Nat. Hist.*, ser. 11, **1**, 148–155.
Costello, H. M., and D. P. Costello. 1938. Copulation in the acoelous turbellarian, *Polychoerus carmelensis*. *Biol. Bull.*, **75**, 85–98.
Costello, H. M., and D. P. Costello. 1939. Egg laying in the acoelous turbellarian, *Polychoerus carmelensis*. *Biol. Bull.*, **76**, 80–89.
Mark, E. L. 1892. *Polychoerus caudatus*. *Festschr. 70ten Geburtstag Rudolph Leuckarts*. Pp. 298–309.*

Stenostomum

Child, C. M. 1902. Studies on regulation. 1. Fission and regulation in *Stenostomum*. *Arch. Entwicklungsmech.*, **15**, 187–237; 355–420.
Kepner, W. A., J. S. Carter, and M. Hess. 1933. Observations upon *Stenostomum oesophagium*. *Biol. Bull.*, **64**, 405–417.
Kepner, W. A., and J. R. Cash. 1915. Ciliated pits of *Stenostomum*. *Jour. Morph.*, **26**, 235–245.
Nuttycombe, J. W., and A. J. Waters. 1935. Feeding habits and pharyngeal structure in *Stenostomum*. *Biol. Bull.*, **69**, 439–446.
Nuttycombe, J. W., and A. J. Waters. 1938. The American species of the genus *Stenostomum*. *Proc. Amer. Philos. Soc.*, **79**, 213–301.*†

Bdelloura

Wilhelmi, J. 1909. Tricladen. *Fauna und Flora des Golfes von Neapel, Monogr.* 32. 405 pp.*†

Syncoelidium

Wheeler, W. M. 1894. *Syncoelidium pellucidum*, a new marine triclad. *Jour. Morph.*, **9**, 167–194.*

Dugesia

Budington, R. A. 1924. The manner of copulation of a turbellarian worm, *Planaria maculata*. *Biol. Bull.*, **47**, 298–304.
Curtis, W. C. 1902. The life history, the normal fission, and the reproductive organs of *Planaria maculata*. *Proc. Boston Soc. Nat. Hist.*, **30**, 515–559.
Kenk, R. 1935. Studies in Virginian triclads. *Jour. Elisha Mitchell Sci. Soc.*, **51**, 79–125.*†
Kenk, R. 1937. Sexual and asexual reproduction in *Euplanaria tigrina*. *Biol. Bull.*, **73**, 280–294.
Morgan, T. H. 1898. Experimental studies of the regeneration of *Planaria maculata*. *Arch. Entwicklungsmech.*, **7**, 364–397.
Morgan, T. H. 1900. Regeneration in planarians. *Arch. Entwicklungsmech.*, **10**, 58–119.

Silber, R. H., and V. Hamburger. 1939. The production of duplicitas cruciata and multiple heads by regeneration in *Euplanaria tigrina*. *Physiol. Zool.*, **12**, 285–300.

Procotyla

Hyman, L. H. 1928. Studies on the morphology, taxonomy, and distribution of North American triclad Turbellaria. I. *Procotyla fluviatilis*, commonly but erroneously known as *Dendrocoelum lacteum*. *Trans. Amer. Micr. Soc.*, **47**, 222–255.*

Redfield, E. 1915. The grasping organ of *Dendrocoelum lacteum*. *Jour. Animal Behavior*, **5**, 375–380.

Hoploplana and other polyclads

Bock, S. 1913. Studien über Polycladen. *Zool. Bidrag*, **2**, 31–343.

Hyman, L. H. 1939. Some polyclads of the New England coast, especially of the Woods Hole region. *Biol. Bull.*, **76**, 127–152.*

Hyman, L. H. 1940. The polyclad flatworms of the Atlantic coast of the United States and Canada. *Proc. U. S. Nat. Mus.*, **89**, 449–495.*†

Lang, A. 1884. Die Polycladen (Seeplanarien) des Golfes von Neapel. *Fauna und Flora des Golfes von Neapel, Monogr.* 11. 688 pp.*

Wheeler, W. M. 1894. *Planocera inquilina*, a polyclad inhabiting the branchial chamber of *Sycotypus canaliculatus*. *Jour. Morph.*, **9**, 195–201.*

Polystomoides

Gallien, L. 1935. Recherches expérimentales sur le dimorphisme évolutif et la biologie de *Polystomum integerrimum* Fröhl. *Trav. stat. zool. Wimereux*, **12**, 1–181.

Paul, A. A. 1938. Life history studies of North American fresh-water polystomes. *Jour. Parasit.*, **24**, 489–510.

Price, E. W. 1939. North American monogenetic trematodes. IV. The family Polystomatidae (Polystomatoidea). *Proc. Helminth. Soc. Wash.*, **6**, 80–92.

Stunkard, H. W. 1917. Studies on North American Polystomidae, Aspidogastridae and Paramphistomidae. *Ill. Biol. Monogr.*, **3**, 283–395.

Zeller, E. 1876. Weiterer Beitrag zur Kenntniss der Polystomem. *Z. wiss. Zool.*, **27**, 238–274.

Aspidogaster

Bychowsky, Irene, and B. Bychowsky. 1934. Über die Morphologie und die Systematik des *Aspidogaster limacoides* Diesing. *Z. Parasitenk.*, **7**, 125–137.

Stafford, J. 1896. Anatomical structure of *Aspidogaster conchicola*. *Zool. Jahr., Abt. Anat.*, **9**, 477–542.

Van Cleave, H. J., and C. O. Williams. 1943. Maintenance of a trematode, *Aspidogaster conchicola*, outside the body of its natural host. *Jour. Parasit.*, **29**, 127–130.

Williams, C. O. 1942. Observations on the life history and taxonomic relationships of the trematode *Aspidogaster conchicola*. *Jour. Parasit.*, **28**, 467–475.

Cryptocotyle

Jägerskiöld, L. A. 1899. *Distomum lingua* Creplin, ein Genitalnapftragendes Distomum. *Bergens Mus. Årbok*, **2**, 1–18.

Linton, E. 1915. *Tocotrema lingua* (Creplin), the adult stage of a skin parasite of the cunner and other fishes of the Woods Hole region. *Jour. Parasit.*, **1**, 128–134.

Stunkard, H. W., and C. H. Willey. 1929. The development of *Cryptocotyle* (Heterophyidae) in its final host. *Amer. Jour. Trop. Med.*, **9**, 117–128.

Stunkard, H. W. 1930. The life history of *Cryptocotyle lingua* (Creplin), with notes on the physiology of the metacercariae. *Jour. Morph. and Physiol.*, **50**, 143–191.

Opisthorchis

Faust, E. C., O. Khaw, Y. Ke-Fang, C. Yung-An, and B. Walker. 1927. Studies on *Clonorchis sinensis* (Cobbold) with a consideration of the molluscan hosts of *Clonorchis sinensis* (Cobbold) in Japan, China and Southeastern Asia and other species of Mollusca closely related to them. *Amer. Jour. Hyg., Monogr. Ser.*, 8. 284 pp.

Vogel, H. 1934. Der Entwicklungszyklus von *Opisthorchis felineus* (Riv.) nebst Bemerkungen über die Systematik und Epidemiologie. *Zoologica*, Orig. Abh., **33** (86), 1–103.

Gorgodera

Cort, W. W. 1912. North American frog bladder flukes. *Trans. Amer. Micr. Soc.*, **31**, 151–166.

Goodchild, C. G. 1943. The life-history of *Phyllodistomum solidum* Rankin, 1937, with observations on the morphology, development, and taxonomy of the Gorgoderinae (Trematoda). *Biol. Bull.*, **84**, 59–86.

Goodchild, C. G. 1948. Additional observations on the bionomics and life history of *Gorgodera amplicava* Looss, 1899 (Trematoda: Gorgoderidae). *Jour. Parasit.*, **34**, 407–427.

Krull, W. H. 1935. Studies on the life history of a frog bladder fluke, *Gorgodera amplicava* Looss, 1899. *Papers Mich. Acad. Sci., Arts and Letters*, **20**, 697–710.

Stafford, J. 1902. The American representatives of *Distomum cygnoides*. *Zool. Jahr., Abt. Syst.*, **17**, 411–424.

Fasciola

Kawana, H. 1940. Study on the development of the excretory system of *Fasciola hepatica* L., with special reference of its first intermediate host in Central China. *Jour. Shanghai Sci. Inst.*, Sect. IV, **5**, 13–34.

Leuckart, R. 1882. Zur Entwickelungsgeschichte des Leberegels. *Zool. Anzeiger*, **5**, 524–528.

Querner, F. R. v. 1929. Zur Histologie des Exkretionsgefässsystems digenetischer Trematoden. I. Teil. *Z. Parasitenk.*, **1**, 489–561.

Stephenson, W. 1947. Physiological and histochemical observations on the adult liver fluke, *Fasciola hepatica* L. *Parasit.*, **38**, 116–144.

Thomas, A. P. 1883. The life history of the liver-fluke (*Fasciola hepatica*). *Quart. Jour. Micr. Sci.*, **23**, 99–133.

Schistosoma

Cort, W. W. 1919. The cercaria of the Japanese blood fluke, *Schistosoma japonicum* Katsurada. *Univ. Calif. Publ. Zool.*, **18**, 485–507.

Cort, W. W. 1919. Notes on the eggs and miracidia of the human schistosomes. *Univ. Calif. Publ. Zool.*, **18**, 509–519.

Faust, E. C., and H. E. Meleney. 1924. Studies on *Schistosomiasis japonica*. *Amer. Jour. Hyg., Monogr. Ser.* 3. 268 pp.

Looss, A. 1895. Zur Anatomie und Histologie der *Bilharzia haematobia* (Cobbold). *Arch. mikr. Anat. Entwicklungsgesch.*, **46**, 1–108.

Wright, W. H., et al. 1947. Studies on Schistosomiasis. *Nat. Inst. Health Bull.* 189. 212 pp.

Phyllobothrium

Curtis, W. C. 1903. *Crossobothrium laciniatum* and developmental stimuli in the Cestoda. *Biol. Bull.*, **5**, 125–142.

Curtis, W. C. 1906. The formation of proglottids in *Crossobothrium laciniatum* (Linton). *Biol. Bull.*, **11**, 202–228.

Linton, E. 1889. Notes on entozoa of marine fishes of New England, with descriptions of several new species. *Annual Report of the Commissioner of Fish and Fisheries for 1886.* Washington. Pp. 453–511.

Otobothrium

Linton, E. 1907. A cestode parasite in the flesh of the butterfish. *Bull. U. S. Bur. Fish.*, **26**, 111–132.

Pintner, T. 1934. Bruchstücke zur Kenntnis der Rüsselbandwürmer. *Zool. Jahr., Abt. Anat.*, **58**, 1–20.

Diphyllobothrium

Janicki, C., and F. Rosen. 1918. Le cycle évolutif du *Dibothriocephalus latus* L. *Bull. soc. Neuchâteloise sci. nat.*, **42**, 19–53.

Smyth, J. D. 1947. Studies on tapeworm physiology. III. Aseptic cultivation of larval Diphyllobothriidae *in vitro*. *Jour. Exp. Biol.*, **24**, 374–386.

Wardle, R. A. 1935. Fish-tapeworm. *Biol. Board of Canada Bull.* 45, 1–25.

Wardle, R. A. 1941. The rate of growth of the tapeworm, *Diphyllobothrium latum* (L.). *Can. Jour. Research*, **D19**, 245–251.

Taenia

Deffke, O. 1891. Die Entozoen des Hundes. *Arch. wiss. prakt. Thierh.*, **17**, (1–2), 1–60; (4–5), 253–289.

Hall, M. C. 1919. The adult taenioid cestodes of dogs and cats, and of related carnivores in North America. *Proc. U. S. Nat. Mus.*, **55**, No. 2258, 1–94.

Smyth, J. D. 1947. The physiology of tapeworms. *Biol. Rev.*, **22**, 214–238.

Young, R. T. 1908. The histogenesis of *Cysticercus pisiformis*. *Zool. Jahr., Abt. Anat.*, **26**, 183–254.

Young, R. T. 1913. The histogenesis of reproductive organs of *Taenia pisiformis*. *Zool. Jahr., Abt. Anat.*, **35**, 355–418.

Echinococcus

Cameron, T. W. M. 1926. Observations on the genus *Echinococcus* Rudolphi, 1801. *Jour. Helminth.*, **4**, 13–22.

Deffke, O. 1891. Die Entozoen des Hundes. *Arch. wiss. prakt. Thierh.,* **17** (1-2), 1-60; (4-5), 253-289.
Dévé, F. 1927. La cuticulisation des capsules proligères échinococciques. *Ann. parasit. humaine et comparée,* **5,** 310-328.
v. Erlanger, R. S. 1890. Der Geschlechtsapparat der *Taenia echinococcus. Z. wiss. Zool.,* **50,** 555-559.

RHYNCHOCOELA
Amphiporus
Bürger, O. 1895. Die Nemertinen des Golfes von Neapel und der Angrenzenden Meeres-abschnitte. *Fauna und Flora des Golfes von Neapel, Monogr.* 22. 743 pp.
Child, C. M. 1901. The habits and natural history of *Stichostemma. Amer. Naturalist,* **35,** 975-1006.
Coe, W. R. 1943. Biology of the nemerteans of the Atlantic coast of North America. *Trans. Conn. Acad. Arts and Sci.,* **35,** 129-328.

ACANTHOCEPHALA
Neoechinorhynchus
Bieler, W. 1913. Zur Kenntnis des männlichen Geschlechtsapparats einiger Acanthocephalen von Fischen. *Zool. Jahr., Abt. Anat.,* **36,** 525-578.
DeGiusti, D. L. 1949. The life cycle of *Leptorhynchoides thecatus* (Linton), an acanthocephalan of fish. *Jour. Parasit.,* **35,** 437-460.
Lynch, J. E. 1936. New species of *Neoechinorhynchus* from the western sucker, *Catostomus macrocheilus* Girard. *Trans. Amer. Micr. Soc.,* **55,** 21-43.
Van Cleave, H. J. 1919. Acanthocephala from the Illinois River, with descriptions of species and a synopsis of the family Neoechinorhynchidae. *Bull. Ill. Nat. Hist. Surv.* 13, Art. 8, 225-257.
Van Cleave, H. J. 1925. A critical study of the Acanthocephala described and identified by Joseph Leidy. *Proc. Acad. Nat. Sci. Philadelphia,* **76,** 279-334.
Ward, H. L. 1940. Studies on the life history of *Neoechinorhynchus cylindratus* (Van Cleave, 1913) (Acanthocephala). *Trans. Amer. Micr. Soc.,* **59,** 327-347.

ASCHELMINTHES
Hydatina
de Beauchamp, P. M. 1909. Recherches sur les Rotifères: les formations tégumentaires et l'appareil digestif. *Arch. zool. exp. et gén.,* 4th series, **10,** 1-410.
Hudson, C. T., and P. H. Gosse. 1889. *The Rotifera; or wheel animalcules both British and foreign.* 2 vols. Longmans, Green & Co., London.
Martini, E. 1912. Studien über die Konstanz histologischer Elemente. III. *Hydatina senta. Z. wiss. Zool.,* **102,** 425-645.
Plate, L. 1886. Beiträge zur Naturgeschichte der Rotatorien. *Jenaische Z. Med. u. Naturwiss.,* **12,** 1-120.
Wesenberg-Lund, C. 1923. Contributions to the biology of the Rotifera. I. The males of the Rotifera. *Mém. acad. roy. sci. et lettres Danemark,* sec. des sci. ser. 8, **4,** 191-345.

Chaetonotus

Brunson, R. B. 1949. The life history and ecology of two North American gastrotrichs. *Trans. Amer. Micr. Soc.*, **68**, 1–20.

Bütschli, O. 1876. Untersuchungen über freilebende Nematoden und die Gattung *Chaetonotus*. *Z. wiss. Zool.*, **26**, 363–413.

Fernald, C. H. 1883. Notes on the *Chaetonotus larus*. *Amer. Naturalist*, **17**, 1217–1220.

Stokes, A. C. 1887. Observations on *Chaetonotus*. *The Microscope*, **7**, 1-9.

Zelinka, C. 1889. Die Gastrotrichen. Eine monographische Darstellung ihrer Anatomie, Biologie und Systematik. *Z. wiss. Zool.*, **49**, 209–384.

Turbatrix

de Man, J. G. 1910. Beiträge zur Kenntnis der in dem weissen Schleimfluss der Eichen lebenden Anguilluliden. *Zool. Jahr., Abt. Syst.*, **29**, 359–394.

Peters, B. G. 1927. On the nomenclature of the vinegar eelworm. *Jour. Helminth.*, **5**, 133–142.

Peters, B. G. 1927. On the anatomy of the vinegar eelworm. *Jour. Helminth.*, **5**, 183–202.

Rhabditis

Chitwood, B. G. 1930. Studies on some physiological functions and morphological characters of *Rhabditis* (Rhabditidae, Nematodes). *Jour. Morph. and Physiol.*, **49**, 251–275.

Johnson, G. E. 1913. On the nematodes of the common earthworm. *Quart. Jour. Micr. Sci.*, **58**, 605–652.

Otter, G. W. 1933. On the biology and life history of *Rhabditis pellio* (Nematoda). *Parasit.*, **25**, 296–307.

Scott, A. C. 1938. Cleaving nematode eggs as research and classroom material. *Sci.*, **87**, 145–146.

Enterobius

Cram, E. B., M. F. Jones, L. Reardon, and M. O. Nolan. 1937. Studies on oxyuriasis. VI. The incidence of oxyuriasis in 1,272 persons in Washington, D. C., with notes on diagnosis. *U. S. Pub. Health Reports*, **52**, 1480–1504.

Koch, E. W. 1925. Oxyurenfortpflanzung im Darm ohne Reinfektion und Magenpassage. *Centralbl. Bakt. Parasit.*, I. Abt. Orig., **94**, 208–236.

Martini, E. 1916. Die Anatomie des *Oxyuris curvula*. *Z. wiss. Zool.*, **116**, 137–534.

Necator

Chandler, A. C. 1929. *Hookworm disease*. The Macmillan Co., N. Y. 494 pp.

Looss, A. 1905. The anatomy and life history of *Agchylostoma duodenale* Dub. Part I. *Records Egyptian Govt. School Med.*, 3, 1–158.

Looss, A. 1911. The anatomy and life history of *Agchylostoma duodenale* Dub. Part II. The development in the free state. *Records Egyptian Govt. School Med.*, 4, 159–613.

Stiles, C. W. 1902. The significance of the recent American cases of hookworm disease (uncinariasis or anchylostomiasis) in man. *18th Ann. Rep.*, Bureau Animal Industry, U. S. Dept. Agr., Washington, D. C. Pp. 183–219.

Ascaris

Deineka, D. 1908. Das Nervensystem von *Ascaris*. *Z. wiss. Zool.*, **89**, 242–307.
Goldschmidt, R. 1906. Mitteilungen zur Histologie von *Ascaris*. *Zool. Anzeiger*, **29**, 719–737.
Goldschmidt, R. 1908. Das Nervensystem von *Ascaris lumbricoides* und *megalocephala*. Ein versuch, in den Aufbau eines einfachen Nervensystems einzudringen. *Z. wiss. Zool.*, **90**, 73–136.
Mueller, J. F. 1929. Studies on the microscopical anatomy and physiology of *Ascaris lumbricoides* and *Ascaris megalocephala*. *Z. Zellforsch. u. mikr. Anat.*, **8**, 361–403.
Stewart, F. H. 1917. On the development of *Ascaris lumbricoides* Lin. and *Ascaris suilla* Duj. in the rat and mouse. *Parasit.*, **9**, 213–227.

Wuchereria

Feng, L. C. 1933. A comparative study of the anatomy of *Microfilaria malayi* Brug, 1927, and *Microfilaria bancrofti* Cobbold, 1877. *Chinese Med. Jour.*, **47**, 1214–1246.
Fülleborn, F. 1913. Beiträge zur Morphologie und Differentialdiagnose der Mikrofilarien. *Arch. Schiffs- u. Tropen-hyg.*, 17 Beiheft 1, 7–72.
Warren, Virginia G., J. Warren, and G. W. Hunter, III. 1946. Studies on filariasis. I. Serological relationships between antigenic extracts of *Wuchereria bancrofti* and *Dirofilaria immitis*. *Amer. Jour. Hyg.*, **43**, 164–170.

Trichinella

Graham, J. Y. 1897. Beiträge zur Naturgeschichte der *Trichina spiralis*. *Arch. mikr. Anat. Entwicklungsgesch.*, **50**, 219–275.
Hall, M. C. 1937. Studies on Trichinosis. IV. The role of the garbage-fed hog in the production of human trichinosis. *U. S. Pub. Health Reports*, 52 (27), 873–886.
Owen, R. 1835. Description of a microscopic entozoon infesting the muscles of the human body. *Trans. Zool. Soc. Lond.*, **1**, 315–324.
Peres, C. E. 1942. *Trichinella spiralis*. II. Incidence of infection in hogs and rats in the New Orleans area. *Jour. Parasit.*, **28**, 223–226.

Trichuris

Chandler, A. C. 1930. Specific characters in the genus *Trichuris*, with a description of a new species, *Trichuris tenuis*, from a camel. *Jour. Parasit.*, **16**, 198–206.
Clapham, P. A. 1945. On some characters of the genus *Trichuris* and a description of *T. parvispicularis* n. sp. from a cane rat. *Jour. Helminth.*, **21**, 85–89.
Cort, W. W., and G. F. Otto. 1937. *Trichuris trichiura* in the United States. *30th anniversary jubilee volume for K. J. Skrjabin*, Moscow. pp. 81–88.

Entoprocta

Barentsia

Cori, C. I. 1936. Kamptozoa. *Bronn's Klassen und Ordnungen des Tier-Reichs*. Vol. 4 (2), 1–119.
Marcus, E. 1939. Bryozoarios marinhos Brasileiros. III. *Univ. São Paulo, bol. faculd. filos., ciênc. e letr. IV. Zool.*, **3**, 111–354.

Rogick, M. D. 1948. Studies on marine bryozoa. II. *Barentsia laxa* Kirkpatrick 1890. *Biol. Bull.,* **94** (2), 128–142.

ECTOPROCTA

Crisia

Borg, F. 1926. Studies on recent cyclostomatous Bryozoa. *Zool. Bidrag Uppsala,* **10**, 181–507.
Borg, F. 1944. The stenolaematous Bryozoa. *Further Zool. Results Swedish Antarctic Exped. of 1901–1903,* **3** (5). 276 pp.
Robertson, A. 1910. The cyclostomatous Bryozoa of the West Coast of North America. *Univ. Calif. Publ. Zool.,* **6** (12), 225–284.

Bowerbankia

Brien, P., and G. Huysmans. 1937. La croissance et le bourgeonnement du stolon chez les Stolonifera (Bowerbankia). *Ann. soc. roy. zool. Belg.,* **68**, 13–40.
Marcus, E. 1937. Bryozoarios marinhos Brasileiros. I. *Univ. São Paulo, bol. faculd. filos, ciênc. e letr. I. Zool.,* **1**, 5–224.
Marcus, E. 1938. Bryozoarios marinhos Brasileiros. II. *Univ. São Paulo, bol. faculd. filos., ciênc. e letr. IV. Zool.,* **2**, 3–138.

Electra

Marcus, E. 1940. Mosdyr (Bryozóa eller Polyzóa). *Danmarks Fauna.* G. E. C. Gads Forlag, København.
Rogick, M. D., and H. Croasdale. 1949. Studies on marine Bryozoa. III. *Biol. Bull.,* **96**, 32–69.

Bugula

Canu, F., and R. S. Bassler. 1929. Bryozoa of the Philippine Region. *U. S. Nat. Mus. Bull.* 100, **9**, 1–685.
Grave, B. H. 1930. The natural history of *Bugula flabellata* at Woods Hole, Mass. *Jour. Morph. and Physiol.,* **49**, 355–383.
Lynch, W. F. 1947. The behavior and metamorphosis of the larva of *Bugula neritina* (L.) etc. *Biol. Bull.,* **92** (2), 115–150.

Plumatella

Allman, G. 1856. *Monograph of fresh-water Polyzoa.* Ray Society, London. 119 pp.
Rogick, M. D., and C. J. D. Brown. 1942. Studies on fresh-water Bryozoa. XII. *Ann. N. Y. Acad. Sci.,* **43** (3), 123–144.
Toriumi, M. 1942. Studies on freshwater Bryozoa of Japan. III. *Sci. Reports Tôhoku Imper. Univ.,* 4th ser., *Biol.,* **17** (2), 197–205.

ANNELIDA

Neanthes virens

Beadle, L. C. 1937. Adaptation to change of salinity in the Polychaetes. 1. Control of body fluid volume and of body fluid concentration in *Nereis diversicolor. Jour. Exp. Biol.,* **14**, 56–70.

Copeland, M., and F. A. Brown, Jr. 1934. Modification of behavior in *Nereis virens*. *Biol. Bull.*, **67**, 356–364.

Ellis, W. G. 1937. The water and electrolyte exchange of *Nereis diversicolor* (Muller). *Jour. Exp. Biol.*, **14**, 340–350.

Gray, J. 1939. Studies in animal locomotion. VIII. The kinetics of locomotion of *Nereis diversicolor*. *Jour. Exp. Biol.*, **16**, 9–17.

Lindroth, A. 1938. Studien uber die respiratorischen Mechanismen von *Nereis virens* Sars. *Zool. Bidrag Uppsala*, **17**, 367–497.

Turnbull, F. M. 1875. On the anatomy and habits of *Nereis virens*. *Trans. Conn. Acad. Arts and Sci.*, **3**, 265–280.*

Arenicola cristata

Ashworth, J. H. 1904. *Arenicola. Liverpool Marine Biological Committee Memoirs.* Vol. 11. Williams and Norgate, London. 118 pp.*

Barcroft, J., and H. Barcroft. 1924. The blood pigment of *Arenicola*. *Proc. Roy. Soc. Lond.*, **B96**, 28–42.

Borden, M. A. 1931. A study of the respiration and of the function of haemoglobin in *Planorbis corneus* and *Arenicola marina*. *Jour. Mar. Biol. Assoc. Plymouth*, **17**, 709–738.

Wu, K. S. 1939. The action of drugs, especially acetylcholine, on the annelid body wall (*Lumbricus, Arenicola*). *Jour. Exp. Biol.*, **16**, 251–257.

Amphitrite ornata

Scott, J. W. 1911. Egg-laying in *Amphitrite*. *Biol. Bull.*, **20**, 252–265.

Thomas, Joan G. 1940. *Pomatoceros, Sabella, and Amphitrite. Liverpool Marine Biological Committee Memoirs.* Vol. 33. Hodder and Stoughton, London. 88 pp.*

Lumbricus terrestris

Bullock, T. H. 1945. Functional organization of the giant fiber system of *Lumbricus*. *Jour. Neurophysiol.*, **8**, 55–71.

Collier, H. O. J. 1938. The immobilization of locomotory movements in the earthworm, *Lumbricus terrestris*. *Jour. Exp. Biol.*, **15**, 339–357.

Collier, H. O. J. 1939. Central nervous activity in the earthworms. 1. Response to tension and to tactile stimulation. *Jour. Exp. Biol.*, **16**, 286–299.

Gray, J., and H. W. Lissmann. 1938. Studies in animal locomotion. VII. Locomotory reflexes in the earthworm. *Jour. Exp. Biol.*, **15**, 506–517.

Holst, E. von. 1933. Weitere Versuche zum nervosem Mechanismus der Bewegung beim Regenworm (*Lumbricus terr.* L.). *Zool. Jahr. Abt. allg. Zool. u. Physiol.*, **53**, 67–100.

Johnson, M. L. 1941. The respiratory function of the haemoglobin of the earthworm. *Jour. Exp. Biol.*, **18**, 266–277.

Prosser, C. L. 1934. The nervous system of the earthworm. *Quart. Rev. Biol.*, **9**, 181–200.

Robertson, J. D. 1936. The function of the calciferous glands of earthworms. *Jour. Exp. Biol.*, **13**, 279–297.

Swartz, Ruth D. 1929. Modification of behavior in earthworms. *Jour. Comp. Psychol.*, **9**, 17–33.

Wolf, A. V. 1940. Paths of water exchange in the earthworms. *Physiol. Zool.,* **13**, 294–308.

Wu, K. S. 1939. On the physiology and pharmacology of the earthworm gut. *Jour. Exp. Biol.,* **16**, 184–197.

Hirudo medicinalis

Gray, J., H. W. Lissmann, and R. J. Pumphrey. 1938. The mechanism of locomotion in the leech (*Hirudo medicinalis* Ray). *Jour. Exp. Biol.,* **15**, 408–430.

SIPUNCULOIDEA

Phascolosoma

Adolph, E. F. 1936. Differential permeability to water and osmotic exchanges in the marine worm *Phascolosoma*. *Jour. Cell. and Comp. Physiol.,* **9**, 117–135.

Andrews, E. A. 1890. Notes on the anatomy of *Sipunculus gouldii* Pourtales. *Studies Biol. Lab., Johns Hopkins Univ.,* **4**, 389–430.*

Florkin, M. 1933. Recherches sur les hemerythrines. *Arch. Inv. Physiol.,* **36**, 247–328.

Gerould, J. H. 1906. The development of *Phascolosoma*. *Zool. Jahr., Abt. Anat.,* **23**, 77–162.*

Gerould, J. H. 1913. The Sipunculids of the eastern coast of North America. *Proc. U. S. Nat. Mus.,* **44**, 373–437.

Gerould, J. H. 1938. The eyes and nervous system of *Phascolosoma verrillii* and other sipunculids. *Trav. stat. zool. Wimereux,* **13**, 313–325.

MOLLUSCA

General

Johnson, Charles W. 1915. Fauna of New England, 13. List of the Mollusca. *Boston Soc. Nat. Hist. Occasional Papers* VII.

Johnson, Charles W. 1934. List of the marine Mollusca of the Atlantic coast from Labrador to Texas. *Proc. Boston Soc. Nat. Hist.,* **40**, No. 1, 1–204.

Pelseneer, Paul. 1906. Part V. Mollusca. *A treatise on zoology.* Ed. by E. Ray Lankester. London.

Chaetopleura apiculata

Grave, B. H. 1932. Embryology and life history of *Chaetopleura apiculata*. *Jour. Morph.,* **54**, 153–160.

Haller, B. 1882. Die Organisation der Chitonen der Adria. *Arb. zool. Inst. Univ. Wien,* **4**, 323–391.

Yoldia limatula

Drew, G. A. 1899. *Yoldia limatula. Mem. Biol. Lab. Johns Hopkins Univ.* IV. Pp. 1–37.

Pecten irradians

Dakin, W. J. 1909. *Pecten,* the edible scallop. *Proc. and Trans. Liverpool Biol. Soc.,* **XXIII**, 333–468.

Drew, G. A. 1906. Habits, anatomy, and embryology of the giant scallop (*Pecten tenuicostatus*, Mighels). *Univ. Maine Studies*, 6.
Gutsell, J. 1930. Natural history of the bay scallop. *Bull. U. S. Bur. Fish.*, **XLVI**, 569–632.

Venus mercenaria

Field, I. A. 1922. The biology and economic value of the sea mussel, *Mytilus edulis*. *Bull. U. S. Bur. Fish.*, **XXXVIII**, 127–259.
Motley, H. L. 1934. Physiological studies concerning the regulation of heart beat in fresh water mussels. *Physiol. Zool.*, **7**, 62–84.
Orton, J. H. 1937. Oyster breeding and oyster culture. Edward Arnold Co., London.
Prosser, C. L. 1940. Acetylcholine and nervous inhibition in the heart of *Venus mercenaria*. *Biol. Bull.*, **78**, 92–102.
Ridewood, W. G. 1903. On the structure of the gills of the Lamellibranchia. *Phil. Trans. Roy. Soc. Lond.*, **B195**, 147–284.
Schecter, V. 1941. Experimental studies on the eggs of the clam, *Mactra solidissima*. *Jour. Exp. Zoöl.*, **86**, 461–480.
Walzl, E. M. 1937. Actions of ions on the heart of the oyster (*Ostrea virginica*). *Physiol. Zool.*, **10**, 125–140.
Walzl, E. M. 1938. Response of the oyster heart to electrical stimulation and effect of calcium and potassium on its threshold of inhibition. *Jour. Cell. and Comp. Physiol.*, **12**, 237–246.
White, K. M. 1937. *Mytilus*. Liverpool Marine Biological Committee Memoirs. Vol. 31. 177 pp.

Fresh-water Mussels

Baker, F. C. 1928. The fresh water Mollusca of Wisconsin. Part II. Pelecypoda. *Bull. Univ. of Wis.* 70, Serial 1527; Gen. Ser. 1301.
Coker, R. E., A. F. Shira, H. W. Clark, and A. D. Howard. 1921. Natural history and propagation of fresh water mussels. *Bull. U. S. Bur. Fish.*, **XXXVII**, 77–181.
Tucker, M. E. 1928. Studies on the life cycle of two species of fresh water mussels belonging to the genus *Anodonta*. *Biol. Bull.*, **54**, 117–127.

Busycon canaliculatum

Carricker, M. R. 1943. On the structure and function of the proboscis in the common oyster drill *Urosalpinx cinerea*. *Jour. Morph.*, **73**, 441–506.
Conklin, E. G. 1897. The embryology of *Crepidula*. *Jour. Morph.*, **13**, 1–226.
Dakin, W. J. 1912. *Buccinum* (the whelk). Liverpool Marine Biological Committee Memoirs. Vol. 20. Univ. Press, Liverpool.*
Herrick, J. C. 1906. Mechanism of the odontophore apparatus in *Sycotypus canaliculatus*. *Amer. Naturalist*, **XL**, 707–737.

Eolis

Alder, Joshua, and Albany Hancock. 1855. Monograph of the British nudibranchiate Mollusca. Ray Society, London.
Glaser, Otto. 1910. The nematocysts of eolids. *Jour. Exp. Zoöl.*, **9**, 117–142.

Hancock, Albany, and D. Embleton. 1846. Anatomy of *Eolis*. *Ann. Mag. Nat. Hist.*, **15**.

Loligo paeleii

Brooks, W. K. 1880. The development of the squid (*Loligo pealeii*). *Ann. Mem. Boston Soc. Nat. Hist.*

Drew, G. A. 1911. Sexual activties of the squid, *Loligo pealeii* (Lesueur). I. Copulation, egg-laying, and fertilization. *Jour. Morph.*, **22**, 329–359.

Drew, G. A. 1919. Sexual activities of the squid. II. The spermatophore; its structure, ejaculation, and formation. *Jour. Morph.*, **32**, 379–435.

Tompsett, D. H. 1939. Sepia. *Liverpool Marine Biological Committee Memoirs.* Univ. Press, Liverpool.*

Williams, L. W. 1910. *Anatomy of the common squid.* Amer. Mus. Nat. Hist., Leiden, Holland.

BRACHIOPODA

General

Cloud, P. E., Jr. 1948. Notes on recent brachiopods. *Amer. Jour. Sci.*, **246**, 241–250.†

Cooper, G. A. 1947. Brachiopoda. *Encyclopedia Britannica*, **3**, 999–1003.

Percival, E. 1944. A contribution to the life history of a brachiopod, *Terebratella inconspicua* Sowerby. *Trans. Roy. Soc. N. Z.*, **74**, Part I, 1–23.

Thompson, J. Allan. 1927. Brachiopod morphology and genera. *N. Z. Board Sci. and Art. Manual 7.* Dominion Museum, Wellington, N. Z.

Yatsu, N. 1902a. On the habits of the Japanese *Lingula*. *Annatationes Zool. Japan.*, **4**, 61–67.

ARTHROPODA

Xiphosura polyphemus

Benham, W. B. S. 1885. On the testis of *Limulus*. *Trans. Linn. Soc. Lond.* (*Zool.*), (2) **2**, 363–366.*

Gerhardt, U. 1935. 2. Ordnung der Merostoma: Xiphosura-Poecilopoda, Schwertschwänze. *Kükenthal und Krumbach's Handbuch der Zoologie.* Bd. 3, Halfte 2, Lief. 8, 47–96.*†

Hanström, B. 1926. Das Nervensystem und die Sinnesorgane von *Limulus polyphemus*. *Lunds Univ. Årsskr.* (n.F.), (2) **22**, (5), 1–79.

Johansson, Gösta. 1933. Beiträge zur Kenntnis der Morphologie und Entwicklung des Gehirns von *Limulus polyphemus*. *Acta Zool.*, **14**, 1–100.

Milne-Edwards, A. 1873. Recherches sur l'anatomie des limules. *Ann. sci. nat. Zool.* (5), **17** (4), 1–67. (Especially on circulatory system.)*

Munson, J. P. 1898–99. The ovarian egg of *Limulus*. A contribution to the problem of the centrosome and yolk-nucleus. *Jour. Morph.*, **15**, 111–220. (Includes anatomy of female reproductive system.)*

Patten, W., and Annah P. Hazen. 1900. The development of the coxal gland, branchial cartilages, and genital ducts of *Limulus polyphemus*. *Jour. Morph.*, **16**, 459–502. (Includes anatomy of adult coxal gland.)*

Patten, W., and W. A. Redenbaugh. 1899–1900. Studies on *Limulus*. II. The nervous system of *Limulus polyphemus*, with observations upon the general anatomy. *Jour. Morph.*, **16**, 91–200. (Most valuable single paper on anatomy, especially of nerves, heart, arteries, and muscles.)*

Schlottke, E. 1935. Biologische, physiologische, und histologische Untersuchungen über die Verdauung von *Limulus*. *Z. vergl. Physiol.*, **22**, 359–413.*
Schulze, P. 1937. Trilobita, Xiphosura, Acarina. Eine morphologische Untersuchung über Plangleichheit zwischen Trilobiten und Spinnentieren. *Z. Morph. Ökol. Tiere*, **32**, 181–226.
Versluys, J., and R. Demoll. 1922–23. Das *Limulus*-Problem. Die Verwandschaftsbeziehungen der Merostomen und Arachnoideen unter sich und mit anderen Arthropoden. *Ergeb. Fortschr. Zool.*, **5**, 67–388.*†

Argiope

Comstock, John Henry. 1940. *The spider book*. 2d Ed. Comstock Publ. Co., Ithaca, N. Y.*†
Ellis, C. H. 1944. The mechanism of extension in the legs of spiders. *Biol. Bull.*, **86**, 41–50.*†
Ewing, Henry E. 1933. Afield with the spiders. *Nat. Geogr. Mag.*, **64**, 163–194.†
Petrunkevitch, Alexander. 1916. *Morphology of invertebrate types*. The Macmillan Co., N. Y. Pp. 158–173.†
Savory, Theodore H. 1928. *The biology of spiders*. The Macmillan Co., N. Y.*†
Snodgrass, R. E. 1948. The feeding organs of Arachnida, including mites and ticks. *Smithson. Misc. Coll.*, **110**, No. 10, 1–93.*

Artemia

Cannon, H. G. 1933. On the feeding mechanism of the Branchiopoda. With an appendix on the mouth parts of the Branchiopoda by H. Graham Cannon and F. M. C. Leak. *Phil. Trans. Roy. Soc. Lond.*, **B222**, 267–352.*
Heath, H. 1924. The external development of certain phyllopods. *Jour. Morph.*, **38**, 453–483.
Linder, F. 1941. Contributions to the morphology and the taxonomy of the Branchiopoda Anostraca. *Zool. Bidrag. Uppsala*, **20**, 101–302.†
Lochhead, J. H. 1941. *Artemia*, the "brine shrimp." *Turtox News*, **19**, 41–45.†
Lochhead, J. H., and Margaret S. Lochhead. 1941. Studies on the blood and related tissues in *Artemia* (Crustacea Anostraca). *Jour. Morph.*, **68**, 593–632.

Daphnia magna

Anderson, B. G., and J. C. Jenkins. 1942. A time study of events in the life span of *Daphnia magna*. *Biol. Bull.*, **83**, 260–272.
Binder, Gertrude. 1931. Über das Muskelsystem von *Daphnia magna*. *Internat. Rev. ges. Hydrobiol. Hydrog.*, **26**, 54–111.*
Cannon, H. G. 1933. See under *Artemia*.
Claus, C. 1876. Zur Kenntniss der Organisation und des feineren Baues der Daphniden und verwandter Cladoceren. *Z. wiss. Zool.*, **27**, 362–402.*
Coker, R. E. 1939. The problem of cyclomorphosis in *Daphnia*. *Quart. Rev. Biol.*, **14**, 137–148.
v. Dehn, Madeleine. 1930–31. Untersuchungen über die Verdauung bei Daphnien. *Z. vergl. Physiol.*, **13**, 334–358.
v. Dehn, Madeleine. 1937. Experimentelle Untersuchungen über den Generationswechsel der Cladoceren. I. *Zool. Jahrb. Abt. allg. Zool. Physiol. Tiere*, **58**, 241–272.
Eckert, F. 1938. Die positiv phototaktische Orientierung von *Daphnia pulex*. I. Versuche mit "Kältetieren." *Z. vergl. Physiol.*, **25**, 655–702.

Fischel, A. 1908. Untersuchungen über vitale Färbung an Süsswassertieren, insbesondere bei Cladoceren. *Internat. Rev. ges. Hydrobiol. Hydrog.*, **1**, 73–141.

Fox, H. M. 1948. The haemoglobin of *Daphnia*. *Proc. Roy. Soc. Lond.*, **B135**, 195–212. (Summarized in *Nature*, **160**, 431, 1947.)

Gallistel, Helma. 1936–37. Histochemische Untersuchungen über das Speichern von Glykogen und Fett bei *Daphnia magna*. *Z. Zellforsch. mikr. Anat.*, **25**, 66–82.

Gicklhorn, J. 1931. Elektive Vitalfärbungen im Dienste der Anatomie und Physiologie der Exkretionsorgane von Wirbellosen (Cladoceren als Beispiel), *and* Beobachtungen an den lateralen Frontalorganen von *Daphnia magna* M. nach elektiver Vitalfarbung. *Protoplasma*, **13**, 701–724, *and* 725–739.

Hérouard, E. 1905. La circulation chez les daphnies. *Mem. soc. zool. France*, **18**, 214–232.

Leder, H. 1915. Untersuchungen über den feineren Bau des Nervensystems der Cladoceren. *Arb. zool. Inst. Univ. Wien*, **20**, 297–392.*

v. Scharfenberg, U. 1910–11. Studien und Experimente über die Eibildung und den Generationszyklus von *Daphnia magna*. *Internat. Rev. ges. Hydrobiol. Hydrog.*, **3**, Biol. Suppl., Ser. 1, Heft 2, 1–42.

Schulz, H. 1928. Über die Bedeutung des Lichtes im Leben niederer Krebse (nach Versuchen an Daphniden). *Z. vergl. Physiol.*, **7**, 488–552.

Storch, O. 1925. Cladocera. Wasserflöhe. *Biol. Tiere Deutschlands* (*herausg. P. Schulze*), Lief. 15, Teil 14, 23–102.*†

Cyclops

Cannon, H. G. 1941. A note on fine needles for dissection. *Jour. Roy. Micr. Soc.*, (3) **61**, 58–59.

Ewers, Lela A. 1936. Propagation and rate of reproduction of some freshwater Copepoda. *Trans. Amer. Micr. Soc.*, **55**, 230–238.†

Gicklhorn, J. 1930. Notiz über die sogenannten "Cornealinsen" von *Cyclops strenuus* Fischer. *Zool. Anzeiger*, **90**, 250–258.

Hartog, M. M. 1888. The morphology of *Cyclops* and the relations of the Copepoda. *Trans. Linn. Soc. Lond.* (*Zool.*), (2) **5**, 1–46.*

Spandl, H. 1926. Copepoda. Ruderfusskrebse. *Biol. Tiere Deutschlands* (*herausg. P. Schulze*), Lief, 19, Teil 15, 1–78.*†

Williams, C. M. 1939. A method for studying living mosquito larvae and other small aquatic invertebrates. *Sci.*, **90**, 21–22.

Lepas anatifera

Cannon, H. G. 1947–48. On the anatomy of the pedunculate barnacle *Lithotrya*. *Phil. Trans. Roy. Soc. Lond.*, **B233**, 89–136.

Defner, A. 1910. Der Bau der Maxillardrüse bei Cirripedien. *Arb. zool. Inst. Wien*, **18**, 183–206.

Hoek, P. P. C. 1883 and 1884. Report on the Cirripedia collected by H.M.S. *Challenger* during the years 1873–76. *Rep. sci. Res. Voyage H.M.S. Challenger: Zoology*, **8**, Part 25: 1–169, *and* Anatomical part, **10**, Part 28: 1–47.*

Krüger, P. 1927. Cirripedia. *Tierwelt Nord- u. Ostsee* (*Grimpe u. Wagler*), Lief. 8, Teil X.d, 1–40.

Krüger, P. 1940. Cirripedia. *Bronn's Klassen und Ordnungen des Tier-Reichs* 5. Bd., 1. Abt., 3. Buch, Teil III. 1–272 (incomplete).

Heteromysis formosa

Cannon, H. G., and S. M. Manton. 1927-28. On the feeding mechanism of a mysid crustacean, *Hemimysis lamornae*. *Trans. Roy. Soc. Edinb.*, **55**, 219-253.*

Hansen, H. J. 1925. Studies on Arthropoda. II. On the comparative morphology of the appendages in the Arthropoda. A. Crustacea. Gyldendalske Boghandel, Copenhagen. 176 pp.

Nair, K. B. 1939. The reproduction, oogenesis, and development of *Mesopodopsis orientalis* Tatt. *Proc. Indian Acad. Sci.*, **B9**, 175-223.

Zimmer, C. 1933. Mysidacea. *Tierwelt Nord- u. Ostsee* (Grimpe u. Wagler), Lief. 23, Teil X.g3, 29-69.*†

Crayfish (and Homarus)

Baumann, H. 1919-21. Das Gefässsystem von *Astacus fluviatilis* (*Potamobius astacus* L.). Ein Beitrag zur Morphologie der Decapoden. *Z. wiss. Zool.*, **118**, 246-312.*

Bock, F. 1925. Die Respirationsorgane von *Potamobius astacus* Leach (*Astacus fluviatilis* Fabr.). Ein Beitrag zur Morphologie der Decapoden. *Z. wiss. Zool.*, **124**, 51-117.*

Herrick, F. H. 1911. Natural history of the American lobster. *Bull. U. S. Bur. Fish.*, **29**, 149-408.*†

Huxley, T. H. 1880. *The crayfish. An introduction to the study of zoology.* (Reprinted 1884, as *An introduction to the study of zoology, illustrated by the crayfish.*) International Science Series, Vol. 28. Appleton, N. Y. xiv + 371 pp.

Keim, W. 1915. Das Nervensystem von *Astacus fluviatilis* (*Potamobius astacus* L.). Ein Beitrag zur Morphologie der Dekapoden. *Z. wiss. Zool.*, **113**, 485-545.*

Maluf, R. 1941. Micturition in the crayfish and further observations on the anatomy of the nephron of this animal. *Biol. Bull.*, **81**, 134-148. (See also Maluf's 1939 paper, which he cites.)*

Marburg University Department of Zoology, *Z. wiss. Zool.*, **116**, 1916; **121**, 1924; **123**, 1924; **126**, 1925; **141**, 1932 (well-illustrated papers on different parts of the anatomy of *Astacus*).

Penn, G. H., Jr. 1943. A study of the life history of the Louisiana red-crawfish, *Cambarus clarkii* Girard. *Ecol.*, **24**, 1-18.

Schmidt, W. 1915. Die Muskulatur von *Astacus fluviatilis* (*Potamobius astacus* L.). Ein Beitrag zur Morphologie der Decapoden. *Z. wiss. Zool.*, **113**, 165-251.*

Welsh, J. H. 1941. The sinus glands and 24-hour cycles of retinal pigment migration in the crayfish. *Jour. Exp. Zoöl.*, **86**, 35-49.*

Wiersma, C. A. G., and E. Novitski. 1942. The mechanism of the nervous regulation of the crayfish heart. *Jour. Exp. Biol.*, **19**, 255-265.

Yonge, C. M. 1924. Studies on the comparative physiology of digestion. II. The mechanism of feeding, digestion and assimilation in *Nephrops norvegicus*. (Brit.) *Jour. Exp. Biol.*, **1**, 343-389.*

Callinectes sapidus

Borradaile, L. A. 1922. On the mouth-parts of the shore crab. *Jour. Linn. Soc. Lond.* (*Zool.*), **35**, 115-142.*

Churchill, E. P., Jr. 1919. Life history of the blue crab. *Bull. U. S. Bur. Fish.*, **36**, 91-128.

Cochran, Doris M. 1935. The skeletal musculature of the blue crab, *Callinectes sapidus* Rathbun. *Smithson. Misc. Coll.,* **92** (9) (*Publ.* 3282), 1–76.*† (Endoskeleton also.)
Cronin, L. E. 1947. Anatomy and histology of the male reproductive system of *Callinectes sapidus* Rathbun. *Jour. Morph.,* **81** (2), 209–239.*†
Pearson, J. 1908. *Cancer. Liverpool Marine Biological Committee Memoirs.* Vol. 16. viii + 209 pp.*†
Smith, R. I. 1947. The action of electrical stimulation and of certain drugs on cardiac nerves of the crab, *Cancer irroratus. Biol. Bull.,* **93**, 72–88.

Spirobolus

Attems, Karl. 1926. Myriapoda: Diplopoda. In Vol. 4 of *Kükenthal und Krumbach's Handbuch der Zoologie.**†
Barber, H. S. 1915. Fragmentary notes on the life-history of the myriapod, *Spirobolus marginatus. Proc. Ent. Soc. Wash.,* **17**, 123–126.
Snodgrass, Robert E. 1928. Morphology and evolution of the insect head and its appendages. *Smithson. Misc. Coll.,* **81** (3), 1–158.†
Verhoeff, K. W. 1926–28. Gliederfüssler: Arthropoda. Klasse Diplopoda. In Vol. 5 of *Bronn's Klassen und Ordnungen des Tier-Reichs.**†
Voges, Ernst. 1916. Myriopodenstudien. *Z. wiss. Zool.,* **116**, 73–125.†

Periplaneta

Bordas, L. 1909. Recherches anatomiques histologiques et physiologiques sur les organes appendiculaires de l'appareil reproducteur femelle des Blattes (*Periplaneta orientalis* L.). *Ann. sci. nat.,* IX Sér., *Zool.,* **9**, 71–121.*
Brown, Frank A., Jr., and Alison Meglitsch. 1940. Comparison of the chromatophorotropic activity of insect corpora cardiaca with that of crustacean sinus gland. *Biol. Bull.,* **79**, 409–418.
Carbonell, C. S. 1947. Thoracic muscles of the cockroach *Periplaneta americana* (L.). *Smithson. Misc. Coll.,* **107** (2), 1–23.†
Cleveland, L. R. 1934. The wood-feeding roach *Cryptocercus,* its protozoa, and the symbiosis between protozoa and roach. *Mem. Amer. Acad. Arts and Sci.,* **17**, 185–342.†
Gresson, R. A. R. 1934. The cytology of the mid-gut and hepatic caeca of *Periplaneta orientalis. Quart. Jour. Micr. Sci.,* **77**, 317–334.†
Gupta, P. D. 1947. On copulation and insemination in the cockroach *Periplaneta americana* (Linn.). *Proc. Nat. Inst. Sci. India,* **13**, 65–71.†
Hebard, Morgan. 1917. The *Blattidae* of North America. *Mem. Amer. Entom. Soc.,* No. 2. 284 pp.
Huxley, Thomas H. 1878. *A manual of the anatomy of invertebrated animals.* Appleton, N. Y. Pp. 343–361.†
McIndoo, N. E. 1945. Innervation of insect hearts. *Jour. Comp. Neur.,* **83**, 141–155.
Metcalf, R. L. 1943. The storage and interaction of water soluble vitamins in the Malpighian system of *Periplaneta americana* L. *Arch. Biochem.,* **2**, 55–62.
Miall, L. C., and Alfred Denny. 1886. *The structure and life-history of the cockroach* (*Periplaneta orientalis*). Lovell Reeve, London.*†
Pringle, J. W. S. 1939. The motor mechanism of the insect leg. *Jour. Exp. Biol.,* **16**, 220–231.

Pryor, M. G. M. 1940. On the hardening of the ootheca of *Blatta orientalis*. *Proc. Roy. Soc.*, **B128**, 378–393.
Rehn, James A. G. 1945. Man's uninvited fellow traveler—the cockroach. *Sci. Monthly*, **61**, 265–276.
Roeder, Kenneth D. 1948. Organization of the ascending giant fiber system of the cockroach (*Periplaneta americana*). *Jour. Exp. Zoöl.*, **108**, 243–262.
Snodgrass, R. E. 1933. Morphology of the insect abdomen. Part II. The genital ducts and the ovipositor. *Smithson. Misc. Coll.*, **89** (8), 1–148.*†
Snodgrass, R. E. 1937. The male genitalia of orthopteroid insects. *Smithson. Misc. Coll.*, **96** (5), 1–107.*†
Snodgrass, R. E. 1944. The feeding apparatus of biting and sucking insects affecting man and animals. *Smithson. Misc. Coll.*, **104** (7), 1–113.*†
Turner, C. H. 1913. Behavior of the common roach (*Periplaneta orientalis* L.) on an open maze. *Biol. Bull.*, **25**, 348–361.
Wigglesworth, V. B. 1939. *The principles of insect physiology*. Dutton, N. Y.*
Yeager, J. Franklin, and J. B. Gahan. 1937. Effects of the alkaloid nicotine on the rhythmicity of isolated heart preparations from *Periplaneta americana* and *Prodenia eridania*. *Jour. Agr. Research*, **55**, 1–19.*
Yeager, J. Franklin, and George O. Hendrickson. 1934. Circulation in wings and wing pads of the cockroach, *Periplaneta americana* Linn. *Ann. Entom. Soc. Amer.*, **27**, 257–272.*

Drosophila (larva)

Bodenstein, Dietrich. 1943. Hormones and tissue competence in the development of *Drosophila*. *Biol. Bull.*, **84**, 34–58.*
Demerec, M., Ed. *The biology of Drosophila*. John Wiley, N. Y. (In press.)*
Ephrussi, Boris, and G. W. Beadle. 1936. A technique for transplantation for *Drosophila*. *Amer. Naturalist*, **70**, 218–225.
Hertweck, Heinrich. 1931. Anatomie und Variabilität des Nervensystems und der Sinnesorgane von *Drosophila melanogaster* (Meigen). *Z. wiss. Zool.*, **139**, 559–663.*
Rühle, Hermann. 1932. Das larvale Tracheensystem von *Drosophila melanogaster* Meigen und seine Variabilität. *Z. wiss. Zool.*, **141**, 159–245.*
Strasburger, Eduard H. 1935. *Drosophila melanogaster Meig. Eine Einführung in den Bau und die Entwicklung*. Springer, Berlin.*
Strasburger, Marie. 1932. Bau, Funktion und Variabilität des Darmtraktus von *Drosophila melanogaster* Meigen. *Z. wiss. Zool.*, **140**, 539–649.*

CHAETOGNATHA

Sagitta

Burfield, S. T. 1927. *Sagitta*. *Liverpool Marine Biological Committee Memoirs*. Vol. 28. 104 pp.*†
Clarke, G. L., E. L. Pierce, and D. F. Bumpus. 1943. The distribution and reproduction of *Sagitta elegans* on Georges Bank in relation to the hydrographical conditions. *Biol. Bull.*, **85**, 201–226.
Doncaster, L. 1902. On the development of *Sagitta*; with notes on the anatomy of the adult. *Quart. Jour. Micr. Sci.*, N. S., **46**, 351–398.
John, C. C. 1931. On the anatomy of the head of *Sagitta*. *Proc. Zool. Soc. Lond.*, **1931**, 1307–1319.

Kuhl, W. 1938. Chaetognatha. *Bronn's Klassen und Ordnungen des Tier-Reichs.* Bd. 4, Abt. 4, Bch. 2, Teil 1. Pp. 1–226.*†

Redfield, A. C., and A. Beale. 1940. Factors determining the distribution of populations of chaetognaths in the Gulf of Maine. *Biol. Bull.,* **79,** 459–487.

Russell, F. S. 1933. On the biology of *Sagitta.* IV. Observations on the natural history of *Sagitta elegans* Verrill and *Sagitta setosa* J. Müller in the Plymouth area. *Jour. Mar. Biol. Assoc. U. K.,* **18,** 559–574.

Schmidt, W. J. 1938. Ueber den feineren Bau der Muskulatur der Körperwand von Sagitta. *Z. Zellforsch.,* **28,** 674–696.

Schmidt, W. J. 1940. Zur Morphologie, Polarisationsoptik und Chemie der Greifhaken von *Sagitta hexaptera. Z. Morph. Ökol. Tiere,* **37,** 62–82.

ECHINODERMATA

Asterias forbesi

Chadwick, H. C. 1923. *Asterias. Liverpool Marine Biological Committee Memoirs.* Vol. 25. 63 pp.*

Galtsoff, P. S., and V. L. Loosanoff. Natural history and method of controlling the starfish (*Asterias forbesi,* Desor). *Bull. U. S. Bur. Fish.,* **49** (31), 75–132.†

Mead, A. D. 1899. The natural history of the star-fish. *Bull. U. S. Bur. Fish.,* **19,** 203–224.

Smith, J. E. 1946. The mechanics and innervation of the starfish tube foot-ampulla system. *Trans. Roy. Soc. Lond.,* **B232,** 279–310.†

Ophioderma brevispinum

Clark, H. L. 1902. Echinoderms of the Woods Hole region. *Bull. U. S. Bur. Fish.,* **22,** 545–576.*

Mortensen, Th. 1927. *Handbook of the echinoderms of the British Isles.* Oxford Univ. Press, London.*

Delage, Yves, and Edgard Hérouard. 1903. *Traité de zoologie concrète.* III. *Les Échinodermes.* Schleicher Frères, Paris.*

Arbacia punctulata

Chadwick, H. C. 1900. *Echinus. Liverpool Marine Biological Committee Memoirs.* Vol. 3. 28 pp.

Jackson, R. T. 1912. Phylogeny of the Echini, with a revision of palaeozoic species. *Boston Jour. Nat. Hist.,* **7,** 1–491.

Petrunkevitch, Alexander. 1916. *Morphology of invertebrate types.* The Macmillan Co., N. Y.

Echinarachnius parma

MacBride, E. W. 1906. Echinodermata. In *Cambridge natural history.* Vol. I. Macmillan, London.

Parker, G. H. 1927. Locomotion and righting movements in echinoderms, especially in *Echinarachnius. Amer. Jour. Psychol.,* **39,** 167–180.

Thyone briareus

Coe, W. R. 1912. Echinoderms of Connecticut. *Conn. State Geol. Nat. Hist. Surv. Bull.,* **19,** 1–152.*†

Kille, F. R. 1935. Regeneration in *Thyone briareus* Leseur following induced autotomy. *Biol. Bull.*, **69**, 82–108.

Kille, F. R. 1939. Regeneration of gonad tubules following extirpation in the sea-cucumber *Thyone briareus* (Leseur). *Biol. Bull.*, **76**, 70–79.

ENTEROPNEUSTA

Saccoglossus

Harmer, S. F. 1904. Hemichordata. *Cambridge natural history.* Vol. 7. Macmillan, London. 760 pp.

Horst, C. J. van der. 1927–39. Hemichordaten. *Bronn's Klassen und Ordnungen des Tier-Reichs.* Bd. 4, Abt. 4, Bch. 2, Teil 2. Pp. 1–737.

CHORDATA

General

Grave, C. 1921. *Amaroucium constellatum* (Verrill). II. The structure and organization of the tadpole larva. *Jour. Morph.*, **36**, 71–101.

Herdman, W. A. 1904. Ascidians and Amphioxus. *Cambridge natural history.* Vol. 7. Macmillan, London. 760 pp.

Huus, J. 1937, 1940. Tunicata. *Kukenthal und Krumbach's Handbuch der Zoologie.* Bd. 5, Halfte 2, Liefer. 6 u. 7. Pp. 545–768.

Seeliger, O., and R. Hartmeyer. 1893–1911. Tunicata (Manteltiere). *Bronn's Klassen und Ordnungen des Tier-Reichs.* Bd. 3 (Supplement), Abt. 2, Bch. 1. Pp. 1–1773.

Van Name, W. G. 1945. The North and South American Ascidians. *Bull. Amer. Mus. Nat. Hist.*, **84**, 1–476.

Index of Scientific and Common Names

Those pages including special accounts of invertebrate types are indicated by bold-faced numbers.

Acanthocephala, 214
Acanthocyclops vernalis, 406
Actinophrys sol, **25–27**
Actinosphaerium eichhorni, **25–27**
Amaroucium constellatum, **549–553**, 561
Amblema, 334–335
Amoeba proteus, **17–19**
Amphileptus, 66, 67
Amphipholis squamata, 527
Amphiporus ochraceus, **209–214**
Amphitrite brunnea, 289
Amphitrite ornata, **289–295**
Ancylostoma caninum, 238
Ancylostoma duodenale, 237
Anguillula aceti, 225
Annelida, 271
Anodonta, 334–335
Anopheles, 32, 36
Arbacia punctulata, 522, **528–538**, 540
Arcella vulgaris, **22–23**
Arenicola cristata, **279–289**
Arenicola marina, 279
Argiope aurantia, **382–394**
Argiope trifasciata, 382
Arrowworm, **505–515**
Artemia, **394–399**, 400, 402
Arthropoda, 360
Ascaris lumbricoides, **243–251**
Aschelminthes, 220
Ascidian larva, **560–563**
Aspidogaster conchicola, **161–165**
Astacus, **422–447**
Asterias forbesi, **515–523**, 535
Asterias glacialis, 517
Asterias rubens, 515
Asterias vulgaris, 515
Astrangia danae, 120, **121**, **127–131**
Aurellia aurita, 92, 111, **112–119**, 126, 138

Barentsia, **261–264**
Bdelloura, **145–148**, 361
Blaberus craniifer, 476, **495–496**
Bladder fluke, *Gorgodera,* **179–185**
Blatella germanica, 30, 476
Blatta orientalis, 30, 476
Blood fluke, *Schistosoma,* **187–190**
Blowfly larva, 496
Blue crab, **447–462**
Bodo, 70
Bowerbankia, **265–266**
Brachiopoda, 357
Brine shrimp, **394–399**
Brittle star, **523–528**
Broad tapeworm, **196–199**
Bryozoa, 261
Bugula, **267–269**
Bugula flabellata, 268
Bugula turrita, 268
Bursaria, 70
Busycon canaliculatum, 157, 319, **336–344**, 347

Callinectes sapidus, **447–462**
Cambarus, **422–447**
Carchesium polypinum, **66–68**
Catenula, 144
Ceratium hirundinella, **5–7**
Chaetognatha, 505
Chaetonotus brevispinosus, **223–225**
Chaetopleura apiculata, **318–319**
Chalina, **78–80**
Chilomonas, 19, 23, 63, 70
Chinese liver fluke, **173–179**
Chironomus, 496
Chiton, **318–319**
Chlamydomonas, 9
Chlorohydra, 105, 109, 130

Chordata, 549
Ciona, 554
Clam, *Venus*, **324–334**
Clams, fresh-water, **334–335**
Clamworm, **271–279**
Cliona, **78–80**
Clonorchis sinensis, **173–179**
Cnidaria, 85
Cockroach, **475–496**
Coelenterata, 85
Coleps hirtus, **40–41**, 70
Colpidium colpoda, 19, 39, **51–52**, 60, 70
Colpoda cucullus, 38, **48–49**, 70
Comb-jellies, **136–140**
Coral, stony, **127–131**
Corymorpha, 90
Crab, *Callinectes*, **447–462**
Crayfish, **422–447**
Crepidula, 344
Crisia, **264–265**
Crossobothrium laciniatum, **190–194**
Cryptocercus, 15, 17
Cryptocotyle lingua, 160, **165–173**
Ctenophora, 136
Cyclonaias tuberculata, 334–335
Cyclops, 104, **406–413**
Cyclops strenuus, 406

Daphnia, 104, 144, 155, 407
Daphnia magna, **399–406**
Dead-man's fingers, **80**
Dendrocoelum lacteum, 154
Didinium nasutum, **38–40**, 70
Dientamoeba, 20
Difflugia oblonga, **23–25**
Diphyllobothrium latum, **196–199**
Dirofilaria immitis, 251
Dolichoglossus kowalevskii, **547–549**
Drosophila melanogaster (larva), **496–504**
Dugesia, **148–154**, 155, 156
Dugesia dorotocephala, 149
Dugesia tigrina, 149

Earthworm, **295–303**
Echinarachnius parma, 522, **538–540**
Echinococcus granulosus, **204–208**
Echinodermata, 515
Echinus, 536

Ectocotyla paguri, 141
Ectoprocta, 264
Electra, **267**
Electra hastingsae, 267
Electra pilosa, 267
Elliptio, 334–335
Endolimax, 20
Entamoeba coli, 20
Entamoeba histolytica, **19–21**
Enterobius vermicularis, **234–237**
Enteropneusta, 547
Entoprocta, 261
Eolis, **344–347**
Epistylis plicatilis, **66–68**
Eubranchipus, 394
Euglena, **1–3**, 26
Euplana gracilis, 158
Euplanaria, 149
Euplectella, **76–77**
Euplotes aediculatus, 62
Euplotes eurystomus, 62
Euplotes patella, **61–63**
Euplotes woodruffi, 62

Fairy shrimp, 394
Fasciola hepatica, **185–187**
Fecampia, 143
Foleyella, 251
Frontonia leucas, 38, **49–50**, 70
Fulgur, **336–344**

Gastrotrich, **223–225**
Glaucoma, 70
Gonionemus murbachii, 92, 98, **100**, **101**, **102–104**
Gonionemus verteus, 102
Goose-neck barnacle, **413–418**
Gorgodera amplicava, **179–185**
Gorgoderina, 179, 182
Gregarina blattarum, **30–32**

Halichondria, **78–80**
Halteria, 70
Heteromysis formosa, **418–422**, 423
Hippospongia, **83–84**
Hirudo medicinalis, **303–309**
Holophrya simplex, **42**
Homarus, **422–447**
Hookworm, **237–243**

INDEX OF SCIENTIFIC AND COMMON NAMES 595

Hoploplana, **156–158**
Horse-shoe crab, **360–381**
Hyalonema, **77**
Hydatina senta, **220–223**
Hydra, **104–112**, 138
Hydractinia echinata, **86**, **87**, **93–96**, 98

Iodamoeba, 20

Jellyfish, *Aurellia*, **112–119**
Jellyfish, *Gonionemus*, **102–104**

Kerona, 105
King crab, **360–381**

Lampsilis, 334–335
Lasmigona, 334–335
Leech, **303–309**
Lepas anatifera, **413–418**
Lepas fascicularis, 414
Lepidometria commensalis, 289
Leucosolenia, **72–74**, 75
Ligumia nasuta, 334–335
Limulus, **360–381**
Lionotus, 66, 67
Littorina, 342
Littorina littorea, 170
Liver fluke, sheep, **185–187**
Lobster, **422–447**
Loligo pealeii, **347–357**
Lugworm, **279–289**
Lumbricus castaneus, 28
Lumbricus rubellus, 28
Lumbricus terrestris, 28, 29, 227, **295–303**

Macrostomum, 143
Mactra, 331
Mactra solidissima, 333
Megacyclops viridis, 406
Mesostoma, 143, 144
Metridium, 89, 97, 108, 110, 111, 113, 129, 130, 133, 138
Metridium dianthus, 119
Metridium fimbriatum, 119
Metridium marginatum, 119
Metridium senile, **119–127**
Microciona, **80**, 105
Microstomum, 143, 144

Millepede, **462–475**
Miranda, **382–394**
Molgula, 550, 551
Molgula, citrina, 560
Molgula manhattensis, **553–560**
Mollusca, 318
Monas, 19
Monocystis lumbrici, **28–30**, 301
Mosquito larva, 496
Musculium partumeium, 184
Mussels, fresh-water, **334–335**
Mya, 331
Mytilus, 90, 123

Neanthes virens, **271–279**
Necator americanus, 234, **237–243**
Nemertea, 209
Nemertine worm, **209–214**
Neoechinorhynchus emydis, **214–219**
Nereis virens, **271–279**
Notoplana atomata, 158
Nudibranch, **344–347**

Obelia geniculata, 92, **99–102**, 129
Opalina obtrigonoidea, **36–38**
Ophioderma brevispinum, **523–528**
Opisthorchis sinensis, **173–179**
Otobothrium crenacolle, **194–196**

Paramecium, 19, 38, 39, 63, 70, 144
Paramecium caudatum, **44–48**
Parazoa, 72
Pecten irradians, **321–324**
Pelmatohydra oligactis, **104–112**, 113
Pennaria, 92, 102, 108, 113, 123, 129, 138
Pennaria tiarella, **96–98**, **100**, **101**
Peranema trichophorum, 1, **3–5**
Periplaneta americana, 30, **475–495**
Phascolosoma gouldii, **309–317**
Phyllobothrium laciniatum, **190–194**
Physa, 344
Pinworm, **234–237**
Planaria, **148–154**
Planaria maculata, 149
Planocera inquilina, 156
Plasmodium falciparum, 32, 34, 35
Plasmodium malariae, 32, 34
Plasmodium ovale, 32, 34
Plasmodium vivax, **32–36**

INDEX OF SCIENTIFIC AND COMMON NAMES

Platyhelminthes, 141
Pleurobrachia pileus, **134, 135, 136–140**
Plumatella, **269–270**
Podocoryne, 94, 95
Podophrya collini, **69–71**
Podophrya fixa, 71
Polychoerus, **141–142**
Polychoerus carmelensis, 141
Polychoerus caudatus, 141
Polystoma integerrimum, 158
Polystomoides oris, **158–161**
Polyzoa, 261
Porifera, 72
Procotyla, 148, **154–156**
Prorodon griseus, 42, **43–44**
Prostoma rubrum, 209, 214
Protozoa, 1

Quadrula, 334–335
Quahog, **324–334**

Renilla amethystina, 131
Renilla köllikeri, 129, **131–136**
Renilla reniformis, 131
Reticulotermes, 15
Rhabditis maupasi, **227–234**
Rhynchocoela, 209
Roach, **475–496**
Rotifer, **220–223**

Saccoglossus kowalevskii, **547–549**
Sagitta elegans, **505–515**
Sand dollar, **538–540**
Schistosoma haemotobium, **187–190**
Schistosoma japonicum, 187–190
Schistosoma mansoni, 187–190
Sciara, 496
Scollap, **321–324**
Scypha, **74–76**
Sea anemone, *Metridium,* **119–127**
Sea cucumber, *Thyone,* **541–546**
Sea pen, *Renilla,* **131–136**
Sea pork, *Amaroucium,* **549–553**
Sea squirt, *Molgula,* **553–560**
Sea urchin, *Arbacia,* **528–538**
Sea walnut, *Pleurobrachia,* **136–140**
Sipunculoidea, 309
Spider, golden garden, **382–394**
Spiny-headed worm, **214–219**

Spirobolus marginatus, **462–475**
Spirostomum ambiguum, 39, **52–55,** 57, 59, 70
Spongilla, **81–83**
Squid, **347–357**
Starfish, common, **515–523**
Stenostomum, **143–144**
Stenostomum tenicauda, 144
Stentor coerulus, 39, **56–58,** 59, 64, 70
Stichocotyle, 161
Strongylocentrotus drobachiensis, 528, 530, 536
Stylactis, 94, 95
Stylochus ellipticus, 158
Stylochus zebra, 158
Stylonychia pustulata, **59–61,** 63
Sycon, **74–76**
Sycopsis, **336–344**
Sycotypus, **336–344**
Syncoelidium pellucidum, 148

Tadpole larva, ascidan, **560–563**
Taenia pisiformis, **199–204**
Tapeworm, fish, **196–199**
Terebratella, **357–359**
Terebratulina septentrionales, **357–359**
Teredo, 139
Tetrahymena, 38, 39, 60
Tetrastemma, 209
Thyone briareus, **541–546**
Trichina worm, **253–256**
Trichinella spiralis, **253–256**
Trichodina, 105
Trichomonas, **13–15,** 16
Trichonympha, **15–17**
Trichuris trichiura, **257–260**
Trypanosoma, **11–12**
Trypanosoma brucei, 11
Trypanosoma cruzi, 11
Trypanosoma gambiense, 11
Tubularia crocea, **85–93,** 94, 95, 96, 98, 103, 108, 109, 110, 113, 116, 123, 129, 133, 138
Turbatrix aceti, **225–227**

Unio, 334–335
Urosalpinx, 342

Venus mercenaria, 79, **324–334**
Venus's flower basket, **76–77**

INDEX OF SCIENTIFIC AND COMMON NAMES

Vinegar eel-worm, **225–227**
Volvox, **7–10**
Volvox aureus, 7
Volvox globator, 7
Vorticella campanula, **63–66**, 68

Water flea, **399–406**
Whelk, *Busycon*, **336–344**

Whipworm, **257–260**
Wuchereria bancrofti, **251–253**

Xiphosura polyphemus, **360–381**

Yoldia limatula, **319–321**

Zootermopsis, 15
Zoothamnium arbuscula, **66–68**
Zygonemertes, 209